SOLUTIONS MANUAL

J. RICHARD CHRISTMAN

U.S. Coast Guard Academy
New London, CT

EDWARD DERRINGH

Wentworth Institute
Boston, MA

to accompany

FUNDAMENTALS OF
PHYSICS

FOURTH EDITION
INCLUDING EXTENDED CHAPTERS

DAVID HALLIDAY

University of Pittsburgh

ROBERT RESNICK

Rensselaer Polytechnic Institute

JEARL WALKER

Cleveland State University

JOHN WILEY & SONS, INC.
New York   Chichester   Brisbane   Toronto   Singapore

PREFACE

This solutions manual is designed for use with the textbook *Fundamentals of Physics*, 4th edition, by David Halliday, Robert Resnick, and Jearl Walker. Its primary purpose is to show students by example how to solve various types of problems given at the ends of chapters in the text.

Most of the solutions start from definitions or fundamental relationships and the final equation is derived. This technique highlights the fundamentals and at the same time gives students the opportunity to review the mathematical steps required to obtain a solution. The mere plugging of numbers into equations derived in the text is avoided. We hope students will learn to examine any assumptions that are made in setting up and solving each problem.

Selection of the problems included in this manual and their solutions are the responsibility of the authors alone.

For their help and encouragement special thanks go to the good people of Wiley, especially Cliff Mills, Catherine Donovan, and Joan Kalkut.

J. Richard Christman
U.S. Coast Guard Academy
15 Mohegan Avenue
New London, CT   06320
March 1993

Edward Derringh
Wentworth Institute of Technology
550 Huntington Avenue
Boston, MA   02115

TABLE OF CONTENTS

Chapter 1

5E

(a) Since $1\,\text{km} = 10^3\,\text{m}$ and $1\,\text{m} = 10^6\,\mu\text{m}$,

$$1\,\text{km} = 10^3\,\text{m} = (10^3\,\text{m})(10^6\,\mu\text{m/m}) = 10^9\,\mu\text{m}.$$

(b) Calculate the number of microns in 1 centimeter. Since $1\,\text{cm} = 10^{-2}\,\text{m}$,

$$1\,\text{cm} = 10^{-2}\,\text{m} = (10^{-2}\,\text{m})(10^6\,\mu\text{m/m}) = 10^4\,\mu\text{m}.$$

Therefore the fraction of a centimeter that equals $1\,\mu\text{m}$ is 10^{-4}.

(c) Since $1\,\text{yard} = (3\,\text{ft})(0.3048\,\text{m/ft}) = 0.9144\,\text{m}$,

$$1.0\,\text{yd} = (0.9144\,\text{m})(10^6\,\mu\text{m/m}) = 9.1 \times 10^5\,\mu\text{m}.$$

6E

(a) The circumference of a sphere of radius R is given by $2\pi R$. Substitute $R = (6.37 \times 10^6\,\text{m})(10^{-3}\,\text{km/m}) = 6.37 \times 10^3\,\text{km}$. Retain 3 significant figures in your answer. You should obtain $4.00 \times 10^4\,\text{km}$.

(b) The surface area of a sphere is given by $4\pi R^2$ and the surface area of the Earth is $4\pi(6.37 \times 10^3\,\text{km})^2 = 5.10 \times 10^8\,\text{km}^2$.

(c) The volume of a sphere is given by $(4\pi/3)R^3$ and the volume of the Earth is $(4\pi/3)(6.37 \times 10^3\,\text{km})^3 = 1.08 \times 10^{12}\,\text{km}^3$.

12P

The volume of ice is given by the product of the area of the semicircle and the thickness. The area is half the area of a circle: $A = \pi R^2/2$, where R is the radius. Thus the volume is given by $V = \pi R^2 T/2$, where T is the thickness. Substitute $R = 2000\,\text{km}$ or, since there are $10^3\,\text{m}$ in $1\,\text{km}$ and $10^2\,\text{cm}$ in a m, $R = (2000\,\text{km})(10^3\,\text{m/km})(10^2\,\text{cm/m}) = 2000 \times 10^5\,\text{cm}$. Also substitute $T = (3000\,\text{m})(10^2\,\text{cm/m}) = 3000 \times 10^2\,\text{cm}$. Thus

$$V = \frac{1}{2}\pi(2000 \times 10^5\,\text{cm})^2(3000 \times 10^2\,\text{cm}) = 1.9 \times 10^{22}\,\text{cm}^3.$$

14P

Use the conversion factors found in Appendix F:

$$1\,\text{acre} \cdot \text{ft} = (43,560\,\text{ft}^2) \cdot \text{ft} = 43,560\,\text{ft}^3.$$

Since 2 in. = 1/6 ft, the volume of water that fell during the storm is

$$V = (26\,\text{km}^2)(1/6\,\text{ft}) = (26\,\text{km}^2)(3281\,\text{ft/km})^2(1/6\,\text{ft}) = 4.66 \times 10^7\,\text{ft}^3.$$

Thus

$$V = \frac{4.66 \times 10^7\,\text{ft}^3}{4.3560 \times 10^4\,\text{ft}^3/\text{acre} \cdot \text{ft}} = 1.1 \times 10^3\,\text{acre} \cdot \text{ft}.$$

Since the area is given to 2 significant figures you are allowed only that many figures in your answer.

15P

(a) Since 1 ft = 0.3048 m, 1 gal = 231 in.^{3}, and 1 in.^{3} = 1.639 × 10^{-2} L,

$$460\,\text{ft}^2/\text{gal} = \left(\frac{460\,\text{ft}^2}{\text{gal}}\right)\left(\frac{0.3048\,\text{m}}{\text{ft}}\right)^2\left(\frac{\text{gal}}{231\,\text{in.}^3}\right)\left(\frac{\text{in.}^3}{1.639 \times 10^{-2}\,\text{L}}\right) = 11.3\,\text{m}^2/\text{L}.$$

(b) Since 1 m^3 = 1000 L,

$$11.3\,\text{m}^2/\text{L} = \left(\frac{11.3\,\text{m}^2}{\text{L}}\right)\left(\frac{1000\,\text{L}}{\text{m}^3}\right) = 1.13 \times 10^4\,\text{m}^{-1}.$$

(c) The reciprocal, 2.17×10^{-3} gal/ft^2, tells us how much paint is required to cover each square foot of surface area. To find the total paint required for a wall, for example, calculate the area of the wall and multiply by this number.

20E

(a) Use the conversion factors given in Appendix F and the definitions of the SI prefixes given in Table 2: 1 m = 3.281 ft and 1 s = 10^9 ns. Thus

$$3.0 \times 10^8\,\text{m/s} = \left(\frac{3.0 \times 10^8\,\text{m}}{\text{s}}\right)\left(\frac{3.281\,\text{ft}}{\text{m}}\right)\left(\frac{\text{s}}{10^9\,\text{ns}}\right) = 0.98\,\text{ft/ns}.$$

(b) Use 1 m = 10^3 mm and 1 s = 10^{12} ps. Thus

$$3.0 \times 10^8\,\text{m/s} = \left(\frac{3.0 \times 10^8\,\text{m}}{\text{s}}\right)\left(\frac{10^3\,\text{mm}}{\text{m}}\right)\left(\frac{\text{s}}{10^{12}\,\text{ps}}\right) = 0.30\,\text{mm/ps}.$$

22E

Use 1 y = 365.25 d, 1 d = 24 h, 1 h = 60 min, and 1 min = 60 s. Thus

$$1\,\text{y} = (365.25\,\text{d})\left(\frac{24\,\text{h}}{\text{d}}\right)\left(\frac{60\,\text{min}}{\text{h}}\right)\left(\frac{60\,\text{s}}{\text{min}}\right) = 3.156 \times 10^7\,\text{s}.$$

27P

You need to convert meters to astronomical units and seconds to minutes. Use $1\,\text{m} = 1 \times 10^{-3}\,\text{km}$, $1\,\text{AU} = 1.50 \times 10^8\,\text{km}$, and $60\,\text{s} = 1\,\text{min}$. Thus

$$3.0 \times 10^8\,\text{m/s} = \left(\frac{3.0 \times 10^8\,\text{m}}{\text{s}}\right)\left(\frac{10^{-3}\,\text{km}}{\text{m}}\right)\left(\frac{\text{AU}}{1.50 \times 10^8\,\text{km}}\right)\left(\frac{60\,\text{s}}{\text{min}}\right) = 0.12\,\text{AU/min}.$$

31P

The last day of the twenty centuries is longer than the first day by

$$(20\,\text{centuries})(0.001\,\text{s/century}) = 0.020\,\text{s}.$$

The average day during the twenty centuries is $(0 + 0.020)/2 = 0.010\,\text{s}$ longer than the first day. Since the increase occurs uniformly, the cumulative effect T is

$$T = (\text{average increase in length of a day})(\text{number of days})$$
$$= \left(\frac{0.010\,\text{s}}{\text{d}}\right)\left(\frac{365.25\,\text{d}}{\text{y}}\right)(2000\,\text{y}) = 7305\,\text{s} = 2.1\,\text{h}$$

32P

Lunar and sidereal months differ because the Earth moves around the sun while the moon moves around the Earth. The relative positions at the beginning (A) and end (B) of a lunar month are shown on the diagram. Since the moon goes around the Earth a little more than once during this time a lunar month is slightly longer than a sidereal month. The moon moves through the angle θ from the end of a sidereal month to the end of a lunar month. The Earth moves through the same angle during a lunar month.

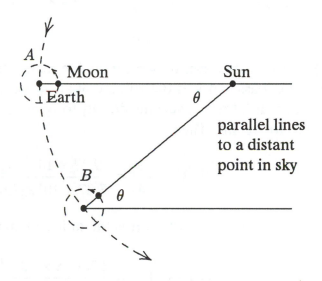

Let T_s be a sidereal month and T_ℓ be a lunar month. Since the moon covers 360° in T_s the angle θ in the diagram is given by $\theta = 360° T_\ell / T_s$. Let T_E be the time for the Earth to go once around the sun (1 y). Then θ is also given by $\theta = 360° T_\ell / T_E$. Equate these two expressions for θ and solve for T_ℓ. You should get

$$T_\ell = \frac{T_s T_E}{T_E - T_s}.$$

Take $T_s = 27.3\,\text{d}$ and $T_E = 365.3\,\text{d}$. The result is $T_\ell = 29.5\,\text{d}$, so $T_\ell - T_s = 2.2\,\text{d}$.

35E

If M_E is the mass of the Earth, m is the average mass of an atom in the Earth, and N is the number of atoms, then $M_E = Nm$ or $N = M_E/m$. The values for M_E and m must have the same units. Convert m to kg. According to Appendix F, $1\,u = 1.661 \times 10^{-27}\,kg$. Thus

$$N = \frac{M_E}{m} = \frac{5.98 \times 10^{24}\,kg}{(40\,u)(1.661 \times 10^{-27}\,kg/u)} = 9.0 \times 10^{49}.$$

37P

(a) Convert grams to kilograms and cubic centimeters to cubic meters: $1\,g = 10^{-3}\,kg$ and $1\,cm^3 = (1 \times 10^{-2}\,m)^3 = 10^{-6}\,m^3$. The density in kg/m^3 is

$$1\,g/cm^3 = \left(\frac{1\,g}{cm^3}\right)\left(\frac{10^{-3}\,kg}{g}\right)\left(\frac{cm^3}{10^{-6}\,m^3}\right) = 10^3\,kg/m^3.$$

(b) Divide the mass (in kilograms) of the water by the time (in seconds) taken to drain it. The mass is the product of the volume of water and its density: $M = (5700\,m^3)(1 \times 10^3\,kg/m^3) = 5.70 \times 10^6\,kg$ and the time is $(10\,h)(3600\,s/h) = 3.6 \times 10^4\,s$, so the mass flow rate R is

$$R = \frac{5.70 \times 10^6\,kg}{3.6 \times 10^4\,s} = 158\,kg/s.$$

40P

(a) The density ρ can be written as the mass per unit volume of a single atom. That is, if M is the mass and V is the volume of an atom, then $\rho = M/V$, or $V = M/\rho$. To obtain the volume in m^3, first convert the density to kg/m^3: $\rho = (7.87\,g/cm^3)(10^{-3}\,kg/g)(10^6\,cm^3/m^3) = 7.87 \times 10^3\,kg/m^3$. Then

$$V = \frac{M}{\rho} = \frac{9.27 \times 10^{-26}\,kg}{7.87 \times 10^3\,kg/m^3} = 1.18 \times 10^{-29}\,m^3.$$

(b) Set $V = 4\pi R^3/3$, where R is the radius of an atom, and solve for R:

$$R = \left(\frac{3V}{4\pi}\right)^{1/3} = \left[\frac{3(1.18 \times 10^{-29}\,m^3)}{4\pi}\right]^{1/3} = 1.41 \times 10^{-10}\,m.$$

The center-to-center distance between atoms is twice the radius or $2.82 \times 10^{-10}\,m$.

Chapter 2

4E

Assume the ball travels with constant velocity and use $\Delta x = v\Delta t$, where Δx is the horizontal distance traveled, Δt is the time, and v is the velocity. Convert v to meters per second. According to Appendix F, $1\,\text{km/h} = 0.2778\,\text{m/s}$, so $160\,\text{km/h} = (160)(0.2778\,\text{m/s}) = 44.45\,\text{m/s}$. Thus

$$\Delta t = \frac{\Delta x}{v} = \frac{18.4\,\text{m}}{44.5\,\text{m/s}} = 0.414\,\text{s}.$$

This may also be written $414\,\text{ms}$.

6E

Calculate the time required to make the trip at each of the two speeds and compute the difference. Use $\Delta x = v\Delta t$, where Δx is the distance, v is the speed, and Δt is the time. At the slower speed the time required is

$$\Delta t = \frac{\Delta x}{v} = \frac{700\,\text{km}}{88.5\,\text{km/h}} = 7.91\,\text{h}$$

and at the faster

$$\Delta t = \frac{\Delta x}{v} = \frac{700\,\text{km}}{105\,\text{km/h}} = 6.67\,\text{h}.$$

The difference is $7.91\,\text{h} - 6.67\,\text{h} = 1.24\,\text{h} = 1\,\text{h}\ 14\,\text{min}$.

8E

(*a*) The average velocity during some time interval is the displacement during that interval divided by the interval: $\bar{v} = \Delta x / \Delta t$, where Δx is the displacement and Δt is the time interval. In this case the interval is divided into two parts. During the first part the displacement is $\Delta x_1 = 40\,\text{km}$ and the time interval is $\Delta t_1 = (40\,\text{km})/(30\,\text{km/h}) = 1.33\,\text{h}$. During the second part the displacement is $\Delta x_2 = 40\,\text{km}$ and the time interval is $\Delta t_2 = (40\,\text{km})/(60\,\text{km/h}) = 0.67\,\text{h}$. Both displacements are in the same direction so the total displacement is $\Delta x = \Delta x_1 + \Delta x_2 = 40\,\text{km} + 40\,\text{km} = 80\,\text{km}$. The total time interval is $\Delta t = \Delta t_1 + \Delta t_2 = 1.33\,\text{h} + 0.67\,\text{h} = 2.00\,\text{h}$. The average velocity is $\bar{v} = (80\,\text{km})/(2.0\,\text{h}) = 40\,\text{km/h}$.

(*b*) The average speed is the total distance traveled divided by the time. In this case the total distance is the magnitude of the total displacement, so the average speed is $40\,\text{km/h}$.

(c) Assume the automobile passes the origin at time $t = 0$. Then its coordinate as a function of time is as shown as the solid lines on the graph to the right. The average velocity is the slope of the dotted line.

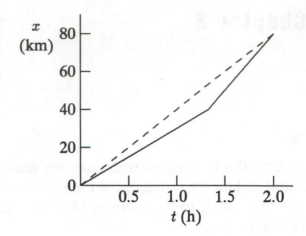

12P

(a) Substitute, in turn, $t = 1, 2, 3,$ and 4 s into the expression $x(t) = 3t - 4t^2 + t^3$:

$$x(1\,\text{s}) = 3 \times 1 - 4 \times 1^2 + 1^3 = 0$$
$$x(2\,\text{s}) = 3 \times 2 - 4 \times 2^2 + 2^3 = -2\,\text{m}$$
$$x(3\,\text{s}) = 3 \times 3 - 4 \times 3^2 + 3^3 = 0$$
$$x(4\,\text{s}) = 3 \times 4 - 4 \times 4^2 + 4^3 = 12\,\text{m}.$$

(b) The displacement during an interval is the coordinate at the end of the interval minus the coordinate at the beginning. For the interval from $t = 0$ to $t = 4$ s, the displacement is $\Delta x = x(4\,\text{s}) - x(0) = 12\,\text{m} - 0 = +12\,\text{m}$. The displacement is in the positive x direction.

(c) The average velocity during an interval is defined as the displacement over the interval divided by the interval: $\bar{v} = \Delta x / \Delta t$. For the interval from $t = 2$ s to $t = 4$ s the displacement is $\Delta x = x(4\,\text{s}) - x(2\,\text{s}) = 12\,\text{m} - (-2\,\text{m}) = 14\,\text{m}$ and the time interval is $\Delta t = 4\,\text{s} - 2\,\text{s} = 2\,\text{s}$. Thus

$$\bar{v} = \frac{\Delta x}{\Delta t} = \frac{14\,\text{m}}{2\,\text{s}} = 7\,\text{m/s}.$$

(d) The solid curve on the graph to the right shows the coordinate x as a function of time. The slope of the dotted line is the average velocity between $t = 2.0$ s and $t = 4.0$ s.

21E

(a) If v_1 is the velocity at the beginning of a time interval (at time t_1) and v_2 is the velocity at the end (at t_2) then the average acceleration in the interval is given by $\bar{a} = (v_2 - v_1)/(t_2 - t_1)$. Take $t_1 = 0$, $v_1 = 18\,\text{m/s}$, $t_2 = 2.4\,\text{s}$, and $v_2 = -30\,\text{m/s}$. Then

$$\bar{a} = \frac{-30\,\text{m/s} - 18\,\text{m/s}}{2.4 - 0\,\text{s} - 0} = -20\,\text{m/s}^2.$$

The minus sign indicates that the acceleration is opposite to the original direction of travel.

(b) Assume the acceleration of the particle is constant. Then the graph of v as a function of t is as shown to the right. The acceleration is the slope of this line. Actually, the graph may have any form whatsoever, as long as the beginning and ending velocities are the same. The average acceleration is the slope of the line from the initial to the final point on the graph.

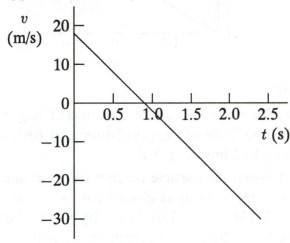

28E

(a) The average velocity during the first 3 s is given by

$$\bar{v} = \frac{x(3\,\text{s}) - x(0)}{\Delta t}.$$

Evaluate the expression $x = 50t + 10t^2$ using $t = 0$ and $t = 3\,\text{s}$: $x(0) = 0$ and $x(3\,\text{s}) = (50)(3) + 10 \times 3^2 = 240\,\text{m}$. Thus

$$\bar{v} = \frac{240\,\text{m} - 0}{3\,\text{s}} = 80\,\text{m/s}.$$

(b) The instantaneous velocity at time t is given by $v = dx/dt = 50 + 20t$, in m/s. At $t = 3.0\,\text{s}$, $v = 50 + 20 \times 3.0 = 110\,\text{m/s}$.

(c) The instantaneous acceleration at time t is given by $a = dv/dt = 20\,\text{m/s}^2$. It is constant, so the acceleration at any time is $20\,\text{m/s}^2$.

(d) and (e) The graph on the left below shows the coordinate x as a function of time (solid curve). The dotted line marked (a) runs from $t = 0$, $x = 0$ to $t = 3.0\,\text{s}$, $x = 240\,\text{m}$. Its slope is the average velocity during the first 3 s of motion. The dotted line marked (b) is tangent to x vs. t at $t = 3.0\,\text{s}$. Its slope is the instantaneous velocity at $t = 3.0\,\text{s}$. The graph on the right shows the instantaneous velocity as a function of time. Its slope is the acceleration.

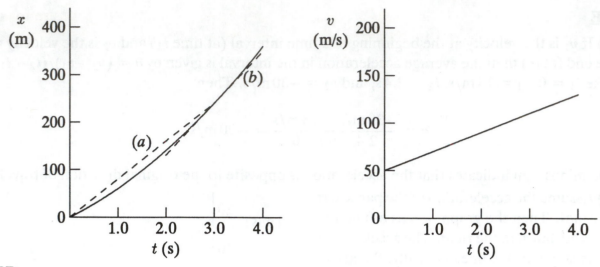

(a) Since the units of et^2 are those of length and the units of t are those of time, the units of e must be those of (length)/(time)^2, or ft/s^2. Since bt^3 has units of length, b must have units of (length)/(time)^3, or ft/s^3.

(b) When the particle reaches its maximum (or its minimum) coordinate its velocity is zero. Since the velocity is given by $v = dx/dt = 2et - 3bt^2$, $v = 0$ occurs for $t = 0$ and for $t = 2e/3b = (2 \times 3.0)/(3 \times 1.0) = 2.0$ s. For $t = 0$, $x = 0$ and for $t = 2.0$ s, $x = 4.0$ ft. Reject the first solution and accept the second.

(c) In the first 4 s the particle moves from the origin to $x = 4.0$ ft, turns around, and goes back to $x(4\,\text{s}) = 3.0 \times 4^2 - 1.0 \times 4^3 = -16$ ft. The total path length it travels is $4.0\,\text{ft} + 4.0\,\text{ft} + 16\,\text{ft} = 24$ ft.

(d) Its displacement is given by $\Delta x = x_2 - x_1$, where $x_1 = 0$ and $x_2 = -16$ ft. Thus $\Delta x = -16$ ft.

(e) The velocity is given by $v = 6.0t - 3.0t^2$. Thus

$$v(1\,\text{s}) = 6.0 \times 1.0 - 3.0 \times 1.0^2 = 3.0\,\text{ft/s}$$
$$v(2\,\text{s}) = 6.0 \times 2.0 - 3.0 \times 2.0^2 = 0.0$$
$$v(3\,\text{s}) = 6.0 \times 3.0 - 3.0 \times 3.0^2 = -9.0\,\text{ft/s}$$
$$v(4\,\text{s}) = 6.0 \times 4.0 - 3.0 \times 4.0^2 = -24\,\text{ft/s}.$$

(f) The acceleration is given by $a = dv/dt = 6.0 - 6.0t$. Thus

$$a(1\,\text{s}) = 6.0 - 6.0 \times 1.0 = 0.0$$
$$a(2\,\text{s}) = 6.0 - 6.0 \times 2.0 = -6.0\,\text{ft/s}^2$$
$$a(3\,\text{s}) = 6.0 - 6.0 \times 3.0 = -12\,\text{ft/s}^2$$
$$a(4\,\text{s}) = 6.0 - 6.0 \times 4.0 = -18\,\text{ft/s}^2.$$

Use $v = v_0 + at$, an equation that is valid for motion with constant acceleration. Take $t = 0$ to be the time when the velocity is $+9.6$ m/s. Then $v_0 = 9.6$ m/s.

(a) Since we wish to calculate the velocity for a time *before* $t = 0$, the value we use for t is negative: $t = -2.5\,\mathrm{s}$. Thus

$$v = (9.6\,\mathrm{m/s}) + (3.2\,\mathrm{m/s^2})(-2.5\,\mathrm{s}) = 1.6\,\mathrm{m/s}.$$

(b) Now $t = +2.5\,\mathrm{s}$ and

$$v = (9.6\,\mathrm{m/s}) + (3.2\,\mathrm{m/s^2})(2.5\,\mathrm{s}) = 18\,\mathrm{m/s}.$$

37E

(a) Solve $v = v_0 + at$ for t: $t = (v - v_0)/a$. Substitute $v = 0.1(3.0 \times 10^8\,\mathrm{m/s}) = 3.0 \times 10^7\,\mathrm{m/s}$, $v_0 = 0$, and $a = 9.8\,\mathrm{m/s^2}$. The result is $t = 3.1 \times 10^6\,\mathrm{s}$.

(b) Evaluate $x = x_0 + v_0 t + \frac{1}{2}at^2$, with $x_0 = 0$. The result is $x = \frac{1}{2}(9.8\,\mathrm{m/s^2})(3.1 \times 10^6\,\mathrm{s})^2 = 4.6 \times 10^{13}\,\mathrm{m}$.

39E

(a) The velocity is given by $v = v_0 + at$. The time to stop can be found by setting $v = 0$ and solving for t: $t = -v_0/a$. Substitute this expression into $x = v_0 t + \frac{1}{2}at^2$ to obtain $x = -\frac{1}{2}v_0^2/a$. (Alternatively, you might use $v^2 = v_0^2 + 2a(x - x_0)$ directly from Table 2.) Use $v_0 = 5.00 \times 10^6\,\mathrm{m/s}$ and $a = -1.25 \times 10^{14}\,\mathrm{m/s^2}$. Notice that since the muon slows the initial velocity and the acceleration must have opposite signs. The result is

$$x = -\frac{1}{2}\frac{(5.00 \times 10^6\,\mathrm{m/s})^2}{(-1.25 \times 10^{14}\,\mathrm{m/s^2})} = 0.100\,\mathrm{m}.$$

(b) Here are the graphs of the coordinate x and velocity v of the muon from the time it enters the field to the time it stops:

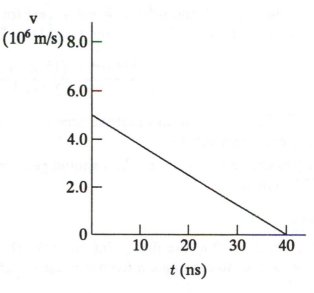

40E

Solve $v^2 = v_0^2 + 2a(x-x_0)$ for a. Take $x_0 = 0$. Then $a = (v^2-v_0^2)/2x$. Use $v_0 = 1.50 \times 10^5$ m/s, $v = 5.70 \times 10^6$ m/s, and $x = 1.0\,\text{cm} = 0.10\,\text{m}$. The result is

$$a = \frac{(5.70 \times 10^6\,\text{m/s})^2 - (1.50 \times 10^5\,\text{m/s})^2}{2(0.010\,\text{m})} = 1.62 \times 10^{15}\,\text{m/s}^2.$$

45E

(a) The word "deceleration" means the car is slowing down; its speed is becoming less. This means the initial velocity and the acceleration must have opposite signs. Assume the acceleration is uniform and solve $v = v_0 + at$ for t: $t(v - v_0)/a$. Substitute $v_0 = 85\,\text{mi/h} = 125\,\text{ft/s}$, $v = 55\,\text{mi/h} = 80.7\,\text{ft/s}$, and $a = -17\,\text{ft/s}^2$. The result is

$$t = \frac{80.7\,\text{ft/s} - 125\,\text{ft/s}}{-17\,\text{ft/s}^2} = 2.6\,\text{s}.$$

(b) Suppose the coordinate of the car is $x = 0$ when the brakes are applied (at time $t = 0$). Then the coordinate of the car as a function of time is given by $x = (125\,\text{ft/s})t - \frac{1}{2}(17\,\text{ft/s}^2)t^2$. This function is plotted from $t = 0$ to $t = 2.6\,\text{s}$ on the graph to the right.

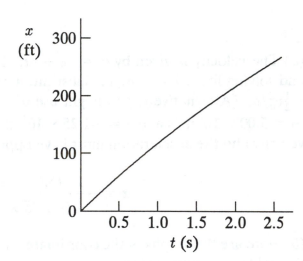

48P

(a) Take $x_0 = 0$, and solve $x = v_0 t + \frac{1}{2}at^2$ for a: $a = 2(x - v_0 t)/t^2$. Substitute $x = 24.0\,\text{m}$, $v_0 = 56.0\,\text{km/h} = 15.55\,\text{m/s}$, and $t = 2.00\,\text{s}$. The result is

$$a = \frac{2[24.0\,\text{m} - (15.55\,\text{m/s})(2.00\,\text{s})]}{(2.00\,\text{s})^2} = -3.56\,\text{m/s}^2.$$

The minus sign indicates that the acceleration is opposite the direction of motion of the car. The car is slowing down.

(b) Evaluate $v = v_0 + at$. You should get $v = 15.55\,\text{m/s} - (3.56\,\text{m/s}^2)(2.00\,\text{s}) = 8.43\,\text{m/s}$ (30.3 km/h).

49P

(a) Let $x = 0$ at the first point and take the time to be zero when the car is there. Use $v = v_0 + at$ to eliminate a from $x = v_0 t + \frac{1}{2}at^2$. The first equation yields $a = (v - v_0)/t$, so

$x = v_0 t + \frac{1}{2}(v - v_0)t = \frac{1}{2}(v + v_0)t$. Solve for v_0: $v_0 = (2x - vt)/t$. Substitute $x = 60.0\,\text{m}$, $v = 15.0\,\text{m/s}$, and $t = 6.00\,\text{s}$. Your result should be

$$v_0 = \frac{2(60.0\,\text{m}) - (15.0\,\text{m/s})(6.00\,\text{s})}{6.00\,\text{s}} = 5.00\,\text{m/s}.$$

(b) Substitute $v = 15.0\,\text{m/s}$, $v_0 = 5.00\,\text{m/s}$, and $t = 6.00\,\text{s}$ into $a = (v - v_0)/t$. The result is $a = (15.0\,\text{m/s} - 5.00\,\text{m/s})/(6.00\,\text{s}) = 1.67\,\text{m/s}^2$.

(c) Substitute $v = 0$, $v_0 = 5.0\,\text{m/s}$, and $a = 1.67\,\text{m/s}^2$ into $v^2 = v_0^2 + 2ax$. Solve for x:

$$x = -\frac{v_0^2}{2a} = -\frac{(5.0\,\text{m/s})^2}{2(1.67\,\text{m/s}^2)} = -7.50\,\text{m}.$$

(d) To draw the graphs you need to know the time at which the car is at rest. Solve $v = v_0 + at = 0$ for t: $t = -v_0/a = -(5.00\,\text{m/s})/(1.67\,\text{m/s}^2) = -3.0\,\text{s}$. Your graphs should look like this:

 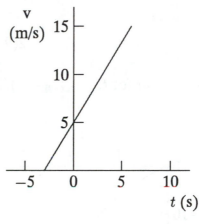

51P

Let t_r be the reaction time and t_b be the braking time. Then the total distance moved by the car is given by $x = v_0 t_r + v_0 t_b + \frac{1}{2}at_b^2$, where v_0 is the initial velocity and a is the acceleration. After the brakes are applied the velocity of the car is given by $v = v_0 + at_b$. Use this equation, with $v = 0$, to eliminate t_b from the first equation. According to the second equation $t_b = -v_0/a$, so $x = v_0 t_r - v_0^2/a + \frac{1}{2}v_0^2/a = v_0 t_r - \frac{1}{2}v_0^2/a$.

Write this equation twice, once for each of the two different initial velocities: $x_1 = v_{01}t_r - \frac{1}{2}v_{01}^2/a$ and $x_2 = v_{02}t_r - \frac{1}{2}v_{02}^2/a$. Solve these equations simultaneously for t_r and a. You should get

$$t_r = \frac{v_{02}^2 x_1 - v_{01}^2 x_2}{v_{01}v_{02}(v_{02} - v_{01})}$$

and

$$a = -\frac{1}{2}\frac{v_{02}v_{01}^2 - v_{01}v_{02}^2}{v_{02}x_1 - v_{01}x_2}.$$

Substitute $x_1 = 186\,\text{ft}$, $v_{01} = 50\,\text{mi/h} = 73.4\,\text{ft/s}$, $x_2 = 80\,\text{ft}$, and $v_{02} = 30\,\text{mi/h} = 44.0\,\text{ft/s}$. The results are

$$t_r = \frac{(44.0\,\text{ft/s})^2(186\,\text{ft}) - (73.4\,\text{ft/s})^2(80\,\text{ft})}{(73.4\,\text{ft/s})(44.0\,\text{ft/s})(73.4\,\text{ft/s} - 44.0\,\text{ft/s})} = 0.74\,\text{s}$$

and

$$a = -\frac{1}{2}\frac{(44.0\,\text{ft/s})(73.4\,\text{ft/s})^2 - (73.4\,\text{ft/s})(44.0\,\text{ft/s})^2}{(44.0\,\text{ft/s})(186\,\text{ft}) - (73.4\,\text{ft/s})(80\,\text{ft})} = -20\,\text{ft/s}^2\,.$$

59E

(a) Take the y axis to be positive in the upward direction and take $t = 0$ and $y = 0$ at the point from which the wrench was dropped. If h is the height from which it was dropped then the ground is at $y = -h$. Solve $v^2 = v_0^2 + 2gh$ for h: $h = (v^2 - v_0^2)/2g$. Substitute $v_0 = 0$, $v = -24\,\text{m/s}$, and $g = 9.8\,\text{m/s}^2$:

$$h = \frac{(24\,\text{m/s})^2}{2(9.8\,\text{m/s}^2)} = 29.4\,\text{m}\,.$$

(b) Solve $v = v_0 - gt$ for t: $t = (v_0 - v)/g = (24\,\text{m/s})/(9.8\,\text{m/s}^2) = 2.45\,\text{s}$.

(c)

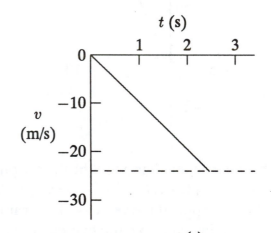

The acceleration is constant until the wrench hits the ground: $a = -9.8\,\text{m/s}^2$. Its graph is as shown on the right.

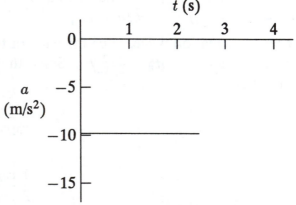

60E

(*a*) At the highest point the velocity of the ball is instantaneously zero. Take the y axis to be upward, set $v = 0$ in $v^2 = v_0^2 - 2gy$, and solve for v_0: $v_0 = \sqrt{2gy}$. Substitute $g = 9.8\,\text{m/s}^2$ and $y = 50\,\text{m}$ to get $v_0 = \sqrt{2(9.8\,\text{m/s}^2)(50\,\text{m})} = 31\,\text{m/s}$.

(*b*) It will be in the air until $y = 0$ again. Solve $y = v_0 t - \frac{1}{2}gt^2$ for t. Since $y = 0$ the two solutions are $t = 0$ and $t = 2v_0/g$. Reject the first and accept the second: $t = 2v_0/g = 2(31\,\text{m/s})/(9.8\,\text{m/s}^2) = 6.4\,\text{s}$.

(*c*)

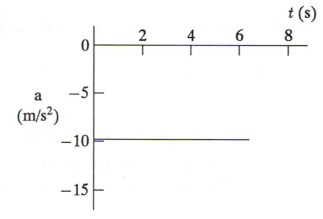

The acceleration is constant while the ball is in flight: $a = -9.8\,\text{m/s}^2$. Its graph is as shown on the right.

66E

Take the y axis to be upward, set $v_0 = 0$, and solve $y = v_0 t - \frac{1}{2}gt^2$ for t: $t = \sqrt{-2y/g}$.

(*a*) For this part $y = -50\,\text{m}$, so

$$t = \sqrt{-\frac{2(-50\,\text{m})}{9.8\,\text{m/s}^2}} = 3.2\,\text{s}.$$

(*b*) For this part set $y = -100\,\text{m}$, so

$$t = \sqrt{-\frac{2(-100\,\text{m})}{9.8\,\text{m/s}^2}} = 4.5\,\text{s}.$$

The difference is the time taken to fall the second 50 m: $4.5\,\text{s} - 3.2\,\text{s} = 1.3\,\text{s}$.

67P

(a) Take the y axis to be positive upward and use $y = v_0 t - \frac{1}{2}gt^2$, with $y = 0.544\,\text{m}$ and $t = 0.200\,\text{s}$. Solve for v_0:

$$v_0 = \frac{y + \frac{1}{2}gt^2}{t} = \frac{0.544\,\text{m} + \frac{1}{2}(9.8\,\text{m/s}^2)(0.200\,\text{s})^2}{0.200\,\text{s}} = 3.70\,\text{m/s}.$$

(b) Use $v = v_0 - gt = 3.70\,\text{m/s} - (9.8\,\text{m/s}^2)(0.200\,\text{s}) = 1.74\,\text{m/s}$.

(c) Use $v^2 = v_0^2 - 2gy$, with $v = 0$. Solve for y:

$$y = \frac{v_0^2}{2g} = \frac{(3.7\,\text{m/s})^2}{2(9.8\,\text{m/s}^2)} = 0.698\,\text{m}$$

It goes $0.698\,\text{m} - 0.544\,\text{m} = 0.154\,\text{m}$ higher.

72P

First find the velocity of the ball just before it hits the ground. During contact with the ground its average acceleration is given by

$$\overline{a} = \frac{\Delta v}{\Delta t},$$

where Δv is the change in its velocity during contact and Δt is the time of contact.

To find the velocity just before contact take the y axis to be positive in the upward direction and put the origin at the point where the ball is dropped. Take the time t to be zero when it is dropped. The ball hits the ground when $y = -15.0\,\text{m}$. It velocity then is given by $v^2 = -2gy$, so

$$v = -\sqrt{-2gy} = -\sqrt{-2(9.8\,\text{m/s}^2)(-15.0\,\text{m})} = -17.1\,\text{m/s}.$$

The negative sign is used since the ball is traveling downward at the time of contact.

The average acceleration during contact with the ground is

$$\overline{a} = \frac{0 - (-17.1\,\text{m/s})}{20.0 \times 10^{-3}\,\text{s}} = 8.57 \times 10^{-2}\,\text{m/s}^2.$$

The positive sign indicates it is upward.

77P

The average acceleration during contact with the floor is given by $\overline{a} = (v_2 - v_1)/\Delta t$, where v_1 is its velocity just before striking the floor and v_2 is its velocity just after leaving the floor. Take the y axis to be upward and place the origin at the point where the ball is dropped. To find the velocity just before hitting the floor, use $v_1^2 = v_0^2 - 2gy$, where $v_0 = 0$ and $y = -4.00\,\text{m}$. The result is

$$v_1 = \sqrt{-2gy} = \sqrt{-2(9.8\,\text{m/s}^2)(-4.00\,\text{m})} = -8.85\,\text{m/s}.$$

The negative square root is used because the ball is traveling downward. To find the velocity just after hitting the floor, use $v^2 = v_2^2 - 2g(y - y_0)$, with $v = 0$, $y = -1.00\,\mathrm{m}$, and $y_0 = -4.00\,\mathrm{m}$. The result is

$$v_2 = \sqrt{2g(y - y_0)} = \sqrt{2(9.8\,\mathrm{m/s^2})(-1.00\,\mathrm{m} + 4.00\,\mathrm{m})} = 7.67\,\mathrm{m/s}\,.$$

The average acceleration is

$$\bar{a} = \frac{v_2 - v_1}{\Delta t} = \frac{7.67\,\mathrm{m/s} + 8.85\,\mathrm{m/s}}{10.0 \times 10^{-3}\,\mathrm{s}} = 1.65 \times 10^3\,\mathrm{m/s^2}\,.$$

80P

Let h be the height of the fall and t be the time of the fall. Take the y axis to be upward and place the origin at the point of release. Then $-h = -\frac{1}{2}gt^2$. Since it falls half the distance down in the last second, it falls the first half in time $t - 1$, where t is in seconds. Thus $-h/2 = -\frac{1}{2}g(t - 1)^2$. Solve these two equations simultaneously for t and h.

(a) First substitute $\frac{1}{2}gt^2$ for h in $h = g(t - 1)^2$ to obtain $t^2 = 2(t - 1)^2$. A little manipulation gives the quadratic equation $t^2 - 4t + 2 = 0$. It has the solutions $t = 2 \pm \sqrt{2} = +3.41\,\mathrm{s}$ and $-0.586\,\mathrm{s}$. You want the positive solution.

(b) To find the height substitute this value for t into $h = \frac{1}{2}gt^2$ to obtain

$$h = \frac{1}{2}(9.8\,\mathrm{m/s^2})(3.41\,\mathrm{s})^2 = 57.1\,\mathrm{m}\,.$$

The negative solution for t corresponds to a ball thrown upward, during the time before it reaches its highest point.

84P

Take the y axis to be upward and place the origin at the point from which the objects are dropped. Suppose the first object is dropped at time $t = 0$. Then its coordinate is given by $y_1 = -\frac{1}{2}gt^2$ and, since the second object starts 1.0 s later, its coordinate is given by $y_2 = -\frac{1}{2}g(t - 1)^2$, where t is in seconds. You want the time for which $y_2 - y_1 = 10\,\mathrm{m}$. Solve $-\frac{1}{2}g(t - 1)^2 + \frac{1}{2}gt^2 = 10$, with $g = 9.8\,\mathrm{m/s^2}$. After writing $(t - 1)^2$ as $t^2 - 2t + 1$ and canceling the terms in t^2 you should get $t = (10/g) + 0.5 = 1.5\,\mathrm{s}$.

86P

(a) Take the y axis to be upward and place the origin on the ground, under the balloon. Since the package is dropped, its initial velocity is the same as the velocity of the balloon, $+12\,\mathrm{m/s}$. The initial coordinate of the package is $y_0 = 80\,\mathrm{m}$; when it hits the ground its coordinate is zero. Solve $y = y_0 + v_0 t - \frac{1}{2}gt^2$ for t:

$$t = \frac{v_0}{g} \pm \sqrt{\frac{v_0^2}{g^2} + \frac{2y_0}{g}} = \frac{12\,\mathrm{m/s}}{9.8\,\mathrm{m/s^2}} + \sqrt{\frac{(12\,\mathrm{m/s})^2}{(9.8\,\mathrm{m/s^2})^2} + \frac{2(80\,\mathrm{m})}{9.8\,\mathrm{m/s^2}}} = 5.4\,\mathrm{s}\,,$$

where the positive solution was used. A negative value for t corresponds to a time before the package was dropped.

(b) Use $v = v_0 - gt = 12\,\text{m/s} - (9.8\,\text{m/s}^2)(5.4\,\text{s}) = -41\,\text{m/s}$. Its speed is $41\,\text{m/s}$.

87P

(a) Take the y axis to be upward and place the origin at the point where then ball is thrown. Take the time to be 0 at the instant the ball is thrown. Then the velocity of the ball is given by $v = v_0 - gt$ and at the highest point $v = 0$. Solve $0 = v_0 - gt$ for t and substitute the result into $y = y_0 + v_0 t - \frac{1}{2}gt^2$ to obtain $y = \frac{1}{2}v_0^2/g$. The initial velocity is $20\,\text{m/s} + 10\,\text{m/s} = 30\,\text{m/s}$. The result is

$$y = \frac{(30\,\text{m/s})^2}{2(9.8\,\text{m/s}^2)} = 46\,\text{m}.$$

It is then $46\,\text{m} + 30\,\text{m} = 76\,\text{m}$ above the ground.

(b) The coordinate of the ball is given by $y = v_0 t - \frac{1}{2}gt^2$. When the ball is thrown the floor of the elevator has coordinate $y_{f0} = -2.0\,\text{m}$ and the elevator is moving upward with constant velocity v_f. Thus the coordinate of the floor at time t is given by $y_f = y_{f0} + v_f t$. You want to solve for the time when the coordinate of the ball and elevator floor are the same: $y = y_f$. This means $v_0 t - \frac{1}{2}gt^2 = y_{f0} + v_f t$. The solution to this quadratic equation is

$$t = \frac{(v_0 - v_f)}{g} \pm \sqrt{\frac{(v_0 - v_f)^2}{g^2} - \frac{2y_{f0}}{g}}$$

$$= \frac{(30\,\text{m/s} - 10\,\text{m/s})}{9.8\,\text{m/s}^2} + \sqrt{\frac{(30\,\text{m/s} - 10\,\text{m/s})^2}{(9.8\,\text{m/s}^2)^2} - \frac{2(-2.0\,\text{m})}{(9.8\,\text{m/s}^2)}} = 4.2\,\text{s},$$

where the positive root was used. The negative root corresponds to a time before the ball is thrown.

Chapter 3

3E

(a) Draw the first vector (**a**) with a length, to scale, of 250 m, 30° east of north and the second vector (**b**) with a length, to the same scale, of 175 m, due east, as shown. The order is not important but the tail of one vector must be at the head of the other. Now draw the resultant vector (**r**), from the tail of the first to the head of the second. Use a rule to measure its length and use the scale factor to find its magnitude. Use a protractor to measure the angle it makes with one of the cardinal compass directions. The magnitude of the resultant is 370 m and the resultant is 36° north of east.

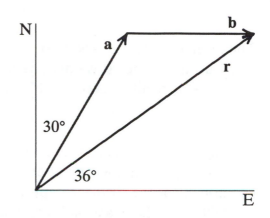

(b) The magnitude of her displacement is 370 m, the distance she walked is $250 + 175 = 425$ m.

6P

The vectors are drawn with the tail of one at the head of the other. The resultant is the vector from the tail of the first to the head of the second. The resultant has a magnitude of 4.2 and is 50.5° north of east.

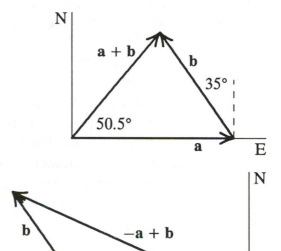

Draw the vector −**a**, opposite in direction to **a**, then draw the vector **b** with its tail at the head of −**a**. The vector −**a** + **b** is from the tail of −**a** to the head of **b**. Its magnitude is 8.0 and it is 24° north of west.

9E

The x component is given by $a_x = 7.3\cos 250° = -2.5$ and the y component is given by $a_y = 7.3\sin 250° = -6.9$. Notice that the vector is 70° below the negative x axis, so the components can also be computed using $a_x = -7.3\cos 70°$ and $a_y = -7.3\sin 70°$. It is also 20° from the negative y axis, so you might also use $a_x = -7.3\sin 20°$ and $a_y = -7.3\cos 20°$. All of these give the same result.

10E

(a) Use $a = \sqrt{a_x^2 + a_y^2}$ to obtain $a = \sqrt{(-25.0)^2 + (+40.0)^2} = 47.2$.

(b) The tangent of the angle between the vector and the positive x axis is

$$\tan\theta = \frac{a_y}{a_x} = \frac{-40.0}{25.0} = -1.6.$$

The inverse tangent is $-58.0°$ or $-58.0° + 180° = 122°$. The first angle has a positive cosine and a negative sine. It is not correct. The second angle has a negative cosine and a positive sine. It is correct for a vector with a negative x component and a positive y component.

17P

The point P is displaced vertically by $2R$, where R is the radius of the wheel. It is displaced horizontally by half the circumference of the wheel, or πR. Since $R = 45.0\,\text{cm}$, the horizontal component of the displacement is $1.41\,\text{m}$ and the vertical component of the displacement is $0.900\,\text{m}$. If the x axis is horizontal and the y axis is vertical, the vector displacement is $\mathbf{r} = (1.41\,\mathbf{i} + 0.900\,\mathbf{j})\,\text{m}$. The displacement has a magnitude of $1.68\,\text{m}$ and it is $32.5°$ above the floor.

19P

(a) The magnitude of the displacement is the distance from one corner to the diametrically opposite corner: $d = \sqrt{(10.0\,\text{ft})^2 + (12.0\,\text{ft})^2 + (14.0\,\text{ft})^2} = 21.0\,\text{ft}$. To see this look at the diagram of the room, with the displacement vector shown. The length of the diagonal across the floor, under the displacement vector, is given by the Pythagorean theorem: $L = \sqrt{\ell^2 + w^2}$, where ℓ is the length and w is the width of the room. Now this diagonal and the room height form a right triangle with the displacement vector as the hypotenuse, so the length of the displacement vector is given by

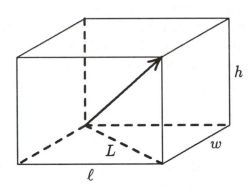

$$d = \sqrt{L^2 + h^2} = \sqrt{\ell^2 + w^2 + h^2}.$$

(b) The displacement vector is along the straight line from the beginning to the end point of the trip. Since a straight line is the shortest distance between two points the length of the path cannot be less than the magnitude of the displacement. It can be greater, however. The fly might, for example, crawl along the edges of the room. Its displacement would be the same but the path length would be $\ell + w + h$. The path length is the same as the magnitude of the displacement if the fly flies along the displacement vector.

(c) Take the x axis to be out of the page, the y axis to be to the right, and the z axis to be upward. Then the x component of the displacement is $w = 12\,\text{ft}$, the y component of the displacement is 14 ft, and the z component is 10 ft. Thus $\mathbf{d} = (12\,\mathbf{i} + 14\,\mathbf{j} + 10\,\mathbf{k})$ ft. You may write an equally correct answer by interchanging the length, width, and height.

(d) Suppose the path of the fly is as shown by the dotted lines and let s be the distance shown. The length of the path is then given by $L = \sqrt{s^2 + w^2} + \sqrt{(\ell - s)^2 + h^2}$. Find the value of s for which this is a minimum. The derivative $dL/ds = 0$ at a minimum so $s[s^2 + w^2]^{-1/2} - (\ell - s)[(\ell - s)^2 + h^2]^{-1/2} = 0$. Solve this equation by moving one of the terms to the right side and squaring both sides. Following a little algebra you should obtain $s^2(w^2 - h^2) - 2\ell w^2 s + w^2\ell^2 = 0$. This quadratic equation has the two solutions $s = \ell w/(w - h)$ and $s = \ell w/(w + h)$. The first leads to a value that is larger than ℓ and must be rejected on physical grounds. Take $s = \ell w/(w + h)$.

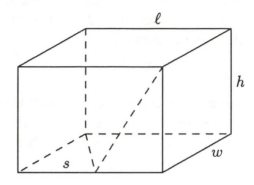

This value is substituted into the equation for L to find the minimum path length:

$$L_{\min} = \sqrt{\ell^2 + w^2 + h^2 + 2wh}$$

$$= \sqrt{(14\,\text{ft})^2 + (12\,\text{ft})^2 + (10\,\text{ft})^2 + 2(12\,\text{ft})(10\,\text{ft})} = 26.1\,\text{ft}.$$

22E

(a) Let $\mathbf{r} = \mathbf{a} + \mathbf{b}$. Then $r_x = a_x + b_x = 4.0 - 13 = -9.0$ and $r_y = a_y + b_y = 3.0 + 7.0 = 10$. Thus $\mathbf{r} = -9.0\,\mathbf{i} + 10\,\mathbf{j}$.

(b) The magnitude of the resultant is $r = \sqrt{r_x^2 + r_y^2} = \sqrt{(-9)^2 + (10)^2} = 13$. The angle θ between the resultant and the positive x axis is given by $\tan\theta = r_y/r_x = -10/9 = -1.1$. θ is either $-48°$ or $132°$. The first angle has a positive cosine and a negative sine while the second angle has a negative cosine and positive sine. Since the x component of the resultant is negative and the y component is positive, $\theta = 132°$.

24E

(a) Let $\mathbf{r} = \mathbf{a} + \mathbf{b}$. Then $r_x = a_x + b_x = 4 - 1 = 3$, $r_y = a_y + b_y = -3 + 1 = -2$, and $r_z = a_z + b_z = 1 + 4 = 5$. Thus $\mathbf{r} = 3\,\mathbf{i} - 2\,\mathbf{j} + 5\,\mathbf{k}$.

(b) Let $\mathbf{r} = \mathbf{a} - \mathbf{b}$. Then $r_x = a_x - b_x = 4 + 1 = 5$, $r_y = a_y - b_y = -3 - 1 = -4$, and $r_z = a_z - b_z = 1 - 4 = -3$. Thus $\mathbf{r} = 5\,\mathbf{i} - 4\,\mathbf{j} - 3\,\mathbf{k}$.

(c) Since $\mathbf{a} - \mathbf{b} + \mathbf{c} = 0$, $\mathbf{c} = \mathbf{b} - \mathbf{a} = (-1 - 4)\mathbf{i} + (1 + 3)\mathbf{j} + (4 - 1)\mathbf{k} = -5\,\mathbf{i} + 4\,\mathbf{j} + 3\,\mathbf{k}$.

27P

(a) The vector $\mathbf{a}$ has a magnitude 10.0 and makes the angle $30°$ with the positive x axis, so its components are $a_x = 10\cos 30° = 8.67$ and $a_y = 10.0\sin 30° = 5.0$. The vector $\mathbf{b}$ has a magnitude of 10.0 and makes an angle of $135°$ with the positive x axis, so its components are

$b_x = 10.0\cos 135° = -7.07$ and $b_y = 10.0\sin 135° = 7.07$. The components of the resultant are $r_x = a_x + b_x = 8.67 - 7.07 = 1.60$ and $r_y = a_y + b_y = 5.0 + 7.07 = 12.1$.

(b) The magnitude of $\mathbf{r}$ is $r = \sqrt{r_x^2 + r_y^2} = \sqrt{(1.60)^2 + (12.1)^2} = 12.2$.

(c) The angle θ between $\mathbf{r}$ and the positive x axis is given by $\tan\theta = r_y/r_x = 12.1/1.60 = 7.56$. θ is either $82.5°$ or $262.5°$. The first angle has a positive cosine and a positive sine and so is the correct answer.

32P

Let $\mathbf{a}$ and $\mathbf{b}$ be two vectors such their sum is perpendicular to their difference. If two vectors are perpendicular to each other then their scalar product vanishes. Hence $(\mathbf{a}+\mathbf{b})\cdot(\mathbf{a}-\mathbf{b}) = 0$. Now $(\mathbf{a}+\mathbf{b})\cdot(\mathbf{a}-\mathbf{b}) = a^2 - \mathbf{a}\cdot\mathbf{b} + \mathbf{b}\cdot\mathbf{a} - b^2 = a^2 - b^2$, so $a^2 - b^2 = 0$, and since the magnitude of a vector is positive, $a = b$.

33P

Place the x axis along $\mathbf{a}$ and place the y axis so $\mathbf{a}$ and $\mathbf{b}$ are in the xy plane, with $\mathbf{b}$ making the angle θ with the positive x axis. $\mathbf{a}$ has the components $a_x = a$, $a_y = 0$ and $\mathbf{b}$ has the components $b_x = b\cos\theta$, $b_y = b\sin\theta$. The resultant $\mathbf{r}$ has the components $r_x = a_x + b_x = a + b\cos\theta$, $r_y = a_y + b_y = b\sin\theta$ and so has the magnitude

$$r = \sqrt{r_x^2 + r_y^2} = \sqrt{(a + b\cos\theta)^2 + (b\sin\theta)^2}$$
$$= \sqrt{a^2 + 2ab\cos\theta + b^2\cos^2\theta + b^2\sin^2\theta} = \sqrt{a^2 + b^2 + 2ab\cos\theta}.$$

The trigonometric identity $\cos^2\theta + \sin^2\theta = 1$ was used.

34P

(a) There are 4 such lines, one from each of the corners on the lower face to the diametrically opposite corner on the upper face. One is shown on the diagram to the right. Place the coordinate system as shown, with the origin at the back, lower, left corner. The position vector of the starting point of the diagonal shown is $a\,\mathbf{i}$ and the position vector of the ending point is $a\,\mathbf{j} + a\,\mathbf{k}$ so the vector along the line is $a\,\mathbf{j} + a\,\mathbf{k} - a\,\mathbf{i}$. The point diametrically opposite the origin has position vector $a\,\mathbf{i} + a\,\mathbf{j} + a\,\mathbf{k}$ and this is the vector along the diagonal. Another corner of the bottom face is at $a\,\mathbf{i} + a\,\mathbf{j}$ and the diametrically opposite corner is at $a\,\mathbf{k}$, so another cube diagonal is $a\,\mathbf{k} - a\,\mathbf{i} - a\,\mathbf{j}$. The fourth diagonal runs from $a\,\mathbf{j}$ to $a\,\mathbf{i} + a\,\mathbf{k}$, so the vector along the diagonal is $a\,\mathbf{i} + a\,\mathbf{k} - a\,\mathbf{j}$.

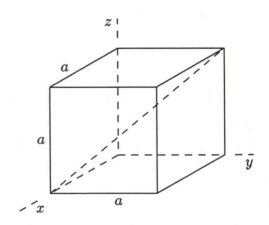

(b) Consider the vector from the back, lower, left corner to the front, upper, right corner. It is $a\mathbf{i} + a\mathbf{j} + a\mathbf{k}$. We may think of it as the sum of the vector $a\mathbf{i}$ parallel to the x axis and the vector $a\mathbf{j} + a\mathbf{k}$ perpendicular to the x axis. The tangent of the angle between the vector and the x axis is the perpendicular component divided by the parallel component. Since the magnitude of the perpendicular component is $\sqrt{a^2 + a^2} = \sqrt{2}a$ and the magnitude of the parallel component is a, $\tan\theta = \sqrt{2}a/a = \sqrt{2}$. Thus $\theta = 54.7°$. The angle between the vector and each of the other two adjacent sides (the y and z axes) is the same as is the angle between any of the other diagonal vectors and any of the cube sides adjacent to them.

(c) The length of any of the diagonals is given by $\sqrt{a^2 + a^2 + a^2} = \sqrt{3}a$.

36E

(a) The vector $\mathbf{a}$ is $56.0°$ above the positive x axis, so its components in the unprimed coordinate system are

$$a_x = (17.0\,\text{m})\cos 56.0° = 9.51\,\text{m}$$

and

$$a_y = (17.0\,\text{m})\sin 56.0° = 14.1\,\text{m}.$$

(b) The vector is $56.0° - 18.0° = 38.0°$ above the positive x' axis, so its components in the primed coordinate system are

$$a_{x'} = (17.0\,\text{m})\cos 38.0° = 13.4\,\text{m}$$

and

$$a_{y'} = (17.0\,\text{m})\sin 38.0° = 10.5\,\text{m}.$$

43E

(a) The scalar product is given by $\mathbf{a}\cdot\mathbf{b} = ab\cos\phi = (10)(6.0)\cos 60° = 30$.

(b) The magnitude of the vector product is given by $|\mathbf{a}\times\mathbf{b}| = ab\sin\phi = (10)(6.0)\sin 60° = 52$.

48P

Since $ab\cos\phi = a_x b_x + a_y b_y + a_z b_z$,

$$\cos\phi = \frac{a_x b_x + a_y b_y + a_z b_z}{ab}.$$

The magnitudes of the vectors given in the problem are $a = \sqrt{(3.0)^2 + (3.0)^2 + (3.0)^2} = 5.2$ and $b = \sqrt{(2.0)^2 + (1.0)^2 + (3.0)^3} = 3.7$. The angle between them is given by

$$\cos\phi = \frac{(3.0)(2.0) + (3.0)(1.0) + (3.0)(3.0)}{(5.2)(3.7)} = 0.926$$

and the angle is $22°$.

55P

Problem 34 showed that each diagonal has a length given by $\sqrt{3}a$, where a is the length of a cube edge. Vectors along two diagonals are $\mathbf{b} = a\,\mathbf{i} + a\,\mathbf{j} + a\,\mathbf{k}$ and $\mathbf{c} = -a\,\mathbf{i} + a\,\mathbf{j} + a\,\mathbf{k}$. Use the technique outlined in Problem 48 to find the angle between them:

$$\cos\phi = \frac{b_x c_x + b_y c_y + b_z c_z}{bc} = \frac{-a^2 + a^2 + a^2}{3a^2} = \frac{1}{3}.$$

The angle is $\phi = \cos^{-1}(1/3) = 70.5°$.

56P

The area of a triangle is half the product of its base and altitude. Take the base to be the side formed by vector $\mathbf{a}$. Then the altitude is $b \sin\phi$ and the area is $A = \frac{1}{2}ab\sin\phi = \frac{1}{2}|\mathbf{a} \times \mathbf{b}|$.

57P

(*a*) $(\mathbf{b} \times \mathbf{a})$ is a vector that is perpendicular to $\mathbf{a}$, so the scalar product of $\mathbf{a}$ with this vector is zero.

(*b*) Let $\mathbf{c} = \mathbf{b} \times \mathbf{a}$. Then the magnitude of $\mathbf{c}$ is $c = ab\sin\phi$. Since $\mathbf{c}$ is perpendicular to $\mathbf{a}$ the magnitude of $\mathbf{a} \times \mathbf{c}$ is ac. The magnitude of $\mathbf{a} \times (\mathbf{b} \times \mathbf{a})$ is therefore $|\mathbf{a} \times (\mathbf{b} \times \mathbf{a})| = ac = a^2 b \sin\phi$. The vector $\mathbf{a} \times (\mathbf{b} \times \mathbf{a})$ is in the plane of $\mathbf{a}$ and $\mathbf{b}$ and is perpendicular to $\mathbf{a}$.

59P

The area of a parallelepiped is equal to the product of its altitude and the area of its base. Take the base to be the parallelogram formed by the vectors $\mathbf{b}$ and $\mathbf{c}$. Its area is $bc \sin\phi$, where ϕ is the angle between $\mathbf{b}$ and $\mathbf{c}$. This is just the magnitude of the vector product $\mathbf{b} \times \mathbf{c}$. See Problem 56 and Fig. 3–30. The area of the parallelogram shown there is just twice the area of the triangle.

The altitude of the parallelepiped is $a \cos\theta$, where θ is the angle between $\mathbf{a}$ and the normal to the plane of $\mathbf{b}$ and $\mathbf{c}$. Since $\mathbf{b} \times \mathbf{c}$ is normal to that plane, θ is the angle between $\mathbf{a}$ and $\mathbf{b} \times \mathbf{c}$. Thus the volume of the parallelepiped is $V = a|\mathbf{b} \times \mathbf{c}| \cos\theta = \mathbf{a} \cdot (\mathbf{b} \times \mathbf{c})$.

Chapter 4

6E

The average velocity is the total displacement divided by the time interval. The total displacement $\mathbf{r}$ is the sum of 3 displacements, each calculated as the product of a velocity and a time interval. The first has a magnitude of $(60.0\,\text{km/h})(40.0\,\text{min})/(60.0\,\text{min/h}) = 40.0\,\text{km}$. Its direction is east. If we take the x axis to be toward the east and the y axis to be toward the north, then this displacement is $\mathbf{r}_1 = (40.0\,\mathbf{i})\,\text{km}$.

The second has a magnitude of $(60.0\,\text{km/h})(20.0\,\text{min})/(60.0\,\text{min/h}) = 20.0\,\text{km}$. Its direction is $50.0°$ east of north, so it may be written $\mathbf{r}_2 = (20.0\sin 50.0°\,\mathbf{i} + 20.0\cos 50.0°\,\mathbf{j})\,\text{km} = (15.3\,\mathbf{i} + 12.9\,\mathbf{j})\,\text{km}$.

The third has a magnitude of $(60.0\,\text{km/h})(50.0\,\text{min})/(60.0\,\text{min/h}) = 50.0\,\text{km}$. Its direction is west, so the displacement may be written $\mathbf{r}_3 = (-50\,\mathbf{i})\,\text{km}$. The total displacement is

$$\mathbf{r} = \mathbf{r}_1 + \mathbf{r}_2 + \mathbf{r}_3 = (40.0\,\mathbf{i})\,\text{km} + (15.3\,\mathbf{i} + 12.9\,\mathbf{j})\,\text{km} - (50\,\mathbf{i})\,\text{km}$$
$$= (5.3\,\mathbf{i} + 12.9\,\mathbf{j})\,\text{km}\,.$$

The total time for the trip is $40\,\text{min} + 20\,\text{min} + 50\,\text{min} = 110\,\text{min} = 1.83\,\text{h}$. Divide $\mathbf{r}$ by this interval to obtain an average velocity of $\bar{\mathbf{v}} = 2.89\,\mathbf{i} + 7.04\,\mathbf{j}$, in km/h. The average velocity has a magnitude of $7.61\,\text{km/h}$ and is directed $67.7°$ north of east.

11E

(a) The velocity is the derivative of the position vector with respect to time:

$$\mathbf{v} = \frac{\mathrm{d}}{\mathrm{d}t}\,(\mathbf{i} + 4t^2\,\mathbf{j} + t\,\mathbf{k}) = 8t\,\mathbf{j} + \mathbf{k}\,.$$

(b) The acceleration is the derivative of the velocity with respect to time:

$$\mathbf{a} = \frac{\mathrm{d}}{\mathrm{d}t}\,(8t\,\mathbf{j} + \mathbf{k}) = 8\,\mathbf{j}\,.$$

The velocity is in m/s and the acceleration is in m/s^2.

15P

(a) The velocity of the particle at any time t is given by $\mathbf{v} = \mathbf{v}_0 + \mathbf{a}t$, where $\mathbf{v}_0$ is the initial velocity and $\mathbf{a}$ is the acceleration. The x component is $v_x = v_{0x} + a_x t = 3.00 - 1.00t$ and the y component is $v_y = v_{0y} + a_y t = -0.500t$. When the particle reaches its maximum x coordinate $v_x = 0$. This means $3.00 - 1.00t = 0$ or $t = 3.00\,\text{s}$. The y component of the velocity at this time is $v_y = (-0.50)(3.00) = -1.50\,\text{m/s}$.

(b) The coordinates of the particle at any time t are $x = v_{0x}t + \frac{1}{2}a_x t^2$ and $y = v_{0y}t + \frac{1}{2}a_y t^2$. At $t = 3.00\,$s their values are

$$x = (3.00)(3.00) - \frac{1}{2}(1.00)(3.00)^2 = 4.50\,\text{m}$$

and

$$y = -\frac{1}{2}(0.50)(3.00)^2 = -2.25\,\text{m}.$$

18E

(a) Take the positive y axis to be downward and place the origin at the firing point. Then the y coordinate of the bullet is given by $y = \frac{1}{2}gt^2$. If t is the time of flight and y is the distance the bullet hits below the target, then

$$t = \sqrt{2y/g} = \sqrt{\frac{2(0.75\,\text{in.})}{(12\,\text{in.}/\text{ft})(32\,\text{ft/s}^2)}} = 6.25 \times 10^{-2}\,\text{s}.$$

Notice that $0.75\,$in. was converted to $(0.75/12)\,$ft.

(b) The muzzle velocity is the initial velocity of the bullet. It is horizontal. If x is the horizontal distance to the target, then $x = v_0 t$ and

$$v_0 = \frac{x}{t} = \frac{100\,\text{ft}}{6.25 \times 10^{-2}\,\text{s}} = 1.6 \times 10^3\,\text{ft/s}.$$

20E

(a) The horizontal component of the velocity is constant. If ℓ is the length of a plate and t is the time an electron is between the plates, then $\ell = v_0 t$, where v_0 is the initial speed. Thus

$$t = \frac{\ell}{v_0} = \frac{2.0\,\text{cm}}{1.0 \times 10^9\,\text{cm/s}} = 2.0 \times 10^{-9}\,\text{s}.$$

(b) The vertical displacement of the electron is

$$y = \frac{1}{2}at^2 = \frac{1}{2}(1.0 \times 10^{17}\,\text{cm/s}^2)(2.0 \times 10^{-9}\,\text{s})^2 = 0.20\,\text{cm},$$

where down was taken to be positive.

(c) The x component of velocity is $v_x = v_0 = 1.0 \times 10^9\,$cm/s and the y component is $v_y = a_y t = (1.0 \times 10^{17}\,\text{cm/s}^2)(2.0 \times 10^{-9}\,\text{s}) = 2.0 \times 10^8\,$cm/s.

25E

Take the y axis to be upward and the x axis to be horizontal and to the right. Place the origin at the throwing point and suppose the stone is thrown at time $t = 0$. The x component of its initial velocity is given by $v_{0x} = v_0 \cos\theta_0$ and the y component is given by $v_{0y} = v_0 \sin\theta_0$, where v_0 is the initial speed and θ_0 is the launch angle.

(a) At $t = 1.10\,\mathrm{s}$ its x coordinate is $x = v_0 t \cos\theta_0 = (20.0\,\mathrm{m/s})(1.10\,\mathrm{s})\cos 40.0° = 16.9\,\mathrm{m}$ and its y coordinate is $y = v_0 t \sin\theta_0 - \frac{1}{2}gt^2 = (20.0\,\mathrm{m/s})(1.10\,\mathrm{s})\sin 40° - \frac{1}{2}(9.80\,\mathrm{m/s^2})(1.10\,\mathrm{s})^2 = 8.21\,\mathrm{m}$.

(b) At $t = 1.80\,\mathrm{s}$, its x coordinate is $x = (20.0\,\mathrm{m/s})(1.80\,\mathrm{s})\cos 40.0° = 27.6\,\mathrm{m}$ and its y coordinate is $y = (20.0\,\mathrm{m/s})(1.80\,\mathrm{s})\sin 40° - \frac{1}{2}(9.80\,\mathrm{m/s^2})(1.80\,\mathrm{s})^2 = 7.26\,\mathrm{m}$.

(c) The stone hits the ground earlier than $t = 5.0\,\mathrm{s}$. To find the time when it hits the ground solve $y = v_0 t \sin\theta_0 - \frac{1}{2}gt^2 = 0$ for t. You should get

$$t = \frac{2v_0}{g}\sin\theta_0 = \left[\frac{2(20.0\,\mathrm{m/s})}{9.8\,\mathrm{m/s^2}}\right]\sin 40° = 2.62\,\mathrm{s}.$$

Its x coordinate on landing is

$$x = v_0 t \cos\theta_0 = (20.0\,\mathrm{m/s})(2.62\,\mathrm{s})\cos 40° = 40.1\,\mathrm{m}.$$

Assuming it stays where it lands its coordinates at $t = 5.00\,\mathrm{s}$ are $x = 40.1\,\mathrm{m}$ and $y = 0$.

33P

Take the y axis to be upward and the x axis to the horizontal. Place the origin at the firing point, let the time be zero at firing, and let θ_0 be the firing angle. If the target is a distance d away then its coordinates are $x = d$, $y = 0$. The kinematic equations are $d = v_0 t \cos\theta_0$ and $0 = v_0 t \sin\theta_0 - \frac{1}{2}gt^2$. Eliminate t and solve for θ_0. The first equation gives $t = d/v_0\cos\theta_0$. This expression is substituted into the second equation to obtain $2v_0^2 \sin\theta_0 \cos\theta_0 - gd = 0$. Use the trigonometric identity $\sin\theta_0 \cos\theta_0 = \frac{1}{2}\sin(2\theta_0)$ to obtain $v_0^2 \sin(2\theta_0) = gd$ or

$$\sin(2\theta_0) = \frac{gd}{v_0^2} = \frac{(32\,\mathrm{ft/s^2})(150\,\mathrm{ft})}{(1500\,\mathrm{ft/s})^2} = 2.13 \times 10^{-3}.$$

The firing angle is $0.0611°$. If the gun is aimed at a point a distance ℓ above the target then $\tan\theta_0 = \ell/d$ or $\ell = d\tan\theta_0 = (150\,\mathrm{ft})\tan 0.0611° = 0.16\,\mathrm{ft} = 1.9\,\mathrm{in}$.

34P

Take the y axis to be upward and place the origin at the firing point. Then the y coordinate is given by $y = v_0 t \sin\theta_0 - \frac{1}{2}gt^2$ and the y component of the velocity is given by $v_y = v_0 \sin\theta_0 - gt$. The maximum height occurs when $v_y = 0$. This means $t = (v_0/g)\sin\theta_0$ and

$$y = v_0\left(\frac{v_0}{g}\right)\sin\theta_0 \sin\theta_0 - \frac{1}{2}\frac{g(v_0\sin\theta_0)^2}{g^2} = \frac{(v_0\sin\theta_0)^2}{g} - \frac{1}{2}\frac{(v_0\sin\theta_0)^2}{g} = \frac{(v_0\sin\theta_0)^2}{2g}.$$

42P

Take the y axis to be upward and the x axis to be horizontal. Place the origin at point where the ball is kicked and take the time to be zero when it is kicked. $x\ (= 50\,\mathrm{yd} = 150\,\mathrm{ft})$ and

y (= -5.0 ft) are coordinates of the landing point. The ball lands at time t (= 4.5 s). Since $x = v_{0x}t$,

$$v_{0x} = \frac{x}{t} = \frac{150\,\text{ft}}{4.5\,\text{s}} = 33.3\,\text{ft/s}.$$

Since $y = v_{0y}t - \frac{1}{2}gt^2$,

$$v_{0y} = \frac{y + \frac{1}{2}gt^2}{t} = \frac{(-5.0\,\text{ft}) + \frac{1}{2}(32\,\text{ft/s}^2)(4.5\,\text{s})^2}{4.5\,\text{s}} = 70.9\,\text{ft/s}.$$

The magnitude of the initial velocity is $v_0 = \sqrt{v_{0x}^2 + v_{0y}^2} = \sqrt{(33.3\,\text{ft/s})^2 + (70.9\,\text{ft/s})^2} = 78.3\,\text{ft/s}$. The firing angle satisfies $\tan\theta_0 = v_{0y}/v_{0x} = (70.9\,\text{ft/s})/(33.3\,\text{ft/s}) = 2.13$. The angle is $\theta_0 = 64.8°$.

49P

(a) Since the projectile is released its initial velocity is the same as the velocity of the plane at the time of release. Take the y axis to be upward and the x axis to be horizontal. Place the origin at the point of release and take the time to be zero at release. Let x and y (= -730 m) be the coordinates of the point on the ground where the projectile hits and let t be the time when it hits. Then

$$y = -v_0 t \cos\theta_0 - \frac{1}{2}gt^2,$$

where $\theta_0 = 53.0°$. This equation gives

$$v_0 = -\frac{y + \frac{1}{2}gt^2}{t\cos\theta_0} = -\frac{-730\,\text{m} + \frac{1}{2}(9.80\,\text{m/s}^2)(5.00\,\text{s})^2}{(5.00\,\text{s})\cos(-53.0°)} = 202\,\text{m/s}.$$

(b) The horizontal distance traveled is $x = v_0 t \sin\theta_0 = (202\,\text{m/s})(5.00\,\text{s})\sin(-53.0°) = 806$ m.

(c) The x component of the velocity is $v_x = v_0 \sin\theta_0 = (202\,\text{m/s})\sin(53.0°) = 161$ m/s and the y component is $v_y = -v_0 \cos\theta_0 - gt = -(202\,\text{m/s})\cos(53°) - (9.80\,\text{m/s}^2)(5.00\,\text{s}) = -171$ m/s.

54P

You want to know how high the ball is from the ground when it is 320 ft from home plate. To calculate this quantity you need to know the initial speed v_0 of the ball. Use the range information to obtain v_0. Put the origin at the point where the ball is hit, take the y axis to be upward and the x axis to be horizontal. If x (= 350 ft) and y (= -4.0 ft) are the coordinates of the ball when it lands, then $x = v_0 t \cos\theta_0$ and $y = v_0 t \sin\theta_0 - \frac{1}{2}gt^2$, where t is the time of flight of the ball. The first equation gives $t = x/v_0\cos\theta_0$ and when this is substituted into the second, the result is

$$y = x\tan\theta_0 - \frac{1}{2}\frac{gx^2}{v_0^2\cos^2\theta_0}.$$

Solve for v_0. You should obtain

$$v_0 = \sqrt{\frac{gx^2}{2(x\tan\theta_0 - y)\cos^2\theta_0}} = \sqrt{\frac{(32\,\text{ft/s}^2)(350\,\text{ft})^2}{2(354\,\text{ft})\cos^2 45°}} = 105\,\text{ft/s}\,.$$

Now take x and y to be the coordinates when the ball is at the fence. Again $x = v_0 t\cos\theta_0$ and $y = v_0 t\sin\theta_0 - \frac{1}{2}gt^2$. The time to reach the fence is given by $t = x/v_0\cos\theta_0$. When this is substituted into the second equation the result is

$$y = x\tan\theta_0 - \frac{gx^2}{2v_0^2\cos^2\theta_0} = (320\,\text{ft})\tan 45° - \frac{(32\,\text{ft/s}^2)(320\,\text{ft})^2}{2(105\,\text{ft/s})^2\cos^2 45°} = 24.1\,\text{ft}\,.$$

Since the ball started 4.0 ft above the ground it is 28.1 ft above the ground when it gets to the fence and it is 4.1 ft above the top of the fence. It goes over the fence.

55P

Take the y axis to be upward and the x axis to be horizontal. Place the origin at the point where the ball is kicked, on the ground, and take the time to be zero at the instant it is kicked. x and y are the coordinates of ball at the goal post. You want to find the kicking angle θ_0 so that $y = 3.44$ m when $x = 50$ m. Write the kinematic equations for projectile motion: $x = v_0 t\cos\theta_0$ and $y = v_0 t\sin\theta_0 - \frac{1}{2}gt^2$. The first equation gives $t = x/v_0\cos\theta_0$ and when this is substituted into the second the result is

$$y = x\tan\theta_0 - \frac{1}{2}\frac{gx^2}{(v_0\cos\theta_0)^2}\,.$$

You may solve this by trial and error: systematically try values of θ_0 until you find the two that satisfy the equation. A little manipulation, however, will give you an algebraic solution. Use the trigonometric identity $1/\cos^2\theta_0 = 1 + \tan^2\theta_0$ to obtain

$$\frac{1}{2}\frac{gx^2}{v_0^2}\tan^2\theta_0 - x\tan\theta_0 + y + \frac{1}{2}\frac{gx^2}{v_0^2} = 0\,.$$

This is a quadratic equation for $\tan\theta_0$. To simplify writing the solution, let $c = \frac{1}{2}gx^2/v_0^2 = \frac{1}{2}(9.80\,\text{m/s}^2)(50\,\text{m})^2/(25\,\text{m/s})^2 = 19.6$ m. Then the quadratic equation becomes $c\tan^2\theta_0 - x\tan\theta_0 + y + c = 0$. It has the solution

$$\tan\theta_0 = \frac{x \pm \sqrt{x^2 + 4(y+c)c}}{2c}$$

$$= \frac{50\,\text{m} \pm \sqrt{(50\,\text{m})^2 - 4(3.44\,\text{m} + 19.6\,\text{m})(19.6\,\text{m})}}{2(19.6\,\text{m})}\,.$$

The two solutions are $\tan\theta_0 = 1.95$ and $\tan\theta_0 = 0.605$. The corresponding angles are $\theta_0 = 63°$ and $\theta_0 = 31°$. If kicked at any angle between these two, the ball will travel above the cross bar on the goalposts.

56P

Write the kinematic equations for the shell and for the tank, with the firing time for the shell delayed from the starting time of the tank. Take the y axis to be upward and the x axis to be horizontal. Place the origin at the firing point and take the time of firing to be t_0. The coordinates of the shell are given by $x = v_0(t - t_0) \cos \theta_0$ and $y = v_0(t - t_0) \sin \theta_0 - \frac{1}{2}g(t - t_0)^2$, where $v_0 = 240 \, \text{m/s}$ and $\theta_0 = 10°$. Let x_0 (= 2.2 km) and y_0 (= −60 m) be the initial coordinates of the tank. Then the tank coordinates are $x = x_0 + \frac{1}{2}at^2$ and $y = y_0$. At the time the shell hits the tank the coordinates of the shell must be the same as the coordinates of the tank: $x_0 + \frac{1}{2}at^2 = v_0(t - t_0) \cos \theta_0$ and $y_0 = v_0(t - t_0) \sin \theta_0 - \frac{1}{2}g(t - t_0)^2$.

The second equation is a quadratic equation in $(t - t_0)$ and can be solved for that quantity:

$$t - t_0 = \frac{v_0 \sin \theta_0 \pm \sqrt{(v_0 \sin \theta_0)^2 - 2gy_0}}{g}$$

$$= \frac{(240 \, \text{m/s}) \sin 10° \pm \sqrt{[(240 \, \text{m/s}) \sin 10°]^2 + 2(9.8 \, \text{m/s}^2)(60 \, \text{m})}}{9.8 \, \text{m/s}^2} = 9.76 \, \text{s} .$$

The positive square root was used. Discard the other solution, using the negative square root, because it is negative and corresponds to a time before the gun is fired. Now solve $x_0 + \frac{1}{2}at^2 = v_0(t - t_0) \cos \theta_0$ for t:

$$t = \sqrt{\frac{2v_0(t - t_0) \cos \theta_0 - 2x_0}{a}}$$

$$= \sqrt{\frac{2(240 \, \text{m/s})(9.76 \, \text{s}) \cos 10° - 2(2.2 \times 10^3 \, \text{m})}{0.90 \, \text{m/s}^2}} = 15.4 \, \text{s} .$$

Now $t_0 = t - (t - t_0) = (15.4 \, \text{s}) - (9.8 \, \text{s}) = 5.6 \, \text{s}$.

62E

(a) The circumference of the circular orbit is given by $2\pi r$, where r is the radius. Since the radius of the Earth is 6.37×10^6 m, the radius of the satellite orbit is 6.37×10^6 m + 640 × 10^3 m = 7.01×10^6 m. The satellite travels a distance equal to one circumference in a time interval equal to one period, so its speed is

$$v = \frac{2\pi r}{T} = \frac{2\pi(7.01 \times 10^6 \, \text{m})}{(98.0 \, \text{min})(60 \, \text{s/min})} = 7.49 \times 10^3 \, \text{m/s} .$$

(b) The magnitude of the acceleration is given by $a = v^2/r = (7.49 \times 10^3 \, \text{m/s})^2/(7.01 \times 10^6 \, \text{m}) = 8.00 \, \text{m/s}^2$.

67E

(a) Let r be the radius of the orbit and solve $a = v^2/r$ for v:

$$v = \sqrt{ra} = \sqrt{(5.0 \, \text{m})(7.0)(9.8 \, \text{m/s}^2)} = 19 \, \text{m/s} .$$

(b) If the astronaut goes around N times in time t then his speed is $v = 2\pi Nr/t$. Solve for N/t, then substitute $v = 19$ m/s and $r = 5.0$ m: $N/t = v/2\pi r = (19 \text{ m/s})(60 \text{ s/min})/2\pi(5.0 \text{ m}) = 35$ rev/min.

68P

The magnitude of the acceleration is given by $a = v^2/R$, where R is the radius of the wheel, and v is the speed of the passenger. Since the passenger goes a distance $2\pi R$ for each revolution, his speed is $5(2\pi)(15 \text{ m})/(60 \text{ s}) = 7.85$ m/s and his acceleration is

$$a = \frac{(7.85 \text{ m/s})^2}{15 \text{ m}} = 4.1 \text{ m/s}^2.$$

(a) At the highest point the acceleration is downward and (b) at the lowest point it is upward, always toward the center of the circle. It has the same magnitude at every position.

69P

(a) The speed of a person at the Earth's equator is $v = 2\pi R/T$, where R is the radius of the Earth (6.37×10^6 m) and T is the length of a day (8.64×10^4 s): $v = 2\pi(6.37 \times 10^6 \text{ m})/(8.64 \times 10^4 \text{ s}) = 463$ m/s. The magnitude of the acceleration is given by $a = v^2/R = (463 \text{ m/s})^2/(6.37 \times 10^6 \text{ m}) = 0.034 \text{ m/s}^2$.
(b) If T is the period, then $v = 2\pi R/T$ is the speed and $a = v^2/R = 4\pi^2 R^2/T^2 R = 4\pi^2 R/T^2$ is the magnitude of the acceleration. Thus

$$T = 2\pi\sqrt{\frac{R}{a}} = 2\pi\sqrt{\frac{6.37 \times 10^6 \text{ m}}{9.8 \text{ m/s}^2}} = 5.1 \times 10^3 \text{ s} = 84 \text{ min}.$$

72P

To calculate the centripetal acceleration of the stone you need to know its speed while it is being whirled around. This the same as its initial speed when it flies off. Use the kinematic equations of projectile motion to find that speed. Take the y axis to be upward and the x axis to be horizontal. Place the origin at the point where the stone leaves its circular orbit and take the time to be zero when this occurs. Then the coordinates of the stone when it is a projectile are given by $x = v_0 t$ and $y = -\frac{1}{2}gt^2$. It hits the ground when $x = 10$ m and $y = -2.0$ m. Note that the initial velocity is horizontal. Solve the second equation for the time: $t = \sqrt{-2y/g}$. Substitute this expression into the first equation and solve for v_0:

$$v_0 = x\sqrt{-\frac{g}{2y}} = (10 \text{ m})\sqrt{-\frac{9.8 \text{ m/s}^2}{2(-2.0 \text{ m})}} = 15.7 \text{ m/s}.$$

The magnitude of the centripetal acceleration is $a = v^2/r = (15.7 \text{ m/s})^2/(1.5 \text{ m}) = 160 \text{ m/s}^2$.

74E

When the escalator is stalled the speed of the person is $v_p = \ell/t$, where ℓ is the length of the escalator and t is the time the person takes to walk up it. This is $v_p = (15 \text{ m})/(90 \text{ s}) =$

0.167 m/s. The escalator moves at $v_e = (15\,\text{m})/(60\,\text{s}) = 0.250\,\text{m/s}$. The speed of the person walking up the moving escalator is $v = v_p + v_e = 0.167\,\text{m/s} + 0.250\,\text{m/s} = 0.417\,\text{m/s}$ and the time taken to move the length of the escalator is $t = \ell/v = (15\,\text{m})/(0.417\,\text{m/s}) = 36\,\text{s}$.

If the various times given are independent of the escalator length then the answer does not depend on that length either. In terms of ℓ (in meters) the speed (in meters per second) of the person walking on the stalled escalator is $\ell/90$, the speed of the moving escalator is $\ell/60$, and the speed of the person walking on the moving escalator is $v = (\ell/90) + (\ell/60) = 0.0278\ell$. The time taken is $t = \ell/v = \ell/0.0278\ell = 36\,\text{s}$ and is independent of ℓ.

80E

Relative to the car the velocity of the snowflakes have a vertical component of 8.0 m/s and a horizontal component of $50\,\text{km/h} = 13.9\,\text{m/s}$. The angle θ from the vertical is given by $\tan\theta = v_h/v_v = (13.9\,\text{m/s})/(8.0\,\text{m/s}) = 1.74$. The angle is 60°.

83P

Since the raindrops fall vertically relative to the train, the horizontal component of the velocity of a raindrop is $v_h = 30\,\text{m/s}$, the same as the speed of the train. If v_v is the vertical component of the velocity and θ is the angle between the direction of motion and the vertical then $\tan\theta = v_h/v_v$. Thus $v_v = v_h/\tan\theta = (30\,\text{m/s})/\tan 22° = 74.3\,\text{m/s}$. The speed of a raindrop is
$$v = \sqrt{v_h^2 + v_v^2} = \sqrt{(30\,\text{m/s})^2 + (74.3\,\text{m/s})^2} = 80\,\text{m/s}.$$

86P

Let $\mathbf{v}_{PG}$ be the velocity of the plane relative to the ground, $\mathbf{v}_{AG}$ be the velocity of the air relative to the ground, and $\mathbf{v}_{PA}$ be the velocity of the plane relative to the air.
The vector diagram is shown to the right: $\mathbf{v}_{PG} = \mathbf{v}_{PA} + \mathbf{v}_{AG}$. Since the magnitudes v_{PG} and v_{PA} are equal the triangle is isosceles, with two sides of equal length. Consider either one of the right triangles formed when the bisector of θ is drawn (the dotted line). It bisects $\mathbf{v}_{AG}$, so

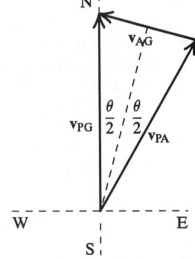

$$\sin(\theta/2) = \frac{v_{AG}}{2v_{PG}} = \frac{70.0\,\text{mi/h}}{2(135\,\text{mi/h})} = 0.259\,.$$

This means $\theta = 30.1°$. Now $\mathbf{v}_{AG}$ makes the same angle with the E-W line as the dotted line does with the N-S line. The wind is blowing in the direction 15° north of west.

(b) The plane is headed in the direction 30° east of north.

There is another solution, with the plane headed 30° west of north and the wind blowing 15° north of east.

Chapter 5

7E

Label the two forces $\mathbf{F}_1$ and $\mathbf{F}_2$. According to Newton's second law $\mathbf{F}_1 + \mathbf{F}_2 = m\mathbf{a}$, so $\mathbf{F}_2 = m\mathbf{a} - \mathbf{F}_1$. In unit vector notation $\mathbf{F}_1 = (20.0\,\mathbf{i})\,\mathrm{N}$ and $\mathbf{a} = (-12\sin 30°\,\mathbf{i} - 12\cos 30°\,\mathbf{j})\,\mathrm{m/s^2} = (-6.0\,\mathbf{i} - 10.4\,\mathbf{j})\,\mathrm{m/s^2}$. Thus $\mathbf{F}_2 = (2.0\,\mathrm{kg})(-6.0\,\mathbf{i}\,\mathrm{m/s^2}) + (2.0\,\mathrm{kg})(-10.4\,\mathbf{j}\,\mathrm{m/s^2}) - 20.0\,\mathbf{i}\,\mathrm{N} = (-32\,\mathbf{i} - 21\,\mathbf{j})\,\mathrm{N}$.

(b) To find the magnitude use $F_2 = \sqrt{F_{2x}^2 + F_{2y}^2} = \sqrt{(32)^2 + (21)^2} = 38\,\mathrm{N}$. The angle that $\mathbf{F}_2$ makes with the positive x axis is given by $\tan\theta = F_{2y}/F_{2x} = (21\,\mathrm{N})/(32\,\mathrm{N}) = 0.656$. The angle is either $33°$ or $33° + 180° = 213°$. Since both the x and y components are negative the correct result is $213°$.

11E

Use the relationship $W = mg$ between the weight W and mass m of an object. (a) The weight is $1400\,\mathrm{lb}$ and the mass is $m = W/g = (1400\,\mathrm{lb})/(32\,\mathrm{ft/s^2}) = 44\,\mathrm{slug}$.

(b) The mass is $421\,\mathrm{kg}$ and the weight is $W = mg = (421\,\mathrm{kg})(9.8\,\mathrm{m/s^2}) = 4100\,\mathrm{N}$.

14E

(a) The mass is $m = W/g = (22\,\mathrm{N})/(9.8\,\mathrm{m/s^2}) = 2.2\,\mathrm{kg}$. At a place where $g = 4.9\,\mathrm{m/s^2}$ the mass is still $2.2\,\mathrm{kg}$ but the weight is $W = mg = (2.2\,\mathrm{kg})(4.9\,\mathrm{m/s^2}) = 11\,\mathrm{N}$.

(b) At a place where $g = 0$ the mass is still $2.2\,\mathrm{kg}$ but the weight is zero.

18E

In all three cases the scale is not accelerating, which means that the two cords exert forces of equal magnitude on it. The scale reads the magnitude of either of these forces. In each case the tension in the cord attached to the salami must be the same in magnitude as the weight of the salami because the salami is not accelerating. Thus the scale reading is mg, where m is the mass of the salami. Its value is $(11.0\,\mathrm{kg})(9.8\,\mathrm{m/s^2}) = 108\,\mathrm{N}$.

21E

Use $F = ma$, where F is the magnitude of the force, a is the magnitude of the acceleration, and m is the mass. The acceleration can be found using constant acceleration kinematics. Solve $v = at$ for a: $a = v/t$. The velocity is $v = (1600\,\mathrm{km/h})(1000\,\mathrm{m/km})/(3600\,\mathrm{s/h}) = 444\,\mathrm{m/s}$, so $a = (444\,\mathrm{m/s})/(1.8\,\mathrm{s}) = 247\,\mathrm{m/s^2}$ and the force is $F = (500\,\mathrm{kg})(247\,\mathrm{m/s^2}) = 1.2 \times 10^5\,\mathrm{N}$.

23E

According to Newton's second law the magnitude of the force is given by $F = ma$, where a is the acceleration of the neutron. Use kinematics to find the acceleration that brings the neutron to rest in a distance d. Assume the acceleration is constant and solve $v^2 = v_0^2 + 2ad$ for a:

$$a = \frac{(v^2 - v_0^2)}{2d} = \frac{-(1.4 \times 10^7 \, \text{m/s})^2}{2(1.0 \times 10^{-14} \, \text{m})} = -9.8 \times 10^{27} \, \text{m/s}^2 \, .$$

The magnitude of the force is $F = ma = (1.67 \times 10^{-27} \, \text{kg})(9.8 \times 10^{27} \, \text{m/s}^2) = 16 \, \text{N}$.

28E

(a) The free-body diagram is shown in Fig. 27 of the text. Since the acceleration of the block is zero, Newton's second law in component form is $T - mg \sin \theta = 0$ and $N - mg \cos \theta = 0$. Solve the first equation for the tension in the string: $T = mg \sin \theta = (8.5 \, \text{kg})(9.8 \, \text{m/s}^2) \sin 30° = 42 \, \text{N}$.

(b) Solve the second equation for N: $N = mg \cos \theta = (8.5 \, \text{kg})(9.8 \, \text{m/s}^2) \cos 30° = 72 \, \text{N}$.

(c) When the string is cut it no longer exerts a force on the block and the block accelerates. The x component of the second law becomes $-mg \sin \theta = ma$, so $a = -g \sin \theta = -(9.8 \, \text{m/s}^2) \sin 30° = -4.9 \, \text{m/s}^2$. The minus sign indicates the acceleration is down the plane.

33E

The acceleration of the electron is vertical and for all practical purposes the only force acting on it is the electric force. The force of gravity is much smaller. Take the x axis to be in the direction of the initial velocity and the y axis to be in the direction of the electrical force. Place the origin at the initial position of the electron. Since the force and acceleration are constant the kinematic equations are $x = v_0 t$ and $y = \frac{1}{2}at^2 = \frac{1}{2}(F/m)t^2$, where $F = ma$ was used to substitute for the acceleration a. The time taken by the electron to travel a distance x ($= 30 \, \text{mm}$) horizontally is $t = x/v_0$ and its deflection in the direction of the force is

$$y = \frac{1}{2}\frac{F}{m}\left(\frac{x}{v_0}\right)^2 = \frac{1}{2}\left(\frac{4.5 \times 10^{-16} \, \text{N}}{9.11 \times 10^{-31} \, \text{kg}}\right)\left(\frac{30 \times 10^{-3} \, \text{m}}{1.2 \times 10^7 \, \text{m/s}}\right)^2 = 1.5 \times 10^{-3} \, \text{m} \, .$$

38P

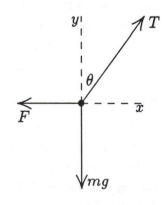

The free-body diagram is shown to the right, with the tension T in the string, the force of gravity mg, and the force F of the air labeled. Take the coordinate system to be as shown. The x component of the net force is $T \sin \theta - F$ and the y component is $T \cos \theta - mg$, where $\theta = 37°$. Since the sphere is motionless the net force on it is zero. Answer the questions in the reverse order to that given. (b) Solve $T \cos \theta - mg = 0$ for the tension in the string: $T = mg/\cos \theta = (3.0 \times 10^{-4} \, \text{kg})(9.8 \, \text{m/s}^2)/\cos 37° = 3.7 \times 10^{-3} \, \text{N}$. (a) Solve $T \sin \theta - F = 0$ for the force of the air: $F = T \sin \theta = (3.7 \times 10^{-3} \, \text{N}) \sin 37° = 2.2 \times 10^{-3} \, \text{N}$.

39P

(a) Since friction is negligible the force of the girl on the sled is the only horizontal force on the sled. The vertical forces, the force of gravity and the normal force of the ice, sum to zero.

The acceleration of the sled is

$$a_s = \frac{F}{m_s} = \frac{5.2\,\text{N}}{8.4\,\text{kg}} = 0.62\,\text{m/s}^2\,.$$

(b) According to Newton's third law, the force of the sled on the girl is also 5.2 N. Her acceleration is

$$a_g = \frac{F}{m_g} = \frac{5.2\,\text{N}}{40\,\text{kg}} = 0.13\,\text{m/s}^2\,.$$

(c) The accelerations of the sled and girl are in opposite directions. Suppose the girl starts at the origin and moves in the positive x direction. Her coordinate is given by $x_g = \frac{1}{2}a_g t^2$. The sled starts at $x = x_0$ (= 1.5 m) and moves in the negative x direction. Its coordinate is given by $x_s = x_0 - \frac{1}{2}a_s t^2$. They meet when $x_g = x_s$ or $\frac{1}{2}a_g t^2 = x_0 - \frac{1}{2}a_s t^2$. This occurs at time

$$t = \sqrt{\frac{2x_0}{a_g + a_s}}\,.$$

By that time the girl has gone the distance

$$x_g = \frac{1}{2}a_g t^2 = \frac{x_0 a_g}{a_g + a_s} = \frac{(15\,\text{m})(0.13\,\text{m/s}^2)}{0.13\,\text{m/s}^2 + 0.62\,\text{m/s}^2} = 2.6\,\text{m}\,.$$

40P

(a) The free-body diagrams are shown to the right. **F** is the applied force and **f** is the force m_1 exerts on m_2. Note that **F** is applied only to m_1 and that m_2 exerts the force $-\mathbf{f}$ on m_1. Newton's third law has been taken into account.

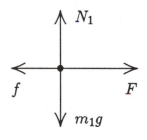

Newton's second law for m_1 is $F - f = m_1 a$, where a is the acceleration. The second law for m_2 is $f = m_2 a$. Since the blocks move together they have the same acceleration and the same symbol is used in both equations. Use the second equation to obtain an expression for a: $a = f/m_2$. Substitute into the first equation to get $F - f = m_1 f/m_2$. Solve for f:

$$f = \frac{Fm_2}{m_1 + m_2} = \frac{(3.2\,\text{N})(1.2\,\text{kg})}{2.3\,\text{kg} + 1.2\,\text{kg}} = 1.1\,\text{N}\,.$$

(b) If **F** is applied to m_2 instead of m_1 the force of contact is

$$f = \frac{Fm_1}{m_1 + m_2} = \frac{(3.2\,\text{N})(2.3\,\text{kg})}{1.2\,\text{kg} + 2.3\,\text{kg}} = 2.1\,\text{N}\,.$$

The acceleration of the blocks is the same in the two cases. Since the contact force is the only horizontal force applied to one of the blocks it must be just right to give that block the same acceleration as the block to which **F** is applied. In the second case the contact force accelerates a more massive block than in the first, so it must be larger.

44P

The free-body diagram is shown to the right. T is the tension in the cable and mg is the force of gravity. If the upward direction is positive then Newton's second law is $T - mg = ma$, where a is the acceleration. Solve for the tension: $T = m(g + a)$. Use constant acceleration kinematics to find the acceleration. If v $(= 0)$ is the final velocity, v_0 $(= -12\,\text{m/s})$ is the initial velocity, and y $(= -42\,\text{m})$ is the coordinate at the stopping point, then $v^2 = v_0^2 + 2ay$. Solve for a: $a = -v_0^2/2y = -(-12\,\text{m/s})^2/2(-42\,\text{m}) = 1.71\,\text{m/s}^2$. Now go back to calculate the tension: $T = m(g+a) = (1600\,\text{kg})(9.8\,\text{m/s}^2 + 1.71\,\text{m/s}^2) = 1.8 \times 10^4\,\text{N}$.

52P

(*a*) The first diagram on the right is the free-body diagram for the person and parachute, considered as a single object with a mass of $80\,\text{kg} + 5\,\text{kg} = 85\,\text{kg}$. F_a is the force of the air on the parachute and mg is the force of gravity. Take the downward direction to be positive. Then Newton's second law is $mg - F_a = ma$, where a is the acceleration. Solve for F_a: $F_a = m(g - a) = (85\,\text{kg})(9.8\,\text{m/s}^2 - 2.5\,\text{m/s}^2) = 620\,\text{N}$.

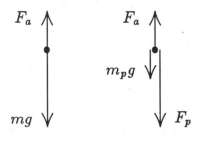

(*b*) The second diagram is the free-body diagram for the parachute alone. F_a is the force of the air, $m_p g$ is the force of gravity, and F_p is the force of the person. Newton's second law is $m_p g + F_p - F_a = m_p a$. Solve for F_p: $F_p = m_p(a - g) + F_a = (5.0\,\text{kg})(2.5\,\text{m/s}^2 - 9.8\,\text{m/s}^2) + 620\,\text{N} = 580\,\text{N}$.

55P

Let g be the acceleration due to gravity near the surface of Callisto, m be the mass of the landing craft, a be the acceleration of the landing craft, and F be the rocket thrust. Take the downward direction to be positive. Then Newton's second law is $mg - F = ma$. If the thrust is F_1 $(= 3260\,\text{N})$ the acceleration is zero, so $mg - F_1 = 0$. If the thrust is F_2 $(= 2200\,\text{N})$ the acceleration is a_2 $(= 0.39\,\text{m/s}^2)$, so $mg - F_2 = ma_2$. (*a*) The first equation gives the weight of the landing craft: $W = mg = F_1 = 3260\,\text{N}$.

(*b*) The second equation gives the mass:

$$m = \frac{mg - F_2}{a_2} = \frac{3260\,\text{N} - 2200\,\text{N}}{0.39\,\text{m/s}^2} = 2.7 \times 10^3\,\text{kg}\,.$$

(*c*) The weight divided by the mass gives the acceleration due to gravity: $g = (3260\,\text{N})/(2.7 \times 10^3\,\text{kg}) = 1.2\,\text{m/s}^2$.

57P

(a) Number the links from bottom to top. The forces acting on the bottom link are the force of gravity mg, downward, and the force $\mathbf{F}_{2\,\text{on}\,1}$ of link 2, upward. Take the positive direction to be upward. Then Newton's second law for this link is $F_{2\,\text{on}\,1} - mg = ma$. Thus $F_{2\,\text{on}\,1} = m(a + g) = (0.100\,\text{kg})(2.50\,\text{m/s}^2 + 9.8\,\text{m/s}^2) = 1.23\,\text{N}$.

The forces acting on the second link are the force of gravity mg, downward, the force $\mathbf{F}_{1\,\text{on}\,2}$ of link 1, downward, and the force $\mathbf{F}_{3\,\text{on}\,2}$ of link 3, upward. $\mathbf{F}_{1\,\text{on}\,2}$ has the same magnitude as $\mathbf{F}_{2\,\text{on}\,1}$. Newton's second law for the second link is $F_{3\,\text{on}\,2} - F_{1\,\text{on}\,2} - mg = ma$, so $F_{3\,\text{on}\,2} = m(a + g) + F_{1\,\text{on}\,2} = (0.100\,\text{kg})(2.50\,\text{m/s}^2 + 9.8\,\text{m/s}^2) + 1.23\,\text{N} = 2.46\,\text{N}$.

For link 3, $F_{4\,\text{on}\,3} - F_{2\,\text{on}\,3} - mg = ma$ and $F_{4\,\text{on}\,3} = m(a+g) + F_{2\,\text{on}\,3} = (0.100\,\text{N})(2.50\,\text{m/s}^2 + 9.8\,\text{m/s}^2) + 2.46\,\text{N} = 3.69\,\text{N}$, where $F_{2\,\text{on}\,3} = F_{3\,\text{on}\,2}$ was used.

For link 4, $F_{5\,\text{on}\,4} - F_{3\,\text{on}\,4} - mg$ and $F_{5\,\text{on}\,4} = m(a + g) + F_{3\,\text{on}\,4} = (0.100\,\text{kg})(2.50\,\text{m/s}^2 + 9.8\,\text{m/s}^2) + 3.69\,\text{N} = 4.92\,\text{N}$, where $F_{3\,\text{on}\,4} = F_{4\,\text{on}\,3}$ was used.

(b) For the top link $F - F_{4\,\text{on}\,5} - mg = ma$ and $F = m(a+g) + F_{4\,\text{on}\,5} = (0.100\,\text{kg})(2.50\,\text{m/s}^2 + 9.8\,\text{m/s}^2) + 4.92\,\text{N} = 6.15\,\text{N}$, where $F_{4\,\text{on}\,5} = F_{5\,\text{on}\,4}$ was used.

(c) Each link has the same mass and the same acceleration, so the same net force acts on each of them: $F_{\text{net}} = ma = (0.100\,\text{kg})(2.50\,\text{m/s}^2) = 0.25\,\text{N}$.

58P

The free-body diagram for each block is shown to the right. T is the tension in the cord and θ (= 30°) is the angle of the incline. For m_1 take the x axis to be up the plane and the y axis to be in the direction of the normal force of the plane on the block. For m_2 take the y axis to be downward. Then the accelerations of the two blocks can be represented by the same symbol a. The x component of Newton's second law for m_1 is $T - m_1 g \sin\theta = m_1 a$ and the y component is $N - mg \cos\theta = 0$, where N is the normal force of the plane on the block. Newton's second law for m_2 is $m_2 g - T = m_2 a$.

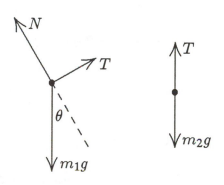

These two equations are to be solved for a and T. The first equation gives $T = m_1 a + m_1 g \sin\theta$. When this is substituted into the third equation the result is $m_2 g - m_1 a - m_1 g \sin\theta = m_2 a$. Solve for a:

$$a = \frac{(m_2 - m_1 \sin\theta)g}{m_1 + m_2} = \frac{[2.30\,\text{kg} - (3.70\,\text{kg})(\sin 30.0°)]\,(9.8\,\text{m/s}^2)}{3.70\,\text{kg} + 2.30\,\text{kg}} = 0.735\,\text{m/s}^2.$$

(b) The result is positive, indicating that the acceleration of m_1 is the up the plane and the acceleration of m_2 is downward.

(c) The tension in the cord is

$$T = m_1 a + m_1 g \sin\theta = (3.70\,\text{kg})(0.735\,\text{m/s}^2) + (3.70\,\text{kg})(9.8\,\text{m/s}^2)\sin 30° = 20.8\,\text{N}.$$

60P

The free-body diagram is shown at the right. N is the normal force of the plane on the block and m is the mass of the block. Take the positive x axis to be down the plane, in the direction of the acceleration, and the positive y axis to be in the direction of the normal force. The x component of Newton's second law is then $mg \sin \theta = ma$, so the acceleration is $a = g \sin \theta$.

(a) Place the origin at the bottom of the plane. The kinematic equations for motion along the x axis are $x = v_0 t + \frac{1}{2}at^2$ and $v = v_0 + at$. The block stops when $v = 0$.

According to the second equation this is at the time $t = -v_0/a$. The coordinate when it stops is

$$x = v_0 \left(\frac{-v_0}{a} \right) + \frac{1}{2}a \left(\frac{-v_0}{a} \right)^2 = -\frac{1}{2} \frac{v_0^2}{a} = -\frac{1}{2} \frac{v_0^2}{g \sin \theta}$$

$$= -\frac{1}{2} \left[\frac{(-3.50 \,\text{m/s})^2}{(9.8 \,\text{m/s}^2) \sin 32.0°} \right] = -1.18 \,\text{m} .$$

(b) The time is

$$t = -\frac{v_0}{a} = -\frac{v_0}{g \sin \theta} = -\frac{-3.50 \,\text{m/s}}{(9.8 \,\text{m/s}^2) \sin 32.0°} = 0.674 \,\text{s} .$$

(c) Now set $x = 0$ and solve $x = v_0 t + \frac{1}{2}at^2$ for t. The result is

$$t = -\frac{2v_0}{a} = -\frac{2v_0}{g \sin \theta} = -\frac{2(-3.50 \,\text{m/s})}{(9.8 \,\text{m/s}^2) \sin 32.0°} = 1.35 \,\text{s} .$$

The velocity is

$$v = v_0 + at = v_0 + gt \sin \theta = -3.50 \,\text{m/s} + (9.8 \,\text{m/s}^2)(1.35 \,\text{s}) \sin 32° = 3.50 \,\text{m/s} .$$

It is down the plane.

63P

(a) Take the positive direction to be upward for both the monkey and the package. Suppose the monkey pulls downward on the rope with a force of magnitude F. According to Newton's third law, the rope pulls upward on the monkey with a force of the same magnitude, so Newton's second law for the monkey is $F - m_m g = m_m a_m$, where m_m is the mass of the monkey and a_m is its acceleration. Since the rope is massless F is the tension in the rope. The rope pulls upward on the package with a force of magnitude F, so Newton's second law for the package is $F + N - m_p g = m_p a_p$, where m_p is the mass of the package, a_p is its acceleration, and N is the normal force of the ground on it.

Now suppose F is the minimum force required to lift the package. Then $N = 0$ and $a_p = 0$. According to the second law equation for the package, this means $F = m_p g$. Substitute $m_p g$ for F in the second law equation for the monkey, then solve for a_m. You should obtain

$$a_m = \frac{F - m_m g}{m_m} = \frac{(m_p - m_m)g}{m_m} = \frac{(15\,\text{kg} - 10\,\text{kg})(9.8\,\text{m/s}^2)}{10\,\text{kg}} = 4.9\,\text{m/s}^2 .$$

(b) Newton's second law equations are $F - m_p g = m_p a_p$ for the package and $F - m_m g = m_m a_m$ for the monkey. If the acceleration of the package is downward then the acceleration of the monkey is upward, so $a_m = -a_p$. Solve the first equation for F: $F = m_p(g + a_p) = m_p(g - a_m)$. Substitute the result into the second equation and solve for a_m:

$$a_m = \frac{(m_p - m_m)g}{m_p + m_m} = \frac{(15\,\text{kg} - 10\,\text{kg})(9.8\,\text{m/s}^2)}{15\,\text{kg} + 10\,\text{kg}} = 2.0\,\text{m/s}^2 .$$

(c) Solve the second law equation for the package to obtain

$$F = m_p(g - a_m) = (15\,\text{kg})(9.8\,\text{m/s}^2 - 2.0\,\text{m/s}^2) = 120\,\text{N} .$$

67P

(a) The force diagram (not to scale) for the block is shown to the right. N is the normal force exerted by the floor and mg is the force of gravity. The x component of Newton's second law is $F \cos\theta = ma$, where m is the mass of block and a is the x component of its acceleration. Solve for a:

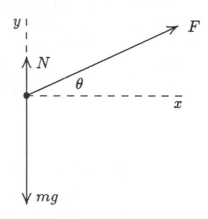

$$a = \frac{F \cos\theta}{m} = \frac{(12.0\,\text{N})\cos 25.0°}{5.00\,\text{kg}} = 2.18\,\text{m/s}^2 .$$

This is its acceleration provided it remains in contact with the floor. Assume it does and find the value of N. If N is positive then the assumption is true; if N is negative then the block leaves the floor. The y component of Newton's second law is $N + F \sin\theta - mg = 0$, so $N = mg - F \sin\theta = (5.00\,\text{kg})(9.8\,\text{m/s}^2) - (12.0\,\text{N})\sin 25.0° = 43.9\,\text{N}$. Hence the block remains on the floor and its acceleration is $a = 2.18\,\text{m/s}^2$.

(b) Suppose F is the minimum force for which the block leaves the floor. Then $N = 0$ and the y component of the acceleration vanishes. The y component of the second law becomes $F \sin\theta - mg = 0$, so

$$F = \frac{mg}{\sin\theta} = \frac{(5.00\,\text{kg})(9.8\,\text{m/s}^2)}{\sin 25.0°} = 116\,\text{N} .$$

(c) The acceleration is still in the x direction and is still given by the equation developed in part (a):

$$a = \frac{F \cos\theta}{m} = \frac{(116\,\text{N})(9.8\,\text{m/s}^2)}{5.00\,\text{kg}} = 21.0\,\text{m/s}^2 .$$

70P

The forces acting on the balloon are the force of gravity $m\mathbf{g}$, down, and the force of the air $\mathbf{F}_a$, up. Take the positive direction to be up. When the mass is M (before the ballast is thrown out) the acceleration is downward and Newton's second law is $F_a - Mg = -Ma$. After the ballast is thrown out the mass is $M - m$, where m is the mass of the ballast, and the acceleration is upward. Newton's second law is $F_a - (M - m)g = (M - m)a$. The first equation gives $F_a = M(g - a)$ and the second gives $M(g - a) - (M - m)g = (M - m)a$. Solve for m: $m = 2Ma/(g + a)$.

72P

(a) Consider a small segment of the rope. It has mass and is pulled down by the gravitational force of the Earth. Equilibrium is reached because neighboring portions of the rope pull up on it. Since the tension is a force along the rope at least one of the neighboring portions must slope up away from the segment we are considering. Then the tension has an upward component. This means the rope sags.

(b) The only force acting with a horizontal component is the applied force $\mathbf{F}$. Consider the block and rope as a single object and write Newton's second law for it: $F = (M + m)a$, where a is the acceleration and the positive direction is taken to be to the right. The acceleration is given by $a = F/(M + m)$.

(c) The force of the rope is the only force with a horizontal component acting on the block. Let this force be F_r. Then Newton's second law for the block gives

$$F_r = Ma = \frac{MF}{M + m},$$

where the expression found above for a was substituted.

(d) Consider the block and half the rope to be a single object, with mass $M + \frac{1}{2}m$. The horizontal force acting on it is the tension at the midpoint of the rope. Let this force be T_m and use Newton's second law to find its value:

$$T_m = (M + \frac{1}{2}m)a = \frac{(M + \frac{1}{2}m)F}{(M + m)} = \frac{(2M + m)F}{2(M + m)}.$$

Chapter 6

1E

(a) The free-body diagram for the bureau is shown to the right. **F** is the applied force, **f** is the force of friction, **N** is the normal force of the floor, and $m\mathbf{g}$ is the force of gravity. Take the x axis to be horizontal and the y axis to be vertical. Assume the bureau does not move and write Newton's second law. The x component is $F - f = 0$ and the y component is $N - mg = 0$. The force of friction is then equal in magnitude to the applied force: $f = F$. The normal force is equal in magnitude to the force of gravity: $N = mg$. As F increases, f increases until $f = \mu_s N$. Then the bureau starts to move. The minimum force that must be applied to start the bureau moving is $F = \mu_s N = \mu_s mg = (0.45)(45\,\text{kg})(9.8\,\text{m/s}^2) = 200\,\text{N}$.

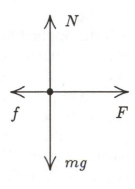

(b) The equation for F is the same but the mass is now $45\,\text{kg} - 17\,\text{kg} = 28\,\text{kg}$. Thus $F = \mu_s mg = (0.45)(28\,\text{kg})(9.8\,\text{m/s}^2) = 120\,\text{N}$.

2E

The free-body diagram for the player is shown to the right. **N** is the normal force of the ground on the player, $m\mathbf{g}$ is the force of gravity, and **f** is the force of friction. The force of friction is related to the normal force by $f = \mu_k N$. Use Newton's second law to find the normal force. The vertical component of the acceleration is zero, so the vertical component of Newton's second law is $N - mg = 0$ and the normal force equals the weight in magnitude: $N = mg$. Thus

$$\mu_k = \frac{f}{N} = \frac{f}{mg} = \frac{470\,\text{N}}{(79\,\text{kg})(9.8\,\text{m/s}^2)} = 0.61\,.$$

8E

(a) The free-body diagram for the crate is shown to the right. **F** is the force of the person on the crate, **f** is the force of friction, **N** is the normal force of the floor, and $m\mathbf{g}$ is the force of gravity. The magnitude of the force of friction is given by $f = \mu_k N$, where μ_k is the coefficient of kinetic friction. The vertical component of Newton's second law is used to find the normal force. Since the vertical component of the acceleration is zero, $N - mg = 0$ and $N = mg$. Thus $f = \mu_k N = \mu_k mg = (0.35)(55\,\text{kg})(9.8\,\text{m/s}^2) = 189\,\text{N}$.

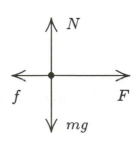

(b) Use the horizontal component of Newton's second law to find the acceleration. Since $F - f = ma$,

$$a = \frac{(F-f)}{m} = \frac{(220\,\text{N} - 189\,\text{N})}{55\,\text{kg}} = 0.56\,\text{m/s}^2 .$$

11E

(a) The free-body diagram for the block is shown to the right. **F** is the applied force, **N** is the normal force of the wall on the block, **f** is the force of friction, and $m\mathbf{g}$ is the force of gravity. To determine if the block falls, find the magnitude f of the force of friction required to hold it without accelerating and also find the normal force of the wall on the block. Compare f and $\mu_s N$. If $f < \mu_s N$ the block does not slide on the wall but if $f > \mu_s N$ the block does slide. The horizontal component of Newton's second law is $F - N = 0$, so $N = F = 12\,\text{N}$ and $\mu_s N = (0.60)(12\,\text{N}) = 7.2\,\text{N}$. The vertical component is $f - mg = 0$, so $f = mg = 5.0\,\text{N}$. Since $f < \mu_s N$ the block does not slide.

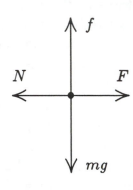

(b) Since the block does not move $f = 5.0\,\text{N}$ and $N = 12\,\text{N}$. The force of the wall on the block is

$$\mathbf{F}_w = -N\mathbf{i} + f\mathbf{j} = (-12\mathbf{i} + 5.0\mathbf{j})\,\text{N},$$

where the axes are as shown on Fig. 18 of the text.

15E

(a) The free-body diagram for the puck is shown to the right. **N** is the normal force of the ice on the puck, **f** is the force of friction, and $m\mathbf{g}$ is the force of gravity. The horizontal component of Newton's second law is $-f = ma$, where the positive direction was taken to be the direction of motion of the puck (to the right in the diagram). Constant acceleration kinematics can be used to find the acceleration. Solve $v^2 = v_0^2 + 2ax$ for a. Since the final velocity is zero, $a = -v_0^2/2x$. This result is substituted into the second law equation to obtain

$$f = \frac{mv_0^2}{2x} = \frac{(0.110\,\text{kg})(6.0\,\text{m/s})^2}{2(15\,\text{m})} = 0.13\,\text{N} .$$

(b) Use $f = \mu_k N$. The vertical component of Newton's second law is $N - mg = 0$, so $N = mg$ and $f = \mu_k mg$. Solve for μ_k:

$$\mu_k = \frac{f}{mg} = \frac{0.13\,\text{N}}{(0.110\,\text{kg})(9.8\,\text{m/s}^2)} = 0.12 .$$

17P

A cross section of the cone of sand is shown to the right. To pile the most sand without extending the radius, sand is added to make the height h as great as possible. Eventually, however, the sides become so steep that sand at the surface begins to slide. You want to find the greatest height (or greatest slope) for which the sand does not slide.

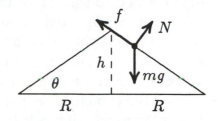

A grain of sand is shown on the diagram and the forces acting on it are labeled. $\mathbf{N}$ is the normal force of the surface, $m\mathbf{g}$ is the force of gravity, and $\mathbf{f}$ is the force of friction. Take the x axis to be down the plane and the y axis to be in the direction of the normal force. Assume the grain does not slide, so its acceleration is zero. Then the x component of Newton's second law is $mg\sin\theta - f = 0$ and the y component is $N - mg\cos\theta = 0$. The first equation gives $f = mg\sin\theta$ and the second gives $N = mg\cos\theta$. If the grain does not slide the condition $f < \mu_s N$ must hold. This means $mg\sin\theta < \mu_s mg\cos\theta$ or $\tan\theta < \mu_s$. The surface of the cone has the greatest slope (and the height of the cone is the greatest) if $\tan\theta = \mu_s$.

Now $h = R\tan\theta$ so we may replace h with $\mu_s R$ in the equation $V = Ah/3$ for the volume of the cone. Also replace the area A of the base with πR^2. The result is $V = \pi\mu_s R^3/3$.

22P

(a) The free-body diagram for the crate is shown to the right. $\mathbf{T}$ is the tension in the rope, $\mathbf{N}$ is the normal force of the floor on the crate, $m\mathbf{g}$ is the force of gravity, and $\mathbf{f}$ is the force of friction. Take the x axis to be horizontal to the right and the y axis to be vertically upward. Assume the crate is motionless. The x component of Newton's second law is then $T\cos\theta - f = 0$ and the y component is $T\sin\theta + N - mg = 0$, where $\theta\ (= 15°)$ is the angle between the rope and the horizontal. The first equation gives $f = T\cos\theta$ and the second gives $N = mg - T\sin\theta$. If the crate is to remain at rest

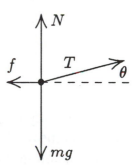

f must be less than $\mu_s N$, or $T\cos\theta < \mu_s(mg - T\sin\theta)$. When the tension is sufficient to just start the crate moving $T\cos\theta = \mu_s(mg - T\sin\theta)$. Solve for T:

$$T = \frac{\mu_s mg}{\cos\theta + \mu_s\sin\theta} = \frac{(0.50)(68\,\text{kg})(9.8\,\text{m/s}^2)}{\cos 15° + 0.50\sin 15°} = 304\,\text{N}.$$

(b) The second law equations for the moving crate are $T\cos\theta - f = ma$ and $N + T\sin\theta - mg = 0$. Now $f = \mu_k N$. The second equation gives $N = mg - T\sin\theta$, as before, so $f = \mu_k(mg - T\sin\theta)$. This expression is substituted for f in the first equation to obtain $T\cos\theta - \mu_k(mg - T\sin\theta) = ma$, so the acceleration is

$$a = \frac{T(\cos\theta + \mu_k\sin\theta)}{m} - \mu_k g$$

$$= \frac{(304\,\text{N})(\cos 15° + 0.35\sin 15°)}{68\,\text{kg}} - (0.35)(9.8\,\text{m/s}^2) = 1.3\,\text{m/s}^2.$$

24P

(a) Free-body diagrams for the blocks A and C, considered as a single object, and for the block B are shown to the right. T is the tension in the rope, N is the magnitude of the normal force of the table on block A, f is the magnitude of the force of friction, W_{AC} is the combined weight of blocks A and C, and W_B is the weight of block B. Assume the blocks are not moving. For the blocks on the table take the x axis to be to the right and the y axis to be upward. The x

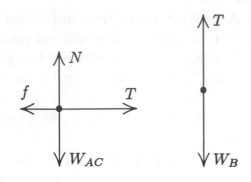

component of Newton's second law is then $T - f = 0$ and the y component is $N - W_{AC} = 0$. For block B take the downward direction to be positive. Then Newton's second law for that block is $W_B - T = 0$. The third equation gives $T = W_B$ and the first gives $f = T = W_B$. The second equation gives $N = W_{AC}$. If sliding is not to occur f must be less than $\mu_s N$, or $W_B < \mu_s W_{AC}$. The smallest that W_{AC} can be with the blocks still at rest is $W_{AC} = W_B/\mu_s = (22\,\text{N})/(0.20) = 110\,\text{N}$. Since the weight of block A is 44 N, the least weight for C is $110\,\text{N} - 44\,\text{N} = 66\,\text{N}$.

(b) The second law equations become $T - f = (W_A/g)a$, $N - W_A = 0$, and $W_B - T = (W_B/g)a$. In addition, $f = \mu_k N$. The second equation gives $N = W_A$, so $f = \mu_k W_A$. The third gives $T = W_B - (W_B/g)a$. Substitute these two expressions into the first equation to obtain $W_B - (W_B/g)a - \mu_k W_A = (W_A/g)a$. Solve for a:

$$a = \frac{g(W_B - \mu_k W_A)}{W_A + W_B} = \frac{(9.8\,\text{m/s}^2)\,[22\,\text{N} - (0.15)(44\,\text{N})]}{44\,\text{N} + 22\,\text{N}} = 2.3\,\text{m/s}^2 \,.$$

30P

The free-body diagrams for block B and for the knot just above block A are shown to the right. T_1 is the tension in the rope pulling on block B, T_2 is the tension in the other rope, f is the magnitude of the force of friction exerted by the horizontal surface on block B, N is the magnitude of the normal force exerted by the surface on block B, W_A is the weight of block A, and W_B is the weight of block B. $\theta\,(=30°)$ is the angle between the second rope and the horizontal.

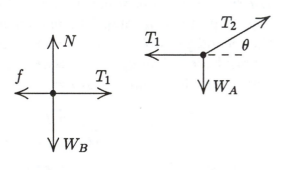

For each object take the x axis to be horizontal and the y axis to be vertical. The x component of Newton's second law for block B is then

$$T_1 - f = 0$$

and the y component is

$$N - W_B = 0\,.$$

The x component of Newton's second law for the knot is

$$T_2 \cos \theta - T_1 = 0$$

and the y component is

$$T_2 \sin \theta - W_A = 0.$$

Eliminate the tensions and find expressions for f and N in terms of W_A and W_B, then select W_A so $f = \mu_s N$. The second Newton's law equation gives $N = W_B$ immediately. The third gives $T_2 = T_1 / \cos \theta$. Substitute this expression into the fourth equation to obtain $T_1 = W_A / \tan \theta$. Substitute $T_1 = W_A / \tan \theta$ into the first equation to obtain $f = W_A / \tan \theta$. For the blocks to remain stationary f must be less than $\mu_s N$ or $W_A / \tan \theta < \mu_s W_B$. The greatest that W_A can be is the value for which $W_A / \tan \theta = \mu_s W_B$. Solve for W_A:

$$W_A = \mu_s W_B \tan \theta = (0.25)(711\,\mathrm{N}) \tan 30° = 100\,\mathrm{N}.$$

31P

(a) First check to see if the bodies start to move. Assume they remain at rest, compute the force of friction that holds them at rest, and compare its magnitude with $\mu_s N$.

The free-body diagrams are shown on the right. T is the tension in the string, f is the magnitude of the force of friction on body B, N is the magnitude of the normal force of the plane on body B, $m_A g$ is the force of gravity on body A, and $m_B g$ is the force of gravity on body B. θ is the angle of incline of the plane ($40°$).

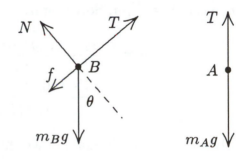

We do not know the direction of the frictional force but we assume it is down the plane. If we obtain a negative result for f then we know the force is actually up the plane.

For B take the x axis to be up the plane and the y axis to be in the direction of the normal force. The x component of Newton's second law is then

$$T - f - m_B g \sin \theta = 0$$

and the y component is

$$N - m_B g \cos \theta = 0.$$

For A take the positive direction to be downward. The second law equation for this object is

$$m_A g - T = 0.$$

The third equation gives $T = m_A g$. Substitute this into the first equation to obtain $m_A g - f - m_B g \sin \theta = 0$. Thus $f = m_A g - m_B g \sin \theta = 32\,\mathrm{N} - (102\,\mathrm{N}) \sin 40° = -34\,\mathrm{N}$. The force of friction is up the plane. The second equation gives $N = m_B g \cos \theta = (102\,\mathrm{N}) \cos 40° = 78\,\mathrm{N}$, so $\mu_s N = (0.56)(78\,\mathrm{N}) = 44\,\mathrm{N}$. Since the magnitude of the force of friction that hold the bodies motionless is less than $\mu_s N$ the bodies remain at rest. Their accelerations are zero.

(b) Since B is moving up the plane the force of friction is down the plane and has the magnitude $f = \mu_k N$. The second law equations become $T - \mu_k N - m_B g \sin\theta = m_B a$, $N - m_B g \cos\theta = 0$, and $m_A g - T = m_A a$, where a is the acceleration. Because the objects move together and because the coordinate axis was chosen to be up the plane for B and downward for A the same symbol can be used for the two accelerations. Substitute $N = m_B g \cos\theta$, from the second equation, and $T = m_A g - m_A a$, from the third, into the first to obtain $m_A g - m_A a - \mu_k m_B g \cos\theta - m_B g \sin\theta = m_B a$. Solve for a:

$$a = \frac{m_A g - m_B g \sin\theta - \mu_k m_B g \cos\theta}{m_A + m_B}$$

$$= \frac{32\,\mathrm{N} - (102\,\mathrm{N})\sin 40° - (0.25)(102\,\mathrm{N})\cos 40°}{(32\,\mathrm{N} + 102\,\mathrm{N})/(9.8\,\mathrm{m/s}^2)} = -3.9\,\mathrm{m/s}^2 \,.$$

The acceleration is down the plane. The objects are slowing down. Notice that $m = W/g$ was used to calculate the masses in the denominator.

(c) Now B is moving down the plane, so the force of friction is up the plane and has a magnitude that is given by $\mu_k N$. The second law equations become $T + \mu_k N - m_B g \sin\theta = m_B a$, $N - m_B g \cos\theta = 0$, and $m_A g - T = m_A a$. Substitute $N = m_B g \cos\theta$, from the second equation, and $T = m_A g - m_A a$, from the third, into the first to obtain $m_A g - m_A a + \mu_k m_B g \cos\theta - m_B g \sin\theta = m_B a$. Solve for a:

$$a = \frac{m_A g - m_B g \sin\theta + \mu_k m_B g \cos\theta}{m_A + m_B}$$

$$= \frac{32\,\mathrm{N} - (102\,\mathrm{N})\sin 40° + (0.25)(102\,\mathrm{N})\cos 40°}{(32\,\mathrm{N} + 102\,\mathrm{N})/(9.8\,\mathrm{m/s}^2)} = -0.98\,\mathrm{m/s}^2 \,.$$

The acceleration is again down the plane. The objects are speeding up.

34P

(a) The free-body diagrams for the two blocks are shown to the right. T is the tension in the string, N_A is the magnitude of the normal force on block A, N_B is the normal force on block B, f_A is the magnitude of the friction force on block A, f_B is the magnitude of the friction force on block B, m_A is the mass of block A, and m_B is the mass of block B. θ is the angle of the incline (30°). We have assumed that the incline goes down from right to left and that block A is leading. It is the 8.0 lb block.

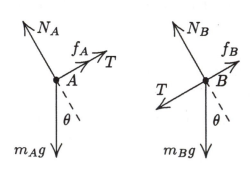

For each block take the x axis to be down the plane and the y axis to be in the direction of the normal force. For block A the x component of Newton's second law is

$$m_A g \sin\theta - f_A - T = m_A a_A$$

and the y component is
$$N - m_{Ag} \cos \theta = 0.$$

Here a_A is the acceleration of the block. The magnitude of the frictional force is

$$f_A = \mu_{kA} N_A = \mu_{kA} m_{Ag} \cos \theta,$$

where $N_A = m_{Ag} \cos \theta$, from the second equation, was substituted. μ_{kA} is the coefficient of kinetic friction for block A. When the expression for f_A is substituted into the first equation the result is

$$m_{Ag} \sin \theta - \mu_{kA} m_{Ag} \cos \theta - T = m_A a_A.$$

The same analysis applied to block B leads to

$$m_{Bg} \sin \theta - \mu_{kB} m_{Bg} \cos \theta + T = m_B a_B.$$

We must first find out if the rope is taut or slack. Assume the blocks are not joined by a rope and calculate the acceleration of each. If the acceleration of A is greater than the acceleration of B then when the rope is attached it is taut. If the acceleration of B is greater than the acceleration of A then even when the rope is attached B gains speed at a greater rate than A and the rope is slack.

Set $T = 0$ in the equation you derived above and solve for a_A and a_B. The results are

$$a_A = g \sin \theta - \mu_{kA} g \cos \theta = (32 \, \text{ft/s}^2)(\sin 30° - 0.10 \cos 30°) = 13 \, \text{ft/s}^2$$

and

$$a_B = g \sin \theta - \mu_{kB} g \cos \theta = (32 \, \text{ft/s}^2)(\sin 30° - 0.20 \cos 30°) = 10 \, \text{ft/s}^2.$$

We have learned that when the blocks are joined, the rope is taut, the tension is not zero, and the two blocks have the same acceleration.

Now go back to $m_{Ag} \sin \theta - \mu_{kA} m_{Ag} \cos \theta - T = m_A a$ and $m_{Bg} \sin \theta - \mu_{kB} m_{Bg} \cos \theta + T = m_B a$, where a has been substituted for both a_A and a_B. Solve the first expression for T, substitute the result into the second, and solve for a. The result is

$$a = g \sin \theta - \frac{\mu_{kA} m_A + \mu_{kB} m_B}{m_A + m_B} g \cos \theta$$

$$= (32 \, \text{ft/s}^2) \sin 30° - \left(\frac{(0.10)(8.0 \, \text{lb}) + (0.20)(16 \, \text{lb})}{8.0 \, \text{lb} + 16 \, \text{lb}} \right) (32 \, \text{ft/s}^2) \cos 30°$$

$$= 11.4 \, \text{ft/s}^2.$$

Strictly speaking, values of the masses rather than weights should be substituted, but the factor g cancels from the numerator and denominator.

(b) Use $m_{Ag} \sin \theta - \mu_{kA} m_{Ag} \cos \theta - T = m_A a$ to find the tension in the rope: $T = m_{Ag} \sin \theta - \mu_{kA} m_{Ag} \cos \theta - m_A a = (8.0 \, \text{lb}) \sin 30° - (0.10)(8.0 \, \text{lb}) \cos 30° - (8.0 \, \text{lb}/32 \, \text{ft/s}^2)(11.4 \, \text{ft/s}^2) = 0.46 \, \text{lb}.$

(c) If the blocks are reversed then the following block has a greater acceleration than the leading block and the rope is slack. The blocks move independently of each other.

37P

The free-body diagrams for the slab and block are shown to the right. F is the magnitude of the force applied to the block, N_s is the magnitude of the normal force of the floor on the slab, N_b is the magnitude of the normal force of the slab on the box, f is the magnitude of the force of friction between the slab and the block, m_s is the mass of the slab, and m_b is the mass of the block. Notice that Newton's third law has been taken into account: the forces of friction on the two objects have the same magnitude but are in opposite directions; the block pushes down on the slab with a force of magnitude N_b.

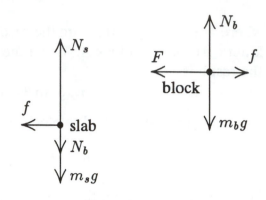

Take the x axis to be to the left and the y axis to be upward. The x component of Newton's second law for the slab is

$$f = m_s a_s$$

and the y component is

$$N_s - N_b - m_s g = 0,$$

where a_s is the acceleration of the slab. The x component of Newton's second law for the block is

$$F - f = m_b a_b$$

and the y component is

$$N_b - m_b g = 0,$$

where a_b is the acceleration of the block.

First check to see if the block slides on the slab. Assume it does not. Then $a_s = a_b$. Use a to denote both these accelerations. Use $f = m_s a$ to eliminate a from $F - f = m_b a$, then solve for f. The result is

$$f = \frac{m_s F}{m_s + m_b} = \frac{(40\,\text{kg})(100\,\text{N})}{40\,\text{kg} + 10\,\text{kg}} = 80\,\text{N}.$$

According to the last of the second law equations $N_b = m_b g$, so

$$\mu_s N_b = \mu_s m_b g = (0.60)(10\,\text{kg})(9.8\,\text{m/s}^2) = 59\,\text{N}.$$

Since $f > \mu_s N_b$ the block slides on the slab. They have different accelerations. Carefully note that N_b, not N_s is used to compute the upper limit of the static frictional force. The force of friction is exerted by the slab and block on each other so the normal force these objects exert on each other is used to compute the upper limit.

Since the block slides on the slab, the magnitude of the frictional force is given by $f = \mu_k N_b$. The x component of Newton's second law for the slab is now $\mu_k N_b = m_s a_s$, the x component of Newton's second law for the block is $F - \mu_k N_b = m_b a_b$, and the y component is $N_b - m_b g = 0$. The last of these gives $N_b = m_b g$. Substitute into the first equation and solve for a_s:

$$a_s = \frac{\mu_k m_b g}{m_s} = \frac{(0.40)(10\,\text{kg})(9.8\,\text{m/s}^2)}{40\,\text{kg}} = 0.98\,\text{m/s}^2 \,.$$

Substitute $N_b = m_b g$ into the second equation and solve for a_b:

$$a_b = \frac{F - \mu_k m_b g}{m_b} = \frac{100\,\text{N} - (0.40)(10\,\text{kg})(9.8\,\text{m/s}^2)}{10\,\text{kg}} = 6.1\,\text{m/s}^2 \,.$$

39P

Each side of the trough exerts a normal force on the crate. The first diagram shows the view looking in toward a cross section. The resultant is along the dotted line. Since each of the normal forces makes an angle of 45° with the dotted line the magnitude of the resultant is given by $N_r = 2N \cos 45° = \sqrt{2}N$, where $\cos 45° = 1/\sqrt{2}$ was used. The second diagram is the free-body diagram for the crate. The force of gravity has magnitude mg, where m is the mass of the crate, and the magnitude of the force of friction is denoted by f.

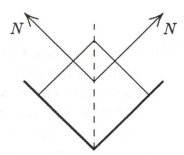

Take the x axis to be down the incline and the y axis to be in the direction of $\mathbf{N}_r$. Then the x component of Newton's second law is $mg \sin\theta - f = ma$ and the y component is $N_r - mg \cos\theta = 0$, where θ is the angle of the incline. Since the crate is moving each side of the trough exerts a force of kinetic friction, so the total frictional force has magnitude $f = 2\mu_k N = 2\mu_k N_r/\sqrt{2} = \sqrt{2}\mu_k N_r$. Substitute this expression and $N_r = mg \cos\theta$, from the y component of the second law, into the x component equation to obtain $mg \sin\theta - \sqrt{2}mg \cos\theta = ma$. Finally, solve for a: $a = g(\sin\theta - \sqrt{2}\mu_k \cos\theta)$.

42P

Let the magnitude of the frictional force be αv, where $\alpha = 70\,\text{N} \cdot \text{s/m}$. Take the direction of the boat's motion to be positive. Newton's second law is then $-\alpha v = m\,dv/dt$. Thus

$$\int_{v_o}^{v} \frac{dv}{v} = -\frac{\alpha}{m} \int_{0}^{t} dt \,,$$

where v_0 is the velocity at time zero and v is the velocity at time t. The integrals can be evaluated, with the result

$$\ln \frac{v}{v_0} = -\frac{\alpha t}{m}.$$

Take $v = v_0/2$ and solve for t:

$$t = \frac{m}{\alpha} \ln 2 = \frac{1000\,\text{kg}}{70\,\text{N} \cdot \text{s/m}} \ln 2 = 9.9\,\text{s}.$$

43E

Use Eq. 18 of the text: $D = \frac{1}{2}C\rho Av^2$, where ρ is the air density, A is the cross-sectional area of the missile, v is the speed of the missile, and C is the drag coefficient. The area is given by $A = \pi R^2$, where R ($= 26.5\,\text{cm} = 0.265\,\text{m}$) is the radius of the missile. Thus $D = \frac{1}{2}(0.75)(1.2\,\text{kg/m}^3)\pi(0.265\,\text{m})^2(250\,\text{m/s})^2 = 6.2 \times 10^3\,\text{N}$.

47E

The acceleration of the car as it rounds the curve is given by v^2/R, where v is the speed of the car and R is radius of the curve. Since the road is horizontal, only the frictional force of the road on the tires provides the force to produce this acceleration. The horizontal component of Newton's second law is $f = mv^2/R$. If N is the normal force of the road on the car and m is the mass of the car, the vertical component of the second law is $N - mg = 0$. Thus $N = mg$ and $\mu_s N = \mu_s mg$. If the car does not slip, $f < \mu_s mg$. This means $v^2/R < \mu_s g$, or $v < \sqrt{\mu_s R g}$. The maximum speed with which the car can round the curve without slipping is

$$v_{\text{max}} = \sqrt{\mu_s R g} = \sqrt{(0.25)(47.5\,\text{m})(9.8\,\text{m/s}^2)} = 11\,\text{m/s}.$$

55E

(a) Let N be the number of revolutions the electron makes in time ΔT. In each revolution the electron goes a distance $2\pi R$, where R is the radius of the orbit. This distance is the circumference of a circle of radius R. The speed of the electron is

$$v = \frac{N(2\pi R)}{\Delta T} = \frac{(6.6 \times 10^{15}\,\text{rev})(2\pi \times 5.3 \times 10^{-11}\,\text{m/rev})}{1.0\,\text{s}} = 2.2 \times 10^6\,\text{m/s}.$$

(b) The acceleration is

$$a = \frac{v^2}{R} = \frac{(2.2 \times 10^6\,\text{m/s})^2}{5.3 \times 10^{-11}\,\text{m}} = 9.1 \times 10^{22}\,\text{m/s}^2.$$

At any instant it is inward from the position of the electron at that instant toward the center of the orbit.

(c) The force is $F = ma = (9.11 \times 10^{-31}\,\text{kg})(9.1 \times 10^{22}\,\text{m/s}^2) = 8.3 \times 10^{-8}\,\text{N}$.

56E

For M to remain at rest the tension T in the cord must equal the gravitational force Mg on M. The tension supplies the centripetal force that keeps m in its circular orbit, so $T = mv^2/r$. Thus $Mg = mv^2/r$. Solve for v: $v = \sqrt{Mgr/m}$.

62E

(a) At the highest point the seat pushes up on the student with a force of magnitude F_t (= 125 lb). The Earth pulls down with a force of magnitude W (= 150 lb). The net force toward the center of the circular orbit is $W - F_t$ and, according to Newton's second law, this must be mv^2/R, where v is the speed of the student and R is the radius of the orbit. Thus $mv^2/R = W - F_t = 150\,\text{lb} - 125\,\text{lb} = 25\,\text{lb}$. Let F_b be the magnitude of the force of the seat when the student is at the bottom. The force is upward so the net force toward the center of the circle is $F_b - W$ and $F_b - W = mv^2/R$. Solve for F_b:

$$F_b = \frac{mv^2}{R} + W = 25\,\text{lb} + 150\,\text{lb} = 175\,\text{lb}\,.$$

(b) At the top $W - F_t = mv^2/R$, so $F_t = W - mv^2/R$. If the speed is doubled mv^2/R increases by a factor of 4, to 100 lb. Then $F_t = 150\,\text{lb} - 100\,\text{lb} = 50\,\text{lb}$.

65P

The free-body diagram for the plane is shown to the right. F is the magnitude of the lift on the wings and m is the mass of the plane. Since the wings are tilted by 40° to the horizontal, the angle θ is 50°. The center of the circular orbit is to the right of the plane, the dotted line being a portion of the radius. Take the x axis to be to the right and the y axis to be upward. Then the x component of Newton's second law is $F\cos\theta = mv^2/R$ and the y component is $F\sin\theta - mg = 0$, where R is the radius of the orbit. The first equation gives $F = mv^2/R\cos\theta$ and when this is substituted into the second, $(mv^2/R)\tan\theta = mg$ results. Solve for R:

$$R = \frac{v^2}{g}\tan\theta\,.$$

The speed of the plane is $v = 480\,\text{km/h} = 133\,\text{m/s}$, so

$$R = \frac{(133\,\text{m/s})^2}{9.8\,\text{m/s}^2}\tan 50° = 2.2 \times 10^3\,\text{m}\,.$$

70P

(a) The free-body diagram for the ball is shown to the right. T_u is the tension in the upper string, T_ℓ is the tension in the lower string, and m is the mass of the ball. Note that the tension in the upper string is greater than the tension in the lower string. It must balance the downward pull of gravity and the force of the lower string.

(b) Take the x axis to be to the left, toward the center of the circular orbit, and the y axis to be upward. The x component of Newton's second law is

$$T_u \cos\theta + T_\ell \cos\theta = \frac{mv^2}{R},$$

where v is the speed of the ball and R is the radius of its orbit. The y component is

$$T_u \sin\theta - T_\ell \sin\theta - mg = 0.$$

The second equation gives the tension in the lower string: $T_\ell = T_u - mg/\sin\theta$. Since the triangle is equilateral $\theta = 30°$. Thus

$$T_\ell = 35\,\text{N} - \frac{(1.34\,\text{kg})(9.8\,\text{m/s}^2)}{\sin 30°} = 8.74\,\text{N}.$$

(c) The net force is toward the left and has magnitude $F_\text{net} = (T_u + T_\ell)\cos\theta = (35\,\text{N} + 8.74\,\text{N})\cos 30° = 37.9\,\text{N}$.

(d) Use $F_\text{net} = mv^2/R$. The radius of the orbit is $[(1.70\,\text{m})/2)]\tan 30° = 1.47\,\text{m}$. Thus

$$v = \sqrt{\frac{RF_\text{net}}{m}} = \sqrt{\frac{(1.47\,\text{m})(37.9\,\text{N})}{1.34\,\text{kg}}} = 6.45\,\text{m/s}.$$

Chapter 7

2E

(a) The force of the worker on the crate is constant so the work it does is given by $W_F = \mathbf{F} \cdot \mathbf{d} = Fd\cos\phi$, where $\mathbf{F}$ is the force, $\mathbf{d}$ is the displacement of the crate, and ϕ is the angle between the force and the displacement. Here $F = 210\,\text{N}$, $d = 3.0\,\text{m}$, and $\phi = 20°$. Thus $W_F = (210\,\text{N})(3.0\,\text{m})\cos 20° = 590\,\text{J}$.

(b) The force of gravity is downward, perpendicular to the displacement of the crate. The angle between this force and the displacement is $90°$ and $\cos 90° = 0$, so the work done by the force of gravity is zero.

(c) The normal force of the floor on the crate is also perpendicular to the displacement, so the work done by this force is also zero.

(d) These are the only forces acting on the crate so the total work done on it is $590\,\text{J}$.

9P

(a) Since the rope is light (massless) the tension is the same throughout its length. It pulls upward on the two lower pulleys with the same force F at four points, so the net force of the rope on the lower pulleys is $4F$. If F is the minimum force that will lift the load, then the load moves upward with constant velocity and Newton's second law is $4F - Mg = 0$, where M is the total mass of the load and lower pulleys. Thus $Mg = 840\,\text{lb} + 20\,\text{lb} = 860\,\text{lb}$ and $F = (860\,\text{lb})/4 = 215\,\text{lb}$.

(b) The work done by the rope is $W = 4Fd = Mgd$, where d is the distance moved by the load. The substitution $4F = Mg$ was made. Hence $W = (860\,\text{lb})(12\,\text{ft}) = 10{,}300\,\text{ft} \cdot \text{lb}$.

(c) For every foot that the load goes up, each segment of rope between the upper and lower sets of pulley shortens by one foot. Thus the free end of the rope goes down 4 feet. The free end moves a total of $4(12\,\text{ft}) = 48\,\text{ft}$.

(d) The work done by the person pulling on the free end of the rope is $W = Fd = Mgd/4$, where d is the distance the free end moves. The substitution $F = Mg/4$ was made. Thus $W = (860\,\text{lb})(48\,\text{ft})/4 = 10{,}300\,\text{ft} \cdot \text{lb}$.

12P

(a) The force of the rope on the block is constant so the work it does is given by $W = \mathbf{F} \cdot \mathbf{d} = Fd\cos\phi$, where $\mathbf{F}$ is the force exerted by the rope, $\mathbf{d}$ is the displacement, and ϕ is the angle between the force and the displacement. Since $F = 7.68\,\text{N}$, $d = 4.06\,\text{m}$, and $\phi = 15.0°$, the work is $W = (7.68\,\text{N})(4.06\,\text{m})\cos 15.0° = 30.1\,\text{J}$.

(b) The free-body diagram for the block is shown to the right. F is the magnitude of the force of the rope, f is the magnitude of the force of friction, N is the magnitude of the normal force of the floor on the block, and mg is the magnitude of the force of gravity. Take the x axis to be to the right and the y axis to be upward. The x component of Newton's second law is $F\cos\phi - f = 0$ and the y component is $N + F\sin\phi - mg = 0$.

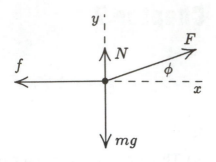

The second equation gives $N = mg - F\sin\phi$. The magnitude of the force of friction is $f = \mu_k N = \mu_k(mg - F\sin\phi)$, so the first equation becomes $F\cos\phi - \mu(mg - F\sin\phi)$. Solve for μ_k:

$$\mu_k = \frac{F\cos\phi}{mg - F\sin\phi} = \frac{(7.68\,\text{N})\cos 15.0°}{(3.57\,\text{kg})(9.8\,\text{m/s}^2) - (7.68\,\text{N})\sin 15.0°} = 0.22\,.$$

16P

(a) The graph shows F as a function of x if x_0 is positive. The work is negative as the object moves from $x = 0$ to $x = x_0$ and positive as it moves from $x = x_0$ to $x = 2x_0$. Since the area of a triangle is $\frac{1}{2}$(base)(altitude), the work done from $x = 0$ to $x = x_0$ is $-\frac{1}{2}(x_0)(F_0)$ and the work done from $x = x_0$ to $x = 2x_0$ is $\frac{1}{2}(2x_0 - x_0)(F_0) = \frac{1}{2}(x_0)(F_0)$. The total work is the sum, which is zero.

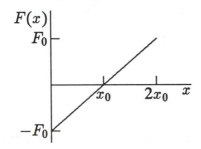

(b) The integral for the work is

$$W = \int_0^{2x_0} F_0\left(\frac{x}{x_0} - 1\right)\,dx = F_0\left(\frac{x^2}{2x_0} - x\right)\Bigg|_0^{2x_0} = 0\,.$$

18E

(a) As the cage moves from $x = x_1$ to $x = x_2$ the work done by the spring is given by

$$W = \int_{x_1}^{x_2}(-kx)\,dx = -\frac{1}{2}kx^2\Bigg|_{x_1}^{x_2} = -\frac{1}{2}k(x_2^2 - x_1^2)\,,$$

where k is the force constant for the spring. Substitute $x_1 = 0$ and $x_2 = 7.6 \times 10^{-3}\,\text{m}$. The result is

$$W = -\frac{1}{2}(1500\,\text{N/m})(7.6 \times 10^{-3}\,\text{m})^2 = -0.043\,\text{J}\,.$$

To use SI units consistently throughout, $15\,\text{N/cm}$ was converted to $1500\,\text{N/m}$.

(b) Now substitute $x_1 = 7.6 \times 10^{-3}\,\text{m}$ and $x_2 = 15.2 \times 10^{-3}\,\text{m}$. The result is now

$$W = -\frac{1}{2}(1500\,\text{N/m})\left[(15.2 \times 10^{-3}\,\text{m})^2 - (7.6 \times 10^{-3}\,\text{m})^2\right] = -0.13\,\text{J}\,.$$

Notice this is more than twice the work done in the first interval. Although the displacements have the same magnitude the force is larger throughout the second interval.

22E

The kinetic energy is given by $K = \frac{1}{2}mv^2$, where m is the mass and v is the speed of the electron. The speed is therefore

$$v = \sqrt{\frac{2K}{m}} = \sqrt{\frac{2(6.7 \times 10^{-19}\,\text{J})}{9.11 \times 10^{-31}\,\text{kg}}} = 1.2 \times 10^6\,\text{m/s}\,.$$

29E

(a) The initial kinetic energy of the car is $K_i = \frac{1}{2}mv_i^2$, where m is the mass and v_i is the initial speed of the car. The initial speed is $60\,\text{km/h} = 16.7\,\text{m/s}$, so the initial kinetic energy is $K_i = \frac{1}{2}(1000\,\text{kg})(16.7\,\text{m/s})^2 = 1.39 \times 10^5\,\text{J}$. After the brakes reduce the kinetic energy by $50\,\text{kJ}$ it is $K_f = 1.39 \times 10^5\,\text{J} - 50 \times 10^3\,\text{J} = 8.9 \times 10^4\,\text{J}$. The final speed of the car is

$$v_f = \sqrt{\frac{2K_f}{m}} = \sqrt{\frac{2(8.9 \times 10^4\,\text{J})}{1000\,\text{kg}}} = 13\,\text{m/s}$$

or $48\,\text{km/h}$.

(b) Since the final kinetic energy of the car is zero, the kinetic energy must be reduced by $8.9 \times 10^4\,\text{J}$ to stop the car.

35E

(a) Let F be the magnitude of the force exerted by the cable on the astronaut. The force of the cable is upward and the weight is mg is downward. Furthermore the acceleration of the astronaut is $g/10$, upward. According to Newton's second law $F - mg = mg/10$, so $F = 11mg/10$. Since the force $\mathbf{F}$ and the displacement $\mathbf{d}$ are in the same direction the work done by $\mathbf{F}$ is

$$W_F = Fd = \frac{11mgd}{10} = \frac{11(72\,\text{kg})(9.8\,\text{m/s}^2)(15\,\text{m})}{10} = 1.16 \times 10^4\,\text{J}\,.$$

(b) The weight has magnitude mg and is opposite in direction to the displacement. It does work

$$W_g = -mgd = -(72\,\text{kg})(9.8\,\text{m/s}^2)(15\,\text{m}) = -1.06 \times 10^4\,\text{J}\,.$$

(c) The total work done is $W_T = 1.16 \times 10^4\,\text{J} - 1.06 \times 10^4\,\text{J} = 1.0 \times 10^3\,\text{J}$. Since the astronaut started from rest the work-energy theorem tells us that this must be her final kinetic energy.

(d) Since $K = \frac{1}{2}mv^2$ her final speed is

$$v = \sqrt{\frac{2K}{m}} = \sqrt{\frac{2(1.0 \times 10^3\,\text{J})}{72\,\text{kg}}} = 5.3\,\text{m/s}\,.$$

36E

(a) Let F be the magnitude of the force of the cord on the block. This force is upward, while the force of gravity, with magnitude Mg, is downward. The acceleration is $g/4$, down. Take the downward direction to be positive. Then Newton's second law is $Mg - F = Mg/4$, so $F = 3Mg/4$. The force is directed opposite to the displacement so the work it does is $W_F = -Fd = -3Mgd/4$.

(b) The force of gravity is in the same direction as the displacement so it does work $W_g = Mgd$.

(c) The total work done on the block is $W_T = -3Mgd/4 + Mgd = Mgd/4$. Since the block starts from rest this is its kinetic energy K after it is lowered a distance d.

(d) Since $K = \frac{1}{2}Mv^2$, where v is the speed, $v = \sqrt{2K/M} = \sqrt{gd/2}$ after the block is lowered a distance d. The result found in (c) was used.

43P

First find the force of the winch. The free-body diagram for the block is shown to the right. F is the magnitude of the force of the winch, f is the magnitude of the force of friction of the incline, N is the magnitude of the normal force of the incline, and mg is the magnitude of the force of gravity. $\theta\ (= \tan^{-1}(30\,\text{m}/40\,\text{m}) = 37°)$ is the angle of the incline. Take the x axis to be up the incline and the y axis to be in the direction of the normal force. The acceleration is zero so the x component of Newton's second law is $F - f - mg\sin\theta = 0$ and the y component is $N - mg\cos\theta = 0$.

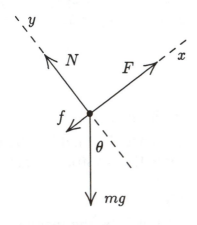

The second equation gives $N = mg\cos\theta$, so $f = \mu_k N = \mu_k mg\cos\theta$. Substitute this expression into the first equation and solve for F: $F = mg(\sin\theta + \mu_k\cos\theta)$. The force of the winch is in the same direction as the velocity of the block so the power of the winch is

$$P = Fv = mgv(\sin\theta + \mu_k\cos\theta)$$
$$= (1400\,\text{kg})(9.8\,\text{m/s}^2)(1.34\,\text{m/s})(\sin 37° + 0.40\cos 37°) = 1.7 \times 10^4\,\text{W}.$$

47P

(a) The power is given by $P = Fv$ and the work done by **F** from time t_1 to time t_2 is given by

$$W = \int_{t_1}^{t_2} P\,dt = \int_{t_1}^{t_2} Fv\,dt.$$

Since **F** is the total force the magnitude of the acceleration is $a = F/m$ and the velocity as a function of time is given by $v = at = (F/m)t$. Thus

$$W = \int_{t_1}^{t_2} (F^2/m)t\,dt = \frac{1}{2}(F^2/m)(t_2^2 - t_1^2).$$

For $t_1 = 0$ and $t_2 = 1.0\,\text{s}$,

$$W = \frac{1}{2}\left[\frac{(5.0\,\text{N})^2}{15\,\text{kg}}\right](1.0\,\text{s})^2 = 0.83\,\text{J}.$$

For $t_1 = 1.0\,\text{s}$ and $t_2 = 2.0\,\text{s}$,

$$W = \frac{1}{2}\left[\frac{(5.0\,\text{N})^2}{15\,\text{kg}}\right]\left[(2.0\,\text{s})^2 - (1.0\,\text{s})^2\right] = 2.5\,\text{J}.$$

For $t_1 = 2.0\,\text{s}$ and $t_2 = 3.0\,\text{s}$,

$$W = \frac{1}{2}\left[\frac{(5.0\,\text{N})^2}{15\,\text{kg}}\right]\left[(3.0\,\text{s})^2 - (2.0\,\text{s})^2\right] = 4.2\,\text{J}.$$

(b) Substitute $v = (F/m)t$ into $P = Fv$ to obtain $P = F^2 t/m$ for the power at any time t. At the end of the third second

$$P = \frac{(5.0\,\text{N})^2(3.0\,\text{s})}{15\,\text{kg}} = 5.0\,\text{W}.$$

48P

The total work is the sum of the works done by gravity on the elevator, the work done by gravity on the counterweight, and the work done by the motor on the system: $W_T = W_e + W_c + W_m$. Since the elevator moves at constant velocity, its kinetic energy does not change and according to the work-energy theorem the total work done it is zero. This means $W_e + W_c + W_m = 0$. The elevator moves upward through 54 m so the work done by gravity on it is

$$W_e = -m_e g d = -(1200\,\text{kg})(9.8\,\text{m/s}^2)(54\,\text{m}) = -6.35 \times 10^5\,\text{J}.$$

The counterweight moves downward the same distance so the work done by gravity on it is

$$W_c = m_c g d = (950\,\text{kg})(9.8\,\text{m/s}^2)(54\,\text{m}) = 5.03 \times 10^5\,\text{J}.$$

Since $W_T = 0$, the work done by the motor is

$$W_m = -W_e - W_c = 6.35 \times 10^5\,\text{J} - 5.03 \times 10^5\,\text{J} = 1.32 \times 10^5\,\text{J}.$$

This work is done in a time interval of $\Delta t = 3.0\,\text{min} = 180\,\text{s}$ so the power supplied by the motor to lift the elevator is

$$P = \frac{W_m}{\Delta t} = \frac{1.32 \times 10^5\,\text{J}}{180\,\text{s}} = 7.35 \times 10^2\,\text{W}.$$

This is $(7.35 \times 10^2\,\text{W})/(746\,\text{W/hp}) = 0.99\,\text{hp}$.

49P

Let the force be $F = \alpha v$, where v is the speed and α is a constant of proportionality. The power required is $P = Fv = \alpha v^2$. Let P_1 be the power required for speed v_1 and P_2 be the power required for speed v_2. Divide $P_2 = \alpha v_2^2$ by $P_1 = \alpha v_1^2$, then solve for P_2. You should obtain $P_2 = (v_2/v_1)^2 P_1$. Since $P_1 = 10\,\text{hp}$ and $v_2 = 3v_1$, $P_2 = (3)^2(10\,\text{hp}) = 90\,\text{hp}$.

50E

(a) Its speed is $v = d/t = (5.1 \times 10^{-2}\,\text{m})/(0.25 \times 10^{-9}\,\text{s}) = 2.04 \times 10^8\,\text{m/s}$. Since the speed of light is $c = 2.998 \times 10^8\,\text{m/s}$, $v = (2.04 \times 10^8/2.998 \times 10^8)c = 0.68c$.

(b) Since the speed of the electron is close to the speed of light, use the relativistic expression for the kinetic energy:

$$K = mc^2 \left(\frac{1}{\sqrt{1 - v^2/c^2}} - 1 \right)$$

$$= (9.11 \times 10^{-31}\,\text{kg})(2.998 \times 10^8\,\text{m/s})^2 \left(\frac{1}{\sqrt{1 - (0.68)^2}} - 1 \right)$$

$$= 3.0 \times 10^{-14}\,\text{J}.$$

This is $(3.0 \times 10^{-14}\,\text{J})/(1.60 \times 10^{-19}\,\text{J/eV}) = 1.9 \times 10^5\,\text{eV} = 190\,\text{keV}$.

(c) Classically the kinetic energy is

$$K = \frac{1}{2}mv^2 = \frac{1}{2}(9.11 \times 10^{-31}\,\text{kg})(2.04 \times 10^8\,\text{m/s})^2 = 1.90 \times 10^{-14}\,\text{J}.$$

The fractional error is

$$\frac{3.0 \times 10^{-14}\,\text{J} - 1.9 \times 10^{-14}\,\text{J}}{3.0 \times 10^{-14}\,\text{J}} = 0.37$$

or 37 per cent.

Chapter 8

1E

The potential energy stored by the spring is given by $U = \frac{1}{2}kx^2$, where k is the spring constant and x is the displacement of the end of the spring from its position when the spring is in equilibrium. Thus

$$k = \frac{2U}{x^2} = \frac{2(25\,\text{J})}{(0.075\,\text{m})^2} = 8.9 \times 10^3\,\text{N/m}.$$

6E

The only force that does work as the flake falls is the force of gravity and it is a conservative force. If K_i is the kinetic energy of the flake at the edge of the bowl, K_f is its kinetic energy at the bottom, U_i is the gravitational potential energy with the flake at the top, and U_f is the gravitational potential energy with it at the bottom, then $K_f + U_f = K_i + U_i$. Take the potential energy to be zero at the bottom of the bowl. Then the potential energy at the top is $U_i = mgR$, where R is the radius of the bowl and m is the mass of the flake. $K_i = 0$ since the flake starts from rest. Since the problem asks for the speed at the bottom, write $\frac{1}{2}mv^2$ for K_f. The energy conservation equation becomes $mgR = \frac{1}{2}mv^2$, so

$$v = \sqrt{2gR} = \sqrt{2(9.8\,\text{m/s}^2)(0.22\,\text{m})} = 2.1\,\text{m/s}.$$

8E

Neglect any work done by the force of friction. Then the only force to do work is the force of gravity, a conservative force. Let K_i be the kinetic energy of the truck at the bottom of the ramp and let K_f be its kinetic energy at the top. Let U_i be the gravitational potential energy when the truck is at the bottom and let U_f be the gravitational potential energy when it is at the top. Then $K_f + U_f = K_i + U_i$. If the potential energy is taken to be zero when the truck is at the bottom then $U_f = mgh$, where h is the final height of the truck above its initial position. $K_i = \frac{1}{2}mv^2$, where v is the initial speed of the truck, and $K_f = 0$ since the truck comes to rest. Thus $mgh = \frac{1}{2}mv^2$ and

$$h = \frac{v^2}{2g}.$$

Substitute $v = 80\,\text{mi/h} = 117\,\text{ft/s}$ to obtain

$$h = \frac{(117\,\text{ft/s})^2}{2(32\,\text{ft/s}^2)} = 214\,\text{ft}.$$

If L is the length of the ramp then $L \sin 15° = 214\,\text{ft}$ or $L = (214\,\text{ft})/\sin 15° = 830\,\text{ft}$. If friction does work on the truck then the ramp might be shorter.

10E

(a) If m is the mass of the projectile and v is its speed just after firing, then its kinetic energy just after firing is $K_i = \frac{1}{2}mv^2 = \frac{1}{2}(2.40\,\text{kg})(150\,\text{m/s})^2 = 2.70 \times 10^4\,\text{J}$.

(b) If the potential energy is taken to be zero when the projectile hits the ground and the initial height of the projectile above the ground is taken to be h, then its initial potential energy is $U_i = mgh = (2.40\,\text{kg})(9.8\,\text{m/s}^2)(125\,\text{m}) = 2.94 \times 10^3\,\text{J}$.

(c) Just before it hits the ground the potential energy is zero and the kinetic energy can be written $K_f = \frac{1}{2}mv_f^2$, where v_f is the speed of the projectile. Mechanical energy is conserved during the trip so $K_f = K_i + U_i$, $\frac{1}{2}mv_f^2 = K_i + U_i$, and

$$v_f = \sqrt{\frac{2(K_i + U_i)}{m}} = \sqrt{\frac{2(2.70 \times 10^4\,\text{J} + 2.94 \times 10^3\,\text{J})}{2.40\,\text{kg}}} = 159\,\text{m/s}\,.$$

K_i, K_f, U_i, and U_f all depend on the mass of the projectile, but the final speed v_f does not if air resistance is negligible. To see this, repeat the calculation with a different value for the mass.

12E

(a) Take the gravitational potential energy to be zero at the position of the marble when the spring is compressed. The gravitational potential energy when the marble is at the top of its flight is then $U_g = mgh$, where h is the height of the highest point. This is $h = 20\,\text{m} + 8\,\text{cm} = 20.08\,\text{m}$. Thus $U_g = (5.0 \times 10^{-3}\,\text{kg})(9.8\,\text{m/s}^2)(20.08\,\text{m}) = 0.984\,\text{J}$.

(b) Since mechanical energy is conserved the potential energy of the compressed spring must be the same as the gravitational potential energy at the top of the flight. Thus $\frac{1}{2}kx^2 = mgh$, where k is the spring constant. Hence

$$k = \frac{2mgh}{x^2} = \frac{2(5.0 \times 10^{-3}\,\text{kg})(9.8\,\text{m/s}^2)(20.08\,\text{m})}{(0.08\,\text{m})^2} = 3.1 \times 10^2\,\text{N/m}\,.$$

13E

(a) Take the zero of gravitational potential energy to be at the bottom of the swing. Since the ball is initially a vertical distance L above the bottom the initial potential energy is $U_i = mgL$ and since the ball is finally a distance $2L$ above the bottom, the final potential energy is $2mgL$. The change in the potential energy is $\Delta U = U_f - U_i = 2mgL - mgL = mgL$.

(b) The final kinetic energy is zero. Write the initial kinetic energy as $K_i = \frac{1}{2}mv^2$, where v is the initial speed. The rod does zero work and the force of gravity is conservative, so mechanical energy is conserved. This means $\Delta K = -\Delta U$, or $-\frac{1}{2}mv^2 = -mgL$, so $v = \sqrt{2gL}$.

17P

Information given in the first sentence allows us to compute the spring constant. Solve $F = kx$ for k: $k = F/x = (270\,\text{N})/(0.02\,\text{m}) = 1.35 \times 10^4\,\text{N/m}$.

(a) Now consider the block sliding down the incline. If it starts from rest at a height h above the point where it momentarily comes to rest, its initial kinetic energy is zero and the initial gravitational potential energy is mgh, where m is the mass of the block. We have taken the zero of gravitational potential energy to be at the point where the block comes to rest. We also take the initial potential energy stored in the spring to be zero. Suppose the block compresses the spring a distance x before coming momentarily to rest. Then the final kinetic energy is zero, the final gravitational potential energy is zero, and final spring potential energy is $\frac{1}{2}kx^2$. The incline is frictionless and the normal force it exerts on the block does no work, so mechanical energy is conserved. This means $mgh = \frac{1}{2}kx^2$, so

$$h = \frac{kx^2}{2mg} = \frac{(1.35 \times 10^4 \, \text{N/m})(0.055 \, \text{m})^2}{2(12 \, \text{kg})(9.8 \, \text{m/s}^2)} = 0.174 \, \text{m}.$$

If the block traveled down a length of incline equal to ℓ, then $\ell \sin 30° = h$, so $\ell = h/\sin 30° = (0.174 \, \text{m})/\sin 30° = 0.35 \, \text{m}$.

(b) Just before it touches the spring it is 0.055 m away from the place where it comes to rest and so is a vertical distance $(0.055 \, \text{m}) \sin 30° = 0.0275 \, \text{m}$ above its final position. The potential energy is then $mgh' = (12 \, \text{kg})(9.8 \, \text{m/s}^2)(0.0275 \, \text{m}) = 3.23 \, \text{J}$. On the other hand, its initial potential energy is $mgh = (12 \, \text{kg})(9.8 \, \text{m/s}^2)(0.174 \, \text{m}) = 20.5 \, \text{J}$. The difference is its final kinetic energy: $K_f = 20.5 \, \text{J} - 3.23 \, \text{J} = 17.2 \, \text{J}$. Its final speed is

$$v = \sqrt{\frac{2K_f}{m}} = \sqrt{\frac{2(17.2 \, \text{J})}{12 \, \text{kg}}} = 1.7 \, \text{m/s}.$$

21P

(a) Let m be the mass of the shell and v_0 be its initial speed. Then its initial kinetic energy is $K_i = \frac{1}{2}mv_0^2 = \frac{1}{2}(5.0 \, \text{kg})(100 \, \text{m/s})^2 = 2.5 \times 10^4 \, \text{J}$.

(b) Take the gravitational potential energy to be zero at the firing point and let U_f be the potential energy at the top of the trajectory. U_f is then the same as the change in potential energy from the time of firing to the time the trajectory top is reached. At that point the shell's velocity is horizontal and has the same value as it has initially: $v_h = v_0 \cos \theta_0$, where θ_0 is the firing angle. The kinetic energy at the top is $K_f = \frac{1}{2}mv_h^2 = \frac{1}{2}mv_0^2 \cos^2 \theta_0$. Since mechanical energy is conserved

$$\frac{1}{2}mv_0^2 = U_f + \frac{1}{2}mv_0^2 \cos^2 \theta_0.$$

Thus

$$U_f = \frac{1}{2}mv_0^2(1 - \cos^2 \theta_0) = \frac{1}{2}mv_0^2 \sin^2 \theta_0 = \frac{1}{2}(5.0 \, \text{kg})(100 \, \text{m/s})^2 \sin^2 34° = 7.8 \times 10^3 \, \text{J}.$$

(c) The potential energy at the top of the trajectory is also given by $U_f = mgh$, where h is the height of that point above the firing point. Solve for h:

$$h = \frac{U_f}{mg} = \frac{7.8 \times 10^3 \, \text{J}}{(5.0 \, \text{kg})(9.8 \, \text{m/s}^2)} = 160 \, \text{m}.$$

25P

Let m be the mass of the block, h the height from which it dropped, and x the compression of the spring. Take the zero of potential energy to be at the initial position of the block. It drops a distance $h + x$ and its final gravitational potential energy is $-mg(h + x)$. Here x is taken to be positive for a compression of the spring. The spring potential energy is initially zero and finally $\frac{1}{2}kx^2$. The kinetic energy is zero at both the beginning and end. Since energy is conserved $0 = -mg(h + x) + \frac{1}{2}kx^2$. This is a quadratic equation for x. Its solution is

$$x = \frac{mg \pm \sqrt{(mg)^2 + 2mghk}}{k}.$$

Now $mg = (2.0\,\text{kg})(9.8\,\text{m/s}^2) = 19.6\,\text{N}$ and $hk = (0.40\,\text{m})(1960\,\text{N/m}) = 784\,\text{N}$, so

$$x = \frac{19.6\,\text{N} \pm \sqrt{(19.6\,\text{N})^2 + 2(19.6\,\text{N})(784\,\text{N})}}{1960\,\text{N/m}}$$

$= 0.10\,\text{m}$ or $-0.080\,\text{m}$. Since x must be positive (a compression) we accept the positive solution and reject the negative solution.

27P

The distance the marble travels is determined by its initial speed, which is determined by the original compression of the spring.

Let h be the height of the table, and x the horizontal distance to the point where the marble lands. Then $x = v_0 t$ and $h = \frac{1}{2}gt^2$, where v_0 is the initial speed of the marble and t is the time it is in the air. The second equation gives $t = \sqrt{2h/g}$ so $x = v_0\sqrt{2h/g}$. The distance to the landing point is directly proportional to the initial speed. Let v_{01} be the initial speed of the first shot and x_1 be the horizontal distance to its landing point; let v_{02} be the initial speed of the second shot and x_2 be the horizontal distance to its landing spot. Then $v_{02} = (x_2/x_1)v_{01}$. When the spring is compressed the potential energy is $\frac{1}{2}k\ell^2$, where ℓ is the compression. When the marble leaves the spring the potential energy is zero and the kinetic energy is $\frac{1}{2}mv_0^2$. Since mechanical energy is conserved $\frac{1}{2}mv_0^2 = \frac{1}{2}k\ell^2$, so the initial speed of the marble is directly proportional to the original compression of the spring. If ℓ_1 is the compression for the first shot and ℓ_2 is the compression for the second, then $v_{02} = (\ell_2/\ell_1)v_{01}$. Combining this with the previous result gives $\ell_2 = (x_2/x_1)\ell_1$. Take $x_1 = 2.20\,\text{m} - 0.27\,\text{m} = 1.93\,\text{m}$, $\ell_1 = 1.10\,\text{cm}$, and $x_2 = 2.2\,\text{m}$, then evaluate the expression for ℓ_2:

$$\ell_2 = \left(\frac{2.20\,\text{m}}{1.93\,\text{m}}\right)(1.10\,\text{cm}) = 1.25\,\text{cm}.$$

32P

Use conservation of energy: the mechanical energy must be the same at the top of the swing as it is initially. Use Newton's second law to find the speed, and hence the kinetic energy, at the top. There the tension T in the string and the force of gravity are both downward, toward the center of the circle. Notice that the radius of the circle is $r = L - d$, so the law can be written $T + mg = mv^2/(L - d)$, where v is the speed and m is the mass of the bob. When the bob passes the highest point with the least possible speed the tension is zero. Then $mg = mv^2/(L - d)$ and $v = \sqrt{g(L - d)}$.

Take the zero of gravitational potential energy to be at the bottom of the swing. Then the initial potential energy is mgL. The initial kinetic energy is zero since the bob starts from rest. The final potential energy, at the top of the swing, is $2mg(L - d)$ and the final kinetic energy is $\frac{1}{2}mv^2 = \frac{1}{2}mg(L - d)$. Conservation of energy yields $mgL = 2mg(L - d) + \frac{1}{2}mg(L - d)$. Solve for d: $d = 3L/5$. If d is greater than this value, so the highest point is lower, then the speed of the bob is greater as it reaches that point and the bob passes the point. If d is less the bob cannot go around. Thus the value you found for d is a lower limit.

33P

(a) Take the zero of potential energy to be at the bottom of the swing. Then the initial potential energy is $2mgL$. The initial kinetic energy is zero since the ball is at rest. Write $\frac{1}{2}mv^2$, where v is the speed of the ball, for the final kinetic energy, at the bottom of the swing. Since mechanical energy is conserved $2mgL = \frac{1}{2}mv^2$ and $v = 2\sqrt{gL}$.

(b) At the bottom of the swing the force of gravity pulls downward and the tension in the rod pulls upward. If T is the magnitude of the tension, Newton's second law is $T - mg = mv^2/L$, so $T = mg + mv^2/L = mg + 4mg = 5mg$.

(c) The diagram to the right is the free-body diagram for the ball when the tension in the rod has the same magnitude as the weight of the ball. We wish to solve for θ. The component of the force of gravity along the radial direction is $mg \cos \theta$ and is outward. The net inward force is $T - mg \cos \theta$ and, according to Newton's second law this must equal mv^2/L, where v is the speed of the ball. Thus $T = mv^2/L + mg \cos \theta$.

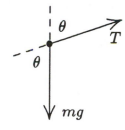

We now need to find the speed of the ball in terms of θ. Take the potential energy to be zero when the rod is horizontal. Since it starts from rest its kinetic energy is also zero. As can be seen on the diagram to the right, when the rod makes the angle θ with the vertical, the ball has dropped through a vertical distance $L \cos \theta$. The

potential energy is then $-mgL \cos \theta$. Write the kinetic energy as $\frac{1}{2}mv^2$ and the conservation of energy equation as $0 = -mgL \cos \theta + \frac{1}{2}mv^2$. Thus $mv^2 = 2mgL \cos \theta$. Substitute this expression into the equation developed above for T: $T = 2mg \cos \theta + mg \cos \theta = 3mg \cos \theta$. According to the condition of the problem this must be equal to mg, so $3mg \cos \theta = mg$, or $\cos \theta = 1/3$. This means $\theta = 71°$.

35P

The work required is the change in the gravitational potential energy as the chain is pulled onto the table. Take the potential energy to be zero when the whole chain is on the table.

Divide the hanging chain into a large number of infinitesimal segments, each of length dy. The mass of a segment is $(m/L)\,dy$ and the potential energy of the segment a distance y below the table top is $dU = -(m/L)gy\,dy$. The total potential energy is

$$U = -(m/L)g \int_0^{L/4} y\,dy = -\frac{1}{2}(m/L)g(L/4)^2 = -mgL/32.$$

The work required to pull the chain onto the table is $-U = mgL/32$.

37P

The free-body diagram for the boy is shown to the right. N is the normal force of the ice on him and m is his mass. The net inward force is $mg\cos\theta - N$ and, according to Newton's second law, this must be equal to mv^2/R, where v is the speed of the boy. At the point where the boy leaves the ice $N = 0$, so $g\cos\theta = v^2/R$. We wish to find his speed. If the gravitational potential energy is taken to be zero when he is at the top of the ice mound then his potential energy at the time shown is $U = -mgR(1-\cos\theta)$. He starts from rest and his kinetic energy at the time shown is $\frac{1}{2}mv^2$. Thus conservation of energy gives $0 = \frac{1}{2}mv^2 - mgR(1-\cos\theta)$, or $v^2 = 2gR(1-\cos\theta)$. Substitute this expression into the equation developed from the second law to obtain $g\cos\theta = 2g(1-\cos\theta)$. This gives $\cos\theta = 2/3$. The height of the boy above the bottom of the mound is $R\cos\theta = 2R/3$.

39P

(a) The force is radial (along the line joining the atoms) and is given by the derivative of U with respect to r:

$$F(r) = -\frac{dU}{dr} = \frac{12A}{r^{13}} - \frac{6B}{r^7}.$$

The equilibrium separation r_0 is the separation for which $F = 0$. This means $12A - 6Br_0^6 = 0$, where the equation above was multiplied by r_0^{13}. The equilibrium separation is given by $r_0 = (2A/B)^{1/6} = 1.12(A/B)^{1/6}$.

(b) The derivative of the force with respect to r, evaluated at the equilibrium separation, is

$$\frac{dF}{dr} = -\frac{12\cdot 13A}{r_0^{14}} + \frac{42B}{r_0^8} = -\frac{(156A - 42Br_0^6)}{r_0^{14}} = -\frac{72A}{r_0^{14}},$$

where $r_0^6 = 2A/B$ was substituted to obtain the final form. The derivative is negative so the force is positive if r is slightly less than r_0, indicating that it is a force of repulsion.

(c) If r is slightly greater than r_0 the force is negative, indicating it is a force of attraction.

49E

Since a joule is a watt·second, $1.0\,\text{kW}\cdot\text{h}$ is $(1.0\,\text{kW}\cdot\text{h})(10^3\,\text{W/kW})(3600\,\text{s/h}) = 3.6\times 10^6\,\text{J}$. This is the energy that can be obtained from $(3.6\times 10^6\,\text{J})/(140\times 10^6\,\text{J/gal}) = 0.0257\,\text{gal}$ of gasoline. At $30\,\text{mi/gal}$ you would go $(30\,\text{mi/gal})(0.0257\,\text{gal}) = 0.77\,\text{mi}$.

(b) You are expending energy at the rate

$$P = \frac{dE}{dt} = \frac{dE}{dx}\frac{dx}{dt} = \frac{dE}{dx}v = \left(\frac{1.0\,\mathrm{kW}\cdot\mathrm{h}}{0.77\,\mathrm{mi}}\right)(55\,\mathrm{mi/h}) = 71\,\mathrm{kW}.$$

50E

(a) If the mass of the boy is m and the distance climbed is h then the change in the gravitational potential energy is $\Delta U = mgh = (51\,\mathrm{kg})(9.8\,\mathrm{m/s}^2)(6.0\,\mathrm{m}) = 3.0 \times 10^3\,\mathrm{J}$.

(b) The boy must supply energy at the same rate as his potential energy is increasing, or $P = \Delta U/\Delta t = (3.0 \times 10^3\,\mathrm{J})/(10\,\mathrm{s}) = 3.0 \times 10^2\,\mathrm{W}$.

55E

To swim at constant velocity the swimmer must push backward against the water with a force of 110 N. Relative to him the water is going at 0.22 m/s toward his rear, in the same direction as his force. His power output is $P = \mathbf{F} \cdot \mathbf{v} = Fv = (110\,\mathrm{N})(0.22\,\mathrm{m/s}) = 24\,\mathrm{W}$.

64E

(a) Take the initial gravitational potential energy to be $U_i = 0$. Then the final gravitational potential energy is $U_f = -mgL$, where L is the length of the tree. The change is $U_f - U_i = -mgL = -(25\,\mathrm{kg})(9.8\,\mathrm{m/s}^2)(12\,\mathrm{m}) = -2.94 \times 10^3\,\mathrm{J}$.

(b) The kinetic energy is $K = \frac{1}{2}mv^2 = \frac{1}{2}(25\,\mathrm{kg})(5.6\,\mathrm{m/s})^2 = 3.92 \times 10^2\,\mathrm{J}$.

(c) According to Eq. 8–26, the change in the mechanical energy is equal to $-fL$, where f is the average frictional force. Thus

$$f = -\frac{\Delta K + \Delta U}{L} = -\frac{3.92 \times 10^2\,\mathrm{J} - 2.94 \times 10^3\,\mathrm{J}}{12\,\mathrm{m}} = 210\,\mathrm{N}.$$

66E

(a) The magnitude of the force of friction is $f = \mu_k N$, where μ_k is the coefficient of kinetic friction and N is the normal force of the surface on the block. The only vertical forces acting on the block are the normal force, upward, and the force of gravity, downward. Since the vertical component of the block's acceleration is zero, Newton's second law tells us that $N = mg$, where m is the mass of the block. Thus $f = \mu_k mg$. The mechanical energy dissipated is given by $\Delta E = f\ell = \mu_k mg\ell$, where ℓ is the distance the block moves before coming to rest. Its value is $\Delta E = (0.25)(3.5\,\mathrm{kg})(9.8\,\mathrm{m/s}^2)(7.8\,\mathrm{m}) = 67\,\mathrm{J}$.

(b) The block had its maximum kinetic energy just as it left the spring and entered the part of the surface where friction acts. The maximum kinetic energy equals the mechanical energy dissipated by friction, 67 J.

(c) The energy that appears as kinetic energy was originally stored as the potential energy of the compressed spring. Thus $\Delta E = \frac{1}{2}kx^2$, where k is the spring constant and x is the compression. Solve for x:

$$x = \sqrt{\frac{2\Delta E}{k}} = \sqrt{\frac{2(67\,\mathrm{J})}{640\,\mathrm{N/m}}} = 0.46\,\mathrm{m}.$$

69P

(a) Take the zero of gravitational potential energy to be at the bottom of the peaks. Then the initial potential energy is $U_i = mgh_i$, where m is the mass of the skier, and h_i is the height of the higher peak. The final potential energy is $U_f = mgh_f$, where h_f is the height of the lower peak. The skier initially has a kinetic energy of $K_i = 0$. Write the final kinetic energy as $K_f = \frac{1}{2}mv^2$, where v is the speed of the skier at the top of the lower peak. The normal force of the slope on the skier does no work and friction is negligible, so mechanical energy is conserved: $U_i + K_i = U_f + K_f$, or $mgh_i = mgh_f + \frac{1}{2}mv^2$. Solve for v:

$$v = \sqrt{2g(h_i - h_f)} = \sqrt{2(9.8\,\text{m/s}^2)(850\,\text{m} - 750\,\text{m})} = 44\,\text{m/s}.$$

(b) As you know from your study of objects sliding down inclined planes, the normal force of the slope on the skier is given by $N = mg\cos\theta$, where θ is the angle of the slope from the horizontal, 30° for each of the slopes shown. The magnitude of the force of friction is given by $f = \mu_k N = \mu_k mg\cos\theta$. The mechanical energy dissipated by the force of friction is $f\ell = \mu_k mg\ell\cos\theta$, where ℓ is the total length of the path. Since the skier gets to the top of the lower peak with no kinetic energy, the mechanical energy dissipated by friction is equal to the difference in the potential energy at the beginning and end of the trip. That is, $\mu_k mg\ell\cos\theta = mg(h_i - h_f)$. Solve for μ_k:

$$\mu_k = \frac{(h_i - h_f)}{\ell\cos\theta}.$$

Thus

$$\mu_k = \frac{(850\,\text{m} - 750\,\text{m})}{(3.2 \times 10^3\,\text{m})\cos 30°} = 0.036.$$

74P

(a) To stretch the spring an external force, equal in magnitude to the force of the spring but opposite to its direction, is applied. Since a spring stretched in the positive x direction exerts a force in the negative x direction, the applied force must be $F = 52.8x + 38.4x^2$, in the positive x direction. The work it does is

$$W = \int_{0.50}^{1.00} (52.8x + 38.4x^2)\,dx$$

$$= \left[\frac{52.8}{2}x^2 + \frac{38.4}{3}x^3\right]_{0.50}^{1.00} = 31.0\,\text{J}.$$

(b) The spring does 31.0 J of work and this must be the increase in the kinetic energy of the particle. It speed is then

$$v = \sqrt{\frac{2K}{m}} = \sqrt{\frac{2(31.0\,\text{J})}{2.17\,\text{kg}}} = 5.35\,\text{m/s}.$$

(c) The force is conservative since the work it does as the particle goes from any point x_1 to any other point x_2 depends only on x_1 and x_2, not on details of the motion between x_1 and x_2.

(a)Let h be the maximum height reached. The mechanical energy dissipated by the air as the stone rises to this height is $\Delta E = -fh$. Use $\Delta E = (K_f + U_f) - (K_i + U_i)$, where K_i and K_f are the initial and final kinetic energies and U_i and U_f are the initial and final gravitational potential energies. Take the potential energy to be zero at the throwing point. The initial kinetic energy is $K_i = \frac{1}{2}mv_0^2$, the initial potential energy is $U_i = 0$, the final kinetic energy is $K_f = 0$, and the final potential energy is $U_f = wh$. Thus $-fh = wh - \frac{1}{2}mv_0^2$. Solve for h:

$$h = \frac{mv_0^2}{2(w+f)} = \frac{wv_0^2}{2g(w+f)} = \frac{v_0^2}{2g(1+f/w)} .$$

Here w/g was substituted for m and both the numerator and denominator were divided by w.

(b) Notice that the force of the air is downward on the trip up and upward on the trip down. It is always opposite to the direction of the velocity. Over the entire trip the energy dissipated by the air is $\Delta E = -2fh$. The final kinetic energy is $K_f = \frac{1}{2}mv^2$, where v is the speed of the stone just before it hits the ground. The final potential energy is $U_f = 0$. Thus $-2fh = \frac{1}{2}mv^2 - \frac{1}{2}mv_0^2$. Substitute the expression you found for h to obtain

$$-\frac{2fv_0^2}{2g(1+f/w)} = \frac{1}{2}mv^2 - \frac{1}{2}mv_0^2 .$$

This leads to

$$v^2 = v_0^2 - \frac{2fv_0^2}{mg(1+f/w)} = v_0^2 - \frac{2fv_0^2}{w(1+f/w)} = v_0^2\left[1 - \frac{2f}{w+f}\right] = v_0^2\frac{w-f}{w+f} .$$

Here w was substituted for mg and some algebraic manipulations were carried out. Thus

$$v = v_0\left(\frac{w-f}{w+f}\right)^{1/2} .$$

84P

The kinetic energy of the automobile may change, its gravitational potential energy may change, and energy is dissipated in friction and air resistance. All this energy must be supplied by the engine. Let ΔE_e be the energy supplied by the engine, ΔK be the change in kinetic energy, ΔU be the change in potential energy, and ΔE_d be the energy dissipated, all in time Δt. Energy balance requires that $\Delta E_e = \Delta K + \Delta U + \Delta E_d$. Divide by Δt and replace $\Delta E_e/\Delta t$ with P (the power supplied by the engine). Thus

$$P = \frac{\Delta K}{\Delta t} + \frac{\Delta U}{\Delta t} + \frac{\Delta E_d}{\Delta t} .$$

(a) The speed of the automobile does not change, so $\Delta K = 0$. It is traveling on a level road, so $\Delta U = 0$. If it goes a distance Δx in time Δt and f is the magnitude of the effective retarding force, then $\Delta E_d/\Delta t = f\Delta x/\Delta t = fv$, where v is the speed of the automobile. Thus $P = fv$ and

$$f = \frac{P}{v} = \frac{16.0 \times 10^3 \text{ W}}{15.0 \text{ m/s}} = 1.07 \times 10^3 \text{ N}.$$

(b) Now the potential energy increases. If the automobile goes a distance Δx the change in potential energy is $\Delta U = mg\Delta x \sin\theta$, where θ is the angle of the incline. The rate at which the potential energy increases is $\Delta U/\Delta t = mgv \sin\theta$ and $P = mgv \sin\theta + fv$. For an 8.00% grade, $\tan\theta = 0.0800$, $\theta = 4.574°$, and $\sin\theta = 0.0797$. Thus

$$P = (1710 \text{ kg})(9.8 \text{ m/s}^2)(15.0 \text{ m/s})(0.0797) + (1.07 \times 10^3 \text{ N})(15.0 \text{ m/s}) = 3.6 \times 10^4 \text{ W}.$$

(c) Now $P = 0$ and $\Delta K = 0$ since the automobile is coasting with constant velocity. Thus $0 = mgv \sin\theta + fv$ and

$$\sin\theta = -\frac{f}{mg} = -\frac{1.07 \times 10^3 \text{ N}}{(1710 \text{ kg})(9.8 \text{ m/s}^2)} = -0.064.$$

Thus $\theta = \sin^{-1}(-0.064) = -3.67°$ and $\tan\theta = -0.064$. The incline has a 6.4% grade.

92E

(a) Use $E = mc^2$, where m is the mass and c is the speed of light. Convert 102 g to 0.102 kg and use $c = 2.998 \times 10^8$ m/s. Thus $E = (0.102 \text{ kg})(2.998 \times 10^8 \text{ m/s})^2 = 9.17 \times 10^{15}$ J.

(b) Use $E = Pt$, where P is the rate of energy consumption and t is the time. Thus

$$t = \frac{E}{P} = \frac{9.17 \times 10^{15} \text{ J}}{1.00 \times 10^3 \text{ W}} = 9.17 \times 10^{12} \text{ s} = 2.91 \times 10^5 \text{ y}.$$

96E

Use $E = mc^2$, where m is the mass and c is the speed of light (2.998×10^8 m/s). Convert 2.31×10^{12} kW·h to $(2.31 \times 10^{12} \text{ kW·h})(10^3 \text{ W/kW})(3600 \text{ s/h}) = 8.32 \times 10^{18}$ J. Thus

$$m = \frac{E}{c^2} = \frac{8.32 \times 10^{18} \text{ J}}{(2.998 \times 10^8 \text{ m/s})^2} = 92.5 \text{ kg}.$$

Chapter 9

1E

(a) Put the origin at the center of the Earth. Then the distance r_{cm} of the center of mass of the Earth-Moon system is given by

$$r_{cm} = \frac{m_M r_M}{m_M + m_E},$$

where m_M is the mass of the Moon, m_E is the mass of the Earth, and r_M is their separation. These values are given in Appendix C. The numerical result is

$$r_{cm} = \frac{(7.36 \times 10^{22}\,\text{kg})(3.82 \times 10^8\,\text{m})}{7.36 \times 10^{22}\,\text{kg} + 5.98 \times 10^{24}\,\text{kg}} = 4.64 \times 10^6\,\text{m}.$$

(b) The radius of the Earth is $R_E = 6.37 \times 10^6$ m so $r_{cm}/R_E = 0.73$.

3E

(a) Let x_1 (= 0), y_1 (= 0) be the coordinates of one particle, x_2 (= 1.0 m), y_2 (= 2.0 m) be the coordinates of the second particle, and x_3 (= 2.0 m), y_3 (= 1.0 m) be the coordinates of the third. Designate the corresponding masses by m_1 (= 3.0 kg), m_2 (= 8.0 kg), and m_3 (= 4.0 kg). Then the x coordinate of the center of mass is

$$\begin{aligned} x_{cm} &= \frac{m_1 x_1 + m_2 x_2 + m_3 x_3}{m_1 + m_2 + m_2} \\ &= \frac{0 + (8.0\,\text{kg})(1.0\,\text{m}) + (4.0\,\text{kg})(2.0\,\text{m})}{3.0\,\text{kg} + 8.0\,\text{kg} + 4.0\,\text{kg}} = 1.1\,\text{m} \end{aligned}$$

and the y coordinate is

$$\begin{aligned} y_{cm} &= \frac{m_1 y_1 + m_2 y_2 + m_3 y_3}{m_1 + m_2 + m_3} \\ &= \frac{0 + (8.0\,\text{kg})(2.0\,\text{m}) + (4.0\,\text{kg})(1.0\,\text{m})}{3.0\,\text{kg} + 8.0\,\text{kg} + 4.0\,\text{kg}} = 1.3\,\text{m}. \end{aligned}$$

(b) As the mass of the topmost particle is increased the center of mass shifts toward that particle. In the limit as the topmost particle is much more massive than the others, the center of mass is at the position of that particle.

12P*

(a) Since the can is uniform its center of mass is at its geometrical center, a distance $H/2$ above its base. The center of mass of the pop alone is at its geometrical center, a distance $x/2$ above the base of the can. When the can is full this is $H/2$. Thus the center of mass of the can and the pop it contains is a distance

$$h = \frac{M(H/2) + m(H/2)}{M + m} = \frac{H}{2}$$

above the base, on the cylinder axis.

(b) We now consider the can alone. The center of mass is $H/2$ above the base, on the cylinder axis.

(c) As x decreases the center of mass of the pop in the can at first drops, then rises to $H/2$ again.

(d) When the top surface of the pop is a distance x above the base of the can the mass of the pop in the can is $m_p = m(x/H)$, where m is the mass when the can is full ($x = H$). The center of mass of the pop alone is a distance $x/2$ above the base of the can. Hence

$$h = \frac{M(H/2) + m_p(x/2)}{M + m_p} = \frac{M(H/2) + m(x/H)(x/2)}{M + (mx/H)} = \frac{MH^2 + mx^2}{2(MH + mx)}.$$

Find the lowest position of the center of mass of the can and pop by setting the derivative of h with respect to x equal to 0 and solving for x. The derivative is

$$\frac{dh}{dx} = \frac{2mx}{2(MH + mx)} - \frac{(MH^2 + mx^2)m}{2(MH + mx)^2} = \frac{m^2x^2 + 2MmHx - MmH^2}{2(MH + mx)^2}.$$

The solution to $m^2x^2 + 2MmHx - MmH^2 = 0$ is

$$x = \frac{MH}{m}\left[-1 + \sqrt{1 + \frac{m}{M}}\right].$$

The positive root was used since x must be positive.

Now substitute the expression you found for x into $h = (MH^2 + mx^2)/2(MH + mx)$. After some algebraic manipulation you should obtain

$$h = \frac{HM}{m}\left(\sqrt{1 + \frac{m}{M}} - 1\right).$$

14E

Let m_c be the mass of the Chrysler and v_c be its velocity. Let m_f be the mass of the Ford and v_f be its velocity. Then the velocity of the center of mass is

$$v_{cm} = \frac{m_c v_c + m_f v_f}{m_c + m_f} = \frac{(2400\,\text{kg})(80\,\text{km/h}) + (1600\,\text{kg})(60\,\text{km/h})}{2400\,\text{kg} + 1600\,\text{kg}} = 72\,\text{km/h}.$$

Notice that the two velocities are in the same direction, so the two terms in the numerator have the same sign.

20P

You need to find the coordinates of the point where the shell explodes and the velocity of the fragment that does not fall straight down. These become the initial conditions for a projectile motion problem to determine where it lands.

Consider first the motion of the shell from firing to the time of the explosion. Place the origin at the firing point, take the x axis to be horizontal, and take the y axis to be vertically upward. The y component of the velocity is given by $v = v_{0y} - gt$ and this is zero at time $t = v_{0y}/g = (v_0/g)\sin\theta_0$, where v_0 is the initial speed and θ_0 is the firing angle. The coordinates of the highest point are

$$x = v_{0x}t = v_0 t \cos\theta_0 = \frac{v_0^2}{g}\sin\theta_0 \cos\theta_0 = \frac{(20\,\text{m/s})^2}{9.8\,\text{m/s}^2}\sin 60°\cos 60° = 17.7\,\text{m}$$

and

$$y = v_{0y}t - \frac{1}{2}gt^2 = \frac{1}{2}\frac{v_0^2}{g}\sin^2\theta_0 = \frac{1}{2}\frac{(20\,\text{m/s})^2}{9.8\,\text{m/s}^2}\sin^2 60° = 15.3\,\text{m}.$$

Since the no horizontal forces act, the horizontal component of the momentum is conserved. Since one fragment has a velocity of zero after the explosion the momentum of the other equals the momentum of the shell before the explosion. The velocity of the shell is $v_0 \cos\theta_0$. Let M be the mass of the shell and let V_0 be the velocity of the fragment. Then $Mv_0 \cos\theta_0 = MV_0/2$, since the mass of the fragment is $M/2$. This means

$$V_0 = 2v_0 \cos\theta_0 = 2(20\,\text{m/s})\cos 60° = 20\,\text{m/s}.$$

Now consider a projectile launched horizontally at time $t = 0$ with a speed of 20 m/s from the point with coordinates $x_0 = 17.7$ m, $y_0 = 15.3$ m. Its y coordinate is given by $y = y_0 - \frac{1}{2}gt^2$, and when it lands this is zero. The time of landing is $t = \sqrt{2y_0/g}$ and the x coordinate of the landing point is

$$x = x_0 + V_0 t = x_0 + V_0\sqrt{\frac{2y_0}{g}} = 17.7\,\text{m} + (20\,\text{m/s})\sqrt{\frac{2(15.3\,\text{m})}{9.8\,\text{m/s}^2}} = 53\,\text{m}.$$

21P

(a) Place the origin of a coordinate system at the center of the pulley, with the x axis horizontal and to the right and with the y axis downward. The center of mass is halfway between the containers, at $x = 0$ and $y = \ell$, where ℓ is the vertical distance from the pulley center to either of the containers.

(b) Suppose 20 g is transferred from the container on the left to the container on the right. The container on the left has mass $m_1 = 480$ g and is at $x_1 = -25$ mm. The container on the right has mass $m_2 = 520$ g and is at $x_2 = +25$ mm. The x coordinate of the center of mass is then

$$x_{\text{cm}} = \frac{m_1 x_1 + m_2 x_2}{m_1 + m_2} = \frac{(480\,\text{g})(-25\,\text{mm}) + (520\,\text{g})(25\,\text{mm})}{480\,\text{g} + 520\,\text{g}} = 1.0\,\text{mm}.$$

The y coordinate is still ℓ. The center of mass is 26 mm from the lighter container, along the line that joins the bodies.

(c) When they are released the heavier container moves downward and the lighter container moves upward, so the center of mass, which must remain closer to the heavier container, moves downward.

(d) Because the containers are connected by the string, which runs over the pulley, their accelerations have the same magnitude but are in opposite directions. If a is the acceleration of m_2 then $-a$ is the acceleration of m_1. The acceleration of the center of mass is

$$a_{cm} = \frac{m_1(-a) + m_2 a}{m_1 + m_2} = a\frac{m_2 - m_1}{m_1 + m_2}.$$

We must resort to Newton's second law to find the acceleration of each container. The force of gravity $m_1 g$, down, and the tension in the string T, up, act on the lighter container. The second law for it is $m_1 g - T = -m_1 a$. The minus sign appears because a is the acceleration of the heavier container. The same forces act on the heavier container and for it the second law is $m_2 g - T = m_2 a$. The first equation gives $T = m_1 g + m_1 a$. This is substituted into the second equation to obtain $m_2 g - m_1 g - m_1 a = m_2 a$, so $a = (m_2 - m_1)g/(m_1 + m_2)$. Thus

$$a_{cm} = \frac{g(m_2 - m_1)^2}{(m_1 + m_2)^2} = \frac{(9.8\,\text{m/s}^2)(520\,\text{g} - 480\,\text{g})^2}{(480\,\text{g} + 520\,\text{g})^2} = 1.6 \times 10^{-2}\,\text{m/s}^2.$$

The acceleration is downward.

22P

Take the x axis to be to the right in the figure, with the origin at the shore. Let m_b be the mass of the boat and x_{bi} its initial coordinate. Let m_d be the mass of the dog and x_{di} his initial coordinate. The coordinate of the center of mass is

$$x_{cm} = \frac{m_b x_{bi} + m_d x_{di}}{m_b + m_d}.$$

Now the dog walks a distance d to the left on the boat. The new coordinates x_{bf} and x_{df} are related by $x_{bf} = x_{df} + d$, so the coordinate of the center of mass can be written

$$x_{cm} = \frac{m_b x_{bf} + m_d x_{df}}{m_b + m_d} = \frac{m_b x_{df} + m_b d + m_d x_{bf}}{m_b + m_d}.$$

Since the net external force on the boat-dog system is zero the velocity of the center of mass does not change. Since the boat and dog were initially at rest the velocity of the center of mass is zero. The center of mass remains at the same place and the two expressions we have written for x_{cm} must equal each other. This means $m_b x_{bi} + m_d x_{di} = m_b x_{df} + m_b d + m_d x_{df}$. Solve for x_{df}:

$$x_{df} = \frac{m_b x_{bi} + m_d x_{di} - m_b d}{m_b + m_d}$$

$$= \frac{(40.0\,\text{lb})(20.0\,\text{ft}) + (10.0\,\text{lb})(20.0\,\text{ft}) - (40.0\,\text{lb})(8.00\,\text{ft})}{40.0\,\text{lb} + 10.0\,\text{lb}} = 13.6\,\text{ft}.$$

Strictly speaking, we should substitute values of the masses rather than the weights, but the factor g cancels from the equation.

23E

The magnitude of the linear momentum is given by $p = mv$. The mass of the automobile is $(3200\,\text{lb})/(32\,\text{ft/s}^2) = 100\,\text{slug}$ and its speed is $(55\,\text{mi/h})(5280\,\text{ft/mi})/(3600\,\text{s/h}) = 80.7\,\text{ft/s}$. Thus $p = (100\,\text{slug})(80.7\,\text{ft/s}) = 8100\,\text{slug}\cdot\text{ft/s}$.

33E

No external forces with horizontal components act on the man-stone system so the total momentum of the system is conserved. Since the man and the stone are initially at rest the total momentum is zero both before and after the stone is kicked. Let m_s be the mass of the stone and v_s be its velocity after it is kicked; let m_m be the mass of the man and v_m be his velocity after he kicks the stone. Then $m_s v_s + m_m v_m = 0$ and $v_m = -m_s v_s/m_m$. Take the axis to be positive in the direction the stone travels. Then $v_m = -(0.15\,\text{lb})(13\,\text{ft/s})/(200\,\text{lb}) = -9.8 \times 10^{-3}\,\text{ft/s}$. Notice that the ratio of the masses is the same as the ratio of the weights, so we did not need to calculate the masses. The minus sign indicates that the man moves in the direction opposite to the direction of motion of the stone.

36E

No external forces with horizontal components act on the cart-man system, so the total momentum of the system is conserved. Let m_c be the mass of the cart, v be its initial velocity, and v_c be its final velocity (after the man jumps off). Let m_m be the mass of the man. His initial velocity is the same as that of the cart and his final velocity is zero. Conservation of momentum yields $(m_m + m_c)v = m_c v_c$. The final speed of the cart is

$$v_c = \frac{v(m_m + m_c)}{m_c} = \frac{(2.3\,\text{m/s})(75\,\text{kg} + 39\,\text{kg})}{39\,\text{kg}} = 6.7\,\text{m/s}.$$

The cart speeds up by $6.7\,\text{m/s} - 2.3\,\text{m/s} = 4.4\,\text{m/s}$. In order to slow himself the man gets the cart to push backward on him by pushing forward on it, so the cart speeds up.

38P

(a) Assume no external forces act on the system composed of the two parts of the last stage. The total momentum of the system is conserved. Let m_c be the mass of the rocket case and m_p be the mass of the payload. At first they are traveling together with velocity v. After the clamp is released m_c has velocity v_c and m_p has velocity v_p. Conservation of momentum yields $(m_c + m_p)v = m_c v_c + m_p v_p$. After the clamp is released the payload, having the lesser mass, will be traveling at the greater speed. Write $v_p = v_c + v_{\text{rel}}$, where v_{rel} is the relative velocity. When this expression is substituted into the conservation of momentum equation the result is $(m_c + m_p)v = m_c v_c + m_p v_c + m_p v_{\text{rel}}$, so

$$
\begin{aligned}
v_c &= \frac{(m_c + m_p)v - m_p v_{\text{rel}}}{m_c + m_p} \\
&= \frac{(290.0\,\text{kg} + 150.0\,\text{kg})(7600\,\text{m/s}) - (150.0\,\text{kg})(910.0\,\text{m/s})}{290.0\,\text{kg} + 150.0\,\text{kg}} = 7290\,\text{m/s}.
\end{aligned}
$$

The final speed of the payload is $v_p = v_c + v_{\text{rel}} = 7290\,\text{m/s} + 910.0\,\text{m/s} = 8200\,\text{m/s}$.

(b) The total kinetic energy before the clamp is released is

$$K_i = \frac{1}{2}(m_c + m_p)v^2 = \frac{1}{2}(290.0\,\text{kg} + 150.0\,\text{kg})(7600\,\text{m/s})^2 = 1.271 \times 10^{10}\,\text{J}.$$

The total kinetic energy after the clamp is released is

$$K_f = \frac{1}{2}m_c v_c^2 + \frac{1}{2}m_p v_p^2 = \frac{1}{2}(290.0\,\text{kg})(7290\,\text{m/s})^2 + \frac{1}{2}(150.0\,\text{kg})(8200\,\text{m/s})^2$$
$$= 1.275 \times 10^{10}\,\text{J}.$$

The total kinetic energy increased slightly. Energy originally stored in the spring was converted to kinetic energy of the rocket parts.

39P

Assume no external forces act, so the total momentum of the three-piece system is conserved. Since the total momentum before the break-up was zero, it is also zero after the break-up. This means the velocity vectors of the three pieces lie in the same plane, as shown in the diagram to the right. Conservation of the y component of momentum leads to $mv \sin\theta_1 - mv \sin\theta_2 = 0$, where v is the speed of each of the smaller pieces. Thus $\theta_1 = \theta_2$ and, since $\theta_1 + \theta_2 = 90°$, both θ_1 and θ_2 are 45°.

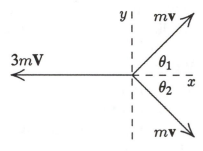

Conservation of the x component of momentum leads to $3mV = 2mv \cos\theta_1$. Thus $V = (2/3)v \cos\theta_1 = (2/3)(30\,\text{m/s})\cos 45° = 14\,\text{m/s}$. The angle between the velocity vector of the large piece and either of the smaller pieces is $180° - 45° = 135°$.

48E

Ignore the gravitational pull of Jupiter and use Eq 47 of the text. If v_i is the initial velocity, M_i is the initial mass, v_f is the final velocity, M_f is the final mass, and u is the speed of the exhaust gas, then

$$v_f = v_i + u \ln \frac{M_i}{M_f}.$$

For this problem $M_i = 6090\,\text{kg}$ and $M_f = 6090 - 80.0 = 6010\,\text{kg}$. Thus

$$v_f = 105\,\text{m/s} + (253\,\text{m/s}) \ln \frac{6090\,\text{kg}}{6010\,\text{kg}} = 108\,\text{m/s}.$$

49E

(a) The thrust of the rocket is given by $T = Ru$, where R is the rate of fuel consumption and u is the speed of the exhaust gas. For this problem $R = 480\,\text{kg/s}$ and $u = 3.27 \times 10^3\,\text{m/s}$, so $T = (480\,\text{kg/s})(3.27 \times 10^3\,\text{m/s}) = 1.57 \times 10^6\,\text{N}.$

(b) The mass of fuel ejected is given by $M_{fuel} = R\Delta t$, where Δt is the time interval of the burn. Thus $M_{fuel} = (480\,\text{kg/s})(250\,\text{s}) = 1.20 \times 10^5\,\text{kg}$. The mass of the rocket after the burn is $M_f = M_i - M_{fuel} = 2.55 \times 10^5\,\text{kg} - 1.20 \times 10^5\,\text{kg} = 1.35 \times 10^5\,\text{kg}$.

(c) Since the initial speed is zero, the final speed is given by

$$v_f = u \ln \frac{M_i}{M_f} = (3.27 \times 10^3\,\text{m/s}) \ln \frac{2.55 \times 10^5\,\text{kg}}{1.35 \times 10^5\,\text{kg}} = 2.08 \times 10^3\,\text{m/s}\,.$$

56P

Take the x axis to be positive to the right in the Fig. 9-38 of the text and take the y axis to be perpendicular to that direction. Consider first the slow barge and suppose the mass of coal shoveled in time Δt is ΔM. If $\mathbf{v}_s$ is the velocity of the barge and $\mathbf{u}$ is the velocity of the coal as it leaves the barge then the change in the momentum of the coal-barge system during this interval is $\Delta \mathbf{P} = \Delta M(\mathbf{u} - \mathbf{v}_s)$. The momentum of the coal changed from $\mathbf{v}_s \Delta M$ to $\mathbf{u}\Delta M$ and the momentum of the barge did not change. The force that must be exerted on the barge to keep its velocity constant is $\mathbf{F}_s = \Delta \mathbf{P}/\Delta t = (\Delta M/\Delta t)(\mathbf{u} - \mathbf{v}_s)$. Now $\mathbf{v}_s = v_s \mathbf{i}$ and if the coal is shoveled perpendicularly to the length of the boat then $\mathbf{u} = v_s \mathbf{i} + u_y \mathbf{j}$. Thus $\mathbf{F}_s = (\Delta M/\Delta t)u_y \mathbf{j}$. u_y is the slight transverse speed the coal must be given to get it from barge to the other. It is not given in the problem so we assume it is so small it may be neglected. The force that must be applied to the slower barge is essentially zero.

Now consider the faster barge, which receives coal with mass ΔM. Initially the coal has velocity $\mathbf{u}$ but after it comes to rest relative to the barge its velocity is $\mathbf{v}_f$, the same as the velocity of the barge. The momentum of the coal changes from $\Delta M \mathbf{u}$ to $\Delta M \mathbf{v}_f$ and the momentum of the barge does not change. The force that must be applied to the barge is $\mathbf{F}_f = (\Delta M/\Delta t)(\mathbf{v}_f - \mathbf{u})$. Now $\mathbf{v}_f = v_f \mathbf{i}$ and $\mathbf{u} = v_s \mathbf{i} + u_y \mathbf{j}$, so the x component of the force is

$$F_{fx} = \frac{\Delta M}{\Delta t}(v_f - v_s) = \left(\frac{1000\,\text{kg}}{60\,\text{s}}\right)(20\,\text{km/h} - 10\,\text{km/h})\left(\frac{1000\,\text{m/km}}{3600\,\text{s/h}}\right) = 46\,\text{N}$$

The rate coal is shoveled is converted from kg/min to kg/s in the first factor and the barge speeds are converted from km/h to m/s by the last factor.

The y component of the force that applied to the faster barge is $F_{fy} = -(\Delta M/\Delta t)u_y$. If u_y is small it is essentially zero.

57P

(a) Let ΔM_a be the mass of air taken in and ejected and let ΔM_f be the mass of fuel ejected in time Δt. The initial velocity of the air is zero and its final velocity is $v - u$, where u is the exhaust speed and v is the velocity of the plane. The initial velocity of the fuel is v and its final velocity is $v - u$. The velocity of the plane changes from v to $v + \Delta v$ over this time interval. The change in the total momentum of the plane-fuel-air system is $\Delta P = M_p \Delta v + \Delta M_f(-u) + \Delta M_a(v - u)$ so the net external force is

$$F = \frac{\Delta P}{\Delta t} = M_p \frac{\Delta v}{\Delta t} - u \frac{\Delta M_f}{\Delta t} + (v - u)\frac{\Delta M_a}{\Delta t}$$

and

$$M_p \frac{\Delta v}{\Delta t} = F + u \frac{\Delta M_f}{\Delta t} - (v - u) \frac{\Delta M_a}{\Delta t}.$$

The thrust is

$$T = u \frac{\Delta M_f}{\Delta t} - (v - u) \frac{\Delta M_a}{\Delta t}$$
$$= (490 \, \text{m/s})(2.9 \, \text{kg/s}) - (180 \, \text{m/s} - 490 \, \text{m/s})(70 \, \text{kg/s}) = 2.3 \times 10^4 \, \text{N} .$$

(b) The power is $P = Tv = (2.3 \times 10^4 \, \text{N})(180 \, \text{m/s}) = 4.2 \times 10^6 \, \text{W}$.

60E

(a) The force of the floor does no work on her because her feet do not move while they in contact with the floor. Nevertheless, the scalar product of the net external force and the displacement of her center of mass gives the change in her center of mass kinetic energy. Once her feet leave the floor this kinetic energy is converted to gravitational potential energy. If F is the magnitude of the force of the floor and d_1 ($= 0.90 \, \text{m} - 0.40 \, \text{m} = 0.50 \, \text{m}$) is the distance her center of mass moves while her feet are on the floor then $(F - mg)d_1 = K$, where K is her center of mass kinetic energy when her feet leave the floor. Mechanical energy is conserved during the rest of the trip, so $K = mgd_2$, where d_2 ($= 1.20 \, \text{m} - 0.90 \, \text{m} = 0.30 \, \text{m}$) is the distance her center of mass rises from the time her feet leave the floor to the time she reaches the top of her leap. Solve $(F - mg)d_1 = mgd_2$ for F:

$$F = \frac{mg(d_1 + d_2)}{d_1} = \frac{(55 \, \text{kg})(9.8 \, \text{m/s}^2)(0.50 \, \text{m} + 0.30 \, \text{m})}{0.50 \, \text{m}} = 860 \, \text{N} .$$

(b) She has her maximum speed at the time her feet leave the floor. Then $\frac{1}{2}mv^2 = (F - mg)d_1$, so

$$v = \sqrt{\frac{2(F - mg)d_1}{m}}$$
$$= \sqrt{\frac{2\left[860 \, \text{N} - (55 \, \text{kg})(9.8 \, \text{m/s}^2)\right](0.50 \, \text{m})}{55 \, \text{kg}}} = 2.4 \, \text{m/s} .$$

Chapter 10

3E

If F is the magnitude of the average force then the magnitude of the impulse is $J = F\Delta t$, where Δt is the time interval over which the force is exerted (see Eq. 10–8). This equals the magnitude of the change in the momentum of the ball and since the ball is initially at rest it equals the magnitude of the final momentum mv. When $F\Delta t = mv$ is solved for v the result is $v = F\Delta t/m = (50\,\text{N})(10 \times 10^{-3}\,\text{s})/(0.20\,\text{kg}) = 2.5\,\text{m/s}$.

9E

Take the initial direction of motion to be positive and let F be the magnitude of the average force, Δt be the duration of the force, m be the mass of the ball, v_i be the initial velocity of the ball, and v_f be the final velocity of the ball. Then the force is in the negative direction and the impulse-momentum theorem yields $-F\Delta t = mv_f - mv_i$. Solve for v_f to obtain

$$v_f = \frac{mv_i - F\Delta t}{m} = \frac{(0.40\,\text{kg})(14\,\text{m/s}) - (1200\,\text{N})(27 \times 10^{-3}\,\text{s})}{0.40\,\text{kg}} = -67\,\text{m/s}.$$

The final speed of the ball is $67\,\text{m/s}$.

12P

(a) The initial momentum of the car is $\mathbf{p}_i = m\mathbf{v}_i = (1400\,\text{kg})(5.3\,\text{m/s})\mathbf{j} = (7400\,\text{kg}\cdot\text{m/s})\mathbf{j}$ and the final momentum is $\mathbf{p}_f = (7400\,\text{kg}\cdot\text{m/s})\mathbf{i}$. The impulse acting on it equals the change in its momentum: $\mathbf{J} = \mathbf{p}_f - \mathbf{p}_i = (7400\,\text{kg}\cdot\text{m/s})(\mathbf{i} - \mathbf{j})$.

(b) The initial momentum of the car is $\mathbf{p}_i = (7400\,\text{kg}\cdot\text{m/s})\mathbf{i}$ and the final momentum is $\mathbf{p}_f = 0$. The impulse acting on it is $\mathbf{J} = \mathbf{p}_f - \mathbf{p}_i = -(7400\,\text{kg}\cdot\text{m/s})\mathbf{i}$.

(c) The average force acting on the car is

$$\mathbf{F}_{\text{av}} = \frac{\Delta \mathbf{p}}{\Delta t} = \frac{\mathbf{J}}{\Delta t} = \frac{(7400\,\text{kg}\cdot\text{m/s})(\mathbf{i} - \mathbf{j})}{4.6\,\text{s}} = (1600\,\text{N})(\mathbf{i} - \mathbf{j})$$

and its magnitude is $F_{\text{av}} = (1600\,\text{N})\sqrt{2} = 2300\,\text{N}$.

(d) The average force is

$$\mathbf{F}_{\text{av}} = \frac{\mathbf{J}}{\Delta t} = \frac{(-7400\,\text{kg}\cdot\text{m/s})\mathbf{i}}{350 \times 10^{-3}\,\text{s}} = (-2.1 \times 10^4\,\text{N})\mathbf{i}$$

and its magnitude is $F_{\text{av}} = 2.1 \times 10^4\,\text{N}$.

(e) The average force is given above in unit vector notation. Its x and y components have equal magnitudes. The x component is positive and the y component is negative, so the force is 45° below the positive x axis.

13P

Take the magnitude of the force to be $F = At$, where A is a constant of proportionality. The condition that $F = 50\,\text{N}$ when $t = 4.0\,\text{s}$ leads to $A = (50\,\text{N})/(4.0\,\text{s}) = 12.5\,\text{N/s}$. The magnitude of the impulse exerted on the object is

$$J = \int_0^{4.0} F\,dt = \int_0^{4.0} At\,dt = \frac{1}{2}At^2\Big|_0^{4.0} = \frac{1}{2}(12.5\,\text{N/s})(4.0\,\text{s})^2 = 100\,\text{N}\cdot\text{s}.$$

This equals the magnitude of the change in the momentum of the ball or since the ball started from rest it equals the magnitude of the final momentum: $J = mv_f$. Thus $v_f = J/m = (100\,\text{N}\cdot\text{s})/(10\,\text{kg}) = 10\,\text{m/s}$.

14P

(a) If m is the mass of a pellet and v is its velocity as it hits the wall, then its momentum is $p = mv = (2.0\times10^{-3}\,\text{kg})(500\,\text{m/s}) = 1.00\,\text{kg}\cdot\text{m/s}$, toward the wall.

(b) The kinetic energy of a pellet is $K = \frac{1}{2}mv^2 = \frac{1}{2}(2.0\times10^{-3}\,\text{kg})(5000\,\text{m/s})^2 = 2.5\times10^2\,\text{J}$.

(c) The force on the wall is given by the rate at which momentum is transferred from the pellets to the wall. Since the pellets do not rebound, each pellet that hits transfers $p = 1.00\,\text{kg}\cdot\text{m/s}$. If ΔN pellets hit in time Δt then the average rate at which momentum is transferred is

$$F_{\text{av}} = \frac{p\,\Delta N}{\Delta t} = (1.0\,\text{kg}\cdot\text{m/s})(10\,\text{s}^{-1}) = 10\,\text{N}.$$

The force on the wall is in the direction of the initial velocity of the pellets.

(d) If Δt is the time interval for a pellet to be brought to rest by the wall, then the average force exerted on the wall by a pellet is

$$F_{\text{av}} = \frac{p}{\Delta t} = \frac{1.0\,\text{kg}\cdot\text{m/s}}{0.6\times10^{-3}\,\text{s}} = 1.7\times10^3\,\text{N}.$$

The force is in the direction of the initial velocity of the pellet.

(e) In part (d) the force was averaged over the time a pellet is in contact with the wall, while in part (c) it was averaged over the time for many pellets to hit the wall. Most of this time no pellet is in contact with the wall, so the average force in part (c) is much less than the average force in (d).

26P

Consider first the lighter part. Suppose the impulse has magnitude J and is in the positive direction. Let m_1 be the mass of the part and v_1 be its velocity after the bolts are exploded. Assume both parts are at rest before the explosion. Then $J = m_1v_1$, so $v_1 = J/m_1 = (300\,\text{N}\cdot\text{s})/(1200\,\text{kg}) = 0.25\,\text{m/s}$. The impulse on the heavier part has the same magnitude but is in the opposite direction, so $-J = m_2v_2$, where m_2 is the mass and v_2 is the velocity of the part. Thus $v_2 = -J/m_2 = -(300\,\text{N}\cdot\text{s})/(1800\,\text{kg}) = -0.167\,\text{m/s}$. The relative speed of the parts after the explosion is $0.25\,\text{m/s} - (-0.167\,\text{m/s}) = 0.417\,\text{m/s}$.

29E

(a) Let m_1 be the mass of the block on the left, v_{1i} be its initial velocity, and v_{1f} be its final velocity. Let m_2 be the mass of the block on the right, v_{2i} be its initial velocity, and v_{2f} be its final velocity. The momentum of the two-block system is conserved, so $m_1v_{1i} + m_2v_{2i} = m_1v_{1f} + m_2v_{2f}$ and

$$v_{1f} = \frac{m_1v_{1i} + m_2v_{2i} - m_2v_{2f}}{m_1}$$
$$= \frac{(1.6\,\text{kg})(5.5\,\text{m/s}) + (2.4\,\text{kg})(2.5\,\text{m/s}) - (2.4\,\text{kg})(4.9\,\text{m/s})}{1.6\,\text{kg}} = 1.9\,\text{m/s}.$$

The block continues going to the right after the collision.

(b) To see if the collision is elastic compare the total kinetic energy before the collision with the total kinetic energy after the collision. The total kinetic energy before is

$$K_i = \frac{1}{2}m_1v_{1i}^2 + \frac{1}{2}m_2v_{2i}^2$$
$$= \frac{1}{2}(1.6\,\text{kg})(5.5\,\text{m/s})^2 + \frac{1}{2}(2.4\,\text{kg})(2.5\,\text{m/s})^2 = 31.7\,\text{J}.$$

The total kinetic energy after is

$$K_f = \frac{1}{2}m_1v_{1f}^2 + \frac{1}{2}m_2v_{2f}^2$$
$$= \frac{1}{2}(1.6\,\text{kg})(1.9\,\text{m/s})^2 + \frac{1}{2}(2.4\,\text{kg})(4.9\,\text{m/s})^2 = 31.7\,\text{J}.$$

Since $K_i = K_f$ the collision is elastic.

(c) Now $v_{2i} = -2.5\,\text{m/s}$ and

$$v_{1f} = \frac{m_1v_{1i} + m_2v_{2i} - m_2v_{2f}}{m_1}$$
$$= \frac{(1.6\,\text{kg})(5.5\,\text{m/s}) - (2.4\,\text{kg})(2.5\,\text{m/s}) - (2.4\,\text{kg})(4.9\,\text{m/s})}{1.6\,\text{kg}} = -5.6\,\text{m/s}.$$

The velocity must be opposite to the direction shown.

33E

(a) Let m_1 be the mass of the cart that is originally moving, v_{1i} be its velocity before the collision, and v_{1f} be its velocity after the collision. Let m_2 be the mass of the cart that is originally at rest and v_{2f} be its velocity after the collision. Then, according to Eq. 10–18,

$$v_{1f} = \frac{m_1 - m_2}{m_1 + m_2}v_{1i}.$$

Solve for m_2 to obtain

$$m_2 = \frac{v_{1i} - v_{1f}}{v_{1f} + v_{1i}}m_1 = \left(\frac{1.2\,\text{m/s} - 0.66\,\text{m/s}}{1.2\,\text{m/s} + 0.66\,\text{m/s}}\right)(340\,\text{g}) = 99\,\text{g}.$$

(b) The velocity of the second cart is given by Eq. 10–19:

$$v_{2f} = \frac{2m_1}{m_1 + m_2}v_{1i} = \left[\frac{2(0.340\,\text{kg})}{0.340\,\text{kg} + 0.099\,\text{kg}}\right](1.2\,\text{m/s}) = 1.9\,\text{m/s}.$$

34E

Let m_1 be the mass of the body that is originally moving, v_{1i} be its velocity before the collision, and v_{1f} be its velocity after the collision. Let m_2 be the mass of the body that is originally at rest and v_{2f} be its velocity after the collision. Then, according to Eq. 10–18,

$$v_{1f} = \frac{m_1 - m_2}{m_1 + m_2}v_{1i}.$$

Solve for m_2 to obtain

$$m_2 = \frac{v_{1i} - v_{1f}}{v_{1f} + v_{1i}}m_1.$$

Substitute $v_{1f} = v_{1i}/4$ to obtain $m_2 = 3m_1/5 = 3(2.0\,\text{kg})/5 = 1.2\,\text{kg}$.

37P

Let m_1 be the mass of one sphere, v_{1i} be its velocity before the collision, and v_{1f} be its velocity after the collision. Let m_2 be the mass of the other sphere, v_{2i} be its velocity before the collision, and v_{2f} be its velocity after the collision. Then, according to Eq. 10–28,

$$v_{1f} = \frac{m_1 - m_2}{m_1 + m_2}v_{1i} + \frac{2m_2}{m_1 + m_2}v_{2i}.$$

Suppose sphere 1 is originally traveling in the positive direction and is at rest after the collision. Sphere 2 is originally traveling in the negative direction. Replace v_{1i} with v, v_{2i} with $-v$, and v_{1f} with zero to obtain $0 = m_1 - 3m_2$. Thus $m_2 = m_1/3 = (300\,\text{g})/3 = 100\,\text{g}$.

40P

(a) First use free-fall kinematics to find the place where the ball collides with the elevator roof and the velocity of the ball just before it collides. Then use the equations for an elastic collision between an object with small mass and another object with large mass to find the velocity with which the ball rebounds. Finally use free-fall kinematics find how high the ball rebounds.

Take the y axis to be upward and place the origin at the top of the elevator shaft. The coordinate of the ball is given by $y = -\frac{1}{2}gt^2$ and the coordinate of the elevator roof is given by $y = y_0 + v_e t$, where $y_0 = -60\,\text{ft}$ and $v_e = 6.0\,\text{ft/s}$. The collision occurs when $-\frac{1}{2}gt^2 = y_0 + v_e t$, or

$$t = \frac{-v_e \pm \sqrt{v_e^2 - 2y_0 g}}{g} = \frac{-6.0\,\text{ft/s} + \sqrt{(-6.0\,\text{ft/s})^2 - 2(-60\,\text{ft})(32\,\text{ft/s}^2)}}{32\,\text{ft/s}^2} = 1.76\,\text{s}.$$

The positive solution was selected since the collision occurs after the ball is dropped. The velocity of the ball just before the collision is $v = -gt = -(32\,\text{ft/s}^2)(1.76\,\text{s}) = -56.3\,\text{ft/s}$ and the collision takes place at $y = y_0 + v_e t = -60\,\text{ft} + (6.0\,\text{ft/s})(1.76\,\text{s}) = -49.5\,\text{ft}$.

Now analyze the collision. Let object 1 be the ball and object 2 be the elevator. Since the elevator is much more massive than the ball the final velocity of the ball is given by $v_{1f} = -v_{1i} + 2v_{2i}$. Substitute $v_{1i} = -56.3\,\text{ft/s}$ and $v_{2i} = 6.0\,\text{ft/s}$ to find $v_{1f} = +56.3\,\text{ft/s} + 2(6.0\,\text{ft/s}) = 68.3\,\text{ft/s}$.

Use $v^2 = v_0^2 - 2g(y - y_0)$ to find the coordinate y of the highest point on the ball's trajectory (where $v = 0$):

$$y = y_0 + \frac{v_0^2}{2g} = -49.5\,\text{ft} + \frac{(68.3\,\text{ft/s})^2}{2(32\,\text{ft/s}^2)} = 23\,\text{ft}.$$

The highest point is 23 ft above the top of the shaft.

(b) Now $v_e = -6.0\,\text{ft/s}$. The time of the collision is

$$t = \frac{-v_e \pm \sqrt{v_e^2 - 2y_0 g}}{g} = \frac{6.0\,\text{ft/s} + \sqrt{(-6.0\,\text{ft/s})^2 - 2(-60\,\text{ft})(32\,\text{ft/s}^2)}}{32\,\text{ft/s}^2} = 2.13\,\text{s}.$$

The positive root was used since the solution must be positive. The velocity of the ball just before the collision is $v = -gt = -(32\,\text{ft/s}^2)(2.13\,\text{s}) = -68.3\,\text{ft/s}$ and collision takes place at $y = y_0 + v_e t = -60\,\text{ft} - (6.0\,\text{ft/s})(2.13\,\text{s}) = -72.8\,\text{ft}$. The velocity of the ball just after the collision is $v_{1f} = -v_{1i} + 2v_{2i} = +68.2\,\text{ft/s} - 2(6.0\,\text{ft/s}) = 56.2\,\text{ft/s}$. Finally, the highest point on the ball's trajectory has coordinate

$$y = y_0 + \frac{v_0^2}{2g} = -72.8\,\text{ft} + \frac{(56.2\,\text{ft/s})^2}{2(32\,\text{ft/s}^2)} = -23\,\text{ft}.$$

The highest point is 23 ft below the top of the shaft.

41E

Let m_m be the mass of the meteor and m_e be the mass of the Earth. Let v_m be the velocity of the meteor just before the collision and let v be the velocity of the Earth (with the meteor) just after the collision. The momentum of the Earth-meteor system is conserved during the collision. Thus, in the reference frame of the Earth before the collision, $m_m v_m = (m_m + m_e)v$, so

$$v = \frac{v_m m_m}{m_m + m_e} = \frac{(7200\,\text{m/s})(5 \times 10^{10}\,\text{kg})}{5.98 \times 10^{24}\,\text{kg} + 5 \times 10^{10}\,\text{kg}} = 6 \times 10^{-11}\,\text{m/s}.$$

This is about 2 mm/y, as you should verify

42E

We need to consider only the horizontal components of the momenta of the package and sled. Let m_s be the mass of the sled and v_s be its initial velocity. Let m_p be the mass of the package and let v be the final velocity of the sled and package together. The horizontal component of the total momentum of the sled-package system is conserved, so $m_s v_s = (m_s + m_p)v$ and

$$v = \frac{v_s m_s}{m_s + m_p} = \frac{(9.0\,\text{m/s})(6.0\,\text{kg})}{6.0\,\text{kg} + 12\,\text{kg}} = 3.0\,\text{m/s}.$$

52P

We want to calculate the force that the scale exerts on the marbles. This force is the sum of the force that holds the marbles already on the scale against the downward force of gravity and the force that brings the falling marbles to rest. At the end of time t the number of marbles on the scale is Rt, their weight is $Rtmg$, and the upward force of the scale that holds them is $F_1 = Rtmg$. Just before it hits the scale a marble that falls from height h has a velocity of $v = \sqrt{2gh}$ and a momentum of $p = m\sqrt{2gh}$. To stop the falling marbles the scale must exert an upward force $F_2 = Rp = Rm\sqrt{2gh}$. The total force of the scale on the marbles is

$$F = F_1 + F_2 = Rtmg + Rm\sqrt{2gh} = Rm\left(tg + \sqrt{2gh}\right).$$

For the data given

$$F = (100\,\text{s}^{-1})(4.50 \times 10^{-3}\,\text{kg})\left[(10.0\,\text{s})(9.8\,\text{m/s}^2) + \sqrt{2(9.8\,\text{m/s}^2)(7.60\,\text{m})}\right]$$

$$= 49.6\,\text{N}.$$

If the scale is calibrated to read in terms of the equivalent mass its reading would be $F/g = (49.6\,\text{N})/(9.8\,\text{m/s}^2) = 5.06\,\text{kg}$.

53P

Let m_F be the mass of the freight car and v_F be its initial velocity. Let m_C be the mass of the caboose and v be the common final velocity of the two when they are coupled. Conservation of the total momentum of the two-car system leads to $m_F v_F = (m_F + m_C)v$, so $v = v_F m_F/(m_F + m_C)$. The initial kinetic energy of the system is

$$K_i = \frac{1}{2}m_F v_F^2$$

and the final kinetic energy is

$$K_f = \frac{1}{2}(m_F + m_C)v^2 = \frac{1}{2}(m_F + m_C)\frac{m_F^2 v_F^2}{(m_F + m_C)^2} = \frac{1}{2}\frac{m_F^2 v_F^2}{(m_F + m_C)}.$$

Since 27% of the original kinetic energy is lost $K_f = 0.73K_i$, or

$$\frac{1}{2}\frac{m_F^2 v_F^2}{(m_F + m_C)} = (0.73)\left(\frac{1}{2}m_F v_F^2\right).$$

Following some obvious cancellations this becomes $m_F/(m_F + m_C) = 0.73$. Solve for m_C: $m_C = (0.27/0.73)m_F = 0.37m_F$. The ratio of the weights is the same as the ratio of the masses: $W_C = 0.37W_F = 0.37(35\,\text{tons}) = 13\,\text{tons}$.

54P

(a) Let v be the final velocity of the ball-gun system. Since the total momentum of the system is conserved $mv_i = (m + M)v$ and $v = mv_i/(m + M)$.

(b) The initial kinetic energy is $K_i = \frac{1}{2}mv_i^2$ and the final kinetic energy is $K_f = \frac{1}{2}(m+M)v^2 = \frac{1}{2}m^2v_i^2/(m + M)$. The difference is the energy U_s stored in the spring:

$$U_s = \frac{1}{2}mv_i^2 - \frac{1}{2}\frac{m^2v_i^2}{(m + M)} = \frac{1}{2}mv_i^2\left(1 - \frac{m}{m + M}\right) = \frac{1}{2}mv_i^2\frac{M}{m + M}\,.$$

The fraction of the original energy that is stored in the spring is $U_s/K_i = M/(m + M)$.

63E

(a) Use Fig. 10–20 of the text and take the cue ball to be body 1, the other ball to be body 2. Conservation of the x component of the total momentum of the two-ball system leads to $mv_{1i} = mv_{1f}\cos\theta_1 + mv_{2f}\cos\theta_2$ and conservation of the y component leads to $0 = -mv_{1f}\sin\theta_1 + mv_{2f}\sin\theta_2$. Notice that the masses are the same and cancel from the equations. Solve the second equation for $\sin\theta_2$:

$$\sin\theta_2 = \frac{v_{1f}}{v_{2f}}\sin\theta_1 = \left(\frac{3.50\,\text{m/s}}{2.00\,\text{m/s}}\right)\sin 22.0° = 0.656\,.$$

The angle is $\theta_2 = 41.0°$.

(b) Solve the first momentum conservation equation for v_{1i}: $v_{1i} = v_{1f}\cos\theta_1 + v_{2f}\cos\theta_2 = (3.50\,\text{m/s})\cos 22.0° + (2.00\,\text{m/s})\cos 41.0° = 4.75\,\text{m/s}$.

(c) The initial kinetic energy is

$$K_i = \frac{1}{2}mv_i^2 = \frac{1}{2}m(4.75\,\text{m/s})^2 = 11.3m$$

and the final kinetic energy is

$$K_f = \frac{1}{2}mv_{1f}^2 + \frac{1}{2}mv_{2f}^2 = \frac{1}{2}m\left[(3.50\,\text{m/s})^2 + (2.00\,\text{m/s})^2\right] = 8.1m\,.$$

Kinetic energy is not conserved.

66P

(a) Let m be the mass and $v_i\,\mathbf{i}$ be the velocity of the body before the explosion. Let m_1, m_2, and m_3 be the masses of the fragments. Write $v_1\,\mathbf{j}$ for the velocity of fragment 1, $-v_2\,\mathbf{i}$ for the velocity of fragment 2, and $v_{3x}\,\mathbf{i} + v_{3y}\,\mathbf{j}$ for the velocity of fragment 3. Since the original body and two of the fragments all move in the xy plane the third fragment must also move in that plane. Conservation of momentum leads to $mv_i\,\mathbf{i} = m_1v_1\,\mathbf{j} - m_2v_2\,\mathbf{i} + m_3v_{3x}\,\mathbf{i} + m_3v_{3y}\,\mathbf{j}$ or $(mv_i + m_2v_2 - m_3v_{3x})\,\mathbf{i} - (m_1v_1 + m_3v_{3y})\,\mathbf{j} = 0$. The x component of this equation gives

$$v_{3x} = \frac{mv_i + m_2v_2}{m_3} = \frac{(20.0\,\text{kg})(200\,\text{m/s}) + (4.00\,\text{kg})(500\,\text{m/s})}{6.0\,\text{kg}} = 1000\,\text{m/s}\,.$$

The y component gives

$$v_{3y} = -\frac{m_1v_1}{m_3} = -\frac{(10.0\,\text{kg})(100\,\text{m/s})}{6.0\,\text{kg}} = -167\,\text{m/s}\,.$$

Thus $\mathbf{v_3} = (1000\,\mathbf{i} - 167\,\mathbf{j})\,\text{m/s}$. The velocity has a magnitude of $1010\,\text{m/s}$ and is $9.48°$ below the x axis.

(b) The initial kinetic energy is

$$K_i = \frac{1}{2}mv_i^2 = \frac{1}{2}(20.0\,\text{kg})(200\,\text{m/s})^2 = 4.00 \times 10^5\,\text{J}.$$

The final kinetic energy is

$$\begin{aligned} K_f &= \frac{1}{2}m_1v_1^2 + \frac{1}{2}m_2v_2^2 + \frac{1}{2}m_3v_3^2 \\ &= \frac{1}{2}\left[(10.0\,\text{kg})(100\,\text{m/s})^2 + (4.0\,\text{kg})(500\,\text{m/s})^2 + (6.0\,\text{kg})(1010\,\text{m/s})^2\right] \\ &= 3.61 \times 10^6\,\text{J}. \end{aligned}$$

The energy released in the explosion was $3.61 \times 10^6\,\text{J} - 4.00 \times 10^5\,\text{J} = 3.21 \times 10^6\,\text{J}$.

68P

Let m_n be the mass of the neutron and m_d be the mass of the deuteron. Write $m_d = 2m_n$. Suppose the neutron enters the collision along the positive x axis with speed v_{ni} and leaves the collision along the positive y axis with speed v_{nf}. The deuteron leaves the collision along a line that is an angle θ below the positive x axis and has speed v_{df}. Conservation of the x component of the total momentum of the two-particle system leads to

$$m_n v_{ni} = m_d v_{df} \cos\theta,$$

conservation of the y component leads to

$$0 = m_n v_{nf} - m_d v_{df} \sin\theta,$$

and conservation of kinetic energy leads to

$$\frac{1}{2}m_n v_{ni}^2 = \frac{1}{2}m_n v_{nf}^2 + \frac{1}{2}m_d v_{df}^2.$$

Once $m_d = 2m_n$ is used and the factors $\frac{1}{2}$ are canceled these equations become

$$v_{ni} = 2v_{df} \cos\theta,$$

$$0 = v_{nf} - 2v_{df} \sin\theta,$$

and

$$v_{ni}^2 = v_{nf}^2 + 2v_{df}^2.$$

Use these equation to eliminate θ and v_{df}, then solve for v_{nf} in terms of v_{ni}. This relationship can be used to find the loss of kinetic energy by the neutron. First use the conservation of momentum equations to eliminate θ. The equation for conservation of the x component of momentum gives $\cos\theta = v_{ni}/2v_{df}$ and the equation for conservation of the y component gives $\sin\theta = v_{nf}/2v_{df}$. Square these and make use of the trigonometric identity $\sin^2\theta + \cos^2\theta = 1$.

The result is $v_{ni}^2 + v_{nf}^2 = 4v_{df}^2$. Now use the conservation of energy equation to eliminate v_{df}. It gives $2v_{df}^2 = v_{ni}^2 - v_{nf}^2$, so

$$v_{ni}^2 + v_{nf}^2 = 2v_{ni}^2 - 2v_{nf}^2.$$

A little rearrangement shows that $v_{nf}^2 = v_{ni}^2/3$. The loss in kinetic energy by the neutron is $\Delta K_n = \frac{1}{2}m_n v_{ni}^2 - \frac{1}{2}m_n v_{nf}^2 = \frac{1}{2}m_n v_{ni}^2 - \frac{1}{2}m_n v_{ni}^2/3 = (2/3)\left(\frac{1}{2}m_n v_{ni}^2\right)$. This is two-thirds its original kinetic energy.

69P

Suppose the objects enter the collision along lines that make the angles θ and ϕ with the x axis, as shown in the diagram to the right. Both have the same mass m and the same initial speed v. We suppose that after the collision the combined objects move in the positive x direction with speed V. Since the y component of the total momentum of the two-object system is conserved $mv\sin\theta - mv\sin\phi = 0$. This means $\phi = \theta$. Since the x component is conserved $2mv\cos\theta = 2mV$. Now use $V = v/2$ to find that $\cos\theta = 1/2$. This means $\theta = 60°$. The angle between the initial velocities is $120°$.

72P

The diagram on the right shows the situation as the incident ball (the left most ball) makes contact with the other two. It exerts an impulse of the same magnitude on each ball, along the line that joins the centers of the incident ball and the target ball. The target balls leave the collision along those lines, while the incident ball leaves the collision along the x axis. The three dotted lines that join the centers of the balls in contact form an equilateral triangle, so both of the angles marked θ are $30°$. Let v_0 be the velocity of the incident ball before the collision, V be its velocity afterward, and v be the speed of each of the target balls after the collision. Each ball has mass m.

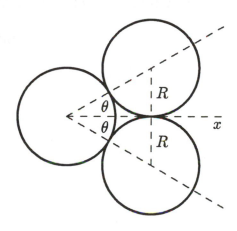

Since the x component of the total momentum of the three-ball system is conserved

$$mv_0 = mV + 2mv\cos\theta$$

and since the total kinetic energy is conserved

$$\frac{1}{2}mv_0^2 = \frac{1}{2}mV^2 + 2\left(\frac{1}{2}mv^2\right).$$

We know the directions in which the target balls leave the collision so we first eliminate V and solve for v. The momentum equation gives $V = v_0 - 2v \cos \theta$, so $V^2 = v_0^2 - 4v_0 v \cos \theta + 4v^2 \cos^2 \theta$ and the energy equation becomes $v_0^2 = v_0^2 - 4v_0 v \cos \theta + 4v^2 \cos^2 \theta + 2v^2$. Solve for v:

$$v = \frac{2v_0 \cos \theta}{1 + 2 \cos^2 \theta} = \frac{2(10 \, \text{m/s}) \cos 30°}{1 + 2 \cos^2 30°} = 6.93 \, \text{m/s} \,.$$

Now use the momentum equation to find a value for V:

$$V = v_0 - 2v \cos \theta = 10 \, \text{m/s} - 2(6.93 \, \text{m/s}) \cos 30° = -2.0 \, \text{m/s} \,.$$

The target ball bounces back in the negative x direction.

74E

Q is defined by $Q = (M_i - M_f)c^2$, where M_i is the sum of the initial masses, M_f is the sum of the final masses, and c is the speed of light. For the reaction given

$$M_i = m_p + m_F = 1.007825 \, \text{u} + 18.998405 \, \text{u} = 20.006230 \, \text{u}$$

and

$$M_f = m_\alpha + m_O = 4.002603 \, \text{u} + 15.994915 \, \text{u} = 19.997518 \, \text{u} \,.$$

Thus

$$M_i - M_f = 20.006230 \, \text{u} - 19.997518 \, \text{u} = 0.008712 \, \text{u}$$

Since 1 u is equivalent to 931.5 MeV,

$$Q = (0.008712 \, \text{u})(931.5 \, \text{MeV/u}) = 8.12 \, \text{MeV} \,.$$

Chapter 11

3E

(a) The time for one revolution is the circumference of the orbit divided by the speed v of the sun: $T = 2\pi R/v$, where R is the radius of the orbit. Since $R = 2.3 \times 10^4 \, \text{ly} = (2.3 \times 10^4 \, \text{ly})(9.460 \times 10^{12} \, \text{km/ly}) = 2.18 \times 10^{17} \, \text{km}$,

$$T = \frac{2\pi(2.18 \times 10^{17} \, \text{km})}{250 \, \text{km/s}} = 5.5 \times 10^{15} \, \text{s}.$$

(b) The number of revolutions is the total time t divided by the time T for one revolution: $N = t/T$. Convert the total time from years to seconds. The result for the number of revolutions is

$$N = \frac{(4.5 \times 10^9 \, \text{y})(3.16 \times 10^7 \, \text{s/y})}{5.5 \times 10^{15} \, \text{s}} = 26.$$

5E

(a) Evaluate $\theta(t) = 2 + 4t^2 + 2t^3$ for $t = 0$ to obtain $\theta(0) = 2 \, \text{rad}$.

(b) The angular velocity (in rad/s) is given by $\omega(t) = d\theta/dt = 8t + 6t^2$. Evaluate this expression for $t = 0$ to obtain $\omega(0) = 0$.

(c) For $t = 4.0 \, \text{s}$, $\omega = 8(4.0) + 6(4.0)^2 = 130 \, \text{rad/s}$.

(d) The angular acceleration (in rad/s^2) is given by $\alpha = d\omega/dt = 8 + 12t$. For $t = 2.0 \, \text{s}$, $\alpha = 8 + 12(2.0) = 32 \, \text{rad/s}^2$.

(e) The angular acceleration, given by $\alpha = 8 + 12t$, depends on the time and so is not constant.

12E

(a) For constant angular acceleration $\omega = \omega_0 + \alpha t$, so $\alpha = (\omega - \omega_0)/t$. Take $\omega = 0$ and to obtain the units requested use $t = (30 \, \text{s})/(60 \, \text{s/min}) = 0.50 \, \text{min}$. Then

$$\alpha = -\frac{33.33 \, \text{rev/min}}{0.50 \, \text{min}} = -66.7 \, \text{rev/min}^2.$$

(b) Use $\theta = \omega_0 t + \frac{1}{2}\alpha t^2 = (33.33 \, \text{rev/min})(0.50 \, \text{min}) + \frac{1}{2}(-66.7 \, \text{rev/min}^2)(0.50 \, \text{min})^2 = 8.3 \, \text{rev}$.

19P

Take $t = 0$ at the start of the interval. Then at the end of the interval $t = 4.0 \, \text{s}$, and the angle of rotation is $\theta = \omega_0 t + \frac{1}{2}\alpha t^2$. Solve for ω_0:

$$\omega_0 = \frac{\theta - \frac{1}{2}\alpha t^2}{t} = \frac{120 \, \text{rad} - \frac{1}{2}(3.0 \, \text{rad/s}^2)(4.0 \, \text{s})^2}{4.0 \, \text{s}} = 24 \, \text{rad/s}.$$

Now use $\omega = \omega_0 + \alpha t$ to find the time when the wheel was at rest ($\omega = 0$):

$$t = -\frac{\omega_0}{\alpha} = -\frac{24 \, \text{rad/s}}{3.0 \, \text{rad/s}^2} = -8.0 \, \text{s}.$$

That is, the wheel started from rest 8.0 s before the start of the 4.0 s interval.

25P

(*a*) A complete revolution is a revolution by $\theta = 2\pi$ rad so the angular velocity in rad/s is given by $\omega = \theta/T = 2\pi/T$.

(*b*) The angular acceleration is given by

$$\alpha = \frac{d\omega}{dt} = -\frac{2\pi}{T^2}\frac{dT}{dt}.$$

For the pulsar described

$$\frac{dT}{dt} = \frac{1.26 \times 10^{-5} \, \text{s/y}}{3.16 \times 10^7 \, \text{s/y}} = 4.00 \times 10^{-13},$$

so

$$\alpha = -\left[\frac{2\pi}{(0.033 \, \text{s})^2}\right](4.00 \times 10^{-13}) = -2.3 \times 10^{-9} \, \text{rad/s}^2.$$

The minus sign indicates that the angular acceleration is opposite the angular velocity and the pulsar is slowing down.

(*c*) Solve $\omega = \omega_0 + \alpha t$ for the time t when $\omega = 0$:

$$t = -\frac{\omega_0}{\alpha} = -\frac{2\pi}{\alpha T} = -\frac{2\pi}{(-2.3 \times 10^{-9} \, \text{rad/s}^2)(0.033 \, \text{s})} = 8.3 \times 10^{10} \, \text{s}.$$

This is about 2600 years.

(*d*) The pulsar was born $1992 - 1054 = 938$ years ago. This is equivalent to $(938 \, \text{y})(3.16 \times 10^7 \, \text{s/y}) = 2.96 \times 10^{10}$ s. It angular velocity was then

$$\omega = \omega_0 + \alpha t = \frac{2\pi}{T} + \alpha t = \frac{2\pi}{0.033 \, \text{s}} + (-2.3 \times 10^{-9} \, \text{rad/s}^2)(-2.96 \times 10^{10} \, \text{s}) = 258 \, \text{rad/s}.$$

Its period was

$$T = \frac{2\pi}{\omega} = \frac{2\pi}{258 \, \text{rad/s}} = 2.4 \times 10^{-2} \, \text{s}.$$

26E

The magnitude of the acceleration is given by $a = \omega^2 r$, where r is the distance from the center of rotation and ω is the angular velocity. You must convert the given angular velocity to rad/s: $\omega = (33.33 \, \text{rev/min})(2\pi \, \text{rad/rev})/(60 \, \text{s/min}) = 3.49 \, \text{rad/s}$. Thus

$$a = (0.15 \, \text{m})(3.49 \, \text{rad/s}^2)^2 = 1.8 \, \text{m/s}^2.$$

The acceleration vector is toward the center of the record.

28E

Use $v = r\omega$. First convert 50 km/h to m/s: (50 km/h)(1000 m/km)/(3600 s/h) = 13.9 m/s. Then $\omega = v/r = (13.9\,\text{m/s})/(110\,\text{m}) = 0.13\,\text{rad/s}$.

35P

(a) The Earth makes one rotation per day and 1 d is $(24\,\text{h})(3600\,\text{s/h}) = 8.64 \times 10^4$ s, so the angular velocity of the Earth is $(2\pi\,\text{rad})/(8.64 \times 10^4\,\text{s}) = 7.27 \times 10^{-5}\,\text{rad/s}$.

(b) Use $v = r\omega$, where r is the radius of its orbit. A point on the Earth at a latitude of 40° goes around a circle of radius $r = R\cos 40°$, where R is the radius of the Earth (6.37×10^6 m). So its speed is $v = (R\cos 40°)\omega = (6.37 \times 10^6\,\text{m})(\cos 40°)(7.27 \times 10^{-5}\,\text{rad/s}) = 355\,\text{m/s}$.

(c) At the equator the value of ω is the same but the latitude is 0° and the speed is $v = R\omega = (6.37 \times 10^6\,\text{m})(7.27 \times 10^{-5}\,\text{rad/s}) = 463\,\text{m/s}$.

39P

(a) In the time light takes to go from the wheel to the mirror and back again, the wheel turns through an angle of $\theta = 2\pi/500 = 1.26 \times 10^{-2}$ rad. That time is

$$t = \frac{2\ell}{c} = \frac{2(500\,\text{m})}{2.998 \times 10^8\,\text{m/s}} = 3.34 \times 10^{-6}\,\text{s}$$

so the angular velocity of the wheel is

$$\omega = \frac{\theta}{t} = \frac{1.26 \times 10^{-2}\,\text{rad}}{3.34 \times 10^{-6}\,\text{s}} = 3.8 \times 10^3\,\text{rad/s}.$$

(b) If r is the radius of the wheel the linear speed of a point on its rim is

$$v = r\omega = (0.05\,\text{m})(3.8 \times 10^3\,\text{rad/s}) = 190\,\text{m/s}.$$

41P

Since the belt does not slip a point on the rim of wheel C has the same tangential acceleration as a point on the rim of wheel A. This means that $r_A\alpha_A = r_C\alpha_C$, where α_A is the angular acceleration of wheel A and α_C is the angular acceleration of wheel C. Thus

$$\alpha_C = \left(\frac{r_A}{r_C}\right)\alpha_A = \left(\frac{10\,\text{cm}}{25\,\text{cm}}\right)(1.6\,\text{rad/s}^2) = 0.64\,\text{rad/s}^2.$$

Since the angular velocity of wheel C is given by $\omega_C = \alpha_C t$, the time for it to reach an angular velocity of 100 rev/min (= 10.5 rad/s) is

$$t = \frac{\omega_C}{\alpha_C} = \frac{10.5\,\text{rad/s}}{0.64\,\text{rad/s}^2} = 16\,\text{s}.$$

45E

The kinetic energy is given by $K = \frac{1}{2}I\omega^2$, where I is the rotational inertia and ω is the angular velocity. Use

$$\omega = \frac{(602 \text{ rev/min})(2\pi \text{ rad/rev})}{60 \text{ s/min}} = 63.0 \text{ rad/s}.$$

Then

$$I = \frac{2K}{\omega^2} = \frac{2(24400 \text{ J})}{(63.0 \text{ rad/s})^2} = 12.3 \text{ kg} \cdot \text{m}^2.$$

47E

Since the rotational inertia of a cylinder of mass M and radius R is $I = \frac{1}{2}MR^2$, the kinetic energy of a cylinder when it rotates with angular velocity ω is $K = \frac{1}{2}I\omega^2 = \frac{1}{4}MR^2\omega^2$. For the first cylinder $K = \frac{1}{4}(1.25 \text{ kg})(0.25 \text{ m})^2(235 \text{ rad/s})^2 = 1.1 \times 10^3 \text{ J}$. For the second $K = \frac{1}{4}(1.25 \text{ kg})(0.75 \text{ m})^2(235 \text{ rad/s})^2 = 9.7 \times 10^3 \text{ J}$.

54E

Use the parallel axis theorem: $I = I_{cm} + Mh^2$, where I_{cm} is the rotational inertia about a parallel axis through the center of mass, M is the mass, and h is the distance between the two axes. In this case the axis through the center of mass is at the 0.50 m mark so $h = 0.50 \text{ m} - 0.20 \text{ m} = 0.30 \text{ m}$. Now $I_{cm} = M\ell^2/12 = (0.56 \text{ kg})(1.0 \text{ m})^2/12 = 4.67 \times 10^{-2} \text{ kg} \cdot \text{m}^2$, so $I = 4.67 \times 10^{-2} \text{ kg} \cdot \text{m}^2 + (0.56 \text{ kg})(0.30 \text{ m})^2 = 9.7 \times 10^{-2} \text{ kg} \cdot \text{m}^2$.

57P

Use the parallel-axis theorem. According to Table 11–2(*f*) the rotational inertia of a uniform slab about an axis through the center and perpendicular to the large faces is given by

$$I_{cm} = \frac{M}{12}(a^2 + b^2).$$

A parallel axis through a corner is a distance $h = \sqrt{(a/2)^2 + (b/2)^2}$ from the center, so

$$I = I_{cm} + Mh^2 = \frac{M}{12}(a^2 + b^2) + \frac{M}{4}(a^2 + b^2) = \frac{M}{3}(a^2 + b^2).$$

58P

(*a*) According to Table 11–2(*c*) the rotational inertia of a uniform solid cylinder about its central axis is given by

$$I_C = \frac{1}{2}MR^2,$$

where M is its mass and R is its radius. For a hoop with mass M and radius R_H Table 11–2(*a*) gives

$$I_H = MR_H^2.$$

If the two bodies have the same mass then they will have the same rotational inertia if $R^2/2 = R_H^2$, or $R_H = R/\sqrt{2}$.

(b) You want the rotational inertia to be given by $I = Mk^2$, where M is the mass of the arbitrary body and k is the radius of the equivalent hoop. Thus

$$k = \sqrt{\frac{I}{M}}.$$

60E

Two forces act on the ball, the force of the rod and the force of gravity. No torque about the pivot point is associated with the force of the rod since that force is along the line from the pivot point to the ball. As can be seen from the diagram, the component of the force of gravity that is perpendicular to the rod is $mg \sin \theta$, so if ℓ is the length of the rod then the torque associated with this force has magnitude $\tau = mg\ell \sin \theta = (0.75\,\text{kg})(9.8\,\text{m/s}^2)(1.25\,\text{m}) \sin 30° = 4.6\,\text{N} \cdot \text{m}$. For the position of the ball shown the torque is counterclockwise.

63P

(a) Take a torque that tends to cause a counterclockwise rotation from rest to be positive and a torque that tends to cause a clockwise rotation from rest to be negative. Thus a positive torque of magnitude $r_1 F_1 \sin \theta_1$ is associated with $\mathbf{F}_1$ and a negative torque of magnitude $r_2 F_2 \sin \theta_2$ is associated with $\mathbf{F}_2$. Both of these are about O. The net torque about O is

$$\tau = r_1 F_1 \sin \theta_1 - r_2 F_2 \sin \theta_2.$$

(b) Substitute the given values to obtain

$$\tau = (1.30\,\text{m})(4.20\,\text{N}) \sin 75.0° - (2.15\,\text{m})(4.90\,\text{N}) \sin 60.0° = -3.85\,\text{N} \cdot \text{m}.$$

66E

(a) Use the kinematic equation $\omega = \omega_0 + \alpha t$, where ω_0 is the initial angular velocity, ω is the final angular velocity, α is the angular acceleration, and t is the time. This gives

$$\alpha = \frac{\omega - \omega_0}{t} = \frac{6.20\,\text{rad/s}}{220 \times 10^{-3}\,\text{s}} = 28.2\,\text{rad/s}^2.$$

(b) If I is the rotational inertia of the diver then the magnitude of the torque acting on her is $\tau = I\alpha = (12.0\,\text{kg} \cdot \text{m}^2)(28.2\,\text{rad/s}^2) = 3.38 \times 10^2\,\text{N} \cdot \text{m}$.

69E

(a) Use $\tau = I\alpha$, where τ is the net torque acting on the shell, I is the rotational inertia of the shell, and α is its angular acceleration. This gives

$$I = \frac{\tau}{\alpha} = \frac{960\,\text{N} \cdot \text{m}}{6.20\,\text{rad/s}^2} = 155\,\text{kg} \cdot \text{m}^2.$$

(b) The rotational inertia of the shell is given by $I = (2/3)MR^2$ (see Table 2 of the text). This means

$$M = \frac{3I}{2R^2} = \frac{3(155\,\text{kg} \cdot \text{m}^2)}{2(1.90\,\text{m})^2} = 64.4\,\text{kg}.$$

73P

(a) Use constant acceleration kinematics. If down is taken to be positive and a is the acceleration of the heavier block, then its coordinate is given by $y = \frac{1}{2}at^2$, so

$$a = \frac{2y}{t^2} = \frac{2(0.750\,\text{m})}{(5.00\,\text{s})^2} = 6.00 \times 10^{-2}\,\text{m/s}^2.$$

The lighter block has an acceleration of $6.00 \times 10^{-2}\,\text{m/s}^2$, upward.

(b) For the heavier block $m_h g - T_h = m_h a$, where m_h is its mass and T_h is the tension in the part of the cord that is attached to it. Thus $T_h = m_h(g - a) = (0.500\,\text{kg})(9.8\,\text{m/s}^2 - 6.00 \times 10^{-2}\,\text{m/s}^2) = 4.87\,\text{N}$.

(c) For the lighter block $m_l g - T_l = -m_l a$, so $T_l = m_l(g + a) = (0.460\,\text{kg})(9.8\,\text{m/s}^2 + 6.00 \times 10^{-2}\,\text{m/s}^2) = 4.54\,\text{N}$.

(d) Since the cord does not slip on the pulley, the tangential acceleration of a point on the rim of the pulley must be the same as the acceleration of the blocks, so $\alpha = a/R = (6.00 \times 10^{-2}\,\text{m/s}^2)/(5.00 \times 10^{-2}\,\text{m}) = 1.20\,\text{rad/s}^2$.

(e) The net torque acting on the pulley is $\tau = (T_h - T_l)R$. Equate this to $I\alpha$ and solve for I:

$$I = \frac{(T_h - T_l)R}{\alpha} = \frac{(4.87\,\text{N} - 4.54\,\text{N})(5.00 \times 10^{-2}\,\text{m})}{1.20\,\text{rad/s}^2} = 1.38 \times 10^{-2}\,\text{kg} \cdot \text{m}^2.$$

77P

(a) If ℓ is the length of the chimney then the radial component of the acceleration of the top is given by $a_r = \ell\omega^2$, where ω is the angular velocity. Use conservation of mechanical energy to find an expression for ω^2 as a function of the angle θ that the chimney makes with the vertical. The potential energy of the chimney is given by $U = Mgh$, where M is its mass and h is the altitude of its center of mass above the ground. When the chimney makes the angle θ with the vertical $h = (\ell/2)\cos\theta$. Initially the potential energy is $U_i = Mg(\ell/2)$ and the kinetic energy is zero. Write $\frac{1}{2}I\omega^2$ for the kinetic energy when the chimney makes the angle θ with the vertical. Here I is its rotational inertia. Conservation of energy then leads to $Mg\ell/2 = Mg(\ell/2)\cos\theta + \frac{1}{2}I\omega^2$, so $\omega^2 = (Mg\ell/I)(1 - \cos\theta)$. Thus

$$a_r = \ell\omega^2 = \left(\frac{Mg\ell^2}{I}\right)(1 - \cos\theta).$$

The chimney is rotating about its base, so $I = Mg\ell^2/3$ and $a_r = 3g(1 - \cos\theta)$.

(b) The tangential component of the acceleration of the chimney top is given by $a_t = \ell\alpha$, where α is the angular acceleration. Differentiate $\omega^2 = (Mg\ell/I)(1 - \cos\theta)$ with respect to time, replace $d\omega/dt$ with α, and replace $d\theta/dt$ with ω to obtain $2\omega\alpha = (Mg\ell/I)\omega\sin\theta$ or $\alpha = (Mg\ell/2I)\sin\theta$. Thus

$$a_t = \ell\alpha = \frac{Mg\ell^2}{2I}\sin\theta = \frac{3g}{2}\sin\theta\,,$$

where $I = M\ell^2/3$ was used to obtain the last result.

(c) The angle θ for which $a_t = g$ is the solution to $(3/2)\sin\theta = g$ or $\sin\theta = 2/3$. It is $\theta = 41.8°$.

83P

Use conservation of mechanical energy. The center of mass is at the midpoint of the cross bar of the H and it drops by $\ell/2$, where ℓ is the length of any one of the rods. The gravitational potential energy decreases by $Mg\ell/2$, where M is the mass of the body. The initial kinetic energy is zero and the final kinetic energy may be written $\frac{1}{2}I\omega^2$, where I is the rotational inertia of the body and ω is its angular velocity when it is vertical. Thus $0 = -Mg\ell/2 + \frac{1}{2}I\omega^2$ and $\omega = \sqrt{Mg\ell/I}$. Since the rods are thin the one along the axis of rotation does not contribute to the rotational inertia. All points on the other leg are the same distance from the axis of rotation so that leg contributes $(M/3)\ell^2$, where $M/3$ is its mass. The cross bar is a rod that rotates around one end so its contribution is $(M/3)\ell^2/3 = M\ell^2/9$. The total rotational inertia is $I = (M\ell^2/3) + (M\ell^2/9) = 4M\ell^2/9$. The angular velocity is

$$\omega = \sqrt{\frac{Mg\ell}{I}} = \sqrt{\frac{Mg\ell}{4M\ell^2/9}} = \sqrt{\frac{9g}{4\ell}}\,.$$

87P

(a) Constant angular acceleration kinematics can be used to compute the angular acceleration α. If ω_0 is the initial angular velocity and t is the time to come to rest, then $0 = \omega_0 + \alpha t$, or $\alpha = -\omega_0/t = -(39.0\,\text{rev/s})(2\pi\,\text{rad/rev})/(32.0\,\text{s}) = -7.66\,\text{rad/s}^2$.

(b) Use $\tau = I\alpha$, where τ is the torque and I is the rotational inertia. The contribution of the rod to I is $M\ell^2/12$, where M is its mass and ℓ is its length. The contribution of each ball is $m(\ell/2)^2$, where m is the mass of a ball. The total rotational inertia is

$$I = \frac{M\ell^2}{12} + 2\frac{m\ell^2}{4} = \frac{(6.40\,\text{kg})(1.20\,\text{m})^2}{12} + \frac{(1.06\,\text{kg})(1.20\,\text{m})^2}{2} = 1.53\,\text{kg}\cdot\text{m}^2\,.$$

The torque is

$$\tau = -(1.53\,\text{kg}\cdot\text{m}^2)(7.66\,\text{rad/s}^2) = -11.7\,\text{N}\cdot\text{m}\,.$$

(c) Since the system comes to rest the change in the mechanical energy is simply the negative of the initial kinetic energy $-K_i = -\frac{1}{2}I\omega_0^2 = -\frac{1}{2}(1.53\,\text{kg}\cdot\text{m}^2)(2\pi\times 39\,\text{rad/s})^2 = -4.59\times10^4\,\text{J}$.

(d) Use $\theta = \omega_0 t + \frac{1}{2}\alpha t^2 = (2\pi\times 39.0\,\text{rad/s})(32.0\,\text{s}) + \frac{1}{2}(-7.66\,\text{rad/s}^2)(32.0\,\text{s})^2 = 3920\,\text{rad}$. This is $3920/2\pi = 624\,\text{rev}$.

(e) Only the loss in mechanical energy can still be computed without additional information. It is 4.59×10^4 J no matter how τ varies with time, as long as the system comes to rest.

88P

(a) Use conservation of energy. At the top of the hill the kinetic energy is $K_i = 0$ and the gravitational potential energy is $U_i = mgh = mgd\sin\theta$, where m is the total mass of the car, d is the length of the hill, and θ is its slope. At the bottom of the hill the kinetic energy is $\frac{1}{2}mv^2 + \frac{1}{2}I\omega^2$, where v is the speed of the car, I is the rotational inertia of the flywheel, and ω is its angular velocity. Conservation of energy leads to $mgd\sin\theta = \frac{1}{2}mv^2 + \frac{1}{2}I\omega^2$. Since the flywheel is geared to the driveshaft the velocity of the car and the angular velocity of the flywheel are proportional to each other. Write $\omega = Av$, where A is the constant of proportionality. Thus $mgd\sin\theta = \frac{1}{2}mv^2 + \frac{1}{2}IA^2v^2$ and

$$v = \sqrt{\frac{2mgd\sin\theta}{m + IA^2}}.$$

Use the fact that $\omega = 240\,\text{rev/s} = 1.51 \times 10^3\,\text{rad/s}$ when $v = 80\,\text{km/h} = 22.2\,\text{m/s}$ to find the value of A: $A = \omega/v = (1.51\times 10^3\,\text{rad/s})/(22.2\,\text{m/s}) = 68.0\,\text{rad/m}$. The rotational inertia of the flywheel (assumed to be a uniform disk) is $I = \frac{1}{2}MR^2 = \frac{1}{2}(200\,\text{N}/9.8\,\text{m/s}^2)(0.55\,\text{m})^2 = 3.09\,\text{kg}\cdot\text{m}^2$. Then

$$v = \sqrt{\frac{2(800\,\text{kg})(9.8\,\text{m/s}^2)(1500\,\text{m})\sin 5°}{800\,\text{kg} + (3.09\,\text{kg}\cdot\text{m}^2)(68.0\,\text{rad/s})^2}} = 11.7\,\text{m/s}.$$

This is $42.1\,\text{km/h}$.

(b) Differentiate $v^2 = 2mgd\sin\theta/(m + IA^2)$ with respect to time, replace dv/dt with the acceleration a, and replace dd/dt with v. The result is

$$2va = \frac{2mgv\sin\theta}{m + IA^2}.$$

so

$$a = \frac{mg\sin\theta}{m + IA^2}.$$

Since $\omega = Av$,

$$\alpha = Aa = \frac{Amg\sin\theta}{m + IA^2} = \frac{(68.0\,\text{rad/s})(800\,\text{kg})(9.8\,\text{m/s}^2)\sin 5°}{800\,\text{kg} + (3.09\,\text{kg}\cdot\text{m}^2)(68.0\,\text{rad/m})^2} = 3.08\,\text{rad/s}^2.$$

(c)The energy stored in the flywheel is $E = \frac{1}{2}I\omega^2$ and the rate at which it is changing is $dE/dt = I\omega\alpha = IAv\alpha = (3.09\,\text{kg}\cdot\text{m}^2)(68.0\,\text{rad/m})(11.7\,\text{m/s})(3.08\,\text{rad/s}^2) = 7.57 \times 10^3\,\text{W}.$

Chapter 12

2E

The work required to stop the hoop is the negative of the initial kinetic energy of the hoop. The initial kinetic energy is given by $K = \frac{1}{2}I\omega^2 + \frac{1}{2}mv^2$, where I is its rotational inertia, m is its mass, ω is its angular velocity about its center of mass, and v is the speed of its center of mass. The rotational inertia of the hoop is given by $I = mr^2$, where r is its radius. Since the hoop rolls without slipping the angular velocity and the speed of the center of mass are related by $\omega = v/r$. Thus

$$K = \frac{1}{2}mr^2\left(\frac{v^2}{r^2}\right) + \frac{1}{2}mv^2 = mv^2 = (140\,\text{kg})(0.150\,\text{m/s})^2 = 3.15\,\text{J}.$$

The work required is $W = -3.15\,\text{J}$.

4E

Let M be the mass of the car and v be its speed. Let I be the rotational inertia of one wheel and ω be the angular velocity of each wheel. The total kinetic energy is given by

$$K = \frac{1}{2}Mv^2 + 4\left(\frac{1}{2}I\omega^2\right),$$

where the factor 4 appears because there are 4 wheels. The kinetic energy of rotation is

$$K_r = 4\left(\frac{1}{2}I\omega^2\right),$$

and the fraction of the total energy that is due to rotation is

$$f = \frac{K_r}{K} = \frac{4I\omega^2}{Mv^2 + 4I\omega^2}.$$

For a uniform wheel $I = \frac{1}{2}mr^2$, where r is the radius of a wheel and m is its mass. Since the wheels roll without slipping $\omega = v/r$. Thus $I\omega^2 = \frac{1}{2}mr^2v^2/r^2 = \frac{1}{2}mv^2$ and

$$f = \frac{2mv^2}{Mv^2 + 2mv^2} = \frac{2m}{M + 2m} = \frac{2(10\,\text{kg})}{1000\,\text{kg} + 2(10\,\text{kg})} = 0.020.$$

Notice that the radius of the wheel cancels from the equations. The rotational inertia is proportional to r^2 and when it is multiplied by $\omega^2 = v^2/r^2$ the result is independent of r.

11P

To find where the sphere lands you need to know its speed as it leaves the track. Use conservation of energy to find that speed. Let M be its mass and I be its rotational inertia. Its initial kinetic energy is $K_i = 0$ and its initial potential energy is $U_i = MgH$. Its final kinetic energy (as it leaves the track) is $K_f = \frac{1}{2}Mv^2 + \frac{1}{2}I\omega^2$ and its final potential energy is Mgh. Here v is the speed of its center of mass and ω is its angular velocity about the center of mass, both as it leaves the track. Conservation of energy yields

$$MgH = \frac{1}{2}Mv^2 + \frac{1}{2}I\omega^2 + Mgh.$$

For a uniform sphere $I = \frac{2}{5}Mr^2$, where r is its radius. Since the sphere rolls without slipping $\omega = v/r$. Thus $I\omega^2 = \frac{2}{5}Mv^2$ and the conservation of energy equation becomes

$$MgH = \frac{1}{2}Mv^2 + \frac{2}{10}Mv^2 + Mgh.$$

Carry out the addition: $\frac{1}{2}Mv^2 + \frac{2}{10}Mv^2 = \frac{7}{10}Mv^2$. Notice that the mass M cancels from the energy equation. Solve for v:

$$v = \sqrt{\frac{10}{7}g(H-h)} = \sqrt{\frac{10}{7}(9.8\,\mathrm{m/s^2})(60\,\mathrm{m} - 20\,\mathrm{m})} = 23.7\,\mathrm{m/s}.$$

Now put the origin at the position of the center of mass when the sphere leaves the track. Take the x axis to be positive toward the right and the y axis to be positive upward. Then the projectile motion equations become

$$x = vt$$

and

$$y = h - \frac{1}{2}gt^2.$$

You want to solve for x at the time when $y = 0$. The second equation gives $t = \sqrt{2h/g}$. This is substituted into the first to obtain

$$x = v\sqrt{\frac{2h}{g}} = (23.7\,\mathrm{m/s})\sqrt{\frac{2(20\,\mathrm{m})}{(9.8\,\mathrm{m/s^2})}} = 48\,\mathrm{m}.$$

16E

(a) An expression for the acceleration is derived in the text and appears as Eq. 12–19:

$$a = -\frac{g}{1 + I/MR_0^2},$$

where M is the mass of the yo-yo, I is its rotational inertia, and R_0 is the radius of its axle. The upward direction was taken to be positive. Substitute $I = 950\,\mathrm{g \cdot cm^2}$, $M = 120\,\mathrm{g}$, $R_0 = 0.32\,\mathrm{cm}$, and $g = 980\,\mathrm{cm/s^2}$ to obtain

$$a = \frac{980\,\mathrm{cm/s^2}}{1 + (950\,\mathrm{g \cdot cm^2})/(120\,\mathrm{g})(0.32\,\mathrm{cm})^2} = 12.5\,\mathrm{cm/s^2}.$$

(b) Solve the kinematic equation $y = \frac{1}{2}at^2$ for t and substitute $y = 120\,\text{cm}$:

$$t = \sqrt{\frac{2y}{a}} = \sqrt{\frac{2(120\,\text{cm})}{12.5\,\text{cm/s}^2}} = 4.38\,\text{s}.$$

(c) As it reaches the end of the string its linear speed is $v = at = (12.5\,\text{cm/s}^2)(4.38\,\text{s}) = 54.8\,\text{cm/s}.$

(d) The translational kinetic energy is $K = \frac{1}{2}mv^2 = \frac{1}{2}(0.120\,\text{kg})(0.548\,\text{m/s})^2 = 1.8 \times 10^{-2}\,\text{J}.$

(e) The angular velocity is given by $\omega = v/R_0$ and the rotational kinetic energy is $K = \frac{1}{2}I\omega^2 = \frac{1}{2}Iv^2/R_0^2 = \frac{1}{2}(9.50 \times 10^{-5}\,\text{kg} \cdot \text{m}^2)(0.548\,\text{m/s})^2/(3.2 \times 10^{-3}\,\text{m})^2 = 1.4\,\text{J}.$

(f) The angular speed is $\omega = v/R_0 = (0.548\,\text{m/s})/(3.2 \times 10^{-3}\,\text{m}) = 1.7 \times 10^2\,\text{rad/s} = 27\,\text{rev/s}.$

30E

(a) We use $\boldsymbol{\ell} = m\mathbf{r} \times \mathbf{v}$, where $\mathbf{r}$ is the position vector of the object, $\mathbf{v}$ is its velocity vector, and m is its mass. The position and velocity vectors have nonvanishing x and z components, so they be written $\mathbf{r} = x\,\mathbf{i} + z\,\mathbf{k}$ and $\mathbf{v} = v_x\,\mathbf{i} + v_z\,\mathbf{k}$. Evaluate the vector product term by term, making sure to keep the order of the factors intact:

$$\mathbf{r} \times \mathbf{v} = (x\,\mathbf{i} + z\,\mathbf{k}) \times (v_x\,\mathbf{i} + v_z\,\mathbf{k}) = xv_x\,\mathbf{i} \times \mathbf{i} + xv_z\,\mathbf{i} \times \mathbf{k} + zv_x\,\mathbf{k} \times \mathbf{i} + zv_z\,\mathbf{k} \times \mathbf{k}.$$

Now use $\mathbf{i} \times \mathbf{i} = 0$, $\mathbf{i} \times \mathbf{k} = -\mathbf{j}$, $\mathbf{k} \times \mathbf{i} = +\mathbf{j}$, and $\mathbf{k} \times \mathbf{k} = 0$ to obtain

$$\mathbf{r} \times \mathbf{v} = (-xv_z + zv_x)\,\mathbf{j}.$$

Thus

$$\boldsymbol{\ell} = m(-xv_z + zv_x)\,\mathbf{j}$$
$$= (0.25\,\text{kg})\,[-(2.0\,\text{m})(5.0\,\text{m/s}) + (-2.0\,\text{m})(-5.0\,\text{m/s})]\,\mathbf{j} = 0.$$

(b) Use $\boldsymbol{\tau} = \mathbf{r} \times \mathbf{F}$, with $\mathbf{F} = F\mathbf{j}$:

$$\boldsymbol{\tau} = (x\,\mathbf{i} + z\,\mathbf{k}) \times (F\mathbf{j}) = xF\,\mathbf{i} \times \mathbf{j} + zf\,\mathbf{k} \times \mathbf{j} = xF\,\mathbf{k} - zF\,\mathbf{i}$$
$$= (2.0\,\text{m})(4.0\,\text{N})\,\mathbf{i} - (-2.0\,\text{m})(4.0\,\text{N})\,\mathbf{k} = (8.0\,\mathbf{i} + 8.0\,\mathbf{k})\,\text{N} \cdot \text{m}.$$

33P

The diagram on the right shows the particles and their lines of motion. The origin is marked O and may be anywhere. The angular momentum of particle 1 has magnitude $\ell_1 = mvr_1 \sin\theta_1 = mv(d + h)$ and it is into the page. The angular momentum of particle 2 has magnitude $\ell_2 = mvr_2 \sin\theta_2 = mvh$ and it is out of the page. The net angular momentum has magnitude $L = mv(d + h) - mvh = mvd$ and is into the page. This result is independent of the location of the origin.

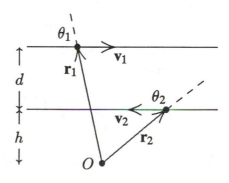

The same result is obtained if O is between the lines of motion of the particles. You should derive the expression for L in this case.

36E

(a) The angular momentum is given by the vector product $\boldsymbol{\ell} = m\mathbf{r} \times \mathbf{v}$, where $\mathbf{r}$ is the position vector of the particle and $\mathbf{v}$ is its velocity. Since the position and velocity vectors are in the xy plane we may write $\mathbf{r} = x\mathbf{i} + y\mathbf{j}$ and $\mathbf{v} = v_x\mathbf{i} + v_y\mathbf{j}$. Thus

$$\mathbf{r} \times \mathbf{v} = (x\mathbf{i} + y\mathbf{j}) \times (v_x\mathbf{i} + v_y\mathbf{j}) = xv_x\,\mathbf{i} \times \mathbf{i} + xv_y\,\mathbf{i} \times \mathbf{j} + yv_x\,\mathbf{j} \times \mathbf{i} + yv_y\,\mathbf{j} \times \mathbf{j}.$$

Use $\mathbf{i} \times \mathbf{i} = 0$, $\mathbf{i} \times \mathbf{j} = \mathbf{k}$, $\mathbf{j} \times \mathbf{i} = -\mathbf{k}$, and $\mathbf{j} \times \mathbf{j} = 0$ to obtain

$$\mathbf{r} \times \mathbf{v} = (xv_y - yv_x)\mathbf{k}.$$

Thus

$$\boldsymbol{\ell} = m(xv_y - yv_x)\mathbf{k}$$
$$= (3.0\,\text{kg})\,[(3.0\,\text{m})(-6.0\,\text{m/s}) - (8.0\,\text{m})(5.0\,\text{m/s})]\,\mathbf{k} = (-174\,\mathbf{k})\,\text{kg} \cdot \text{m}^2/\text{s}.$$

(b) The torque is given by $\boldsymbol{\tau} = \mathbf{r} \times \mathbf{F}$. Since the force has only an x component we may write $\mathbf{F} = F_x\mathbf{i}$ and

$$\boldsymbol{\tau} = (x\mathbf{i} + y\mathbf{j}) \times (F_x\mathbf{i}) = -yF_x\mathbf{k} = -(8.0\,\text{m}) \times (-7.0\,\text{N})\mathbf{k} = (56\,\mathbf{k})\,\text{N} \cdot \text{m}.$$

(c) According to Newton's second law $\boldsymbol{\tau} = d\boldsymbol{\ell}/dt$, so the time rate of change of the angular momentum is $56\,\text{kg} \cdot \text{m}^2/\text{s}^2$, in the positive z direction.

43E

(a) Since $\tau = dL/dt$ the average torque acting during any interval is given by $\overline{\tau} = (L_f - L_i)/\Delta t$, where L_i is the initial angular momentum, L_f is the final angular momentum, and Δt is the time interval. Thus

$$\overline{\tau} = \frac{0.800\,\text{kg} \cdot \text{m}^2/\text{s} - 3.00\,\text{kg} \cdot \text{m}^2/\text{s}}{1.50\,\text{s}} = -1.47\,\text{N} \cdot \text{m}.$$

In this case the minus sign simply indicates that the direction of the torque is opposite the direction of the initial angular momentum, which was taken to be positive.

(b) The angle turned is $\theta = \omega_0 t + \frac{1}{2}\alpha t^2$. If the angular acceleration α is uniform then so is the torque and $\alpha = \tau/I$. Furthermore $\omega_0 = L_i/I$, so

$$\theta = \frac{L_i + \frac{1}{2}\tau t^2}{I} = \frac{(3.00\,\text{kg} \cdot \text{m}^2/\text{s})(1.50\,\text{s}) + \frac{1}{2}(-1.47\,\text{N} \cdot \text{m})(1.50\,\text{s})^2}{0.140\,\text{kg} \cdot \text{m}^2} = 20.3\,\text{rad}.$$

(c) The work done on the wheel is

$$W = \tau\theta = (-1.47\,\text{N} \cdot \text{m})(20.3\,\text{rad}) = -29.8\,\text{J}.$$

(d) The average power is the work done by the flywheel (the negative of the work done on the flywheel) divided by the time interval:

$$\overline{P} = -\frac{W}{\Delta t} = -\frac{-29.8\,\text{J}}{1.50\,\text{s}} = 19.9\,\text{W} .$$

44E

(a) A particle contributes md^2 to the rotational inertia. Here d is the distance from the origin O to the particle. The total rotational inertia is

$$I = m(3\ell)^2 + m(2\ell)^2 + m(\ell)^2 = 14m\ell^2 .$$

(b) The angular momentum of the middle particle is given by $L_m = I_m\omega$, where $I_m = 4m\ell^2$ is its rotational inertia. Thus $L_m = 4m\ell^2\omega$.

(c) The total angular momentum is $L = I\omega = 14m\ell^2\omega$.

48P

Suppose cylinder 1 exerts a uniform force of magnitude F on cylinder 2, tangent to the cylinder's surface at the point of contact. The torque applied to cylinder 2 is $\tau_2 = R_2 F$ and the angular acceleration of that cylinder is $\alpha_2 = \tau_2/I_2 = R_2 F/I_2$. As a function of time its angular velocity is $\omega_2 = \frac{1}{2}\alpha_2 t^2 = \frac{1}{2}R_2 F t^2/I_2$.

The force of either cylinder on the other obeys Newton's third law, so the magnitude of the force of cylinder 2 on cylinder 1 is also F. The torque exerted by cylinder 2 on cylinder 1 is $\tau_1 = R_1 F$ and the angular acceleration of cylinder 1 is $\alpha_1 = \tau_1/I_1 = R_1 F/I_1$. This torque slows the cylinder. As a function of time its angular velocity is $\omega_1 = \omega_0 - R_1 F t/I_1$. The force ceases and the cylinder continue rotating with constant angular speeds when the speeds of points on their rims are the same. This means when $R_1\omega_1 = R_2\omega_2$. Thus

$$R_1\omega_0 - \frac{R_1^2 F t}{I_1} = \frac{R_2^2 F t}{I_2} .$$

When this equation is solved for Ft the result is

$$Ft = \frac{R_1 I_1 I_2}{I_1 R_2^2 + I_2 R_1^2}\,\omega_0 .$$

Substitute this expression for Ft in $\omega_2 = R_2 Ft/I_2$ to obtain

$$\omega_2 = \frac{R_1 R_2 I_1}{I_1 R_2^2 + I_2 R_1^2}\,\omega_0 .$$

51E

(a) No external torques act on the system consisting of the man, weights, and platform, so the total angular momentum of that system is conserved. Let I_i be the initial rotational inertia of the system and let I_f be the final rotational inertia. If ω_i is the initial angular velocity and ω_f is the final angular velocity, then $I_i\omega_i = I_f\omega_f$ and

$$\omega_f = \left(\frac{I_i}{I_f}\right)\omega_i = \left(\frac{6.0\,\text{kg}\cdot\text{m}^2}{2.0\,\text{kg}\cdot\text{m}^2}\right)(1.2\,\text{rev/s}) = 3.6\,\text{rev/s}.$$

(b) The initial kinetic energy is $K_i = \frac{1}{2}I_i\omega_i^2$, the final kinetic energy is $K_f = \frac{1}{2}I_f\omega_f^2$, and their ratio is

$$\frac{K_f}{K_i} = \frac{I_f\omega_f^2}{I_i\omega_i^2} = \frac{(2.0\,\text{kg}\cdot\text{m}^2)(3.6\,\text{rev/s})^2}{(6.0\,\text{kg}\cdot\text{m}^2)(1.2\,\text{rev/s})^2} = 3.0.$$

(c) The man did work in decreasing the rotational inertia by pulling the weights closer to his body.

53E

(a) No external torques act on the system consisting of the two wheels, so its total angular momentum is conserved. Let I_1 be the rotational inertia of the wheel that is originally spinning and I_2 be the rotational inertial of the wheel that is initially at rest. If ω_i is the initial angular velocity of the first wheel and ω_f is common final angular velocity of each wheel, then $I_1\omega_i = (I_1 + I_2)\omega_f$ and

$$\omega_f = \frac{I_1}{I_1 + I_2}\omega_i.$$

Substitute $I_2 = 2I_1$ and $\omega_i = 800\,\text{rev/min}$ to obtain $\omega_f = 267\,\text{rev/min}$.

(b) The initial kinetic energy is $K_i = \frac{1}{2}I_1\omega_i^2$ and the final kinetic energy is $K_f = \frac{1}{2}(I_1 + I_2)\omega_f^2$. The fraction lost is

$$\frac{\Delta K}{K_i} = \frac{K_i - K_f}{K_i} = \frac{I_1\omega_i^2 - (I_1 + I_2)\omega_f^2}{I_1\omega_i^2} = \frac{\omega_i^2 - 3\omega_f^2}{\omega_i^2}$$
$$= \frac{(800\,\text{rev/min})^2 - 3(267\,\text{rev/min})^2}{(800\,\text{rev/min})^2} = 0.67.$$

56E

(a) In terms of the radius of gyration k the rotational inertia of the merry-go-round is $I = Mk^2$ and its value is $(180\,\text{kg})(0.910\,\text{m})^2 = 149\,\text{kg}\cdot\text{m}^2$.

(b) Recall that an object moving along a straight line has angular momentum about any point that is not on the line. Its magnitude is mvd, where m is the mass of the object, v is the speed of the object, and d is the distance from the origin to the line of motion. In particular, the angular momentum of the child about the center of the merry-go-round is $L_c = mvR$, where R is the radius of the merry-go-round. Its value is $(44.0\,\text{kg})(3.00\,\text{m/s})(1.20\,\text{m}) = 158\,\text{kg}\cdot\text{m}^2/\text{s}$.

(c) No external torques act on the system consisting of the child and the merry-go-round, so the total angular momentum of the system is conserved. The initial angular momentum is given by mvR; the final angular momentum is given by $(I + mR^2)\omega$, where ω is the final common angular velocity of the merry-go-round and child. Thus $mvR = (I + mR^2)\omega$ and

$$\omega = \frac{mvR}{I + mR^2} = \frac{158\,\text{kg}\cdot\text{m}^2/\text{s}}{149\,\text{kg}\cdot\text{m}^2 + (44.0\,\text{kg})(1.20\,\text{m})^2} = 0.746\,\text{rad/s}.$$

60P

No external torques act on the system consisting of the train and wheel. The angular momentum of the system is initially zero and remains zero. Let $I\ (= MR^2)$ be the rotational inertial of the wheel. Its final angular momentum is $L_w = I\omega = MR^2\omega$, where M is the mass of the wheel. The speed of the track is $R\omega$ and the speed of the train is $R\omega - v$. The angular momentum of the train is $L_t = m(R\omega - v)R$, where m is its mass. The direction of rotation of the track was taken to be positive. If the train is moving slowly relative to the track its velocity and angular momentum are positive; if it is moving fast its velocity and angular momentum are negative. Conservation of angular momentum yields $0 = MR^2\omega + m(R\omega - v)R$. When this equation is solved for ω the result is

$$\omega = \frac{mvR}{(M + m)R^2} = \frac{mv}{(M + m)R}.$$

65P

(a) If we consider a short time interval from just before the wad hits to just after it hits and sticks, we may use the principle of conservation of angular momentum. The initial angular momentum is the angular momentum of the falling putty wad. The wad initially moves along a line that is $\ell/2$ distant from the axis of rotation, where ℓ is the length of the rod. The angular momentum of the wad is $mv\ell/2$. After the wad sticks, the rod has angular velocity ω and angular momentum $I\omega$, where I is the rotational inertia of the system consisting of the rod with the two balls and the wad at its end. Conservation of angular momentum yields $mv\ell/2 = I\omega$. If M is the mass of one of the balls, $I = (2M + m)(\ell/2)^2$. When $mv\ell/2 = (2M + m)(\ell/2)^2\omega$ is solved for ω the result is

$$\omega = \frac{2mv}{(2M + m)\ell} = \frac{2(0.0500\,\text{kg})(3.00\,\text{m/s})}{[2(2.00\,\text{kg}) + 0.0500\,\text{kg}](0.500\,\text{m})} = 0.148\,\text{rad/s}.$$

(b) The initial kinetic energy is $K_i = \frac{1}{2}mv^2$, the final kinetic energy is $K_f = \frac{1}{2}I\omega^2$, and their ratio is $K_f/K_i = I\omega^2/mv^2$. When $I = (2M + m)\ell^2/4$ and $\omega = 2mv/(2M + m)\ell$ are substituted this becomes

$$\frac{K_f}{K_i} = \frac{m}{2M + m} = \frac{0.0500\,\text{kg}}{2(2.00\,\text{kg}) + 0.0500\,\text{kg}} = 0.0123.$$

(c) As the rod rotates the sum of its kinetic and potential energies is conserved. If one of the balls is lowered a distance h the other is raised the same distance and the sum of the potential energies of the balls does not change. We need consider only the potential energy of the putty wad. It moves through a 90° arc to reach the lowest point on its path, gaining kinetic energy and losing gravitational potential energy as it goes. It then swings up through an angle θ, losing kinetic energy and gaining potential energy, until it momentarily comes to rest. Take the lowest point on the path to be the zero of potential energy. It starts a distance $\ell/2$ above this point, so its initial potential energy is $U_i = mg\ell/2$. If it swings through the angle θ, measured from its lowest point then its final position is $(\ell/2)(1 - \cos\theta)$ above the lowest point and its final potential energy is $U_f = mg(\ell/2)(1 - \cos\theta)$. The initial kinetic energy is the sum of the kinetic energies of the balls and wad: $K_i = \frac{1}{2}I\omega^2 = (2M + m)(\ell/2)^2\omega^2$. At its final position the rod is instantaneously stopped, so the final kinetic energy is $K_f = 0$. Conservation of energy yields $mg\ell/2 + \frac{1}{2}(2M + m)(\ell/2)^2\omega^2 = mg(\ell/2)(1 - \cos\theta)$. When this equation is solved for $\cos\theta$ the result is

$$\cos\theta = -\left(\frac{2M + m}{mg}\right)\left(\frac{\ell}{2}\right)\omega^2$$

$$= -\left[\frac{2(2.00\,\text{kg}) + 0.0500\,\text{kg}}{(0.0500\,\text{kg})(9.8\,\text{m/s}^2)}\right]\left(\frac{0.500\,\text{m}}{2}\right)(0.148\,\text{rad/s})^2 = -0.0453\,.$$

The result for θ is 92.6°. The total angle of the swing is $90° + 92.6° = 182.6°$.

66E

The torque acting on the top is given by $\tau = mg\ell\sin\theta$, where m is the mass of the top, ℓ is the distance from the pivot point to the center of mass, and θ is the angle between the axis of the top and the vertical. Notice that $mg\sin\theta$ gives the component of the force of gravity along a line that is perpendicular to the axis of the top. This is the component that is instrumental in producing a torque on the top; the component along the axis does not produce a torque.

The torque is horizontal and produces a change in the horizontal component of the angular momentum. The tip of the angular momentum vector (along the top axis) moves in a horizontal circle. If it moves through the small angle $\Delta\phi$ in the short time interval Δt, then the change in the horizontal component of the angular momentum is given by $\Delta L_h = L_h\,\Delta\phi$ and the rate of change is given by $dL_h/dt = L_h\,d\phi/dt$ or $dL_h/dt = L_h\Omega$, where $\Omega\ (= d\phi/dt)$ is the rate of precession. Here ϕ must be in radians and Ω is in rad/s.

Since $\tau = dL_h/dt$, $mg\ell\sin\theta = L_h\Omega$ and

$$\Omega = \frac{mg\ell\sin\theta}{L_h}\,.$$

Now $L_h = L\sin\theta$ and $L = I\omega$, where ω is the rate of rotation of the top, in rad/s. Thus

$$\Omega = \frac{mg\ell\sin\theta}{I\omega\sin\theta} = \frac{mg\ell}{I\omega}$$

$$= \frac{(0.50\,\text{kg})(9.8\,\text{m/s}^2)(0.040\,\text{m})}{(5.0\times10^{-4}\,\text{kg}\cdot\text{m}^2)(30\,\text{rev/s})(2\pi\,\text{rad/rev})} = 2.08\,\text{rad/s}\,.$$

Notice that 30 rev/s was multiplied by 2π to convert it to rad/s.

Precession is in the direction of the change in angular momentum, which is in the direction of the torque. This is clockwise when viewed from above.

Chapter 13

5E

(a) The forces are balanced when they sum to zero: $\mathbf{F_1} + \mathbf{F_2} + \mathbf{F_3} = 0$. This means $\mathbf{F_3} = -\mathbf{F_1} - \mathbf{F_2} = -(10\mathbf{i} - 4\mathbf{j})\,\text{N} - (17\mathbf{i} + 2\mathbf{j})\,\text{N} = (-27\mathbf{i} + 2\mathbf{j})\,\text{N}$.

(b) If θ is the angle the vector makes with the x axis then

$$\tan\theta = \frac{F_{3y}}{F_{3x}} = \frac{2\,\text{N}}{-27\,\text{N}} = -0.741.$$

The angle is either $-4.2°$ or $176°$. The second solution yields a negative x component and a positive y component and is therefore the correct solution.

10E

Three forces act on the ball: the tension $\mathbf{T}$ in the rope (acting along the rope), the force of wall $\mathbf{N}$ (acting horizontally away from the wall), and the force of gravity $\mathbf{W}$ (acting downward). Since the ball is in equilibrium they sum to zero. Let θ be the angle between the rope and the vertical. Then the vertical component of Newton's second is $T\cos\theta - W = 0$. The horizontal component is $N - T\sin\theta = 0$.

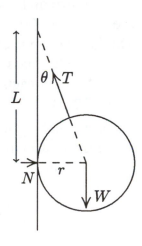

(a) Solve the first equation for T: $T = W/\cos\theta$. Substitute $\cos\theta = L/\sqrt{L^2 + r^2}$ to obtain $T = W\sqrt{L^2 + r^2}/L$.

(b) Solve the second equation for N: $N = T\sin\theta$. Use $\sin\theta = r/\sqrt{L^2 + r^2}$ to obtain

$$N = \frac{Tr}{\sqrt{L^2 + r^2}} = \frac{W\sqrt{L^2 + r^2}}{L}\frac{r}{\sqrt{L^2 + r^2}} = \frac{Wr}{L}.$$

13E

The board is in equilibrium, so the sum of the forces and the sum of the torques acting on it are each zero. Take the upward direction to be positive. Take the force of the left pedestal to be F_1 and suppose its position is $x = x_1$, where the x axis is along the diving board. Take the force of the right pedestal to be F_2 and suppose its position is $x = x_2$. Let W be the weight of the diver, at $x = x_3$. Set the expression for the sum of the forces equal to zero:

$$F_1 + F_2 - W = 0.$$

Set the expression for the torque about x_2 equal to zero:

$$F_1(x_2 - x_1) + W(x_3 - x_2) = 0.$$

(*a*) The second equation gives

$$F_1 = -\frac{x_3 - x_2}{x_2 - x_1} W = -\left(\frac{3.0\,\text{m}}{1.5\,\text{m}}\right)(580\,\text{N}) = -1160\,\text{N}\,.$$

The result is negative, indicating that this force is downward.

(*b*) The first equation gives

$$F_2 = W - F_1 = 580\,\text{N} + 1160\,\text{N} = 1740\,\text{N}\,.$$

The result is positive, indicating that this force is upward.

(*c*) The force of the diving board on the left pedestal is upward (opposite to the direction of the force of the pedestal on the diving board), so this pedestal is being stretched. The force of the diving board on the right pedestal is downward, so this pedestal is being compressed.

14E

Place the x axis along the meter stick, with the origin at the zero position on the scale. The forces acting on it are shown on the diagram to the right. The nickels are at $x = x_1$ ($= 0.120\,\text{m}$) and m is their total mass. The knife edge is at $x = x_2$ ($= 0.455\,\text{m}$) and exerts force **F**. The mass of the meter stick is M and the force of gravity acts at the center of the stick, $x = x_3$ ($= 0.500\,\text{m}$).

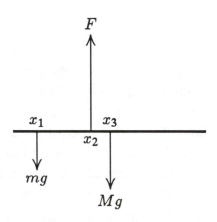

Since the meter stick is in equilibrium the sum of the torques about x_2 must vanish: $Mg(x_3-x_2)-mg(x_2-x_1) = 0$. Thus

$$M = \frac{x_2 - x_1}{x_3 - x_2} m = \left(\frac{0.455\,\text{m} - 0.120\,\text{m}}{0.500\,\text{m} - 0.455\,\text{m}}\right)(10.0\,\text{g}) = 74.0\,\text{g}\,.$$

16E

The forces on the ladder are shown in the diagram to the right. F_1 is the force of the window, horizontal because the window is frictionless. F_2 and F_3 are components of the force of the ground on the ladder. M is the mass of the window cleaner and m is the mass of the ladder. The first acts at a point 3.0 m up the ladder and the second acts at the center of the ladder. Let θ be the angle between the ladder and the ground. You can use $\cos\theta = d/L$ or $\sin\theta = \sqrt{L^2 + d^2}/L$ to find its value. Here L is the length of the ladder (5.0 m) and d is the distance from the wall to the foot of the ladder (2.5 m). In this case $\theta = 60°$.

(*a*) Since the ladder is in equilibrium the sum of the torques about its foot (or any other point) vanishes. Let ℓ be the distance from the foot of the ladder to the position of the window cleaner. Then $Mg\ell \cos\theta + mg(L/2) \cos\theta - F_1 L \sin\theta = 0$ and

$$F_1 = \frac{(M\ell + mL/2)g\cos\theta}{L\sin\theta}$$

$$= \frac{[(75\,\text{kg})(3.0\,\text{m}) + (10\,\text{kg})(2.5\,\text{m})]\,(9.8\,\text{m/s}^2)\cos 60°}{(5.0\,\text{m})\sin 60°} = 280\,\text{N}\,.$$

This force is outward, away from the wall. The force of the ladder on the window has the same magnitude but is in the opposite direction: it is 280 N, inward.

(b) The sum of the horizontal forces and the sum of the vertical forces also vanish: $F_1 - F_3 = 0$ and $F_2 - Mg - mg = 0$. The first of these equations gives $F_3 = F_1 = 280\,\text{N}$ and the second gives $F_2 = (M+m)g = (75\,\text{kg} + 10\,\text{kg})(9.8\,\text{m/s}^2) = 830\,\text{N}$. The magnitude of the force of the ground on the ladder is given by the square root of the sum of the squares of its components: $F = \sqrt{F_2^2 + F_3^2} = \sqrt{(280\,\text{N})^2 + (830\,\text{N})^2} = 880\,\text{N}$. The angle ϕ between the force and the horizontal is given by $\tan\phi = F_2/F_1 = (830\,\text{N})/(280\,\text{N}) = 2.94$, so $\phi = 71°$. The force points to the left and upward, 71° above the horizontal. Note that it is not along the ladder.

20P

(a) The forces acting on bucket are the force of gravity, down, and the tension in cable A, up. Since the bucket is in equilibrium and its weight is 1800 lb, the tension in cable A is $T_A = 1800\,\text{lb}$.

(b) Use the coordinates axes defined in the diagram. Cable A makes an angle of 66° with the negative y axis, cable B makes an angle of 27° with the positive y axis, and cable C is along the x axis. The y components of the forces must sum to zero since the knot is in equilibrium. This means $T_B\cos 27° - T_A\cos 66° = 0$ and

$$T_B = \frac{\cos 66°}{\cos 27°}T_A = \left(\frac{\cos 66°}{\cos 27°}\right)(1800\,\text{lb}) = 822\,\text{lb}\,.$$

The x components must also sum to zero. This means $T_C + T_B\sin 27° - T_A\sin 66° = 0$ and

$$T_C = T_A\sin 66° - T_B\sin 27° = (1800\,\text{lb})\sin 66° - (822\,\text{lb})\sin 27° = 1270\,\text{lb}\,.$$

27P

Consider the wheel as it leaves the lower floor. The floor no longer exerts a force on the wheel and the only forces acting are the force F applied horizontally at the axle, the force of gravity W acting vertically at the center of the wheel, and the force of the step corner, shown as the two components f_h and f_v. If the minimum force is applied the wheel does not accelerate, so both the total force and the total torque acting on it are zero.

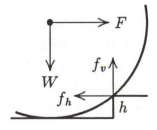

Calculate the torque around the step corner. Look at the
second diagram to see that the distance from the line of F
to the corner is $r - h$, where r is the radius of the wheel
and h is the height of the step. The distance from the line
of W to the corner is $\sqrt{r^2 + (r - h)^2} = \sqrt{2rh - h^2}$. Thus
$F(r - h) - W\sqrt{2rh - h^2} = 0$. The solution for F is

$$F = \frac{\sqrt{2rh - h^2}}{r - h} W.$$

31P

The beam is in equilibrium: the sum of the forces and the sum of the torques acting on it each
vanish. As you can see from Fig. 13–38, the beam makes an angle of 60° with the vertical and
the wire makes an angle of 30° with the vertical.

(a) Calculate the torques around the hinge. Their sum is $TL \sin 30° - W(L/2) \sin 60° = 0$.
Here W is the force of gravity, acting at the center of the beam, and T is the tension in the
wire. Solve for T:

$$T = \frac{W \sin 60°}{2 \sin 30°} = \frac{(50.0 \, \text{lb}) \sin 60°}{2 \sin 30°} = 43.3 \, \text{lb}.$$

(b) Let F_h be the horizontal component of the force exerted by the hinge and take it to be
positive if the force is outward from the wall. Then the vanishing of the horizontal component
of the net force on the beam yields $F_h - T \sin 30° = 0$ or $F_h = T \sin 30° = (43.3 \, \text{lb}) \sin 30° = $
21.7 lb.

(c) Let F_v be the vertical component of the force exerted by the hinge and take it to be positive
if it is upward. Then the vanishing of the vertical component of the net force on the beam
yields $F_v + T \cos 30° - W = 0$ or $F_v = W - T \cos 30° = 50.0 \, \text{lb} - (43.3 \, \text{lb}) \cos 30° = 12.5 \, \text{lb}$.

34P

The bar is in equilibrium, so the forces and the torques acting on it each sum to zero. Let T_l
be the tension in the left-hand cord, T_r be the tension in the right-hand cord, and W be the
weight of the bar. The equations for equilibrium are

$$\text{vertical component of force:} \quad T_l \cos \theta + T_r \cos \phi - W = 0$$
$$\text{horizontal component of force:} \quad -T_l \sin \theta + T_r \sin \phi = 0$$
$$\text{torque:} \quad Wx - T_r L \cos \phi = 0$$

The origin was chosen to be at the left end of the bar for purposes of calculating the torque.
The unknown quantities are T_l, T_r, and x. You want to eliminate T_l and T_r, then solve for
x. The second equation yields $T_l = T_r \sin \phi / \sin \theta$ and when this is substituted into the first
and solved for T_r the result is $T_r = W \sin \theta / (\sin \phi \cos \theta + \cos \phi \sin \theta)$. This expression is
substituted into the third equation and the result is solved for x:

$$x = L \frac{\sin \theta \cos \phi}{\sin \phi \cos \theta + \cos \phi \sin \theta} = L \frac{\sin \theta \cos \phi}{\sin(\theta + \phi)}.$$

The last form was obtained using the trigonometric identity $\sin(A + B) = \sin A \cos B + \cos A \sin B$. For the special case of this problem $\theta + \phi = 90°$ and $\sin(\theta + \phi) = 1$. Thus

$$x = L \sin\theta \cos\phi = (6.10\,\text{m}) \sin 36.9° \cos 53.1° = 2.20\,\text{m}.$$

39P

The diagram on the right shows the forces acting on the plank. Since the roller is frictionless the force it exerts is normal to the plank and makes the angle θ with the vertical. Its magnitude is designated F. W is the magnitude of the force of gravity; this force acts at the center of the plank, a distance $L/2$ from the point where the plank touches the floor. N is the normal force of the floor and f is the force of friction. The distance from the foot of the plank to the wall is denoted by d. This quantity is not given directly but it can be computed using $d = h/\tan\theta$.

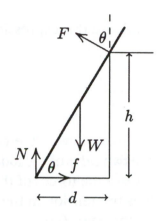

The equations of equilibrium are:

horizontal component of force: $\quad F \sin\theta - f = 0$

vertical component of force: $\quad F \cos\theta - W + N = 0$

torque: $\quad Nd - fh - W\left(d - \dfrac{L}{2}\cos\theta\right) = 0$

The point of contact between the plank and the roller was used as the origin for writing the torque equation.

When $\theta = 70°$ the plank just begins to slip and $f = \mu_s N$, where μ_s is the coefficient of static friction. You want to use the equations of equilibrium to compute N and f for $\theta = 70°$, then use $\mu_s = f/N$ to compute the coefficient of friction.

The second equation gives $F = (W - N)/\cos\theta$ and this is substituted into the first to obtain $f = (W - N)\sin\theta/\cos\theta = (W - N)\tan\theta$. This is substituted into the third equation and the result is solved for N:

$$N = \frac{d - (L/2)\cos\theta + h\tan\theta}{d + h\tan\theta}\,W\,.$$

Now replace d with $h/\tan\theta$ and multiply both numerator and denominator by $\tan\theta$. The result is

$$N = \frac{h(1 + \tan^2\theta) - (L/2)\sin\theta}{h(1 + \tan^2\theta)}\,W\,.$$

Use the trigonometric identity $1 + \tan^2\theta = 1/\cos^2\theta$ and multiply both numerator and denominator by $\cos^2\theta$ to obtain

$$N = W\left(1 - \frac{L}{2h}\cos^2\theta\sin\theta\right)\,.$$

Now substitute the expression for N into $f = (W - N) \tan \theta$ to obtain

$$f = \frac{WL}{2h} \sin^2 \theta \cos \theta.$$

Substitute the expressions for f and N into $\mu_s = f/N$ to obtain

$$\mu_s = \frac{L \sin^2 \theta \cos \theta}{2h - L \sin \theta \cos^2 \theta}.$$

Evaluate this expression for $\theta = 70°$:

$$\mu_s = \frac{(20 \, \text{ft}) \sin^2 70° \cos 70°}{2(10 \, \text{ft}) - (20 \, \text{ft}) \sin 70° \cos^2 70°} = 0.34.$$

40P

The diagrams to the right show the forces on the two sides of the ladder, separated. F_A and F_E are the forces of the floor on the two feet, T is the tension in the tie rod, W is the force of the man (equal to his weight), F_h is the horizontal component of the force exerted by one side of the ladder on the other, and F_v is the vertical component of that force. Note that the forces exerted by the floor are normal to the floor since the floor is frictionless. Also note that the force of the left side on the right and the force of the right side on the left are equal in magnitude and opposite in direction.

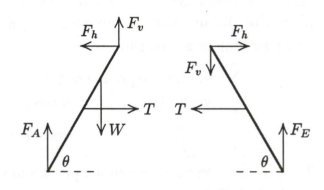

Since the ladder is in equilibrium, the vertical components of the forces on the left side of the ladder must sum to zero: $F_v + F_A - W = 0$. The horizontal components must sum to zero: $T - F_h = 0$. The torques must also sum to zero. Take the origin to be at the hinge and let L be the length of a ladder side. Then $F_A L \cos \theta - W(L/4) \cos \theta - T(L/2) \sin \theta = 0$. Here we recognize that the man is one-fourth the length of the ladder side from the top and the tie rod is at the midpoint of the side.

The analogous equations for the right side are $F_E - F_v = 0$, $F_h - T = 0$, and $F_E L \cos \theta - T(L/2) \sin \theta = 0$.

There are 5 different equations:

$$F_v + F_A - W = 0,$$
$$T - F_h = 0,$$
$$F_A L \cos \theta - W(L/4) \cos \theta - T(L/2) \sin \theta = 0,$$
$$F_E - F_v = 0,$$
$$F_E L \cos \theta - T(L/2) \sin \theta = 0.$$

The unknown quantities are F_A, F_E, F_v, F_h, and T.

(*a*) First solve for T by systematically eliminating the other unknowns. The first equation gives $F_A = W - F_v$ and the fourth gives $F_v = F_E$. Use these to substitute into the remaining 3 equations to obtain

$$T - F_h = 0,$$
$$WL \cos\theta - F_E L \cos\theta - W(L/4)\cos\theta - T(L/2)\sin\theta = 0,$$
$$F_E L \cos\theta - T(L/2)\sin\theta = 0.$$

The last of these gives $F_E = T \sin\theta / 2\cos\theta$. Substitute this expression into the second equation and solve for T. The result is

$$T = \frac{3W \cos\theta}{4\sin\theta}.$$

To find $\sin\theta$ and $\cos\theta$ consider the right triangle formed by the upper half of one side of the ladder, half the tie rod, and the vertical line from the hinge to the tie rod. The lower side of the triangle has a length of 1.25 ft, the hypotenuse has a length of 4.0 ft, and the vertical side has a length of $\sqrt{(4.0\,\text{ft})^2 - (1.25\,\text{ft})^2} = 3.8\,\text{ft}$. This means $\sin\theta = (3.8\,\text{ft})/(4.0\,\text{ft}) = 0.95$ and $\cos\theta = (1.25\,\text{ft})/(4.0\,\text{ft}) = 0.3125$. Thus

$$T = \frac{3(192\,\text{lb})(0.3125)}{4(0.95)} = 47\,\text{lb}.$$

(*b*) Now solve for F_A. Since $F_v = F_E$ and $F_E = T\sin\theta / 2\cos\theta$, $F_v = 3W/8$. Substitute this into $F_v + F_A - W = 0$ and solve for F_A. You should get $F_A = W - F_v = W - 3W/8 = 5W/8 = 5(192\,\text{lb})/8 = 120\,\text{lb}$.

(*c*) You have already obtained an expression for F_E: $F_E = 3W/8$. Evaluate it to obtain $F_E = 72\,\text{lb}$.

43P

(*a*) Examine the box when it is about to tip. Since it will rotate about the lower right edge, that is where the normal force of the floor is exerted. This force is labelled N on the diagram to the right. The force of friction is denoted by f, the applied force by F, and the force of gravity by W. Note that the force of gravity is applied at the center of the box. When the minimum force is applied the box does not accelerate, so the sum of the horizontal force components vanishes: $F - f = 0$, the sum of the vertical force components vanishes: $N - W = 0$, and the sum of the torques vanishes: $FL - WL/2 = 0$. Here L is the length of a side of the box and the origin was chosen to be at the lower right edge.

Solve the torque equation for F:

$$F = \frac{W}{2} = \frac{890\,\text{N}}{2} = 445\,\text{N}.$$

(b) The coefficient of static friction must be large enough that the box does not slip. The box is on the verge of slipping if $\mu_s = f/N$. According to the equations of equilibrium $N = W = 890\,\text{N}$ and $f = F = 445\,\text{N}$, so $\mu_s = (445\,\text{N})/(890\,\text{N}) = 0.50$.

(c) The box can be rolled with a smaller applied force if the force points upward as well as to the right. Let θ be the angle the force makes with the horizontal. The torque equation then becomes $FL\cos\theta + FL\sin\theta - WL/2 = 0$, with the solution

$$F = \frac{W}{2(\cos\theta + \sin\theta)}.$$

You want $\cos\theta + \sin\theta$ to have the largest possible value. This occurs if $\theta = 45°$, a result you can prove by setting the derivative of $\cos\theta + \sin\theta$ equal to zero and solving for θ. The minimum force needed is

$$F = \frac{W}{4\cos 45°} = \frac{890\,\text{N}}{4\cos 45°} = 315\,\text{N}.$$

44P

(a) The force diagram shown on the right depicts the situation just before the crate tips, when the normal force acts at the front edge. However, it may also be used to calculate the angle for which the crate begins to slide. W is the weight of the crate, N is normal force of the plane on the crate, and f is the force of friction. Take the x axis to be down the plane and the y axis to be in the direction of the normal force. Assume the acceleration is zero but the crate is on the verge of sliding. The x component of

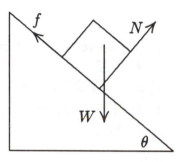

Newton's second law is then $W\sin\theta - f = 0$ and the y component is $N - W\cos\theta = 0$. The y component equation gives $N = W\cos\theta$. Since the crate is about to slide $f = \mu_s N = \mu_s W\cos\theta$, where μ_s is the coefficient of static friction. Substitute into the x component equation to obtain $W\sin\theta - \mu_s W\cos\theta = 0$, or $\tan\theta = \mu_s$. This means $\theta = \tan^{-1}\mu_s = \tan^{-1}0.60 = 31.0°$

Now develop an expression for the total torque about the center of mass when the crate is about to tip. Then the normal force and the force of friction act at the front edge. The torque associated with the force of friction tends to turn the crate clockwise and has magnitude fh, where h is the perpendicular distance from the bottom of the crate to the center of mass. The torque associated with the normal force tends to turn the crate counterclockwise and has magnitude $N\ell/2$, where ℓ is the length of a edge. Since the total torque vanishes, $fh = N\ell/2$. When the crate is about to tip the acceleration of the center of mass vanishes, so $f = W\sin\theta$ and $N = W\cos\theta$. Substitute these expressions into the torque equation and solve for θ. You should get

$$\theta = \tan^{-1}\frac{\ell}{2h} = \tan^{-1}\frac{4.0\,\text{ft}}{2(3.0\,\text{ft})} = 33.7°.$$

As θ is increased from zero the crate slides before it tips. It starts to slide when $\theta = 31.0°$.

(b) The analysis is the same. The crate begins to slide when $\theta = \tan^{-1} \mu_s = \tan^{-1} 0.70 = 35.0°$ and begins to tip when $\theta = 33.7°$. Thus it tips first as the angle is increased. Tipping begins at $\theta = 33.7°$.

47E

(a) If L (= 1500 cm) is the unstretched length of the rope and ΔL (= 2.8 cm) is the amount it stretches then the strain is $\Delta L/L = (2.8 \text{ cm})/(1500 \text{ cm}) = 1.9 \times 10^{-3}$.

(b) The stress is given by F/A where F is the stretching force applied to one end of the rope and A is the cross-sectional area of the rope. Here F is the weight of the rock climber. If m is the mass of the rock climber then $F = mg$. If r is the radius of the rope then $A = \pi r^2$. Thus the stress is

$$\frac{F}{A} = \frac{mg}{\pi r^2} = \frac{(95 \text{ kg})(9.8 \text{ m/s}^2)}{\pi (4.8 \times 10^{-3} \text{ m})^2} = 1.3 \times 10^7 \text{ N/m}^2 \,.$$

(c) Young's modulus is the stress divided by the strain: $E = (1.3 \times 10^7 \text{ N/m}^2)/(1.9 \times 10^{-3}) = 6.9 \times 10^9 \text{ N/m}^2$.

50E

(a) The shear stress is given by F/A, where F is the magnitude of the force applied parallel to one face of the aluminum rod and A is the cross-sectional area of the rod. In this case F is the weight of the object hung on the end: $F = mg$, where m is the mass of the object. If r is the radius of the rod then $A = \pi r^2$. Thus the shear stress is

$$\frac{F}{A} = \frac{mg}{\pi r^2} = \frac{(1200 \text{ kg})(9.8 \text{ m/s}^2)}{\pi (0.024 \text{ m})^2} = 6.5 \times 10^6 \text{ N/m}^2 \,.$$

(b) The shear modulus G is given by

$$G = \frac{F/A}{\Delta x/L} \,,$$

where L is the protrusion of the rod and Δx is its vertical deflection at its end. Thus

$$\Delta x = \frac{(F/A)L}{G} = \frac{(6.5 \times 10^6 \text{ N/m}^2)(0.053 \text{ m})}{3.0 \times 10^{10} \text{ N/m}^2} = 1.1 \times 10^{-5} \text{ m} \,.$$

55P

(a) Let F_A and F_B be the forces exerted by the wires on the log and let m be the mass of the log. Since the log is in equilibrium $F_A + F_B - mg = 0$. Information given about the stretching of the wires allows us to find a relationship between F_A and F_B. If wire A originally had a length L_A and stretches by ΔL_A, then $\Delta L_A = F_A L_A/AE$, where A is the cross-sectional area of the wire and E is Young's modulus for steel ($200 \times 10^9 \text{ N/m}^2$). Similarly, $\Delta L_B = F_B L_B/AE$. If

ℓ is the amount by which B was originally longer than A then, since they have the same length after the log is attached, $\Delta L_A = \Delta L_B + \ell$. This means

$$\frac{F_A L_A}{AE} = \frac{F_B L_B}{AE} + \ell.$$

Solve for F_B:

$$F_B = \frac{F_A L_A}{L_B} - \frac{AE\ell}{L_B}.$$

Substitute into $F_A + F_B - mg = 0$ and solve for F_A:

$$F_A = \frac{mg L_B + AE\ell}{L_A + L_B}.$$

The cross-sectional area of a wire is $A = \pi r^2 = \pi(1.20 \times 10^{-3}\,\text{m})^2 = 4.52 \times 10^{-6}\,\text{m}^2$. Both L_A and L_B may be taken to be 2.50 m without loss of significance. Thus

$$F_A = \frac{(103\,\text{kg})(9.8\,\text{m/s}^2) + (4.52 \times 10^{-6}\,\text{m}^2)(200 \times 10^9\,\text{N/m}^2)(2.0 \times 10^{-3}\,\text{m})}{2.50\,\text{m} + 2.50\,\text{m}}$$

$$= 867\,\text{N}.$$

(b) Solve $F_A + F_B - mg = 0$ for F_B: $F_B = mg - F_A = (103\,\text{kg})(9.8\,\text{m/s}^2) - 867\,\text{N} = 143\,\text{N}$.

(c) The net torque must also vanish. Put the origin on the surface of the log at a point directly above the center of mass. The force of gravity does not exert a torque about this point. Then the torque equation is $F_A d_A - F_B d_B = 0$ and $d_A/d_B = F_B/F_A = (143\,\text{N})/(867\,\text{N}) = 0.165$.

Chapter 14

4E

(a) The motion repeats every $0.500\,\mathrm{s}$ so the period must be $T = 0.500\,\mathrm{s}$.

(b) The frequency is the reciprocal of the period: $f = 1/T = 1/(0.500\,\mathrm{s}) = 2.00\,\mathrm{Hz}$.

(c) The angular frequency ω is $\omega = 2\pi f = 2\pi(2.00\,\mathrm{Hz}) = 12.57\,\mathrm{rad/s}$.

(d) The angular frequency is related to the spring constant k and the mass m by $\omega = \sqrt{k/m}$. Solve for k: $k = m\omega^2 = (0.500\,\mathrm{kg})(12.57\,\mathrm{rad/s})^2 = 79.0\,\mathrm{N/m}$.

(e) Let x_m be the amplitude. The maximum speed is $v_m = \omega x_m = (12.57\,\mathrm{rad/s})(0.350\,\mathrm{m}) = 4.40\,\mathrm{m/s}$.

(f) The maximum force is exerted when the displacement is a maximum and its magnitude is given by $F_m = kx_m = (79.0\,\mathrm{N/m})(0.350\,\mathrm{m}) = 27.6\,\mathrm{N}$.

7E

The magnitude of the maximum acceleration is given by $a_m = \omega^2 x_m$, where ω is the angular frequency and x_m is the amplitude. The angular frequency for which the maximum acceleration is g is given by $\omega = \sqrt{g/x_m}$ and the corresponding frequency is given by

$$ f = \frac{\omega}{2\pi} = \frac{1}{2\pi}\sqrt{\frac{g}{x_m}} = \frac{1}{2\pi}\sqrt{\frac{9.8\,\mathrm{m/s^2}}{1.0 \times 10^{-6}\,\mathrm{m}}} = 500\,\mathrm{Hz}\,. $$

For frequencies greater than $500\,\mathrm{Hz}$ the acceleration exceeds g for some part of the motion.

11E

(a) The angular frequency ω is given by $\omega = 2\pi f = 2\pi/T$, where f is the frequency and T is the period. The relationship $f = 1/T$ was used to obtain the last form. Thus $\omega = 2\pi/(1.00 \times 10^{-5}\,\mathrm{s}) = 6.28 \times 10^5\,\mathrm{rad/s}$.

(b) The maximum speed v_m and maximum displacement x_m are related by $v_m = \omega x_m$, so

$$ x_m = \frac{v_m}{\omega} = \frac{1.00 \times 10^3\,\mathrm{m/s}}{6.28 \times 10^5\,\mathrm{rad/s}} = 1.59 \times 10^{-3}\,\mathrm{m}\,. $$

13E

(a) The amplitude is half the range of the displacement, or $x_m = 1.0\,\mathrm{mm}$.

(b) The maximum speed v_m is related to the amplitude x_m by $v_m = \omega x_m$, where ω is the angular frequency. Since $\omega = 2\pi f$, where f is the frequency, $v_m = 2\pi f x_m = 2\pi(120\,\mathrm{Hz})(1.0 \times 10^{-3}\,\mathrm{m}) = 0.75\,\mathrm{m/s}$.

(c) The maximum acceleration is $a_m = \omega^2 x_m = (2\pi f)^2 x_m = (2\pi \times 120\,\text{Hz})^2 (1.0 \times 10^{-3}\,\text{m}) = 570\,\text{m/s}^2$.

18E

Use $v_m = \omega x_m = 2\pi f x_m$. The frequency is $180/(60\,\text{s}) = 3.0\,\text{Hz}$ and the amplitude is half the stroke, or $0.38\,\text{m}$. Thus $v_m = 2\pi(3.0\,\text{Hz})(0.38\,\text{m}) = 7.2\,\text{m/s}$.

26P

The maximum force that can be exerted by the surface must be less than $\mu_s N$ or else the block will not follow the surface in its motion. Here μ_s is the coefficient of static friction and N is the normal force exerted by the surface on the block. Since the block does not accelerate vertically you know that $N = mg$, where m is the mass of the block. If the block follows the table and moves in simple harmonic motion the magnitude of the maximum force exerted on it is given by $F = ma_m = m\omega^2 x_m = m(2\pi f)^2 x_m$, where a_m is the magnitude of the maximum acceleration, ω is the angular frequency, and f is the frequency. The relationship $\omega = 2\pi f$ was used to obtain the last form.

Substitute $F = m(2\pi f)^2 x_m$ and $N = mg$ into $F < \mu_s N$ to obtain $m(2\pi f)^2 x_m < \mu_s mg$, or

$$x_m < \frac{\mu_s g}{(2\pi f)^2} = \frac{(0.50)(9.8\,\text{m/s}^2)}{(2\pi \times 2.0\,\text{Hz})^2} = 0.031\,\text{m}.$$

A larger amplitude requires a larger force at the end points of the motion. The surface cannot supply the larger force and the block slips.

30P

(a) The object oscillates about its equilibrium point, where the downward force of gravity is balanced by the upward force of the spring. If ℓ is the elongation of the spring at equilibrium, then $k\ell = mg$, where k is the spring constant and m is the mass of the object. Thus $k/m = g/\ell$ and $f = \omega/2\pi = (1/2\pi)\sqrt{k/m} = (1/2\pi)\sqrt{g/\ell}$. Now the equilibrium point is halfway between the points where the object is momentarily at rest. One of these points is where the spring is unstretched and the other is the lowest point, 10 cm below. Thus $\ell = 5.0\,\text{cm} = 0.050\,\text{m}$ and

$$f = \frac{1}{2\pi}\sqrt{\frac{9.8\,\text{m/s}^2}{0.50\,\text{m}}} = 2.24\,\text{Hz}.$$

(b) Use conservation of energy. Take the zero of gravitational potential energy to be at the initial position of the object, where the spring is unstretched. Then both the initial potential and kinetic energies are zero. Take the y axis to be positive in the downward direction and let $y = 0.080\,\text{m}$. The potential energy when the object is at this point is $U = \frac{1}{2}ky^2 - mgy$. The energy equation becomes $0 = \frac{1}{2}ky^2 - mgy + \frac{1}{2}mv^2$. Solve for v:

$$v = \sqrt{2gy - \frac{k}{m}y^2} = \sqrt{2gy - \frac{g}{\ell}y^2}$$

$$= \sqrt{2(9.8\,\text{m/s}^2)(0.080\,\text{m}) - \left(\frac{9.8\,\text{m/s}^2}{0.050\,\text{m}}\right)(0.080\,\text{m})^2} = 0.56\,\text{m/s}.$$

(c) Let m be the original mass and Δm be the additional mass. The new angular frequency is $\omega' = \sqrt{k/(m + \Delta m)}$. This should be half the original angular frequency, or $\frac{1}{2}\sqrt{k/m}$. Solve $\sqrt{k/(m + \Delta m)} = \frac{1}{2}\sqrt{k/m}$ for m. Square both sides of the equation, then take the reciprocal to obtain $m + \Delta m = 4m$. This gives $m = \Delta m/3 = (300\,\mathrm{g})/3 = 100\,\mathrm{g}$.

(d) The equilibrium position is determined by the balancing of the gravitational and spring forces: $ky = (m + \Delta m)g$. Thus $y = (m + \Delta m)g/k$. You will need to find the value of the spring constant k. Use $k = m\omega^2 = m(2\pi f)^2$. Then

$$ y = \frac{(m + \Delta m)g}{m(2\pi f)^2} = \frac{(0.10\,\mathrm{kg} + 0.30\,\mathrm{kg})(9.8\,\mathrm{m/s^2})}{(0.10\,\mathrm{kg})(2\pi \times 2.24\,\mathrm{Hz})^2} = 0.20\,\mathrm{m}. $$

This is measured from the initial position.

31P

(a) Let

$$ x_1 = \frac{A}{2}\cos\left(\frac{2\pi t}{T}\right) $$

be the coordinate as a function of time for particle 1 and

$$ x_2 = \frac{A}{2}\cos\left(\frac{2\pi t}{T} + \frac{\pi}{6}\right) $$

be the coordinate as a function of time for particle 2. Here T is the period. Note that since the range of the motion is A the amplitudes are both $A/2$. The arguments of the cosine functions are in radians.

Particle 1 is at one end of its path ($x_1 = A/2$) when $t = 0$. Particle 2 is at $A/2$ when $2\pi t/T + \pi/6 = 0$ or $t = -T/12$. That is, particle 1 lags particle 2 by one-twelfth a period. We want the coordinates of the particles 0.50 s later; that is, at $t = 0.50\,\mathrm{s}$:

$$ x_1 = \frac{A}{2}\cos\left(\frac{2\pi \times 0.50\,\mathrm{s}}{1.5\,\mathrm{s}}\right) = -0.250A $$

and

$$ x_2 = \frac{A}{2}\cos\left(\frac{2\pi \times 0.50\,\mathrm{s}}{1.5\,\mathrm{s}} + \frac{\pi}{6}\right) = -0.433A. $$

Their separation at that time is $x_1 - x_2 = -0.250A + 0.433A = 0.183A$.

(b) The velocities of the particles are given by

$$ v_1 = \frac{dx_1}{dt} = \frac{\pi A}{T}\sin\left(\frac{2\pi t}{T}\right) $$

and

$$ v_2 = \frac{dx_2}{dt} = \frac{\pi A}{T}\sin\left(\frac{2\pi t}{T} + \frac{\pi}{6}\right). $$

Evaluate these expressions for $t = 0.50\,\text{s}$. You will find they are both negative, indicating that the particles are moving in the same direction.

35P

We wish to find the effective spring constant for the combination of springs shown in Fig. 28. We do this by finding the magnitude F of the force exerted on the mass when the total elongation of the springs is Δx. Then $k_{\text{eff}} = F/\Delta x$.

Suppose the left-hand spring is elongated by Δx_ℓ and the right-hand spring is elongated by Δx_r. The left-hand spring exerts a force of magnitude $k\,\Delta x_\ell$ on the right-hand spring and the right-hand spring exerts a force of magnitude $k\,\Delta x_r$ on the left-hand spring. By Newton's third law these must be equal, so $\Delta x_\ell = \Delta x_r$. The two elongations must be the same and the total elongation is twice the elongation of either spring: $\Delta x = 2\Delta x_\ell$. The left-hand spring exerts a force on the mass and its magnitude is $F = k\,\Delta x_\ell$. Thus $k_{\text{eff}} = k\,\Delta x_\ell/2\Delta x_r = k/2$. The mass behaves as if it were subject to the force of a single spring, with spring constant $k/2$. To find the frequency of its motion replace k_{eff} in $f = (1/2\pi)\sqrt{k_{\text{eff}}/m}$ with $k/2$ to obtain

$$f = \frac{1}{2\pi}\sqrt{\frac{k}{2m}}.$$

37P

(a) First consider a single spring with spring constant k and unstretched length L. One end is attached to a wall and the other is attached to an object. If it is elongated by Δx the magnitude of the force it exerts on the object is $F = k\,\Delta x$. Now consider it to be two springs, with spring constants k_1 and k_2, arranged so spring 1 is attached to the object. If spring 1 is elongated by Δx_1 then the magnitude of the force exerted on the object is $F = k_1\,\Delta x_1$. This must be the same as force of the single spring, so $k\,\Delta x = k_1\,\Delta x_1$. We must determine the relationship between Δx and Δx_1.

The springs are uniform so equal unstretched lengths are elongated by the same amount and the elongation of any portion of the spring is proportional to its unstretched length. This means spring 1 is elongated by $\Delta x_1 = CL_1$ and spring 2 is elongated by $\Delta x_2 = CL_2$, where C is a constant of proportionality. The total elongation is $\Delta x = \Delta x_1 + \Delta x_2 = C(L_1 + L_2) = CL_2(n + 1)$, where $L_1 = nL_2$ was used to obtain the last form. Since $L_2 = L_1/n$, this can also be written $\Delta x = CL_1(n + 1)/n$. Substitute $\Delta x_1 = CL_1$ and $\Delta x = CL_1(n + 1)/n$ into $k\,\Delta x = k_1\,\Delta x_1$ and solve for k_1. The result is $k_1 = k(n + 1)/n$.

Now suppose the object is placed at the other end of the composite spring, so spring 2 exerts a fore on it. Now $k\,\Delta x = k_2\,\Delta x_2$. Substitute $\Delta x_2 = CL_2$ and $\Delta x = CL_2(n + 1)$, then solve for k_2. The result is $k_2 = k(n + 1)$.

(b) To find the frequency when spring 1 is attached to mass m, replace k in $(1/2\pi)\sqrt{k/m}$ with $k(n + 1)/n$ to obtain

$$f_1 = \frac{1}{2\pi}\sqrt{\frac{(n + 1)k}{nm}} = \sqrt{\frac{n + 1}{n}}f,$$

where the substitution $f = (1/2\pi)\sqrt{k/m}$ was made. To find the frequency when spring 2 is attached to the mass, replace k with $k(n + 1)$ to obtain

$$f_2 = \frac{1}{2\pi}\sqrt{\frac{(n + 1)k}{m}} = \sqrt{n + 1}f,$$

where the same substitution was made.

39E

When the block is at the end of its path and is momentarily stopped, its displacement is equal to the amplitude and all the energy is potential in nature. If the spring potential energy is taken to be zero when the block is at its equilibrium position, then

$$E = \frac{1}{2}kx_m^2 = \frac{1}{2}(1.3 \times 10^2\,\text{N/m})(0.024\,\text{m})^2 = 3.7 \times 10^{-2}\,\text{J}.$$

41E

(a) The spring stretches until the magnitude of its upward force on the block equals the magnitude of the downward force of gravity: $ky = mg$, where y is the elongation of the spring at equilibrium, k is the spring constant, and m is the mass of the block. Thus $k = mg/y = (1.3\,\text{kg})(9.8\,\text{m/s}^2)/(0.096\,\text{m}) = 133\,\text{N/m}$.

(b) The period is given by $T = 1/f = 2\pi/\omega = 2\pi\sqrt{m/k} = 2\pi\sqrt{(1.3\,\text{kg})/(133\,\text{N/m})} = 0.62\,\text{s}$.

(c) The frequency is $f = 1/T = 1/0.62\,\text{s} = 1.6\,\text{Hz}$.

(d) The block oscillates in simple harmonic motion about the equilibrium point determined by the forces of the spring and gravity. It is started from rest 5.0 cm below the equilibrium point so the amplitude is 5.0 cm.

(e) Take both the gravitational and spring potential energies to be zero when the spring is neither compressed or extended. The block starts from rest at a point 9.6 cm + 5.0 cm = 14.6 cm below that point. There the spring potential energy is $\frac{1}{2}ky^2 = \frac{1}{2}(133\,\text{N/m})(0.146\,\text{m})^2 = 1.42\,\text{J}$ and the gravitational potential energy is $-mgy = -(1.3\,\text{kg})(9.8\,\text{m/s}^2)(0.146\,\text{m}) = -1.860\,\text{J}$. The total energy is $E = 1.42\,\text{J} - 1.86\,\text{J} = -0.44\,\text{J}$.

(f) The block has maximum speed as it passes the equilibrium point. Then the spring is elongated by $y = 9.6\,\text{cm}$ and the spring potential energy is $\frac{1}{2}ky^2 = \frac{1}{2}(133\,\text{N/m})(0.096\,\text{m})^2 = 0.612\,\text{J}$ and the gravitational potential energy is $-mgy = -(1.3\,\text{kg})(9.8\,\text{m/s}^2)(0.096\,\text{m}) = -1.22\,\text{J}$. The total potential energy is $0.612\,\text{J} - 1.22\,\text{J} = -0.61\,\text{J}$. At the initial position the block is not moving but it has potential energy $U_i = -0.44\,\text{J}$. Write the equation for conservation of energy as $U_i = U_f + \frac{1}{2}mv^2$ and solve for v:

$$v = \sqrt{\frac{2(U_i - U_f)}{m}} = \sqrt{\frac{2(-0.44\,\text{J} + 0.61\,\text{J})}{1.3\,\text{kg}}} = 0.51\,\text{m/s}.$$

44E

(a) and (b) The total energy is given $E = \frac{1}{2}kx_m^2$, where k is the spring constant and x_m is the amplitude. When $x = \frac{1}{2}x_m$ the potential energy is $U = \frac{1}{2}kx^2 = \frac{1}{8}kx_m^2$. The ratio is

$$\frac{U}{E} = \frac{\frac{1}{8}kx_m^2}{\frac{1}{2}kx_m^2} = \frac{1}{4}.$$

The fraction of the energy that is kinetic is

$$\frac{K}{E} = \frac{E - U}{E} = 1 - \frac{U}{E} = 1 - \frac{1}{4} = \frac{3}{4}.$$

(c) Since $E = \frac{1}{2}kx_m^2$ and $U = \frac{1}{2}kx^2$, $U/E = x^2/x_m^2$. Solve $x^2/x_m^2 = 1/2$ for x. You should get $x = x_m/\sqrt{2}$.

49P

(a) Assume the bullet becomes imbedded and moves with the block before the block moves a significant distance. Then the momentum of the bullet-block system is conserved during the collision. Let m be the mass of the bullet, M be the mass of the block, v_0 be the initial speed of the bullet, and v be the final speed of the block and bullet. Conservation of momentum yields $mv_0 = (m + M)v$, so

$$v = \frac{mv_0}{m + M} = \frac{(0.050\,\text{kg})(150\,\text{m/s})}{0.050\,\text{kg} + 4.0\,\text{kg}} = 1.85\,\text{m/s}.$$

After the collision the block with the bullet embedded oscillates with simple harmonic motion. The energy is all kinetic and is given by $E = \frac{1}{2}(m + M)v^2$. When the block is at the end of its path the energy is all potential and is given by $E = \frac{1}{2}ky_m^2$, where k is the spring constant and y_m is the amplitude. Since energy is conserved, $\frac{1}{2}(m + M)v^2 = \frac{1}{2}ky_m^2$ and

$$y_m = \sqrt{\frac{(m + M)v^2}{k}} = \sqrt{\frac{(0.050\,\text{kg} + 4.0\,\text{kg})(1.85\,\text{m/s})^2}{500\,\text{N/m}}} = 0.167\,\text{m}.$$

(b) The original energy of the bullet is $E_0 = \frac{1}{2}mv_0^2 = \frac{1}{2}(0.050\,\text{kg})(150\,\text{m/s})^2 = 563\,\text{J}$. The energy after the collision is $E = \frac{1}{2}(m + M)v^2 = \frac{1}{2}(0.050\,\text{kg} + 4.0\,\text{kg})(1.85\,\text{m/s})^2 = 6.94\,\text{J}$. The ratio is $E/E_0 = (6.94\,\text{J})/(563\,\text{J}) = 0.0123$ or 1.23%.

50P

(a) Use conservation of energy. Initially the energy is all potential in nature and is given by $E = \frac{1}{2}kx_m^2$, where k is the spring constant and x_m is the initial elongation of the spring. When the disk is at the equilibrium point (and the kinetic energy is maximum) the energy is all kinetic in nature and is given by $E = \frac{1}{2}Mv_m^2 + \frac{1}{2}I\omega_m^2$, where M is the mass of the disk, v_m is the speed of its center of mass, I is its rotational inertia, and ω_m is its angular speed (about the center of mass). The first term in the energy is the translational kinetic energy and the second is the rotational kinetic energy. Since the wheel rolls without slipping $\omega_m = v_m/R$, where R is its radius. Furthermore the rotational inertia of a disk is given by $I = \frac{1}{2}MR^2$. When these substitutions are made the expression for the energy at the equilibrium point becomes $E = \frac{1}{2}Mv_m^2 + \frac{1}{4}Mv_m^2 = \frac{3}{4}Mv_m^2$. Conservation of energy yields $\frac{1}{2}kx_m^2 = \frac{3}{4}Mv_m^2$. Thus $Mv_m^2 = \frac{2}{3}kx_m^2$ and the translational kinetic energy is $\frac{1}{2}Mv_m^2 = \frac{1}{3}kx_m^2 = \frac{1}{3}(3.0\,\text{N/m})(0.25\,\text{m})^2 = 0.0625\,\text{J}$.

(b) The rotational kinetic energy at the equilibrium point is $\frac{1}{2}I\omega_m^2$. Replace I with $\frac{1}{2}MR^2$ and ω_m with v_m/R to obtain $\frac{1}{4}Mv_m^2$. Now replace Mv_m^2 with $\frac{2}{3}kx_m^2$ to obtain $\frac{1}{6}kx_m^2 = \frac{1}{6}(3.0\,\text{N/m})(0.25\,\text{m/s})^2 = 0.031\,\text{J}$.

(c) Suppose the center of mass executes simple harmonic motion with angular frequency ω (not to be confused with the angular speed of rotation). Then $v_m = \omega x_m$. Use $\frac{1}{2}Mv_m^2 = \frac{1}{3}kx_m^2$ to find that $v_m = \sqrt{2k/3M}\,x_m$. Thus $\sqrt{2k/3M}\,x_m = \omega x_m$ and $\omega = \sqrt{2k/3M}$. The period is

$$T = \frac{2\pi}{\omega} = 2\pi\sqrt{\frac{3M}{2k}}\,.$$

54P

(a) Take the angular displacement of the wheel to be $\theta = \theta_m \cos(2\pi t/T)$, where θ_m is the amplitude and T is the period. Differentiate with respect to time to find the angular velocity: $\Omega = -(2\pi/T)\theta_m \sin(2\pi t/T)$. The symbol Ω is used for the angular velocity of the wheel so it is not confused with the angular frequency. The maximum angular velocity is

$$\Omega_m = \frac{2\pi\theta_m}{T} = \frac{(2\pi)(\pi\,\text{rad})}{0.500\,\text{s}} = 39.5\,\text{rad/s}\,.$$

(b) When $\theta = \pi/2$, then $\theta/\theta_m = 1/2$, $\cos(2\pi t/T) = 1/2$, and

$$\sin(2\pi t/T) = \sqrt{1 - \cos^2(2\pi t/T)} = \sqrt{1 - (1/2)^2} = \sqrt{3}/2\,,$$

where the trigonometric identity $\cos^2 A + \sin^2 A = 1$ was used. Thus

$$\Omega = -\frac{2\pi}{T}\theta_m \sin\left(\frac{2\pi t}{T}\right) = -\left(\frac{2\pi}{0.500\,\text{s}}\right)(\pi\,\text{rad})\left(\frac{\sqrt{3}}{2}\right) = -34.2\,\text{rad/s}\,.$$

The minus sign is not significant. During another portion of the cycle its angular speed is $+34.2\,\text{rad/s}$ when its angular displacement is $\pi/2\,\text{rad}$.

(c) The angular acceleration is

$$\alpha = \frac{d^2\theta}{dt^2} = -\left(\frac{2\pi}{T}\right)^2 \theta_m \cos(2\pi t/T) = -\left(\frac{2\pi}{T}\right)^2 \theta\,.$$

When $\theta = \pi/4$,

$$\alpha = -\left(\frac{2\pi}{0.500\,\text{s}}\right)^2 \left(\frac{\pi}{4}\right) = -124\,\text{rad/s}^2\,.$$

Again the minus sign is not significant.

57E

The period of a simple pendulum is given by $T = 2\pi\sqrt{L/g}$, where L is its length. Thus $L = T^2 g/4\pi^2 = (2.0\,\text{s})^2(9.8\,\text{m/s}^2)/(4\pi^2) = 0.99\,\text{m}$.

62E

(a) The period of the pendulum is given by $T = 2\pi\sqrt{I/mgd}$, where I is its rotational inertia, m is its mass, and d is the distance from the center of mass to the pivot point. The rotational inertia of a rod pivoted at its center is $mL^2/12$ and, according to the parallel-axis theorem, its rotational inertia when it is pivoted a distance d from the center is $I = mL^2/12 + md^2$. Thus

$$T = 2\pi\sqrt{\frac{m(L^2/12 + d^2)}{mgd}} = 2\pi\sqrt{\frac{L^2 + 12d^2}{12gd}}.$$

(b) $(L^2 + 12d^2)/12gd$, considered as a function of d, has a minimum at $d = L/\sqrt{12}$, so the period increases as d decreases if $d < L/\sqrt{12}$ and decreases as d decreases if $d > L/\sqrt{12}$.

(c) L occurs only in the numerator of the expression for the period, so T increases as L increases.

(d) The period does not depend on the mass of the pendulum, so T does not change when m increases.

66E

(a) A uniform disk pivoted at its center has a rotational inertia of $\frac{1}{2}MR^2$, where M is its mass and R is its radius. The disk of this problem rotates about a point that is displaced from its center by $R + L$, where L is the length of the rod, so, according to the parallel-axis theorem, its rotational inertia is $\frac{1}{2}MR^2 + M(L + R)^2$. The rod is pivoted at one end and has a rotational inertia of $mL^2/3$, where m is its mass. The total rotational inertia of the disk and rod is $I = \frac{1}{2}MR^2 + M(L + R)^2 + \frac{1}{3}mL^2 = \frac{1}{2}(0.500\,\text{kg})(0.100\,\text{m})^2 + (0.500\,\text{kg})(0.500\,\text{m} + 0.100\,\text{m})^2 + \frac{1}{3}(0.270\,\text{kg})(0.500\,\text{m})^2 = 0.205\,\text{kg} \cdot \text{m}^2$.

(b) Put the origin at the pivot. The center of mass of the disk is $\ell_d = L + R = 0.500\,\text{m} + 0.100\,\text{m} = 0.600\,\text{m}$ away and the center of mass of the rod is $\ell_r = L/2 = (0.500\,\text{m})/2 = 0.250\,\text{m}$ away, on the same line. The distance from the pivot point to the center of mass of the disk-rod system is

$$d = \frac{M\ell_d + m\ell_r}{M + m} = \frac{(0.500\,\text{kg})(0.600\,\text{m}) + (0.270\,\text{kg})(0.250\,\text{m})}{0.500\,\text{kg} + 0.270\,\text{kg}} = 0.477\,\text{m}.$$

(c) The period of oscillation is

$$T = 2\pi\sqrt{\frac{I}{(M + m)gd}} = 2\pi\sqrt{\frac{0.205\,\text{kg} \cdot \text{m}^2}{(0.500\,\text{kg} + 0.270\,\text{kg})(9.8\,\text{m/s}^2)(0.447\,\text{m})}} = 1.50\,\text{s}.$$

71P

(a) The frequency for small amplitude oscillations is $f = (1/2\pi)\sqrt{g/L}$, where L is the length of the pendulum. This gives $f = (1/2\pi)\sqrt{(9.80\,\text{m/s}^2)/(2.0\,\text{m})} = 0.35\,\text{Hz}$.

(b) The forces acing on the pendulum are the tension $\mathbf{T}$ in the rod and the force of gravity $m\mathbf{g}$. Newton's second law yields $\mathbf{T} + m\mathbf{g} = m\mathbf{a}$, where m is the mass and $\mathbf{a}$ is the acceleration of the pendulum. Let $\mathbf{a} = \mathbf{a}_e + \mathbf{a}'$, where $\mathbf{a}_e$ is the acceleration of the elevator and $\mathbf{a}'$ is the acceleration of the pendulum relative to the elevator. Newton's second law can then be written $m(\mathbf{g} - \mathbf{a}_e) + \mathbf{T} = m\mathbf{a}'$. Relative to the elevator the motion is exactly the same as it would be in an inertial frame where the acceleration due to gravity is $\mathbf{g} - \mathbf{a}_e$. Since $\mathbf{g}$ and $\mathbf{a}_e$ are along the same line and in opposite directions we can find the frequency for small amplitude oscillations by replacing g with $g + a_e$ in the expression $f = (1/2\pi)\sqrt{g/L}$. Thus

$$f = \frac{1}{2\pi}\sqrt{\frac{g + a_e}{L}} = \frac{1}{2\pi}\sqrt{\frac{9.8\,\text{m/s}^2 + 2.0\,\text{m/s}^2}{2.0\,\text{m}}} = 0.39\,\text{Hz}.$$

(c) Now the acceleration due to gravity and the acceleration of the elevator are in the same direction and have the same magnitude. That is $\mathbf{g} - \mathbf{a}_e = 0$. To find the frequency for small amplitude oscillations, replace g with zero in $f = (1/2\pi)\sqrt{g/L}$. The result is zero. The pendulum does not oscillate.

75P

If the torque exerted by the spring on the rod is proportional to the angle of rotation of the rod and if the torque tends to pull the rod toward its equilibrium orientation, then the rod will oscillate in simple harmonic motion. If $\tau = -C\theta$, where τ is the torque, θ is the angle of rotation, and C is constant of proportionality, then the angular frequency of oscillation is $\omega = \sqrt{C/I}$ and the period is $T = 2\pi/\omega = 2\pi\sqrt{I/C}$, where I is the rotational inertia of the rod. The plan is to find the torque as a function of θ and identify the constant C in terms of given quantities. This immediately gives the period in terms of given quantities.

Let ℓ_0 be the distance from the pivot point to the wall. This is also the equilibrium length of the spring. Suppose the rod turns through the angle θ, with the left end moving away from the wall. This end is now $(L/2)\sin\theta$ further from the wall and has moved $(L/2)(1 - \cos\theta)$ to the right. The length of the spring is now $\sqrt{(L/2)^2(1 - \cos\theta)^2 + [\ell_0 + (L/2)\sin\theta]^2}$. If the angle θ is small we may approximate $\cos\theta$ with 1 and $\sin\theta$ with θ in radians. Then the length of the spring is given by $\ell_0 + L\theta/2$ and its elongation is $\Delta x = L\theta/2$. The force it exerts on the rod has magnitude $F = k\,\Delta x = kL\theta/2$. Since θ is small we may approximate the torque exerted by the spring on the rod by $\tau = -FL/2$, where the pivot point was taken as the origin. Thus $\tau = -(kL^2/4)\theta$. The constant of proportionality C that relates the torque and angle of rotation is $C = kL^2/4$.

The rotational inertia for a rod pivoted at it center is $I = mL^2/12$, where m is its mass. Thus the period of oscillation is

$$T = 2\pi\sqrt{\frac{I}{C}} = 2\pi\sqrt{\frac{4mL^2}{12kL^2}} = 2\pi\sqrt{\frac{m}{3k}}.$$

81E

Let $A = x_m e^{-bt/2m}$. You want to evaluate $A/x_m = e^{-bt/2m}$ for $t = 20T$, where T is the period. Values for b (70 g/s), and m (250 g) are given in Sample Problem 10 and T is found in that problem to be 0.34 s. Thus $bt/2m = (0.070\,\mathrm{kg/s})(6.82\,\mathrm{s})/2(0.250\,\mathrm{kg}) = 0.955$ and

$$\frac{A}{x_m} = e^{-0.955} = 0.385\,.$$

Note the conversions to SI units.

82E

(a) You want to solve $e^{-bt/2m} = 1/3$ for t. Take the natural logarithm of both sides to obtain $-bt/2m = \ln(1/3)$. Now solve for t: $t = -(2m/b)\ln(1/3) = (2m/b)\ln 3$, where the sign was reversed when the argument of the logarithm was replaced by its reciprocal. Thus

$$t = \frac{2(1.50\,\mathrm{kg})}{0.230\,\mathrm{kg/s}}\ln 3 = 14.3\,\mathrm{s}\,.$$

(b) The angular frequency is

$$\omega_d = \sqrt{\frac{k}{m} - \frac{b^2}{4m^2}} = \sqrt{\frac{8.00\,\mathrm{N/m}}{1.50\,\mathrm{kg}} - \frac{(0.230\,\mathrm{kg/s})^2}{4(1.50\,\mathrm{kg})^2}} = 2.31\,\mathrm{rad/s}\,.$$

The period is $T = 2\pi/\omega_d = (2\pi)/(2.31\,\mathrm{rad/s}) = 2.72\,\mathrm{s}$ and the number of oscillations is $t/T = (14.3\,\mathrm{s})/(2.72\,\mathrm{s}) = 5.27$.

Chapter 15

1E

The magnitude of the force of one particle on the other is given by $F = Gm_1m_2/r^2$, where m_1 and m_2 are the masses, r is their separation, and G is the universal gravitational constant. Solve for r:

$$r = \sqrt{\frac{Gm_1m_2}{F}} = \sqrt{\frac{(6.67 \times 10^{-11}\,\text{N} \cdot \text{m}^2/\text{kg}^2)(5.2\,\text{kg})(2.4\,\text{kg})}{2.3 \times 10^{-12}\,\text{N}}} = 19\,\text{m}.$$

4E

Use $F = Gm_sm_m/r^2$, where m_s is the mass of the satellite, m_m is the mass of the meteor, and r is the distance between their centers. The distance between centers is $r = R + d = 15\,\text{m} + 3\,\text{m} = 18\,\text{m}$. Here R is the radius of the satellite and d is the distance from its surface to the center of the meteor. Thus

$$F = \frac{(6.67 \times 10^{-11}\,\text{N} \cdot \text{m}^2/\text{kg}^2)(20\,\text{kg})(7.0\,\text{kg})}{(18\,\text{m})^2} = 2.9 \times 10^{-11}\,\text{N}.$$

6E

At the point where the forces balance $GM_em/r_1^2 = GM_sm/r_2^2$, where M_e is the mass of the Earth, M_s is the mass of the sun, m is the mass of the space probe, r_1 is the distance from the center of the Earth to the probe, and r_2 is the distance from the center of the sun to the probe. Substitute $r_2 = d - r_1$, where d is the distance from the center of the Earth to the center of the sun, to find

$$\frac{M_e}{r_1^2} = \frac{M_s}{(d - r_1)^2}.$$

Take the square root of both sides, then solve for r_1. A little algebra yields

$$r_1 = \frac{d\sqrt{M_e}}{\sqrt{M_s} + \sqrt{M_e}} = \frac{(150 \times 10^9\,\text{m})\sqrt{5.98 \times 10^{24}\,\text{kg}}}{\sqrt{1.99 \times 10^{30}\,\text{kg}} + \sqrt{5.98 \times 10^{24}\,\text{kg}}} = 2.6 \times 10^8\,\text{m}.$$

Values can be found in Appendix C.

15P

If the lead sphere were not hollowed the magnitude of the force it exerts on m would be $F_1 = GMm/d^2$. Part of this force is due to material that is removed. Calculate the force exerted on m by a sphere that just fills the cavity, at the position of the cavity, and subtract it from the force of the solid sphere. The cavity has a radius $r = R/2$. The material that fills it has the same density (mass to volume ratio) as the solid sphere. That is $M_c/r^3 = M/R^3$, where M_c is the mass that fills the cavity. The common factor $4\pi/3$ has been canceled. Thus

$$M_c = \left(\frac{r^3}{R^3}\right) M = \left(\frac{R^3}{8R^3}\right) M = \frac{M}{8}.$$

The center of the cavity is $d - r = d - R/2$ from m, so the force it exerts on m is

$$F_2 = \frac{G(M/8)m}{(d - R/2)^2}.$$

The force of the hollowed sphere on m is

$$F = F_1 - F_2 = GMm \left[\frac{1}{d^2} - \frac{1}{8(d - R/2)^2}\right] = \frac{GMm}{d^2} \left[1 - \frac{1}{8(1 - R/2d)^2}\right].$$

16E

The period of a simple pendulum is given by $T = 2\pi\sqrt{L/g}$, where L is the length of the pendulum. Since g is different at different places on the Earth's surface the period of a pendulum varies as it is carried from place to place. The ratio of the period at the south pole to the period at the equator is $T_p/T_e = \sqrt{g_e/g_p}$, so $T_p = T_e\sqrt{g_e/g_p}$. According to Fig. 7, g_e is about $9.780\,\text{m/s}^2$ and g_p is about $9.835\,\text{m/s}^2$, so

$$T_p = (1.000\,\text{s})\sqrt{\frac{9.780\,\text{m/s}^2}{9.835\,\text{m/s}^2}} = 0.997\,\text{s}.$$

18E

The acceleration due to gravity is given by $g = GM/r^2$, where M is the mass of the Earth and r is the distance from the Earth's center. Substitute $r = R + h$, where R is the radius of the Earth and h is the altitude, to obtain $g = GM/(R + h)^2$. Solve for h. You should get $h = \sqrt{GM/g} - R$. According Appendix C of the text $R = 6.37 \times 10^6$ m and $M = 5.98 \times 10^{24}$ kg, so

$$h = \sqrt{\frac{(6.67 \times 10^{-11}\,\text{m}^3/\text{s}^2 \cdot \text{kg})(5.98 \times 10^{24}\,\text{kg})}{4.9\,\text{m/s}^2}} - 6.37 \times 10^6\,\text{m} = 2.6 \times 10^6\,\text{m}.$$

29P

(a) The forces acting on an object being weighed are the downward force of gravity and the upward force of the spring balance. Let F_g be the magnitude of the force of gravity and let W be the force of the spring balance. The reading on the balance gives the value of W. The object is traveling around a circle of radius R and so has a centripetal acceleration. Newton's second law becomes $F_g - W = mV^2/R$, where V is the speed of the object as measured in an inertial frame and m is the mass of the object. Now $V = R\omega \pm v$, where ω is the angular velocity of the Earth as it rotates and v is the speed of the ship relative to the Earth. Notice that the first term gives the speed of a point fixed to the rotating Earth. The plus sign is used if the ship is traveling in the same direction as the portion of the Earth under it (west to east) and the negative sign is used if the ship is traveling in the opposite direction (east to west).

Newton's second law is now $F_g - W = m(R\omega \pm v)^2/R$. When we expand the parentheses we may neglect the term v^2 since v is much smaller than $R\omega$. Thus $F_g - W = m(R^2\omega^2 \pm 2R\omega v)/R$ and $W = F_g - mR\omega^2 \mp 2m\omega v$. When $v = 0$ the scale reading is $W_0 = F_g - mR\omega^2$, so $W = W_0 \mp 2m\omega v$. Replace m with W_0/g to obtain $W = W_0(1 \mp 2\omega v/g)$.

(b) The upper sign ($-$) is used if the ship is sailing eastward and the lower sign ($+$) is used if the ship is sailing westward.

34P

(a) The magnitude of the force on a particle with mass m at the surface of the Earth is given by $F = GMm/R^2$, where M is the total mass of the Earth and R is the Earth's radius. The acceleration due to gravity is

$$g = \frac{F}{m} = \frac{GM}{R^2} = \frac{(6.67 \times 10^{-11}\,\mathrm{m^3/s^2 \cdot kg})(5.98 \times 10^{24}\,\mathrm{kg})}{(6.37 \times 10^6\,\mathrm{m})^2} = 9.83\,\mathrm{m/s^2}\,.$$

(b) Now $g = GM/R^2$, where M is the total mass contained in the core and mantle together and R is the outer radius of the mantle (6.345×10^6 m, according to Fig. 35). The total mass is $M = 1.93 \times 10^{24}\,\mathrm{kg} + 4.01 \times 10^{24}\,\mathrm{kg} = 5.94 \times 10^{24}\,\mathrm{kg}$. The first term is the mass of the core and the second is the mass of the mantle. Thus

$$g = \frac{(6.67 \times 10^{-11}\,\mathrm{m^3/s^2 \cdot kg})(5.94 \times 10^{24}\,\mathrm{kg})}{(6.345 \times 10^6\,\mathrm{m})^2} = 9.84\,\mathrm{m/s^2}\,.$$

(c) A point 25 km below the surface is at the mantle-crust interface and is on the surface of a sphere with a radius of $R = 6.345 \times 10^6$ m. Since the mass is now assumed to be uniformly distributed the mass within this sphere can be found by multiplying the mass per unit volume by the volume of the sphere: $M = (R^3/R_e^3)M_e$, where M_e is the total mass of the Earth and R_e is the radius of the Earth. Thus

$$M = \left[\frac{6.345 \times 10^6\,\mathrm{m}}{6.37 \times 10^6\,\mathrm{m}}\right]^3 (5.98 \times 10^{24}\,\mathrm{kg}) = 5.91 \times 10^{24}\,\mathrm{kg}\,.$$

The acceleration due to gravity is

$$g = \frac{GM}{R^2} = \frac{(6.67 \times 10^{-11}\,\mathrm{m^3/s^2 \cdot kg})(5.91 \times 10^{24}\,\mathrm{kg})}{(6.345 \times 10^6\,\mathrm{m})^2} = 9.79\,\mathrm{m/s^2}\,.$$

36P

(a) Measure the position of the train with the variable x shown on the diagram to the right. The origin is at the center of the path. If the train is to the left of the path center x it negative; if the train is to the right x is positive. When the train is at the point shown the gravitational force due to all of the Earth's mass outside a sphere of radius r is zero. The mass within the sphere is given by $M = M_e(r/R)^3$, where M_e is the total mass of the Earth and R is the radius of the Earth. The magnitude of the gravitation force on the train is $F = GMm/r^2 = GM_emr/R^3$, where m is the mass of the train.

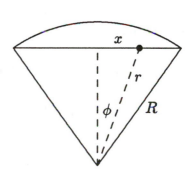

We are interested in the component of the gravitational force along the train's path. This is the component that produces the acceleration of the train. The component perpendicular to the path is balanced by the normal force of the track on the train. The component along the path is $F_x = -F \sin\phi = -(GM_em/R^3)r \sin\phi$. Now $x = r \sin\phi$, so $F_x = -(GM_em/R^3)x$. Note that this force is toward the center of the train's path. That is, it is to the right (positive) if x is negative and to the left (negative) if x is positive. It is a restoring force that is proportional to the displacement. We conclude that the train is in simple harmonic motion. Since the train is at rest at the beginning and end of the path, a one-way trip takes half a period.

(b) Newton's second law yields

$$-\frac{GM_em}{R^3} x = m \frac{d^2x}{dt^2}$$

or, when m is canceled,

$$-\left(\frac{GM_e}{R^3}\right) x = \frac{d^2x}{dt^2}.$$

By comparison with the analogous equation for a spring-mass system, $-\omega^2 x = dx^2/dt^2$, we can identify the angular frequency: $\omega = \sqrt{GM_e/R^3}$. The period is

$$T = \frac{2\pi}{\omega} = 2\pi\sqrt{\frac{R^3}{GM_e}} = 2\pi\sqrt{\frac{(6.37 \times 10^6 \text{ m})^3}{(6.67 \times 10^{-11} \text{ m}^3/\text{s}^2 \cdot \text{kg})(5.98 \times 10^{24} \text{ kg})}} = 5.06 \times 10^3 \text{ s}.$$

The time for a trip is $T/2 = 2.53 \times 10^3 \text{ s} = 42 \text{ min}$.

46P

(a) The work done by you in moving the sphere of mass m_2 equals the change in the potential energy of the three-sphere system. The initial potential energy is

$$U_i = -\frac{Gm_1m_2}{d} - \frac{Gm_1m_3}{L} - \frac{Gm_2m_3}{L-d}$$

and the final potential energy is

$$U_f = -\frac{Gm_1m_2}{L-d} - \frac{Gm_1m_3}{L} - \frac{Gm_2m_3}{d}.$$

The work is

$$W = U_f - U_i = Gm_2\left[m_1\left(\frac{1}{d} - \frac{1}{L-d}\right) + m_3\left(\frac{1}{L-d} - \frac{1}{d}\right)\right]$$

$$= (6.67 \times 10^{-11} \text{ m}^3/\text{s}^2 \cdot \text{kg})(0.10 \text{ kg})\left[(0.80 \text{ kg})\left(\frac{1}{0.040 \text{ m}} - \frac{1}{0.080 \text{ m}}\right)\right.$$

$$\left. + (0.20 \text{ kg})\left(\frac{1}{0.080 \text{ m}} - \frac{1}{0.040 \text{ m}}\right)\right]$$

$$= +5.0 \times 10^{-11} \text{ J}.$$

(b) The work done by the force of gravity is $-(U_f - U_i) = -5.0 \times 10^{-11}$ J.

47P

(a) Use the principle of conservation of energy. Initially the rocket is at the Earth's surface and the potential energy is $U_i = -GMm/R_e = -mgR_e$, where M is the mass of the Earth, m is the mass of the rocket, and R_e is the radius of the Earth. The relationship $g = GM/R_e^2$ was used. The initial kinetic energy is $\frac{1}{2}mv^2 = 2mgR_e$, where the substitution $v = 2\sqrt{gR_e}$ was made. If the rocket can escape then conservation of energy must lead to a positive kinetic energy no matter how far from the Earth it gets. Take the final potential energy to be zero and let K_f be the final kinetic energy. Then $U_i + K_i = U_f + K_f$ leads to $K_f = U_i + K_i = -mgR_e + 2mgR_e = mgR_e$. The result is positive and the rocket has enough kinetic energy to escape the gravitational pull of the Earth.

(b) Write $\frac{1}{2}mv_f^2$ for the final kinetic energy. Then $\frac{1}{2}mv_f^2 = mgR_e$ and $v_f = \sqrt{2gR_e}$.

48P

(a) Use the principle of conservation of energy. Initially the particle is at the surface of the asteroid and has potential energy $U_i = -GMm/R$, where M is the mass of the asteroid, R is its radius, and m is the mass of the particle being fired upward. Take the initial kinetic energy to be $\frac{1}{2}mv^2$. The particle just escapes if its kinetic energy is zero when it is infinitely far from the asteroid. The final potential and kinetic energies are both zero. Conservation of energy yields $-GMm/R + \frac{1}{2}mv^2 = 0$. Replace GM/R with gR, where g is the acceleration due to gravity at the surface. Then the energy equation becomes $-gR + \frac{1}{2}v^2 = 0$. Solve for v:

$$v = \sqrt{2gR} = \sqrt{2(3.0\,\text{m/s}^2)(500 \times 10^3\,\text{m})} = 1.7 \times 10^3\,\text{m/s}.$$

(b) Initially the particle is at the surface; the potential energy is $U_i = -GMm/R$ and the kinetic energy is $K_i = \frac{1}{2}mv^2$. Suppose the particle is a distance h above the surface when it momentarily comes to rest. The final potential energy is $U_f = -GMm/(R + h)$ and the final kinetic energy is $K_f = 0$. Conservation of energy yields

$$-\frac{GMm}{R} + \frac{1}{2}mv^2 = -\frac{GMm}{R + h}.$$

Replace GM with gR^2 and cancel m in the energy equation to obtain

$$-gR + \frac{1}{2}v^2 = -\frac{gR^2}{(R + h)}.$$

The solution for h is

$$h = \frac{2gR^2}{2gR - v^2} - R$$

$$= \frac{2(3.0\,\text{m/s}^2)(500 \times 10^3\,\text{m})^2)}{2(3.0\,\text{m/s}^2)(500 \times 10^3\,\text{m}) - (1000\,\text{m/s})^2} - (500 \times 10^3\,\text{m})$$

$$= 2.5 \times 10^5\,\text{m}.$$

(c) Initially the particle is a distance h above the surface and is at rest. Its potential energy is $U_i = -GMm/(R + h)$ and its initial kinetic energy is $K_i = 0$. Just before it hits the asteroid its potential energy is $U_f = -GMm/R$. Write $\frac{1}{2}mv_f^2$ for the final kinetic energy. Conservation of energy yields

$$-\frac{GMm}{R + h} = -\frac{GMm}{R} + \frac{1}{2}mv^2.$$

Replace GM with gR^2 and cancel m to obtain

$$-\frac{gR^2}{R + h} = -gR + \frac{1}{2}v^2.$$

The solution for v is

$$v = \sqrt{2gR - \frac{2gR^2}{R + h}}$$

$$= \sqrt{2(3.0\,\text{m/s}^2)(500 \times 10^3\,\text{m}) - \frac{2(3.0\,\text{m/s}^2)(500 \times 10^3\,\text{m})^2}{500 \times 10^3\,\text{m} + 1000 \times 10^3\,\text{m}}}$$

$$= 1.4 \times 10^3\,\text{m/s}.$$

51P

(a) The momentum of the two-star system is conserved, and since the stars have the same mass, their speeds and kinetic energies are the same. Use the principle of conservation of energy. The initial potential energy is $U_i = -GM^2/r_i$, where M is the mass of either star and r_i is their initial center-to-center separation. The initial kinetic energy is zero since the stars are at rest. The final potential energy is $U_f = -2GM^2/r_i$ since the final separation is $r_i/2$. Write Mv^2 for the final kinetic energy of the system. This is the sum of two terms, each of which is $\frac{1}{2}Mv^2$. Conservation of energy yields

$$-\frac{GM^2}{r_i} = -\frac{2GM^2}{r_i} + Mv^2.$$

The solution for v is

$$v = \sqrt{\frac{GM}{r_i}} = \sqrt{\frac{(6.67 \times 10^{-11}\,\text{m}^3/\text{s}^2 \cdot \text{kg}) \times (10^{30}\,\text{kg})}{10^{10}\,\text{m}}} = 8.2 \times 10^4\,\text{m/s}.$$

(b) Now the final separation of the centers is $r_f = 2R = 2 \times 10^5\,\text{m}$, where R is the radius of either of the stars. The final potential energy is given by $U_f = -GM^2/r_f$ and the energy equation becomes $-GM^2/r_i = -GM^2/r_f + Mv^2$. The solution for v is

$$v = \sqrt{GM\left(\frac{1}{r_f} - \frac{1}{r_i}\right)}$$

$$= \sqrt{(6.67 \times 10^{-11}\,\text{m}^3/\text{s}^2 \cdot \text{kg})(10^{30}\,\text{kg})\left(\frac{1}{2 \times 10^5\,\text{m}} - \frac{1}{10^{10}\,\text{kg}}\right)}$$

$$= 1.8 \times 10^7\,\text{m/s}.$$

56E

The period T and orbit radius r are related by the law of periods: $T^2 = (4\pi^2/GM)r^3$, where M is the mass of Mars. The period is 7 h 39 min, which is 2.754×10^4 s. Solve for M:

$$M = \frac{4\pi^2 r^3}{GT^2}$$

$$= \frac{4\pi^2(9.4 \times 10^6 \text{ m})^3}{(6.67 \times 10^{-11} \text{ m}^3/\text{s}^2 \cdot \text{kg})(2.754 \times 10^4 \text{ s})^2} = 6.5 \times 10^{23} \text{ kg}.$$

58E

Let N be the number of stars in the galaxy, M be the mass of the sun, and r be the radius of the galaxy. The total mass in the galaxy is NM and the magnitude of the gravitational force acting on the sun is $F = GNM^2/r^2$. The force points toward the galactic center. The magnitude of the sun's acceleration is $a = v^2/R$, where v is its speed. If T is the period of the sun's motion around the galactic center then $v = 2\pi R/T$ and $a = 4\pi^2 R/T^2$. Newton's second law yields $GNM^2/R^2 = 4\pi^2 MR/T^2$. The solution for N is

$$N = \frac{4\pi^2 R^3}{GT^2 M}.$$

The period is 2.5×10^8 y, which is 7.88×10^{15} s, so

$$N = \frac{4\pi^2(2.2 \times 10^{20} \text{ m})^3}{(6.67 \times 10^{-11} \text{ m}^3/\text{s}^2 \cdot \text{kg})(7.88 \times 10^{15} \text{ s})^2(2.0 \times 10^{30} \text{ kg})} = 5.1 \times 10^{10}.$$

60E

(a) If r is the radius of the orbit then the magnitude of the gravitational force acting on the satellite is given by GMm/r^2, where M is the mass of the Earth and m is the mass of the satellite. The magnitude of the acceleration of the satellite is given by v^2/r, where v is its speed. Newton's second law yields $GMm/r^2 = mv^2/R$. Since the radius of the Earth is 6.37×10^6 m the orbit radius is $r = 6.37 \times 10^6$ m $+ 160 \times 10^3$ m $= 6.53 \times 10^6$ m. The solution for v is

$$v = \sqrt{\frac{GM}{r}} = \sqrt{\frac{(6.67 \times 10^{-11} \text{ m}^2/\text{s}^2 \cdot \text{kg})(5.98 \times 10^{24} \text{ kg})}{6.53 \times 10^6 \text{ m}}} = 7.82 \times 10^3 \text{ m/s}.$$

(b) Since the circumference of the circular orbit is $2\pi r$, the period is $T = 2\pi r/v = 2\pi(6.53 \times 10^6$ m$)/(7.82 \times 10^3$ m/s$) = 5.25 \times 10^3$ s. This is 87.4 min.

62E

(a) The greatest distance between the satellite and the Earth's center (the apogee distance) is $R_a = 6.37 \times 10^6$ m $+ 360 \times 10^3$ m $= 6.73 \times 10^6$ m. The least distance (perigee distance) is $R_p = 6.37 \times 10^6$ m $+ 180 \times 10^3$ m $= 6.55 \times 10^6$ m. Here 6.37×10^6 m is the radius of the Earth. Look at Fig. 16 to see that the semimajor axis is $a = (R_a + R_p)/2 = (6.73 \times 10^6$ m $+ 6.55 \times 10^6$ m$)/2 = 6.64 \times 10^6$ m.

(b) The apogee and perigee distances are related to the eccentricity e by $R_a = a(1 + e)$ and $R_p = a(1 - e)$. Add to obtain $R_a + R_p = 2a$ and $a = (R_a + R_p)/2$. Subtract to obtain $R_a - R_p = 2ae$. Thus

$$e = \frac{R_a - R_p}{2a} = \frac{R_a - R_p}{R_a + R_p} = \frac{6.73 \times 10^6 \text{ m} - 6.55 \times 10^6 \text{ m}}{6.73 \times 10^6 \text{ m} + 6.55 \times 10^6 \text{ m}} = 0.0136 \,.$$

74P

Each star is attracted toward each of the other two by a force of magnitude GM^2/L^2, along the line that joins the stars. The net force on each star has magnitude $(GM^2/L^2)\cos 30°$ and is directed toward the center of the triangle. This is a centripetal force and keeps the stars on the same circular orbit if their speeds are appropriate. If R is the radius of the orbit Newton's second law yields $(GM^2/L^2)\cos 30° = Mv^2/R$.

The stars rotate about their center of mass (marked by $\odot$ on the diagram to the right), at the intersection of the perpendicular bisectors of the triangle sides, and the radius of the orbit is the distance from a star to the center of mass of the three-star system. Put a coordinate system as shown in the diagram, with its origin at the left-most star. The altitude of an equilateral triangle is $(\sqrt{3}/2)L$, so the stars are located at $x = 0$, $y = 0$; $x = L$, $y = 0$; and $x = L/2$, $y = \sqrt{3}L/2$. The x coordinate of the center of mass is $x_c = (L + L/2)/3 = L/2$ and the y coordinate is $y_c = (\sqrt{3}L/2)/3 = L/2\sqrt{3}$. The distance from a star to the center of mass is $R = \sqrt{x_c^2 + y_c^2} = \sqrt{(L^2/4) + (L^2/12)} = L/\sqrt{3}$.

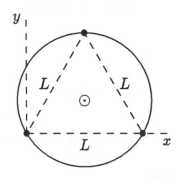

Once the substitution for R is made Newton's second law becomes $(2GM^2/L^2)\cos 30° = \sqrt{3}Mv^2/L$. This can be simplified somewhat by recognizing that $\cos 30° = \sqrt{3}/2$. Also divide the equation by M. Then $GM/L^2 = v^2/L$ and $v = \sqrt{GM/L}$.

76E

(a) Use the law of periods: $T^2 = (4\pi^2/GM)r^3$, where M is the mass of the sun (1.99×10^{30} kg) and r is the radius of the orbit. The radius of the orbit is twice the radius of the Earth's orbit: $r = 2r_e = 2(150 \times 10^9 \text{ m}) = 300 \times 10^9 \text{ m}$. Thus

$$T = \sqrt{\frac{4\pi^2 r^3}{GM}}$$

$$= \sqrt{\frac{4\pi^2(300 \times 10^9 \text{ m})^3}{(6.67 \times 10^{-11} \text{ m}^3/\text{s}^2 \cdot \text{kg})(1.99 \times 10^{30} \text{ kg})}} = 8.96 \times 10^7 \text{ s} \,.$$

Divide by $(365 \text{ d/y})(24 \text{ h/d})(60 \text{ min/h})(60 \text{ s/min})$ to obtain $T = 2.8 \text{ y}$.

(b) The kinetic energy any asteroid or planet in a circular orbit of radius r is given by $K = GMm/2r$, where m is the mass of the asteroid or planet. Notice that it is proportional to m

and inversely proportional to r. The ratio of the kinetic energy of the asteroid to the kinetic energy of the Earth is $K/K_e = (m/m_2)(r_e/r)$. Substitute $m = 2.0 \times 10^{-4} m_e$ and $r = 2r_e$ to obtain $K/K_e = 1.0 \times 10^{-4}$.

79P

The total energy is given by $E = -GMm/2a$, where M is the mass of the central attracting body (the sun, for example), m is the mass of the object (a planet, for example), and a is the semimajor axis of the orbit. If the object is a distance r from the central body the potential energy is $U = -GMm/r$. Write $\frac{1}{2}mv^2$ for the kinetic energy. Then $E = K + U$ becomes $-GMm/2a = \frac{1}{2}mv^2 - GMm/r$. Solve for v^2. The result is

$$v^2 = GM\left(\frac{2}{r} - \frac{1}{a}\right).$$

84P

(a) The force acting on the satellite has magnitude GMm/r^2, where M is the mass of the Earth, m is the mass of the satellite, and r is the radius of the orbit. The force points toward the center of the orbit. Since the acceleration of the satellite is v^2/r, where v is its speed, Newton's second law yields $GMm/r^2 = mv^2/r$ and the speed is given by $v = \sqrt{GM/r}$. The radius of the orbit is the sum of the Earth's radius and the altitude of the satellite: $r = 6.37 \times 10^6 + 640 \times 10^3 = 7.01 \times 10^6$ m. Thus

$$v = \sqrt{\frac{(6.67 \times 10^{-11}\,\text{m}^3/\text{s}^2 \cdot \text{kg})(5.98 \times 10^{24}\,\text{kg})}{7.01 \times 10^6\,\text{m}}} = 7.54 \times 10^3\,\text{m/s}.$$

(b) The period is $T = 2\pi r/v = 2\pi(7.01 \times 10^6\,\text{m})/(7.54 \times 10^3\,\text{m/s}) = 5.84 \times 10^3$ s. This is 97.4 min.

(c) If the total energy is known $E = -GMm/2r$ can be used to calculate the radius of the orbit. If E_0 is the initial energy then the energy after n orbits is $E = E_0 - nC$, where $C = 1.4 \times 10^5$ J/orbit. For a circular orbit the energy and orbit radius are related by $E = GMm/2r$, so the radius after n orbits is given by $r = -GMm/2E$.

The initial energy is

$$E_0 = -\frac{(6.67 \times 10^{-11}\,\text{m}^3/\text{s}^2 \cdot \text{kg})(5.98 \times 10^{24}\,\text{kg})(220\,\text{kg})}{2(7.01 \times 10^6\,\text{m})} = -6.26 \times 10^9\,\text{J},$$

the energy after 1500 orbits is

$$E = E_0 - nC = -6.26 \times 10^9\,\text{J} - (1500\,\text{orbit})(1.4 \times 10^5\,\text{J/orbit}) = -6.47 \times 10^9\,\text{J},$$

and the orbit radius after 1500 orbits is

$$r = -\frac{(6.67 \times 10^{-11}\,\text{m}^3/\text{s}^2 \cdot \text{kg})(5.98 \times 10^{24}\,\text{kg})(220\,\text{kg})}{-6.47 \times 10^9\,\text{m}} = 6.78 \times 10^6\,\text{m}.$$

The altitude is $h = r - R = 6.78 \times 10^6\,\text{m} - 6.37 \times 10^6\,\text{m} = 4.1 \times 10^5\,\text{m}$. Here R is the radius of the Earth.

(d) The speed is

$$v = \sqrt{\frac{GM}{r}} = \sqrt{\frac{(6.67 \times 10^{-11} \, \text{m}^3/\text{s}^2 \cdot \text{kg})(5.98 \times 10^{24} \, \text{kg})}{6.78 \times 10^6 \, \text{m}}} = 7.67 \times 10^3 \, \text{m/s}$$

(e) The period is

$$T = \frac{2\pi r}{v} = \frac{2\pi(6.78 \times 10^6 \, \text{m})}{7.67 \times 10^3 \, \text{m/s}} = 5.6 \times 10^3 \, \text{s}.$$

This is 93 min.

(f) Let F be the magnitude of the average force and s be the distance traveled by the satellite. Then the work done by the force is $W = -Fs$. This is the change in energy: $-Fs = \Delta E$. Thus $F = -\Delta E/s$. Evaluate this expression for the first orbit. For a complete orbit $s = 2\pi r = 2\pi(7.01 \times 10^6 \, \text{m}) = 4.40 \times 10^7 \, \text{m}$ and $\Delta E = -1.4 \times 10^5 \, \text{J}$. Thus

$$F = -\frac{\Delta E}{s} = (1.4 \times 10^5 \, \text{J})/(4.40 \times 10^7 \, \text{m}) = 3.3 \times 10^{-3} \, \text{N}.$$

(g) The resistive force exerts a torque on the satellite, so its angular momentum is not conserved.

Chapter 16

3E

The pressure increase is the applied force divided by the area: $\Delta p = F/A = F/\pi r^2$, where r is the radius of the piston. Thus $\Delta p = (42\,\text{N})/\pi(0.011\,\text{m})^2 = 1.1 \times 10^5\,\text{Pa}$. Notice that the value for the radius was converted to meters.

5E

The air inside pushes outward with a force given by $p_i A$, where p_i is the pressure inside the room and A is the area of the window. Similarly, the air on the outside pushes inward with a force given by $p_o A$, where p_o is the pressure outside. The magnitude of the net force is $F = (p_i - p_o)A$. Since $1\,\text{atm} = 1.013 \times 10^5\,\text{Pa}$,

$$F = (1.0\,\text{atm} - 0.96\,\text{atm})(1.013 \times 10^5\,\text{Pa/atm})(3.4\,\text{m})(2.1\,\text{m}) = 2.9 \times 10^4\,\text{N}.$$

7P

The magnitude F of the force required to pull the lid off is $F = (p_o - p_i)A$, where p_o is the pressure outside the box, p_i is the pressure inside, and A is the area of the lid. This gives

$$p_i = p_o - \frac{F}{A} = 15\,\text{lb/in.}^2 - \frac{108\,\text{lb}}{12\,\text{in.}^2} = 6.0\,\text{lb/in.}^2.$$

Notice that p_o is given in lb/in.^2 and A is given in.^{2}, so no unit conversions are required.

8P

(a) At every point on the surface there is a net inward force, normal to the surface, due to the difference in pressure between the air inside and outside the sphere. The diagram to the right shows half the sphere and some of the force vectors. We suppose a team of horses are pulling to the right. To pull the sphere apart it must exert a force at least as great as the horizontal component of the net force of the air.

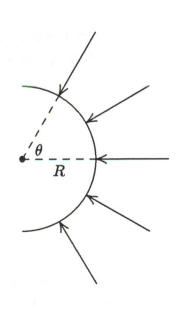

Consider the force acting at the angle θ shown. Its horizontal component is $\Delta p \cos \theta \, dA$, where dA is an infinitesimal area element at the point where the force is applied. We take the area to be that of a ring of constant θ on the surface. The radius of the ring is $R \sin \theta$, where R is the radius of the sphere. If the angular width of the ring is $d\theta$, in radians, then its width is $R\,d\theta$ and its area is $dA = 2\pi R^2 \sin \theta \, d\theta$. Thus the net horizontal component of the force of the air is given by

$$F_h = 2\pi R^2 \Delta p \int_0^{\pi/2} \sin\theta \cos\theta \, d\theta$$

$$= \pi R^2 \Delta p \sin^2\theta \Big|_0^{\pi/2} = \pi R^2 \Delta p \,.$$

This is the force that must be exerted by each team of horses to pull the sphere apart.

(b) Use 1 atm = 14.70 lb/in.^{2} to show that $\Delta p = 0.90$ atm = 13.23 lb/in.^{2} The sphere radius is 12 in., so $F_h = \pi(12\,\text{in.})^2(13.23\,\text{lb/in.}^2) = 6.0 \times 10^3$ lb.

(c) One team of horses could be used if one half of the sphere is attached to a stout tree or building. Two teams were probably used to heighten the dramatic effect.

11E

Consider the sewage in the pipe at any time. The minimum force of the pump just balances the force of gravity on the sewage and the force of the sewage (or air) in the sewer. Then the sewage will be pushed into the sewer without change in its kinetic energy. The force of gravity on the sewage is $\rho g \ell A$ where ρ is its density, ℓ (= 8.2 m − 2.1 m = 6.1 m) is the length of the pipe, and A is the cross-sectional area of the pipe. If p_0 is the pressure in the sewer then $p_0 A$ gives the force pushing down on the sewage in the pipe. If p is the pressure exerted by the pump then the force of the pump on the sewage is pA. The net force on the sewage is given by $(p - p_0)A - \rho g \ell A$ and p is a minimum if this vanishes. The pressure difference that must be maintained by the pump is $(p - p_0) = \rho g \ell = (900\,\text{kg/m}^3)(9.8\,\text{m/s}^2)(6.1\,\text{m}) = 5.4 \times 10^4$ Pa.

16E

The pressure p at the depth d of the hatch cover is $p_0 + \rho g d$, where ρ is the density of ocean water and p_0 is atmospheric pressure. The downward force of the water on the hatch cover is $(p_0 + \rho g d)A$, where A is the area of the cover. If the air in the submarine is at atmospheric pressure then it exerts an upward force of $p_0 A$. The minimum force that must be applied by the crew to open the cover has magnitude $F = (p_0 + \rho g d)A - p_0 A = \rho g d A = (1025\,\text{kg/m}^3)(9.8\,\text{m/s}^2)(100\,\text{m})(1.2\,\text{m})(0.60\,\text{m}) = 7.2 \times 10^5$ N.

18P

When the levels are the same the height of the liquid is $h = (h_1 + h_2)/2$, where h_1 and h_2 are the original heights. Suppose h_1 is greater than h_2. The final situation can then be achieved by taking liquid with volume $A(h_1 - h)$ and mass $\rho A(h_1 - h)$, in the first vessel, and lowering it a distance $h - h_2$. The work done by the force of gravity is $W = \rho A(h_1 - h)g(h - h_2)$. Substitute $h = (h_1 + h_2)/2$ to obtain $W = \frac{1}{4}\rho g A(h_1 - h_2)^2$.

22P

Assume that the pressure is the same at all points a distance $d = 20$ km below the surface. For points on the left side of the figure this pressure is given by $p = p_0 + \rho_o g d_o + \rho_c g d_c + \rho_m g d_m$, where p_0 is atmospheric pressure, ρ_o is the density of ocean water and d_o is the depth of the ocean, ρ_c is the density of the crust and d_c is the thickness of the crust, and ρ_m is the density of the mantle and d_m is the thickness of the mantle (to a depth of 20 km). For points on the right side of the figure p is given by $p = p_0 + \rho_c g d$. Equate the two expressions for p and note that g cancels to obtain $\rho_c d = \rho_o d_o + \rho_c d_c + \rho_m d_m$. Substitute $d_m = d - d_o - d_c$ to obtain

$$\rho_c d = \rho_o d_o + \rho_c d_c + \rho_m d - \rho_m d_o - \rho_m d_c .$$

Solve for d_o:

$$d_o = \frac{\rho_c d_c - \rho_c d + \rho_m d - \rho_m d_c}{\rho_m - \rho_o} = \frac{(\rho_m - \rho_c)(d - d_c)}{\rho_m - \rho_o}$$

$$= \frac{(3.3\,\mathrm{g/cm^3} - 2.8\,\mathrm{g/cm^3})(20\,\mathrm{km} - 12\,\mathrm{km})}{3.3\,\mathrm{g/cm^3} - 1.0\,\mathrm{g/cm^3}}$$

$$= 1.7\,\mathrm{km} .$$

23P

(a) At depth y the gauge pressure of the water is $p = \rho g y$, where ρ is the density of the water. Consider a strip of water with width W and thickness dy, across the dam. Its area is $dA = W\,dy$ and the force it exerts on the dam is $dF = p\,dA = \rho g W y\,dy$. The total force of the water on the dam is

$$F = \int_0^D \rho g W y\,dy = \frac{1}{2}\rho g W D^2 .$$

(b) Again consider the strip of water at depth y. Its moment arm for the torque it exerts about O is $D - y$ so the torque it exerts is $d\tau = dF(D - y) = \rho g W y (D - y)\,dy$ and the total torque of the water is

$$\tau = \int_0^D \rho g W y (D - y)\,dy = \rho g W \left(\frac{1}{2}D^3 - \frac{1}{3}D^3\right) = \frac{1}{6}\rho g W D^3 .$$

(c) Write $\tau = rF$, where r is the effective moment arm. Then

$$r = \frac{\tau}{F} = \frac{\frac{1}{6}\rho g W D^3}{\frac{1}{2}\rho g W D^2} = \frac{D}{3} .$$

31E

Let m_c be the mass of the can and m_ℓ be the mass of the lead shot. The force of gravity on the can and the shot it contains is $(m_c + m_\ell)g$ and the buoyant force of the water is $\rho g V$, where $\rho\ (= 998\,\mathrm{kg/m^3})$ is the density of water and V is the displaced volume. In equilibrium these forces balance each other and $(m_c + m_\ell)g = \rho g V$. The can contains the greatest mass of shot when it about to sink and the displaced volume is the volume of the can ($1200\,\mathrm{cm^3} = 1200 \times 10^{-6}\,\mathrm{m^3}$). Then

$$m_\ell = \rho V - m_c = (998\,\mathrm{kg/m^3})(1200 \times 10^{-6}\,\mathrm{m^3}) - 0.130\,\mathrm{kg} = 1.07\,\mathrm{kg} .$$

This is $1.07 \times 10^3\,\mathrm{g}$.

34E

(a) The anchor is completely submerged. It appears lighter than its actual weight because the water is pushing up on it with a buoyant force of $\rho_w g V$, where ρ_w is the density of water and V is the volume of the anchor. Its effective weight (in water) is $W_{\text{eff}} = W - \rho_w g V$, where W is its actual weight (the force of gravity). Thus

$$V = \frac{W - W_{\text{eff}}}{\rho_w g} = \frac{200\,\text{N}}{(998\,\text{kg/m}^3)(9.8\,\text{m/s}^2)} = 2.045 \times 10^{-2}\,\text{m}^3 .$$

(b) The mass of the anchor is $m = \rho V$, where ρ is the density of iron. Its weight in air is $W = mg = \rho g V = (7870\,\text{kg/m}^3)(9.8\,\text{m/s}^2)(2.045 \times 10^{-2}\,\text{m}^3) = 1.58 \times 10^3\,\text{N}$.

41P

(a) The force of gravity mg is balanced by the buoyant force of the liquid $\rho g V$: $mg = \rho g V_s$. Here m is the mass of the sphere, ρ is the density of the liquid, and V_s is the submerged volume. Thus $m = \rho V_s$. The submerged volume is half the volume enclosed by the outer surface of the sphere, or $V_s = \frac{1}{2}(4\pi/3)r_o^3$, where r_o is the outer radius. This means

$$m = \frac{4\pi}{6}\rho r_o^3 = \left(\frac{4\pi}{6}\right)(800\,\text{kg/m}^3)(0.090\,\text{m})^3 = 1.22\,\text{kg} .$$

Air in the hollow sphere, if any, has been neglected.

(b) The density ρ_m of the material, assumed to be uniform, is given by $\rho_m = m/V$, where m is the mass of the sphere and V is its volume. If r_i is the inner radius, the volume is

$$\frac{4\pi}{3}\left(r_o^3 - r_i^3\right) = \frac{4\pi}{3}\left[(0.090\,\text{m})^3 - (0.080\,\text{m})^3\right] = 9.09 \times 10^{-4}\,\text{m}^3 .$$

The density is

$$\frac{1.22\,\text{kg}}{9.09 \times 10^{-4}\,\text{m}^3} = 1.3 \times 10^3\,\text{kg/m}^3 .$$

43P

The volume V_{cav} of the cavities is the difference between the volume V_{cast} of the casting as a whole and the volume V_{iron} in the casting: $V_{\text{cav}} = V_{\text{cast}} - V_{\text{iron}}$. The volume of the iron is given by $V_{\text{iron}} = W/g\rho_{\text{iron}}$, where W is the weight of the casting and ρ_{iron} is the density of iron. The effective weight in water can be used to find the volume of the casting. It is less than the actual weight W because the water pushes up on it with a force of $g\rho_w V_{\text{cast}}$. That is, $W_{\text{eff}} = W - g\rho_w V_{\text{cast}}$. Thus $V_{\text{cast}} = (W - W_{\text{eff}})/g\rho_w$ and

$$\begin{aligned}
V_{\text{cav}} &= \frac{W - W_{\text{eff}}}{g\rho_w} - \frac{W}{g\rho_{\text{iron}}} \\
&= \frac{6000\,\text{N} - 4000\,\text{N}}{(9.8\,\text{m/s}^2)(0.998 \times 10^3\,\text{kg/m}^3)} - \frac{6000\,\text{N}}{(9.8\,\text{m/s}^2)(7.87 \times 10^3\,\text{kg/m}^3)} \\
&= 0.127\,\text{m}^3 .
\end{aligned}$$

44P

(a) Assume that the top surface of the slab is at the surface of the water and that the force of gravity is balanced by the buoyant force of the water on the slab. Then the area is a minimum. Also assume that the automobile is at the center of the ice surface. Then the ice slab does not tilt. Let M be the mass of the automobile, ρ_i be the density of ice, and ρ_w be the density of water. Suppose the ice slab has area A and thickness h. Since the volume of ice is Ah, the downward force of gravity on the automobile and ice is $(M + \rho_i Ah)g$. The buoyant force of the water is $\rho_w Ahg$, so the condition of equilibrium is $(M + \rho_i Ah)g - \rho_w Ahg = 0$ and

$$A = \frac{M}{(\rho_w - \rho_i)h} = \frac{1100\,\text{kg}}{(0.998 \times 10^3\,\text{kg/m}^3 - 0.917 \times 10^3\,\text{kg/m}^3)(0.30\,\text{m})} = 45\,\text{m}^2.$$

The densities are found in Table 1.

(b) It matters where the car is placed since the ice tilts if the automobile is not at the center of its surface.

51P

(a) We want to show that the net force on the rod is proportional to the displacement of the rod and always pushes it toward its equilibrium position. Then the motion is simple harmonic.

The force of gravity on the rod has magnitude mg, where m is its mass (including the lead). Since ℓ is the submerged length, the buoyant force of the water on the rod has magnitude $\rho g \ell A$, where A is the cross-sectional area of the rod. The force of gravity is downward and the buoyant force is upward, so the net downward force is $F = mg - \rho g \ell A$. The rod is in equilibrium if the net force is zero. This occurs if $\ell = m/\rho A$. If the rod is submerged more than ℓ_0 ($= m/\rho A$) then the net force is upward; if the rod is submerged less than ℓ_0 the net force is downward. That is, the net force is a restoring force that pushed the rod toward its equilibrium position.

Substitute $\ell = \ell_0 + x$, where x is the displacement of the rod from its equilibrium position, and use $\ell_0 = m/\rho A$ to obtain

$$F = mg - \rho g(\ell_0 + x)A = -\rho g A x = -\frac{mg}{\ell_0}x$$

for the net force. It is proportional to the displacement, so the motion of the rod is simple harmonic.

(b) The acceleration of the rod is $a = d^2\ell/dt^2$ and is positive if it is downward. When $\ell = \ell_0 + x$ is used, this becomes $a = d^2x/dt^2$. Thus Newton's second law yields $-(g/\ell_0)x = d^2x/dt^2$. Comparison with the analogous equation for a mass on a spring reveals that the angular frequency of oscillation is $\omega = \sqrt{g/\ell_0}$ and that the period is

$$T = \frac{2\pi}{\omega} = 2\pi\sqrt{\frac{\ell_0}{g}} = 2\pi\sqrt{\frac{2.50\,\text{m}}{9.8\,\text{m/s}^2}} = 3.17\,\text{s}.$$

55E

Use the equation of continuity. Let v_1 be the speed of the water in the hose and v_2 be its speed as it leaves one of the holes. Let A_1 be the cross-sectional area of the hose. If there are n holes you may think of the water in the hose as n tubes of flow, each of which goes through a single hole. The area of each tube of flow is A_1/n. If A_2 is the area of a hole the equation of continuity becomes $v_1 A_1/n = v_2 A_2$. Thus $v_2 = (A_1/nA_2)v_1 = (R^2/nr^2)v_1$, where R is the radius of the hose and r is the radius of a hole. Thus

$$v_2 = \frac{R^2}{nr^2}v_1 = \left[\frac{(0.375\,\text{in.})^2}{24(0.025\,\text{in.})^2}\right](3.0\,\text{ft/s}) = 28\,\text{ft/s}.$$

56P

Suppose that a mass Δm of water is pumped in time Δt. The pump increases the potential energy of the water by Δmgh, where h is the vertical distance through which it is lifted, and increases its kinetic energy by $\frac{1}{2}\Delta mv^2$, where v is its final speed. The work it does is $\Delta W = \Delta mgh + \frac{1}{2}\Delta mv^2$ and its power is

$$P = \frac{\Delta W}{\Delta t} = \frac{\Delta m}{\Delta t}\left(gh + \frac{1}{2}v^2\right).$$

Now the rate of mass flow is $\Delta m/\Delta t = \rho Av$, where ρ is the density of water and A is the area of the hose. The area of the hose is $A = \pi r^2 = \pi(0.010\,\text{m})^2 = 3.14 \times 10^{-4}\,\text{m}^2$ and $\rho Av = (998\,\text{kg/m}^3)(3.14 \times 10^{-4}\,\text{m}^2)(5.0\,\text{m/s}) = 1.57\,\text{kg/s}$. Thus

$$P = \rho Av\left(gh + \frac{1}{2}v^2\right)$$

$$= (1.57\,\text{kg/s})\left[(9.8\,\text{m/s}^2)(3.0\,\text{m}) + \frac{(5.0\,\text{m/s})^2}{2}\right]$$

$$= 66\,\text{W}.$$

58E

(a) Use the equation of continuity: $A_1v_1 = A_2v_2$. Here A_1 is the area of the pipe at the top and v_1 is the speed of the water there; A_2 is the area of the pipe at the bottom and v_2 is the speed of the water there. Thus $v_2 = (A_1/A_2)v_1 = [(4.0\,\text{cm}^2)/(8.0\,\text{cm}^2)](5.0\,\text{m/s}) = 2.5\,\text{m/s}$.

(b) Use the Bernoulli equation: $p_1 + \frac{1}{2}\rho v_1^2 + \rho gh_1 = p_2 + \frac{1}{2}\rho v_2^2 + \rho gh_2$, where ρ is the density of water, h_1 is its initial altitude, and h_2 is its final altitude. Thus

$$p_2 = p_1 + \frac{1}{2}\rho(v_1^2 - v_2^2) + \rho g(h_1 - h_2)$$

$$= 1.5 \times 10^5\,\text{Pa} + \frac{1}{2}(0.998 \times 10^3\,\text{kg/m}^3)\left[(5.0\,\text{m/s})^2 - (2.5\,\text{m/s})^2\right]$$

$$+ (0.998 \times 10^3\,\text{kg/m}^3)(9.8\,\text{m/s}^2)(10\,\text{m})$$

$$= 2.6 \times 10^5\,\text{Pa}.$$

64E

(a) Use the Bernoulli equation: $p_1 + \frac{1}{2}\rho v_1^2 + \rho g h_1 = p_2 + \frac{1}{2}\rho v_2^2 + \rho g h_2$, where h_1 is the height of the water in the tank, p_1 is the pressure there, and v_1 is the speed of the water there; h_2 is the altitude of the hole, p_2 is the pressure there, and v_2 is the speed of the water there. ρ is the density of water. The pressure at the top of the tank and at the hole is atmospheric, so $p_1 = p_2$. Since the tank is large we may neglect the water speed at the top; it is much smaller than the speed at the hole. The Bernoulli equation then becomes $\rho g h_1 = \frac{1}{2}\rho v_2^2 + \rho g h_2$ and

$$v_2 = \sqrt{2g(h_1 - h_2)} = \sqrt{2(32\,\text{ft/s}^2)(1.0\,\text{ft})} = 8.0\,\text{ft/s}.$$

Note that the British system is used and g is taken to be $32\,\text{ft/s}^2$. The flow rate is $A_2 v_2 = (1\,\text{in.}/12\,\text{in.}/\text{ft})^2(8.0\,\text{ft/s}) = 5.6 \times 10^{-2}\,\text{ft}^3/\text{s}$.

(b) Use the equation of continuity: $A_2 v_2 = A_3 v_3$, where $A_3 = A_2/2$ and v_3 is the water speed where the area of the stream is half its area at the hole. Thus $v_3 = (A_2/A_3)v_2 = 2v_2 = 16\,\text{ft/s}$. The water is in free fall and we wish to know how far it has fallen when its speed is doubled to $16\,\text{ft/s}$. Since the pressure is the same throughout the fall, $\frac{1}{2}\rho v_2^2 + \rho g h_2 = \frac{1}{2}\rho v_3^2 + \rho g h_3$. Thus

$$h_2 - h_3 = \frac{v_3^2 - v_2^2}{2g} = \frac{(16\,\text{ft/s})^2 - (8.0\,\text{ft/s})^2}{2(32\,\text{ft/s}^2)} = 3.0\,\text{ft}.$$

67E

Use the Bernoulli equation: $p_\ell + \frac{1}{2}\rho v_\ell^2 = p_u + \frac{1}{2}\rho v_u^2$, where p_ℓ is the pressure at the lower surface, p_u is the pressure at the upper surface, v_ℓ is the air speed at the lower surface, v_u is the air speed at the upper surface, and ρ is the density of air. The two tubes of flow are essentially at the same altitude. We want to solve for v_u such that $p_\ell - p_u = 900\,\text{Pa}$. That is,

$$v_u = \sqrt{\frac{2(p_\ell - p_u)}{\rho} + v_\ell^2} = \sqrt{\frac{2(900\,\text{Pa})}{1.30\,\text{kg/m}^3} + (110\,\text{m/s})^2} = 116\,\text{m/s}.$$

73P

If p_i is the air pressure inside the room and p_o is the air pressure outside the window, then the net force on the window is $(p_i - p_o)A$, where A is the area of the window. Use the Bernoulli equation to find the pressure difference: $p_o + \frac{1}{2}\rho v^2 = p_i$, where v is the air speed outside and ρ is the density of air. The air in the room is assumed to be still. Thus $p_i - p_o = \frac{1}{2}\rho v^2$ and

$$F = \frac{1}{2}\rho v^2 A = \frac{1}{2}(1.23\,\text{kg/m}^3)(30.0\,\text{m/s})^2(4.00\,\text{m})(3.00\,\text{m}) = 1.11 \times 10^4\,\text{N}.$$

76P

Since the pressure is uniform over the surface the torque it exerts may be calculated as if the air acts at the center of mass. The same statement can be made for the force of gravity. Therefore the net torque vanishes if the force of gravity equals the force of the air. Let p_ℓ

be the pressure at the lower surface, p_u be the pressure at the upper surface, v be the air speed over the upper surface, and ρ be the density of air. According to Bernoulli's equation $p_\ell = p_u + \frac{1}{2}\rho v^2$ and $p_\ell - p_u = \frac{1}{2}\rho v^2$. The magnitude of the force of the air is $F = (p_\ell - p_u)A$, where A is the area of the plate. At equilibrium $F = mg$, where m is the mass of the plate. Thus $\frac{1}{2}\rho v^2 A = mg$ and

$$v = \sqrt{\frac{2mg}{\rho A}} = \sqrt{\frac{2(0.500\,\text{kg})(9.8\,\text{m/s}^2)}{(1.23\,\text{kg/m}^3)(80 \times 10^{-4}\,\text{m}^2)}} = 32\,\text{m/s}.$$

81P

Both points are at the same altitude, so Bernoulli's equation is $p_1 + \frac{1}{2}\rho v_1^2 = p_2 + \frac{1}{2}\rho v_2^2$. The equation of continuity is $Av_1 = av_2$, so $v_2 = (A/a)v_1$. Substitute this expression into Bernoulli's equation to obtain $p_1 + \frac{1}{2}\rho v_1^2 = p_2 + \frac{1}{2}\rho(A/a)^2 v_1^2$. Solve for v_1:

$$v_1 = \sqrt{\frac{2(p_1 - p_2)a^2}{\rho(A^2 - a^2)}} = \sqrt{\frac{2a^2\Delta p}{\rho(A^2 - a^2)}},$$

where Δp was used to denote $p_1 - p_2$.

Chapter 17

3E

(a) The period T is the time for one oscillation, or $T = (20\,\text{s})/12 = 1.67\,\text{s}$.

(b) The speed v is the distance d traveled by the wave divided by the time t, or $v = d/t = (12\,\text{m})/(6.0\,\text{s}) = 2.0\,\text{m/s}$.

(c) A sinusoidal wave travels one wavelength in one period: $v = \lambda/T$. So $\lambda = vT = (2.0\,\text{m/s})(1.67\,\text{s}) = 3.3\,\text{m}$.

(d) The amplitude is the height of a wave crest above the level water surface. In this case it is 15 cm.

8E

Substitute $\omega = kv$ to obtain

$$y = y_m \sin(kx - kvt) = y_m \sin k(x - vt).$$

Substitute $k = 2\pi/\lambda$ and $\omega = 2\pi f$ to obtain

$$y = y_m \sin\left(\frac{2\pi x}{\lambda} - 2\pi ft\right) = y_m \sin 2\pi\left(\frac{x}{\lambda} - ft\right).$$

Substitute $k = \omega/v$ to obtain

$$y = y_m \sin\left(\frac{\omega x}{v} - \omega t\right) = y_m \sin\omega\left(\frac{x}{v} - t\right).$$

Substitute $k = 2\pi/\lambda$ and $\omega = 2\pi/T$ to obtain

$$y = y_m \sin\left(\frac{2\pi x}{\lambda} - \frac{2\pi t}{T}\right) = y_m \sin 2\pi\left(\frac{x}{\lambda} - \frac{t}{T}\right).$$

14P

(a) Write the expression for the displacement in the form $y(x,t) = y_m \sin(kx - \omega t)$. A minus sign is used before the ωt term in the argument of the sine function because the wave is traveling in the positive x direction. The angular wave number k is $k = 2\pi/\lambda = 2\pi/(0.10\,\text{m}) = 62.8\,\text{m}^{-1}$ and the angular frequency is $\omega = 2\pi f = 2\pi(400\,\text{Hz}) = 2510\,\text{rad/s}$. Here λ is the wavelength and f is the frequency. The amplitude is $y_m = 0.020\,\text{m}$. Thus $y(x,t) = 0.020\sin(62.8x - 2510t)$, where y and x are in meters and t is in seconds.

(b) The speed of a point on the cord is given by $u(x,t) = \partial y/\partial t = -\omega y_m \cos(kx - \omega t)$ and the maximum speed is $u_m = \omega y_m = (2510\,\text{rad/s})(0.020\,\text{m}) = 50\,\text{m/s}$.

(c) The wave speed is $v = \lambda/T = \omega/k = (2510\,\text{rad/s})/(62.8\,\text{m}^{-1}) = 40\,\text{m/s}$.

17E

The wave speed v is given by $v = \sqrt{\tau/\mu}$, where τ is the tension in the rope and μ is the linear mass density of the rope. The linear mass density is the mass per unit length of rope: $\mu = m/L = (0.0600\,\text{kg})/(2.00\,\text{m}) = 0.0300\,\text{kg/m}$. Thus

$$v = \sqrt{\frac{500\,\text{N}}{0.0300\,\text{kg/m}}} = 129\,\text{m/s}\,.$$

23E

(a) The wave speed is given by $v = \lambda/T = \omega/k$, where λ is the wavelength, T is the period, ω is the angular frequency $(2\pi/T)$, and k is the angular wave number $(2\pi/\lambda)$. The displacement has the form $y = y_m \sin(kx + \omega t)$ so $k = 2.0\,\text{m}^{-1}$ and $\omega = 30\,\text{rad/s}$. Thus $v = (30\,\text{rad/s})/(2.0\,\text{m}^{-1}) = 15\,\text{m/s}$.

(b) Since the wave speed is given by $v = \sqrt{\tau/\mu}$, where τ is the tension in the string and μ is the linear mass density of the string, the tension is $\tau = \mu v^2 = (1.6 \times 10^{-4}\,\text{kg/m})(15\,\text{m/s})^2 = 0.036\,\text{N}$.

25P

Write the string displacement in the form $y = y_m \sin(kx + \omega t)$. The plus sign is used since the wave is traveling in the negative x direction. The frequency is $f = 100\,\text{Hz}$ so the angular frequency is $\omega = 2\pi f = 2\pi(100\,\text{Hz}) = 628\,\text{rad/s}$. The wave speed is given by $v = \sqrt{\tau/\mu}$, where τ is the tension in the string and μ is the linear mass density of the string, so the wavelength is $\lambda = v/f = \sqrt{\tau/\mu}/f$ and the angular wave number is

$$k = \frac{2\pi}{\lambda} = 2\pi f\sqrt{\frac{\mu}{\tau}} = 2\pi(100\,\text{Hz})\sqrt{\frac{0.50\,\text{kg/m}}{10\,\text{N}}} = 141\,\text{m}^{-1}\,.$$

The amplitude is $y_m = 0.12\,\text{mm} = 1.2 \times 10^{-4}\,\text{m}$. Thus

$$y = \left(1.2 \times 10^{-4}\right)\sin(141x + 628t)\,,$$

where x and y are in meters and t is in seconds.

26P

Take the form of the displacement to be $y(x,t) = y_m \sin(kx - \omega t)$. The speed of a point on the cord is $u(x,t) = \partial y/\partial t = -\omega y_m \cos(kx - \omega t)$ and its maximum value is $u_m = \omega y_m$. The wave speed, on the other hand, is given by $v = \lambda/T = \omega/k$. The ratio is

$$\frac{u_m}{v} = \frac{\omega y_m}{\omega/k} = k y_m = \frac{2\pi y_m}{\lambda}\,.$$

This ratio is different for two waves with the same frequency traveling on different cords with different linear mass densities but with the same tension. Since the wave speeds are different and the frequencies are the same, the wavelengths are different.

27P

(a) Read the amplitude from the graph. It is about 5.0 cm.

(b) Read the wavelength from the graph. The curve crosses $y = 0$ at about $x = 15$ cm and again with the same slope at about $x = 55$ cm so $\lambda = 55$ cm $- 15$ cm $= 40$ cm $= 0.40$ m.

(c) The wave speed is $v = \sqrt{\tau/\mu}$, where τ is the tension in the string and μ is the linear mass density of the string. Thus

$$v = \sqrt{\frac{3.6\,\text{N}}{25 \times 10^{-3}\,\text{kg/m}}} = 12\,\text{m/s}.$$

(d) The frequency is $f = v/\lambda = (12\,\text{m/s})/(0.40\,\text{m}) = 30$ Hz and the period is $T = 1/f = 1/(30\,\text{Hz}) = 0.033$ s.

(e) The maximum string speed is $u_m = \omega y_m = 2\pi f y_m = 2\pi(30\,\text{Hz})(5.0\,\text{cm}) = 940$ cm/s $= 9.4$ m/s.

(f) Assume the string displacement has the form $y(x, t) = y_m \sin(kx + \omega t + \phi)$. A plus sign appears in the argument of the trigonometric function because the wave is moving in the negative x direction. The amplitude is $y_m = 5.0 \times 10^{-2}$ m, the angular frequency is $\omega = 2\pi f = 2\pi(30\,\text{Hz}) = 190$ rad/s, and the angular wave number is $k = 2\pi/\lambda = 2\pi/(0.40\,\text{m}) = 16\,\text{m}^{-1}$. According to the graph, the displacement at $x = 0$ and $t = 0$ is about 3.6×10^{-2} m. The formula for the displacement gives $y(0, 0) = y_m \sin\phi$ so we wish to select ϕ so that $5.0 \times 10^{-2} \sin\phi = 3.6 \times 10^{-2}$. The solution is either 0.80 rad or 2.34 rad. In the first case the function has a positive slope at $x = 0$ and matches the graph. In the second case it has negative slope and does not match the graph. We select $\phi = 0.80$ rad. The expression for the displacement is

$$y(x, t) = (5.0 \times 10^{-2})\sin(16x + 190t + 0.80),$$

where x and y are in meters and t is in seconds.

30P

The pulses have the same speed v. Suppose one pulse starts from the left end of the wire at time $t = 0$. Its coordinate at time t is $x_1 = vt$. The other pulse starts from the right end, at $x = L$, where L is the length of the wire, at time $t = 30$ ms. If this time is denoted by t_0 then the coordinate of this wave at time t is $x_2 = L - v(t - t_0)$. They meet when $x_1 = x_2$, or, what is the same, when $vt = L - v(t - t_0)$. Solve for the time they meet: $t = (L + vt_0)/2v$ and the coordinate of the meeting point is $x = vt = (L + vt_0)/2$.

Now calculate the wave speed:

$$v = \sqrt{\frac{\tau}{\mu}} = \sqrt{\frac{(250\,\text{N})(10.0\,\text{m})}{0.100\,\text{kg}}} = 158\,\text{m/s}.$$

Here τ is the tension in the wire and μ is the linear mass density of the wire. The coordinate of the meeting point is

$$x = \frac{10.0\,\text{m} + (158\,\text{m/s})(30 \times 10^{-3}\,\text{s})}{2} = 7.37\,\text{m}.$$

This is the distance from the left end of the wire.

32P

(a) The wave speed at any point on the rope is given by $v = \sqrt{\tau/\mu}$, where τ is the tension at that point and μ is the linear mass density. Because the rope is hanging the tension varies from point to point. Consider a point on the rope a distance y from the bottom end. The forces acting on it are the weight of the rope below it, pulling down, and the tension, pulling up. Since the rope is in equilibrium these balance. The weight of the rope below is given by $\mu g y$ so the tension is $\tau = \mu g y$. The wave speed is $v = \sqrt{\mu g y/\mu} = \sqrt{g y}$.

(b) The time dt for the wave to move past a length dy, a distance y from the bottom end, is $dt = dy/v = dy/\sqrt{g y}$ and the total time for the wave to move the entire length of the rope is

$$t = \int_0^\ell \frac{dy}{\sqrt{g y}} = 2\sqrt{\frac{y}{g}}\Big|_0^\ell = 2\sqrt{\frac{\ell}{g}}.$$

35P

(a) Assume the displacement of the string has the form $y(x,t) = y_m \sin(kx - \omega t)$, where y_m is the amplitude, k is the angular wave number, and ω is the angular frequency. The wave is propagating in the positive x direction and the string oscillates along the y axis. The string velocity is $u(x,t) = \partial y/\partial t = -\omega y_m \cos(kx - \omega t)$ and its maximum value is $u_m = \omega y_m$. For this wave the frequency is $f = 120\,\text{Hz}$ and the angular frequency is $\omega = 2\pi f = 2\pi(120\,\text{Hz}) = 754\,\text{rad/s}$. Since the bar moves through a distance of $1.00\,\text{cm}$, the amplitude is half of that, or $y_m = 5.00 \times 10^{-3}\,\text{m}$. The maximum string speed is $u_m = (754\,\text{rad/s})(5.00 \times 10^{-3}\,\text{m}) = 3.77\,\text{m/s}$.

(b) Consider the string at coordinate x and at time t and suppose it makes the angle θ with the x axis. The tension is along the string and makes the same angle with the x axis. Its transverse component is $\tau \sin\theta$. Now θ is given by $\tan\theta = \partial y/\partial x = k y_m \cos(kx - \omega t)$ and its maximum value is given by $\tan\theta_m = k y_m$.

We must calculate the angular wave number k. It is given by $k = \omega/v$, where v is the wave speed. The wave speed is given by $v = \sqrt{\tau/\mu}$, where τ is the tension in the rope and μ is the linear mass density of the rope. Using the data given,

$$v = \sqrt{\frac{90.0\,\text{N}}{0.120\,\text{kg/m}}} = 27.4\,\text{m/s}$$

and

$$k = \frac{754\,\text{rad/s}}{27.4\,\text{m/s}} = 27.5\,\text{m}^{-1}.$$

Thus
$$\tan\theta_m = (27.5\,\text{m}^{-1})(5.00\times10^{-3}\,\text{m}) = 0.138$$

and $\theta = 7.83°$. The maximum value of the transverse component of the tension in the string is $(90.0\,\text{N})\sin 7.83° = 12.3\,\text{N}$.

Notice that $\sin\theta$ is nearly the same as $\tan\theta$ because θ is small. We can approximate the maximum value of the transverse component of the tension by $\tau k y_m$.

(c) Consider the string at x. The transverse component of the tension pulling on it due to the string to the left is $-\tau\,\partial y/\partial x = -\tau k y_m \cos(kx - \omega t)$ and it reaches its maximum value when $\cos(kx - \omega t) = -1$. The wave speed is $u = \partial y/\partial t = -\omega y_m \cos(kx - \omega t)$ and it also reaches its maximum value when $\cos(kx - \omega t) = -1$. The two quantities reach their maximum values at the same value of the phase. When $\cos(kx - \omega t) = -1$ the value of $\sin(kx - \omega t)$ is zero and the displacement of the string is $y = 0$.

(d) When the string at any point moves through a small displacement Δy, the tension does work $\Delta W = \tau_{\text{trans}}\,\Delta y$. The rate at which it does work is
$$P = \frac{\Delta W}{\Delta t} = \tau_{\text{trans}}\frac{\Delta y}{\Delta t} = \tau_{\text{trans}}\,u\,.$$

P has its maximum value when the transverse component τ_{trans} of the tension and the string speed u have their maximum values. Hence the maximum power is $(12.3\,\text{N})(3.77\,\text{m/s}) = 46.4\,\text{W}$.

(e) As shown above $y = 0$ when the transverse component of the tension and the string speed have their maximum values.

(f) The power transferred is zero when the transverse component of the tension and the string speed are zero.

(g) $P = 0$ when $\cos(kx - \omega t) = 0$ and when it does $\sin(kx - \omega t) = \pm 1$. The string displacement is $y = \pm y_m = \pm 0.50\,\text{cm}$.

36E

The displacement of the string is given by $y = y_m\sin(kx - \omega t) + y_m\sin(kx - \omega t + \phi) = 2y_m\cos(\frac{1}{2}\phi)\sin(kx - \omega t + \frac{1}{2}\phi)$, where $\phi = \pi/2$. The amplitude is $A = 2y_m\cos(\frac{1}{2}\phi) = 2y_m\cos(\pi/4) = 1.41 y_m$.

39P

The string displacement is given by
$$y(x,t) = y_m\sin(kx - \omega t) + \frac{1}{2}y_m\sin\left(kx - \omega t + \frac{\pi}{2}\right) + \frac{1}{3}y_m\sin\left(kx - \omega t + \pi\right).$$

For ease in writing, let $\gamma = kx - \omega t$. Then
$$y = y_m\sin\gamma + \frac{1}{2}y_m\sin\left(\gamma + \frac{\pi}{2}\right) + \frac{1}{3}y_m\sin\left(\gamma + \pi\right).$$

The sum of any number of sinusoidal waves with the same frequency is again a sinusoidal wave with that frequency. That is, $y = A\sin(\gamma + \alpha)$. We want to find A and α.

Use the trigonometric identity $\sin(A + B) = \sin A \cos B + \cos A \sin B$ to expand the original expression for y. You should obtain

$$y = y_m \sin \gamma + \frac{1}{2} y_m \cos \gamma - \frac{1}{3} y_m \sin \gamma$$

$$= \frac{2}{3} y_m \sin \gamma + \frac{1}{2} y_m \cos \gamma.$$

Here we made use of $\cos(\pi/2) = 0$, $\sin(\pi/2) = 1$, $\cos \pi = -1$, and $\sin \pi = 0$. In the last form we collected all terms with the factor $\sin \gamma$ and all terms with the factor $\cos \gamma$.

Now expand $y = A \sin(\gamma + \alpha)$:

$$y = A \sin \gamma \cos \alpha + A \cos \gamma \sin \alpha.$$

Equate the coefficients of $\sin \gamma$ in the two expressions:

$$A \cos \alpha = \frac{2}{3} y_m$$

and equate the coefficients of $\cos \gamma$:

$$A \sin \alpha = \frac{1}{2} y_m.$$

Square these two equations and add them. For the left sides you should get $A^2(\sin^2 \alpha + \cos^2 \alpha) = A^2$ and for the right sides you should get $(2/3)^2 y_m^2 + (1/2)^2 y_m^2 = (5/6)^2 y_m^2$, so $A^2 = (5/6)^2 y_m^2$ and $A = 5 y_m/6$. Divide the second equation by the first to obtain $\tan \alpha = (2y_m/3)/(y_m/2) = 3/4$. This means $\alpha = 0.6435$ rad. The composite wave is

$$y = \frac{5}{6} y_m \sin(\gamma + 0.6435) = \frac{5}{6} y_m \sin(kx - \omega t + 0.6435).$$

The graph below shows the wave at time $t = 0$. As time goes on it moves to the right with speed $v = \omega/k$.

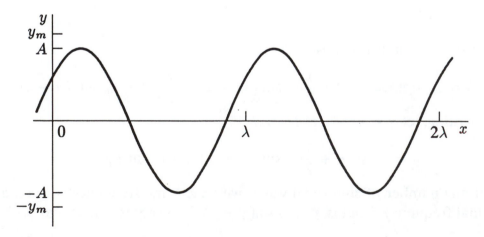

45E

(a) The wave speed is given by $v = \sqrt{\tau/\mu}$, where τ is the tension in the string and μ is the linear mass density of the string. Since the mass density is the mass per unit length, $\mu = M/L$, where M is the mass and L is the length,

$$v = \sqrt{\frac{\tau L}{M}} = \sqrt{\frac{(96.0\,\text{N})(8.40\,\text{m})}{0.120\,\text{kg}}} = 82.0\,\text{m/s}.$$

(b) The longest possible wavelength λ for a standing wave is related to the length by $L = \lambda/2$, so $\lambda = 2L = 2(8.40\,\text{m}) = 16.8\,\text{m}$.

(c) The frequency is $f = v/\lambda = (82.0\,\text{m/s})/(16.8\,\text{m}) = 4.88\,\text{Hz}$.

50E

Possible wavelengths are given by $\lambda = 2L/n$, where L is the length of the wire and n is an integer. The corresponding frequencies are given by $f = v/\lambda = nv/2L$, where v is the wave speed. The wave speed is given by $v = \sqrt{\tau/\mu} = \sqrt{\tau L/M}$, where τ is the tension in the wire, μ is the linear mass density of the wire, and M is the mass of the wire. $\mu = M/L$ was used to obtain the last form. Thus

$$f = \frac{n}{2L}\sqrt{\frac{\tau L}{M}} = \frac{n}{2}\sqrt{\frac{\tau}{LM}} = \frac{n}{2}\sqrt{\frac{250\,\text{N}}{(10.0\,\text{m})(0.100\,\text{kg})}} = 7.91n\,\text{Hz}.$$

For $n = 1$, $f = 7.91\,\text{Hz}$; for $n = 2$, $f = 15.8\,\text{Hz}$; and for $n = 3$, $f = 23.7\,\text{Hz}$.

54P

(a) The angular frequency is $\omega = 8.0\pi/2 = 4.0\pi\,\text{rad/s}$, so the frequency is $f = \omega/2\pi = (4.0\pi\,\text{rad/s})/2\pi = 2.0\,\text{Hz}$. The angular wave number is $k = 2.0\pi/2 = 1.0\pi\,\text{m}^{-1}$ so the wavelength is $\lambda = 2\pi/k = 2\pi/(1.0\pi\,\text{m}^{-1}) = 2.0\,\text{m}$. The wave speed is

$$v = \lambda f = (2.0\,\text{m})(2.0\,\text{Hz}) = 4.0\,\text{m/s}.$$

(b) You need to add 2 cosine functions. First convert them to sine functions using $\cos\alpha = \sin(\alpha + \pi/2)$, then apply Eq. 42. Here are the steps:

$$\cos\alpha + \cos\beta = \sin\left(\alpha + \frac{\pi}{2}\right) + \sin\left(\beta + \frac{\pi}{2}\right) = 2\sin\left(\frac{\alpha + \beta + \pi}{2}\right)\cos\left(\frac{\alpha - \beta}{2}\right)$$

$$= 2\cos\left(\frac{\alpha + \beta}{2}\right)\cos\left(\frac{\alpha - \beta}{2}\right).$$

Let $\alpha = kx$ and $\beta = \omega t$. Then

$$y_m\cos(kx + \omega t) + y_m\cos(kx - \omega t) = 2y_m\cos(kx)\cos(\omega t)$$

Nodes occur where $\cos(kx) = 0$ or $kx = n\pi + \pi/2$, where n is an integer (including zero). Since $k = 1.0\pi\,\text{m}^{-1}$, this means $x = 1.0(n + \frac{1}{2})$, in meters. Nodes occur at $x = 0.50\,\text{m}$, $1.5\,\text{m}$, $2.5\,\text{m}$, etc.

(c) The motion is a maximum where $\cos(kx) = \pm 1$. This means $kx = n\pi$, where n is an integer. Thus $x = 1.0n$, in meters. Maxima occur at $x = 0$, $1.0\,\mathrm{m}$, $2.0\,\mathrm{m}$, $3.0\,\mathrm{m}$, etc.

55P

(a) The amplitude of each of the traveling wave is half the maximum displacement of the string when the standing wave is present, or $0.25\,\mathrm{cm}$. Each traveling wave has an angular frequency of $\omega = 40\pi\,\mathrm{rad/s}$ and an angular wave number of $k = \pi/3\,\mathrm{cm}^{-1}$. The wave speed is $v = \omega/k = (40\pi\,\mathrm{rad/s})/(\pi/3\,\mathrm{cm}^{-1}) = 120\,\mathrm{cm/s}$.

(b) The distance between nodes is half a wavelength: $d = \lambda/2 = \pi/k = \pi/(\pi/3\,\mathrm{cm}^{-1}) = 3.0\,\mathrm{cm}$. Here $2\pi/k$ was substituted for λ.

(c) The string speed is given by $u(x,t) = \partial y/\partial t = -\omega y_m \sin(kx)\sin(\omega t)$. For the given coordinate and time $u = -(40\pi\,\mathrm{rad/s})(0.50\,\mathrm{cm})\sin(1.5\pi/3)\sin(40\pi \times 9/8) = 0$.

56P

(a) The resonant wavelengths are given by $\lambda = 2L/n$, where L is the length of the string and n is an integer, and the resonant frequencies are given by $f = v/\lambda = nv/2L$, where v is the wave speed. Suppose the lower frequency is associated with the integer n. Then, since there are no resonant frequencies between, the higher frequency is associated with $n + 1$. That is, $f_1 = nv/2L$ is the lower frequency and $f_2 = (n + 1)v/2L$ is the higher. The ratio of the frequencies is

$$\frac{f_2}{f_1} = \frac{n + 1}{n}.$$

The solution for n is

$$n = \frac{f_1}{f_2 - f_1} = \frac{315\,\mathrm{Hz}}{420\,\mathrm{Hz} - 315\,\mathrm{Hz}} = 3.$$

The lowest possible resonant frequency is $f = v/2L = f_1/n = (315\,\mathrm{Hz})/3 = 105\,\mathrm{Hz}$.

(b) The longest possible wavelength is $\lambda = 2L$. If f is the lowest possible frequency then $v = \lambda f = 2Lf = 2(0.75\,\mathrm{m})(105\,\mathrm{Hz}) = 158\,\mathrm{m/s}$.

60P

(a) Since the standing wave has 3 loops the string is 3 half-wavelengths long. If L is the length of the string and λ is the wavelength, then $L = 2\lambda/3$, or $\lambda = 2L/3$. If v is the wave speed, then the frequency is $f = v/\lambda = 3v/2L = 3(100\,\mathrm{m/s})/2(3.0\,\mathrm{m}) = 50\,\mathrm{Hz}$.

(b) The waves have the same amplitude, the same angular frequency, and the same angular wave number, but they travel in opposite directions. Take them to be $y_1 = y_m \sin(kx - \omega t)$ and $y_2 = y_m \sin(kx + \omega t)$. The amplitude y_m is half the maximum displacement of the standing wave, or $5.0 \times 10^{-3}\,\mathrm{m}$. The angular frequency is the same as that of the standing wave, or $\omega = 2\pi f = 2\pi(50\,\mathrm{Hz}) = 314\,\mathrm{rad/s}$. The angular wave number is $k = 2\pi/\lambda = 2\pi/(2.0\,\mathrm{m}) = 3.14\,\mathrm{m}^{-1}$. Thus

$$y_1 = (5.0 \times 10^{-3})\sin(3.14x - 314t)$$

$$y_2 = (5.0 \times 10^{-3})\sin(3.14x + 314t),$$

where x, y_1, and y_2 are in meters and t is in seconds.

61P

(a) Since the string has 4 loops its length must be 2 wavelengths. That is, $\lambda = L/2$, where λ is the wavelength and L is the length of the string. The wavelength is related to the frequency f and wave speed v by $\lambda = v/f$, so $L/2 = v/f$ and $L = 2v/f = 2(400\,\text{m/s})/(600\,\text{Hz}) = 1.3\,\text{m}$.

(b) Write the expression for the string displacement in the form $y = y_m \sin(kx)\cos(\omega t)$, where y_m is the maximum displacement, k is the angular wave number, and ω is the angular frequency. The angular wave number is $k = 2\pi/\lambda = 2\pi f/v = 2\pi(600\,\text{Hz})/(400\,\text{m/s}) = 9.4\,\text{m}^{-1}$ and the angular frequency is $\omega = 2\pi f = 2\pi(600\,\text{Hz}) = 3800\,\text{rad/s}$. y_m is $2.0\,\text{mm} = 2.0 \times 10^{-3}\,\text{m}$. The displacement is given by $y(x,t) = (2.0 \times 10^{-3})\sin(9.4x)\cos(3800t)$, where x and y are in meters and t is in seconds.

63P

Consider an infinitesimal segment of a string oscillating in a standing wave pattern. Its length is dx and its mass is $dm = \mu\,dx$, where μ is its linear mass density. If it is moving with speed u its kinetic energy is $dK = \frac{1}{2}u^2\,dm = \frac{1}{2}\mu u^2\,dx$. If the segment is located at x its displacement at time t is $y = 2y_m\sin(kx)\cos(\omega t)$ and its velocity is $u = \partial y/\partial t = -2\omega y_m\sin(kx)\sin(\omega t)$, so its kinetic energy is

$$dK = \left(\frac{1}{2}\right)\left(4\mu\omega^2 y_m^2\right)\sin^2(kx)\sin^2(\omega t) = 2\mu\omega^2 y_m^2 \sin^2(kx)\sin^2(\omega t)\,.$$

Here y_m is the amplitude of either one of the traveling waves that combine to form the standing wave.

The infinitesimal segment has maximum kinetic energy when $\sin^2(\omega t) = 1$ and the maximum kinetic energy is given by

$$dK_m = 2\mu\omega^2 y_m^2 \sin^2(kx)\,.$$

Note that every portion of the string has its maximum kinetic energy at the same time although the values of these maxima are different for different parts of the string.

If the string is oscillating with n loops the kinetic energy in any one of them is given by the integral

$$K_m = 2\mu\omega^2 y_m^2 \int_0^{L/n} \sin^2(kx)\,dx\,.$$

Use the trigonometric identity $\sin^2(kx) = \frac{1}{2}[1 + 2\cos(2kx)]$ to write this

$$K_m = \mu\omega^2 y_m^2 \int_0^{L/n} [1 + 2\cos(2kx)]\,dx = \mu\omega^2 y_m^2 \left[\frac{L}{n} + \frac{1}{2k}\sin\frac{2kL}{n}\right]\,.$$

For a standing wave of n loops the wavelength is $\lambda = 2L/n$ and the angular wave number is $k = 2\pi/\lambda = n\pi/L$, so $2kL/n = 2\pi$ and $\sin(2kL/n) = 0$, no matter what the value of n. Thus

$$K_m = \frac{\mu\omega^2 y_m^2 L}{n}\,.$$

To obtain the expression given in the problem statement, first make the substitutions $\omega = 2\pi f$ and $L/n = \lambda/2$, where f is the frequency and λ is the wavelength. This produces $K_m = 2\pi^2 \mu y_m^2 f^2 \lambda$. Now substitute the wave speed v for $f\lambda$ to obtain $K_m = 2\pi^2 \mu y_m^2 fv$.

64P

(a) The frequency of the wave is the same for both sections of the wire. The wave speed and wavelength, however, are different. Suppose there are n_1 loops in the aluminum section of the wire. Then $\ell_1 = n_1\lambda_1/2 = n_1 v_1/2f$, where λ_1 is the wavelength and v_1 is the wave speed in that section. The substitution $\lambda_1 = v_1/f$, where f is the frequency, was made. Thus $f = n_1 v_1/2\ell_1$. A similar expression holds for the steel section: $f = n_2 v_2/2\ell_2$. Since the frequency is the same for the two sections, $n_1 v_1/\ell_1 = n_2 v_2/\ell_2$.

Now the wave speed in the aluminum section is given by $v_1 = \sqrt{\tau/\mu_1}$, where μ_1 is the linear mass density of the aluminum wire. The mass of aluminum in the wire is given by $m_1 = \rho_1 A\ell_1$, where ρ_1 is the mass density (mass per unit volume) for aluminum and A is the cross-sectional area of the wire. Thus $\mu_1 = \rho_1 A\ell_1/\ell_1 = \rho_1 A$ and $v_1 = \sqrt{\tau/\rho_1 A}$. A similar expression holds for the wave speed in the steel section: $v_2 = \sqrt{\tau/\rho_2 A}$. Note that the cross-sectional area and the tension are the same for the two sections.

The equality of the frequencies for the two sections now leads to $n_1/\ell_1\sqrt{\rho_1} = n_2/\ell_2\sqrt{\rho_2}$, where A has been canceled from both sides. The ratio of the integers is

$$\frac{n_2}{n_1} = \frac{\ell_2\sqrt{\rho_2}}{\ell_1\sqrt{\rho_1}} = \frac{(0.866\,\text{m})\sqrt{7.80 \times 10^3\,\text{kg/m}^3}}{(0.600\,\text{m})\sqrt{2.60 \times 10^3\,\text{kg/m}^3}} = 2.5\,.$$

The smallest integers that have this ratio are $n_1 = 2$ and $n_2 = 5$. The frequency is $f = n_1 v_1/2\ell_1 = (n_1/2\ell_1)\sqrt{\tau/\rho_1 A}$. The tension is provided by the hanging block and is $\tau = mg$, where m is the mass of the block. Thus

$$f = \frac{n_1}{2\ell_1}\sqrt{\frac{mg}{\rho_1 A}} = \frac{2}{2(0.600\,\text{m})}\sqrt{\frac{(10.0\,\text{kg})(9.8\,\text{m/s}^2)}{(2.60 \times 10^3\,\text{kg/m}^3)(1.00 \times 10^{-6}\,\text{m}^2)}} = 324\,\text{Hz}\,.$$

(b) The standing wave pattern has 2 loops in the aluminum section and 5 loops in the steel section, or 7 loops in all. There are 8 nodes, counting the end points.

Chapter 18

4E

(a) The time for the sound to travel from the kicker to a spectator is given by d/v, where d is the distance and v is the speed of sound. The time for light to travel the same distance is given by d/c, where c is the speed of light. The delay between seeing and hearing the kick is $\Delta t = (d/v) - (d/c)$. The speed of light is so much greater than the speed of sound that the delay can be approximated by $\Delta t = d/v$. This means $d = v\,\Delta t$. The distance from the kicker to the first spectator is $d_1 = v\,\Delta t_1 = (343\,\text{m/s})(0.23\,\text{s}) = 79\,\text{m}$. The distance from the kicker to the second spectator is $d_2 = v\,\Delta t_2 = (343\,\text{m/s})(0.12\,\text{s}) = 41\,\text{m}$.

(b) Lines from the kicker to each spectator and from one spectator to the other form a right triangle with the line joining the spectators as the hypotenuse, so the distance between the spectators is $D = \sqrt{d_1^2 + d_2^2} = \sqrt{(79\,\text{m})^2 + (41\,\text{m})^2} = 89\,\text{m}$.

10P

If d is the distance from the location of the earthquake to the seismograph and v_s is the speed of the S waves then the time for these waves to reach the seismograph is $t_s = d/v_s$. Similarly the time for P waves to reach the seismograph is $t_p = d/v_p$. The time delay is $\Delta t = (d/v_s) - (d/v_p) = d(v_p - v_s)/v_s v_p$, so

$$d = \frac{v_s v_p \,\Delta t}{(v_p - v_s)} = \frac{(4.5\,\text{km/s})(8.0\,\text{km/s})(3.0\,\text{min})(60\,\text{s/min})}{8.0\,\text{km/s} - 4.5\,\text{km/s}} = 1900\,\text{km}\,.$$

Notice that values for the speeds were substituted as given, in km/s, but that the value for the time delay was converted from minutes to seconds.

11P

Let t_f be the time for the stone to fall to the water and t_s be the time for the sound of the splash to travel from the water to the top of the well. Then the total time elapsed from dropping the stone to hearing the splash is $t = t_f + t_s$. If d is the depth of the well, then the kinematics of free fall gives $d = \frac{1}{2}gt_f^2$, or $t_f = \sqrt{2d/g}$. The sound travels at a constant speed v_s, so $d = v_s t_s$, or $t_s = d/v_s$. Thus the total time is $t = \sqrt{2d/g} + d/v_s$. This equation is to be solved for d. Rewrite it as $\sqrt{2d/g} = t - d/v_s$ and square both sides to obtain $2d/g = t^2 - 2(t/v_s)d + (1/v_s^2)d^2$. Now multiply by gv_s^2 and rearrange to get $gd^2 - 2v_s(gt + v_s)d + gv_s^2 t^2 = 0$. This is a quadratic equation for d. Its solutions are

$$d = \frac{2v_s(gt + v_s) \pm \sqrt{4v_s^2(gt + v_s)^2 - 4g^2 v_s^2 t^2}}{2g}\,.$$

The physical solution must yield $d = 0$ for $t = 0$, so we take the solution with the negative sign in front of the square root. Once values are substituted the result $d = 40.8\,\text{m}$ is obtained.

14E

(a) Use $\lambda = v/f$, where v is the speed of sound in air and f is the frequency. Thus $\lambda = (343\,\text{m/s})/(4.5 \times 10^6\,\text{Hz}) = 7.62 \times 10^{-5}\,\text{m}$.

(b) Now $\lambda = v/f$, where v is the speed of sound in tissue. The frequency is the same for air and tissue. Thus $\lambda = (1500\,\text{m/s})/(4.5 \times 10^6\,\text{Hz}) = 3.33 \times 10^{-4}\,\text{m}$.

28E

The intensity is the rate of energy flow per unit area perpendicular to the flow. The rate at which energy flows across every sphere centered at the source is the same, regardless of the sphere radius, and is the same as the power output of the source. If P is the power output and I is the intensity a distance r from the source, then $P = IA = 4\pi r^2 I$, where A ($= 4\pi r^2$) is the area of a sphere of radius r. Thus $P = 4\pi(2.50\,\text{m})^2(1.91 \times 10^{-4}\,\text{W/m}^2) = 1.50 \times 10^{-2}\,\text{W}$.

29E

The intensity is given by $I = \frac{1}{2}\rho v\omega^2 s_m^2$, where ρ is the density of air, v is the speed of sound in air, ω is the angular frequency, and s_m is the displacement amplitude for the sound wave. Replace ω with $2\pi f$ and solve for s_m:

$$s_m = \sqrt{\frac{I}{2\pi^2 \rho v f^2}} = \sqrt{\frac{1.00 \times 10^{-6}\,\text{W/m}^2}{2\pi^2(1.21\,\text{kg/m}^3)(343\,\text{m/s})(300\,\text{Hz})^2}} = 3.68 \times 10^{-8}\,\text{m}.$$

31E

(a) Let I_1 be the original intensity and I_2 be the final intensity. The original sound level is $\beta_1 = 10\log(I_1/I_0)$ and the final sound level is $\beta_2 = 10\log(I_2/I_0)$, both in decibels. Here I_0 is the reference intensity. Since $\beta_2 = \beta_1 + 30$, $10\log(I_2/I_0) = 10\log(I_1/I_0) + 30$, or $10\log(I_2/I_0) - 10\log(I_1/I_0) = 30$. Divide by $10\,\text{dB}$ and use $\log(I_2/I_0) - \log(I_1/I_0) = \log(I_2/I_1)$ to obtain $\log(I_2/I_1) = 3$. Now use each side as an exponent of 10 and recognize that $10^{\log(I_2/I_1)} = I_2/I_1$. The result is $I_2/I_1 = 10^3$. The intensity is increased by a factor of 1000.

(b) The pressure amplitude is proportional to the square root of the intensity so it is increased by a factor of $\sqrt{1000} = 32$.

35E

(a) The intensity is given by $I = \frac{1}{2}\rho v\omega^2 s_m^2$, where ρ is the density of the medium, v is the speed of sound, ω is the angular frequency, and s_m is the displacement amplitude. The displacement and pressure amplitudes are related by $\Delta p_m = \rho v\omega s_m$, so $s_m = \Delta p_m/\rho v\omega$ and $I = (\Delta p_m)^2/2\rho v$. For waves of the same frequency the ratio of the intensity for propagation in water to the intensity for propagation in air is

$$\frac{I_w}{I_a} = \left(\frac{\Delta p_{mw}}{\Delta p_{ma}}\right)^2 \frac{\rho_a v_a}{\rho_w v_w},$$

where the subscript a denotes air and the subscript w denotes water.

Since $I_a = I_w$,

$$\frac{\Delta p_{mw}}{\Delta p_{ma}} = \sqrt{\frac{\rho_w v_w}{\rho_a v_a}} = \sqrt{\frac{(0.998 \times 10^3 \,\text{kg/m}^3)(1482 \,\text{m/s})}{(1.21 \,\text{kg/m}^3)(343 \,\text{m/s})}} = 59.7.$$

The speeds of sound are given in Table 1 of this chapter and the densities are given in Table 1 of Chapter 16.

(b) Now $\Delta p_{mw} = \Delta p_{ma}$, so

$$\frac{I_w}{I_a} = \frac{\rho_a v_a}{\rho_w v_w} = \frac{(1.21 \,\text{kg/m}^3)(343 \,\text{m/s})}{(0.998 \times 10^3 \,\text{kg/m}^3)(1482 \,\text{m/s})} = 2.81 \times 10^{-4}.$$

36P

(a) Take the wave to be a plane wave and consider a region formed by the surface of a rectangular solid, with two plane faces of area A perpendicular to the direction of travel and separated by a distance ℓ, along the direction of travel. The energy contained in this region is $U = uA\ell$. If the wave speed is v then all the energy passes through one end of the region in time $t = \ell/v$. The energy passing through per unit time is $U/t = uA\ell v/\ell = uvA$. The intensity is the energy passing through per unit time, per unit area, or $I = U/tA = uv$.

(b) The power output P of the source equals the rate at which energy crosses the surface of any sphere centered at the source. It is related to the intensity I a distance r away by $P = AI = 4\pi r^2 I$, where $A \,(= 4\pi r^2)$ is the area of a sphere of radius r. Substitute $I = uv$ to obtain $P = 4\pi r^2 uv$, then solve for u:

$$u = \frac{P}{4\pi r^2 v} = \frac{50,000 \,\text{W}}{4\pi (480 \times 10^3 \,\text{m})^2 (3.00 \times 10^8 \,\text{m/s})} = 5.76 \times 10^{-17} \,\text{J/m}^3.$$

38P

(a) Let P be the power output of the source. This is the rate at which energy crosses the surface of any sphere centered at the source and is therefore equal to the product of the intensity I at the sphere surface and the area of the sphere. For a sphere of radius r, $P = 4\pi r^2 I$ and $I = P/4\pi r^2$. The intensity is proportional to the square of the displacement amplitude s_m. If we write $I = Cs_m^2$, where C is a constant of proportionality, then $Cs_m^2 = P/4\pi r^2$. Thus $s_m = \sqrt{P/4\pi r^2 C} = \left(\sqrt{P/4\pi C}\right)(1/r)$. The displacement amplitude is proportional to the reciprocal of the distance from the source.

Take the wave to be sinusoidal. It travels radially outward from the source, with points on a sphere of radius r in phase. If ω is the angular frequency and k is the angular wave number then the time dependence is $\sin(kr - \omega t)$. The displacement wave is given by

$$s(r,t) = \sqrt{\frac{P}{4\pi C}}\,\frac{1}{r}\sin(kr - \omega t) = \frac{Y}{r}\sin(kr - \omega t),$$

where $Y = \sqrt{P/4\pi C}$.

(b) Since s and r both have dimensions of length and the trigonometric function is dimensionless, the dimensions of Y must be $(\text{length})^2$.

45P

(a) When the right side of the instrument is pulled out a distance d the path length for sound waves increases by $2d$. Since the interference pattern changes from a minimum to the next maximum, this distance must be half a wavelength of the sound. So $2d = \lambda/2$, where λ is the wavelength. Thus $\lambda = 4d$ and, if v is the speed of sound, the frequency is $f = v/\lambda = v/4d = (343\,\text{m/s})/4(0.0165\,\text{m}) = 5.2 \times 10^3\,\text{Hz}$.

(b) The displacement amplitude is proportional to the square root of the intensity (see Eq. 18–26). Write $\sqrt{I} = Cs_m$, where I is the intensity, s_m is the displacement amplitude, and C is a constant of proportionality. At the minimum interference is destructive and the displacement amplitude is the difference in the amplitudes of the individual waves: $s_m = s_{SAD} - s_{SBD}$, where the subscripts indicate the paths of the waves. At the maximum the waves interfere constructively and the displacement amplitude is the sum of the amplitudes of the individual waves: $s_m = s_{SAD} + s_{SBD}$. Solve $\sqrt{100} = C(s_{SAD} - s_{SBD})$ and $\sqrt{900} = C(s_{SAD} + s_{SBD})$ for s_{SAD} and s_{SBD}. Add the equations to obtain $s_{SAD} = (\sqrt{100} + \sqrt{900})/2C = 20/C$, then subtract them to obtain $s_{SBD} = (\sqrt{900} - \sqrt{100})/2C = 10/C$. The ratio of the amplitudes is $s_{SAD}/s_{SBD} = 2$.

(c) Any energy losses, such as might be caused by frictional forces of the walls on the air in the tubes, result in a decrease in the displacement amplitude.

47P

Let P_s be the power output of the source. Then the intensity at the reflector is $I_r = P_s/4\pi r^2$, where r is the distance from the source to the reflector. The rate at which energy enters the reflector is $P_r = A_r I_r = \pi R^2 I_r$, where A is the area of the circular opening of the reflector and R is its radius. This means $P_r = P_s R^2/4r^2$. All of this energy is reflected into the tube, but only 12% of it reaches the ear of the listener. The rate at which energy reaches the ear is $P_e = 0.12P_r = 0.12P_s R^2/4r^2$. Since the cross-sectional area of the tube is $A_t = \pi r_t^2$, where r_t is its radius, the intensity at the ear is $I_e = P_e/A_e = 0.12P_s R^2/4\pi r^2 r_t^2$. The sound level in dB at the ear is $\beta_e = 10\log(I_e/I_0) = 10\log(0.12P_s R^2/4\pi r^2 r_t^2 I_0)$, where I_0 is the reference intensity.

We must now find the power output of the source. We know that the sound level at a distance $d = 1.0\,\text{m}$ is 20 dB. The intensity there is $P_s/4\pi d^2$, so $20 = 10\log(P_s/4\pi d^2 I_0)$. We could solve this equation for P_s but we shall use it as is to help find the maximum distance r at which a whisper an be heard.

Subtract $20 = 10\log(P_s/4\pi d^2 I_0)$ from $\beta_e = 10\log(0.12P_s R^2/4\pi r^2 r_t^2 I_0)$ and remember that the difference of two logarithms is the logarithm of their quotient. You should obtain $\beta_e - 20 = 10\log(0.12R^2 d^2/r^2 r_t^2)$. Set $\beta_e = 0$ and solve for r. First obtain $\log(0.12R^2 d^2/r^2 r_t^2) = -2$. Now use both sides as exponents of 10 to obtain $0.12R^2 d^2/r^2 r_t^2 = 10^{-2}$. Finally

$$r = \sqrt{\frac{0.12R^2 d^2}{10^{-2} r_t^2}} = \sqrt{\frac{0.12(0.50\,\text{m})^2(1.0\,\text{m})^2}{10^{-2}(5.0 \times 10^{-3}\,\text{m})^2}} = 350\,\text{m}\,.$$

52E

(a) The string is fixed at both ends and when vibrating at its lowest resonant frequency exactly half a wavelength fits between the ends. If L is the length of the string and λ is the wavelength, then $\lambda = 2L$. The frequency is $f = v/\lambda = v/2L$, where v is the speed of waves on the string. Thus $v = 2Lf = 2(0.220\,\text{m})(920\,\text{Hz}) = 405\,\text{m/s}$.

(b) The wave speed is given by $v = \sqrt{\tau/\mu}$, where τ is the tension in the string and μ is the linear mass density of the string. If M is the mass of the string, then $\mu = M/L$ since the string is uniform. Thus $\tau = \mu v^2 = (M/L)v^2 = \left[(800 \times 10^{-6}\,\text{kg})/(0.220\,\text{m})\right](405\,\text{m/s})^2 = 596\,\text{N}$.

(c) The wavelength is $\lambda = 2L = 2(0.220\,\text{m}) = 0.440\,\text{m}$.

(d) The frequency of the sound wave in air is the same as the frequency of oscillation of the string. The wavelength is different because the wave speed is different. If v_a is the speed of sound in air the wavelength in air is $\lambda_a = v_a/f = (343\,\text{m/s})/(920\,\text{Hz}) = 0.373\,\text{m}$.

58P

(a) Since the pipe is open at both ends there are displacement antinodes at both ends and an integer number of half-wavelengths fit into the length of the pipe. If L is the pipe length and λ is the wavelength then $\lambda = 2L/n$, where n is an integer. If v is the speed of sound then the resonant frequencies are given by $f = v/\lambda = nv/2L$. Now $L = (18.0\,\text{in})/(12\,\text{in./ft}) = 1.50\,\text{ft}$, so $f = n(1130\,\text{ft/s})/2(1.50\,\text{ft}) = 37n\,\text{Hz}$. To find the resonant frequencies that lie between 1000 Hz and 2000 Hz, first set $f = 1000\,\text{Hz}$ and solve for n, then set $f = 2000\,\text{Hz}$ and again solve for n. You should get 2.65 and 5.3. This means $n = 3, 4$, and 5 are the appropriate values of n. For $n = 3$, $f = 3(376\,\text{Hz}) = 1130\,\text{Hz}$; for $n = 4$, $f = 4(376\,\text{Hz}) = 1500\,\text{Hz}$; and for $n = 5$, $f = 5(376\,\text{Hz}) = 1880\,\text{Hz}$.

(b) For any integer value of n the displacement has n nodes and $n + 1$ antinodes, counting the ends. The nodes (N) and antinodes (A) are marked on the diagrams below for the 3 resonances found in part a.

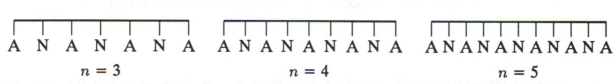

$$\begin{array}{ccc} \text{A N A N A N A} & \text{A N A N A N A N A} & \text{A N A N A N A N A} \\ n = 3 & n = 4 & n = 5 \end{array}$$

59P

The top of the water is a displacement node and the top of the well is a displacement antinode. At the lowest resonant frequency exactly one-fourth of a wavelength fits into the depth of the well. If d is the depth and λ is the wavelength then $\lambda = 4d$. The frequency is $f = v/\lambda = v/4d$, where v is the speed of sound. The speed of sound is given by $v = \sqrt{B/\rho}$, where B is the bulk modulus and ρ is the density of air in the well. Thus $f = (1/4d)\sqrt{B/\rho}$ and

$$d = \frac{1}{4f}\sqrt{\frac{B}{\rho}} = \left[\frac{1}{4(7.00\,\text{Hz})}\right]\sqrt{\frac{1.33 \times 10^5\,\text{Pa}}{1.10\,\text{kg/m}^3}} = 12.4\,\text{m}.$$

62P

(a) We expect the center of the star to be a displacement node. The star has spherical symmetry and the waves are spherical. If matter at the center moved it would move equally in all directions and this is not possible.

(b) Assume the oscillation is at the lowest resonance frequency. Then exactly one-fourth of a wavelength fits the star radius. If λ is the wavelength and R is the star radius then $\lambda = 4R$. The frequency is $f = v/\lambda = v/4R$, where v is the speed of sound in the star. The period is $T = 1/f = 4R/v$.

(c) The speed of sound is $v = \sqrt{B/\rho}$, where B is the bulk modulus and ρ is the density of stellar material. The radius is $R = 9.0 \times 10^{-3}R_s$, where R_s is the radius of the sun (6.96×10^8 m). Thus

$$T = 4R\sqrt{\frac{\rho}{B}} = 4(9.0 \times 10^{-3})(6.96 \times 10^8 \,\text{m})\sqrt{\frac{10^{10} \,\text{kg/m}^3}{1.33 \times 10^{22} \,\text{Pa}}} = 22\,\text{s}\,.$$

63P

The string is fixed at both ends so the resonant wavelengths are given by $\lambda = 2L/n$, where L is the length of the string and n is an integer. The resonant frequencies are given by $f = v/\lambda = nv/2L$, where v is the wave speed on the string. Now $v = \sqrt{\tau/\mu}$, where τ is the tension in the string and μ is the linear mass density of the string. Thus $f = (n/2L)\sqrt{\tau/\mu}$. Suppose the lower frequency is associated with $n = n_1$ and the higher frequency is associated with $n = n_1 + 1$. There are no resonant frequencies between so you know that the integers associated with the given frequencies differ by 1. Thus $f_1 = (n_1/2L)\sqrt{\tau/\mu}$ and

$$f_2 = \frac{n_1 + 1}{2L}\sqrt{\frac{\tau}{\mu}} = \frac{n_1}{2L}\sqrt{\frac{\tau}{\mu}} + \frac{1}{2L}\sqrt{\frac{\tau}{\mu}} = f_1 + \frac{1}{2L}\sqrt{\frac{\tau}{\mu}}\,.$$

This means $f_2 - f_1 = (1/2L)\sqrt{\tau/\mu}$ and

$$\begin{aligned}
\tau &= 4L^2\mu(f_2 - f_1)^2 \\
&= 4(0.300\,\text{m})^2(0.650 \times 10^{-3}\,\text{kg/m})(1320\,\text{Hz} - 880\,\text{Hz})^2 \\
&= 45.3\,\text{N}\,.
\end{aligned}$$

64E

Since the beat frequency equals the difference between the frequencies of the two tuning forks, the frequency of the first fork is either 381 Hz or 387 Hz. When mass is added to this fork its frequency decreases (recall, for example, that the frequency of an mass-spring oscillator is proportional to $1/\sqrt{m}$). Since the beat frequency also decreases the frequency of the first fork must be greater than the frequency of the second. It must be 387 Hz.

67P

Each wire is vibrating in its fundamental mode so the wavelength is twice the length of the wire ($\lambda = 2L$) and the frequency is $f = v/\lambda = (1/2L)\sqrt{\tau/\mu}$, where v ($= \sqrt{\tau/\mu}$) is the wave speed for the wire, τ is the tension in the wire, and μ is the linear mass density of the wire.

Suppose the tension in one wire is τ and the oscillation frequency of that wire is f_1. The tension in the other wire is $\tau + \Delta\tau$ and its frequency is f_2. You want to calculate $\Delta\tau/\tau$ for $f_1 = 600\,\text{Hz}$ and $f_2 = 606\,\text{Hz}$.

Now $f_1 = (1/2L)\sqrt{\tau/\mu}$ and $f_2 = (1/2L)\sqrt{(\tau + \Delta\tau)/\mu}$, so $f_2/f_1 = \sqrt{(\tau + \Delta\tau)/\tau} = \sqrt{1 + (\Delta\tau/\tau)}$. This means $\Delta\tau/\tau = (f_2/f_1)^2 - 1 = [(606\,\text{Hz})/(600\,\text{Hz})]^2 - 1 = 0.020$.

72E

The general expression for the Doppler shifted frequency is

$$f' = f\,\frac{v \pm v_D}{v \mp v_S},$$

where f is the unshifted frequency, v is the speed of sound, v_D is the speed of the detector, and v_S is the speed of the source. All speeds are measured relative to the medium of propagation, the air in this case. The detector (the second plane) is moving toward the source (the first plane). This tends to increase the frequency, so we use the plus sign in the numerator. The source is moving away from the detector. This tends to decrease the frequency, so we use the plus sign in the denominator. Thus

$$f' = f\,\frac{v + v_D}{v + v_S} = (16,000\,\text{Hz})\left(\frac{343\,\text{m/s} + 250\,\text{m/s}}{343\,\text{m/s} + 200\,\text{m/s}}\right) = 17,500\,\text{Hz}.$$

77E

The half angle θ of the Mach cone is given by $\sin\theta = v/v_S$, where v is the speed of light in water and v_S is the speed of the electrons (see Eq. 18–53). Thus

$$v_S = \frac{v}{\sin\theta} = \frac{(3/4)c}{\sin\theta} = \frac{(3/4)(3.0 \times 10^8\,\text{m/s})}{\sin 60^\circ} = 2.6 \times 10^8\,\text{m/s}.$$

Here c is the speed of light in vacuum.

81P

(a) The half angle θ of the Mach cone is given by $\sin\theta = v/v_S$, where v is the speed of sound and v_S is the speed of the plane. Since $v_S = 1.5v$, $\sin\theta = v/1.5v = 1/1.5$. This means $\theta = 42^\circ$.

(b) Let h be the altitude of the plane and suppose the Mach cone intersects the Earth a distance d behind the plane. The situation is shown on the diagram to the right, with P indicating the plane and O indicating the observer. The cone angle is related to h and d by $\tan\theta = h/d$, so $d = h/\tan\theta$. The shock wave reaches O in the time the plane takes to fly the distance d: $t = d/v = h/v\tan\theta = (5000\,\text{m})/1.5(331\,\text{m/s})\tan 42^\circ = 11\,\text{s}$.

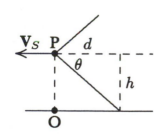

94P

(a) The expression for the Doppler shifted frequency is

$$f' = f\frac{v \pm v_D}{v \mp v_S},$$

where f is the unshifted frequency, v is the speed of sound, v_D is the speed of the detector (the uncle), and v_S is the speed of the source (the locomotive). All speeds are relative to the air. The uncle is at rest with respect to the air, so $v_D = 0$. The speed of the source is $v_S = 10\,\text{m/s}$. Since the locomotive is moving away from the uncle the frequency decreases and we use the plus sign in the denominator. Thus

$$f' = f\frac{v}{v + v_S} = (500.0\,\text{Hz})\left(\frac{343\,\text{m/s}}{343\,\text{m/s} + 10.00\,\text{m/s}}\right) = 485.8\,\text{Hz}.$$

(b) The girl is now the detector. Relative to the air she is moving with speed $v_D = 10.00\,\text{m/s}$ toward the source. This tends to increase the frequency and we use the plus sign in the numerator. The source is moving at $v_S = 10.00\,\text{m/s}$ away from the girl. This tends to decrease the frequency and we use the plus sign in the denominator. Thus $(v + v_D) = (v + v_S)$ and $f' = f = 500.0\,\text{Hz}$.

(c) Relative to the air the locomotive is moving at $v_S = 20.00\,\text{m/s}$ away from the uncle. Use the plus sign in the denominator. Relative to the air the uncle is moving at $v_D = 10.00\,\text{m/s}$ toward the locomotive. Use the plus sign in the numerator. Thus

$$f' = f\frac{v + v_D}{v + v_S} = (500.0\,\text{Hz})\left(\frac{343\,\text{m/s} + 10.00\,\text{m/s}}{343\,\text{m/s} + 20.00\,\text{m/s}}\right) = 486.2\,\text{Hz}.$$

(d) Relative to the air the locomotive is moving at $v_S = 20.00\,\text{m/s}$ away from the girl and the girl is moving at $v_D = 20.00\,\text{m/s}$ toward the locomotive. Use the plus signs in both the numerator and the denominator. Thus $(v + v_D) = (v + v_S)$ and $f' = f = 500.0\,\text{Hz}$.

96E

The Doppler shifted wavelength is given by $\lambda' = \lambda(1 \mp u/c)$, where λ is the unshifted wavelength and u is the radial speed of the galaxy. Since $\lambda' > \lambda$ use the plus sign. Substitute $\lambda' = 1.0040\lambda$ to obtain $1 + u/c = 1.0040$, or $u = 0.0040c = (0.0040)(3.0 \times 10^8\,\text{m/s}) = 1.2 \times 10^6\,\text{m/s}$. The wavelength has increased, which means the frequency has decreased and the galaxy is receding.

Chapter 19

2E

Since the output voltage V is linear in the temperature T we may write $V = A + BT$, where A and B are constants. The two points that are given allow us to solve for their values. Since $V = 0.000$ when $T = 0.000°$ C, $A = 0$. Since $V = 28.0$ mV when $T = 510°$ C, $B = (28.0\,\text{mV})/(510°\,\text{C}) = 0.0549\,\text{mV}/°\text{C}$.

The temperature T when $V = 10.2$ mV is

$$T = \frac{V}{B} = \frac{10.2\,\text{mV}}{0.0549\,\text{mV}/°\text{C}} = 186°\,\text{C}.$$

6P

Take p_3 to be 80 mm of mercury for both thermometers. According to Fig 19–6 the nitrogen thermometer gives 373.35 K for the boiling point of water. Use Eq. 19–5 to compute the pressure:

$$p_N = \frac{T}{273.16\,\text{K}} p_3 = \left(\frac{373.36\,\text{K}}{273.16\,\text{K}} \right) (80\,\text{mm of mercury}) = 109.343\,\text{mm of mercury}.$$

The hydrogen thermometer gives 373.16 K for the boiling point of water and

$$p_H = \left(\frac{373.16\,\text{K}}{273.16\,\text{K}} \right) (80\,\text{mm of mercury}) = 109.287\,\text{mm of mercury}.$$

The pressure in the nitrogen thermometer is higher than the pressure in the hydrogen thermometer by 0.056 mm of mercury.

7P

Let T_L be the temperature and p_L be the pressure in the left-hand thermometer. Let T_R be the temperature and p_R be the pressure in the right-hand thermometer. According to the problem statement the pressure is the same in the two thermometers when they are both at the triple point of water. Take this pressure to be p_3. Write Eq. 19–5 for each thermometer: $T_L = (273.16\,\text{K})(p_L/p_3)$ and $T_R = (273.16\,\text{K})(p_R/p_3)$. Subtract the second equation from the first to obtain

$$T_L - T_R = (273.16\,\text{K}) \frac{p_L - p_R}{p_3}.$$

First take $T_L = 373.125$ K (the boiling point of water) and $T_R = 273.16$ K (the triple point of water). Then $p_L - p_R = 120$ mm of mercury. Solve

$$373.125\,\text{K} - 273.16\,\text{K} = (273.16\,\text{K}) \left(\frac{120\,\text{mm of mercury}}{p_3} \right)$$

for p_3. The result is $p_3 = 328$ mm of mercury. Now let $T_L = 273.16$ K (the triple point of water) and let T_R be the unknown temperature. The pressure difference is $p_L - p_R = 90.0$ mm of mercury. Solve

$$273.16\,\text{K} - T_R = (273.16\,\text{K})\left(\frac{90.0\,\text{mm of mercury}}{328\,\text{mm of mercury}}\right)$$

for T_R. The result is $T_R = 348$ K.

14E

(a) Fahrenheit and Celsius temperatures are related by $T_F = (9/5)T_C + 32°$. T_F is numerically equal to T_C if $T_F = (9/5)T_F + 32°$. The solution to this equation is $T_F = -(5/4)(32°) = -40°$ F.

(b) Fahrenheit and Kelvin temperatures are related by $T_F = (9/5)T_C + 32° = (9/5)(T - 273.15) + 32°$. The Fahrenheit temperature T_F is numerically equal to the Kelvin temperature T if $T_F = (9/5)(T_F - 273.15) + 32°$. The solution to this equation is

$$T_F = \frac{5}{4}\left(\frac{9}{5} \times 273.15 - 32°\right) = 575°\,\text{F}\,.$$

(c) Since $T_C = T - 273.15$ the Kelvin and Celsius temperatures can never have the same numerical value.

17P

(a) Changes in temperature take place by means of radiation, conduction, and convection. A can be reduced by placing the object in isolation, by surrounding it with a vacuum jacket, for example. This reduces conduction and convection. Absorption of radiation can be reduced by polishing the surface to a mirror finish. Clearly A depends on the condition of the surface and on the ability of the environment to conduct or convect energy to or from the object. A has the dimensions of reciprocal time.

(b) Rearrange the equation to obtain

$$\frac{1}{\Delta T}\frac{\mathrm{d}\Delta T}{\mathrm{d}t} = -A\,.$$

Now integrate with respect to time and recognize that

$$\int \frac{1}{\Delta T}\frac{\mathrm{d}\Delta T}{\mathrm{d}t}\,\mathrm{d}t = \int \frac{1}{\Delta T}\,\mathrm{d}(\Delta T)\,.$$

Thus

$$\int_{\Delta T_0}^{\Delta T} \frac{1}{\Delta T}\,\mathrm{d}(\Delta T) = -\int_0^t A\,\mathrm{d}t\,.$$

The integral on the right side yields $-At$ and the integral on the left yields $\ln \Delta T\big|_{\Delta T_0}^{\Delta T} = \ln(\Delta T) - \ln(\Delta T_0) = \ln(\Delta T/\Delta T_0)$, so

$$\ln \frac{\Delta T}{\Delta T_0} = -At\,.$$

Use each side as the exponent of e, the base of the natural logarithms, to obtain

$$\frac{\Delta T}{\Delta T_0} = e^{-At}$$

or

$$\Delta T = \Delta T_0\, e^{-At}.$$

22E

When the temperature changes from T to $T + \Delta T$ the diameter of the mirror changes from D to $D + \Delta D$, where $\Delta D = \alpha D\, \Delta T$. Here α is the coefficient of linear expansion for Pyrex glass $(3.2 \times 10^{-6}/°C$, according to Table 3). The range of values for the diameters can be found by setting ΔT equal to the temperature range. Thus $\Delta D = (3.2 \times 10^{-6}\,/°C)(200\,\text{in.})(60°\,C) = 3.8 \times 10^{-2}$ in.

29E

If the window is L_1 high and L_2 wide at the lower temperature and $L_1 + \Delta L_1$ high and $L_2 + \Delta L_2$ wide at the higher temperature then its area changes from $A_1 = L_1 L_2$ to $A_2 = (L_1 + \Delta L_1)(L_2 + \Delta L_2) \approx L_1 L_2 + L_1 \Delta L_2 + L_2 \Delta L_1$, where the term $\Delta L_1 \Delta L_2$ has been omitted because it is much smaller than the other terms if the changes in the lengths are small. Thus the change is area is $\Delta A = A_2 - A_1 = L_1 \Delta L_2 + L_2 \Delta L_1$. If ΔT is the change in temperature then $\Delta L_1 = \alpha L_1 \Delta T$ and $\Delta L_2 = \alpha L_2 \Delta T$, where α is the coefficient of linear expansion. Thus

$$\Delta A = \alpha(L_1 L_2 + L_1 L_2)\Delta T = 2\alpha L_1 L_2\, \Delta T$$
$$= 2(9 \times 10^{-6}\,/°C)(30\,\text{cm})(20\,\text{cm})(30°\,C) = 0.32\,\text{cm}^2.$$

31E

Since a volume is the product of three lengths, the change in volume due to a temperature change ΔT is given by $\Delta V = 3\alpha V\, \Delta T$, where V is the original volume and α is the coefficient of linear expansion. The proof is similar to the proof given in the solution to Exercise 29. Since $V = (4\pi/3)R^3$, where R is the original radius of the sphere,

$$\Delta V = 3\alpha\left(\frac{4\pi}{3}R^3\right)\Delta T = (23 \times 10^{-6}\,/°C)(4\pi)(10\,\text{cm})^3(100°\,C) = 29\,\text{cm}^3.$$

34E

If V_c is the original volume of the cup, α_a is the coefficient of linear expansion of aluminum, and ΔT is the temperature increase, then the change in the volume of the cup is $\Delta V_c = 3\alpha_a V_c\, \Delta T$. If β is the coefficient of volume expansion for glycerin then the change in the volume of glycerin is $\Delta V_g = \beta V_c\, \Delta T$. Note that the original volume of glycerin is the same as the original volume of the cup. The volume of glycerin that spills is

$$\Delta V_g - \Delta V_c = (\beta - 3\alpha_a)V_c\, \Delta T$$
$$= [(5.1 \times 10^{-4}\,/°C) - 3(23 \times 10^{-6}\,/°C)]\,(100\,\text{cm}^3)(6°C) = 0.26\,\text{cm}^3.$$

36P

After the change in temperature the diameter of the steel rod is $D_s = D_{s0} + \alpha_s D_{s0} \Delta T$ and the diameter of the brass ring is $D_b = D_{b0} + \alpha_b D_{b0} \Delta T$, where D_{s0} and D_{b0} are the original diameters, α_s and α_b are the coefficients of linear expansion, and ΔT is the change in temperature. The rod just fits through the ring if $D_s = D_b$. This means $D_{s0} + \alpha_s D_{s0} \Delta T = D_{b0} + \alpha_b D_{b0} \Delta T$. Solve for ΔT:

$$\Delta T = \frac{D_{s0} - D_{b0}}{\alpha_b D_{b0} - \alpha_s D_{s0}}$$

$$= \frac{3.000\,\text{cm} - 2.992\,\text{cm}}{(19 \times 10^{-6}\,/°\text{C})(2.992\,\text{cm}) - (11 \times 10^{-6}\,/°\text{C})(3.000\,\text{cm})} = 335°\,\text{C}.$$

The temperature is $T = 25°\,\text{C} + 335°\,\text{C} = 360°\,\text{C}$.

41P

The change in volume of the liquid is given by $\Delta V = \beta V \Delta T$. If A is the cross-sectional area of the tube and h is the height of the liquid, then $V = Ah$ is the original volume and $\Delta V = A\Delta h$ is the change in volume. Since the tube does not change the cross-sectional area of the liquid remains the same. Therefore $A\Delta h = \beta Ah \Delta T$ or $\Delta h = \beta h \Delta T$.

43P

The period of the pendulum is given by $P = 2\pi\sqrt{\ell/g}$, where ℓ is its length. When the temperature is increased the pendulum lengthens, the period increases, and the clock runs slow. If the length changes by $\Delta \ell$, the period changes by $\Delta P = (dP/d\ell)\Delta \ell = \pi(1/\sqrt{\ell g})\Delta \ell = \frac{1}{2}(P/\ell)\Delta \ell$. The increase in length is given by $\Delta \ell = \alpha \ell \Delta T$, where α is the coefficient of linear expansion for invar and ΔT is the temperature increase. Thus $\Delta P = \frac{1}{2}\alpha P \Delta T$. Multiply the number of periods in 30 days to find the time keeping error. Let $t = 30\,\text{d}$ and Δt represent the error. Then

$$\Delta t = \frac{1}{2}\alpha t \Delta T = \frac{1}{2}(0.7 \times 10^{-6}\,/°\text{C})(30\,\text{d})(10°\,\text{C}) = 1.05 \times 10^{-4}\,\text{d} = 9.1\,\text{s}.$$

47P

Consider half the bar. Its original length is $\ell_0 = L_0/2$ and its length after the temperature increase is $\ell = \ell_0 + \alpha \ell_0 \Delta T$. The old position of the half-bar, its new position, and the distance x that one end is displaced form a right triangle, with a hypotenuse of length ℓ, one side of length ℓ_0, and the other side of length x. The Pythagorean theorem yields $x^2 = \ell^2 - \ell_0^2 = \ell_0^2(1 + \alpha \Delta T)^2 - \ell_0^2$. Since the change in length is small we may approximate $(1 + \alpha \Delta T)^2$ by $1 + 2\alpha \Delta T$, where the small term $(\alpha \Delta T)^2$ was neglected. Then

$$x^2 = \ell_0^2 + 2\ell_0^2 \alpha \Delta T - \ell_0^2 = 2\ell_0^2 \alpha \Delta T$$

and

$$x = \ell_0\sqrt{2\alpha \Delta T} = \frac{3.77\,\text{m}}{2}\sqrt{2(25 \times 10^{-6}\,/°\text{C})(32°\,\text{C})} = 7.5 \times 10^{-2}\,\text{m}.$$

49P

The height of the liquid in the tube is its volume divided by the cross-sectional area of the tube. When the temperature increases the cross-sectional area and the volume occupied by the liquid both increase. Let V be the original volume of the liquid. Its volume after the temperature increase is $V + \beta V \, \Delta T$, where β is the coefficient of volume expansion for the liquid. Let A be the original cross-sectional area of the tube. Its cross-sectional area after the temperature increase is $A + 2\alpha A \, \Delta T$, where α is the coefficient of linear expansion for the glass. The height of the liquid in the tube after the temperature increase is

$$h = \frac{V(1 + \beta \, \Delta T)}{A(1 + 2\alpha \, \Delta T)}.$$

Now V/A is the original height of the liquid. Denote this by h_0 and write

$$h = \frac{1 + \beta \, \Delta T}{1 + 2\alpha \, \Delta T} h_0.$$

The change in the height is

$$h - h_0 = \left(\frac{1 + \beta \, \Delta T}{1 + 2\alpha \, \Delta T} - 1 \right) h_0 = \frac{(\beta - 2\alpha) \, \Delta T}{1 + 2\alpha \, \Delta T} h_0$$

$$= \left[\frac{[(4.0 \times 10^{-5} \, /°C) - 2(1.0 \times 10^{-5} \, /°C)] \, (10°C)}{1 + 2(1.0 \times 10^{-5} \, /°C)(10° \, C)} \right] (0.64 \, \text{m}) = 1.3 \times 10^{-4} \, \text{m}.$$

52P

Let L be the original length of each rod. After the temperature is increased by ΔT the length of the invar rod is $L_I = L(1 + \alpha_I \, \Delta T)$, the length of the aluminum rod is $L_A = L(1 + \alpha_A \, \Delta T)$, and the length of the steel rod is $L_S = L(1 + \alpha_S \, \Delta T)$. According to the trigonometric law of cosines $L_I^2 = L_A^2 + L_S^2 - 2L_A L_S \cos\theta$, where θ is the angle opposite the invar side (see Appendix G). Substitute for L_I, L_A, and L_S. In each case approximate $(1 + \alpha \, \Delta T)^2$ by $1 + 2\alpha \, \Delta T$, where the term $(\alpha \, \Delta T)^2$ is neglected because the changes in the lengths are small. Also neglect the term that is proportional to $(\Delta T)^2$ when you expand $(1 + \alpha_A \, \Delta T)(1 + \alpha_S \, \Delta T)$. You should get

$$L^2 (1 + 2\alpha_I \Delta T) = L^2 (1 + 2\alpha_A \Delta T) + L^2 (1 + 2\alpha_S \Delta T) - 2L^2 (1 + \alpha_A \Delta T + \alpha_S \Delta T) \cos\theta.$$

Solve for ΔT. The result is

$$\Delta T = \frac{1}{2} \left[\frac{1 - 2\cos\theta}{\alpha_I - (\alpha_A + \alpha_S)(1 - \cos\theta)} \right]$$

$$= \frac{1}{2} \left[\frac{1 - 2\cos 59.95°}{0.7 \times 10^{-6} \, /°C - (23 \times 10^{-6} \, /°C + 11 \times 10^{-6} \, /°C)(1 - \cos 59.95°)} \right]$$

$$= 46° \, C.$$

The temperature is $20.0° \, C + 46° \, C = 66° \, C$.

The force of gravity on the cube is $\rho_A g V$, where V is the volume of the cube and ρ_A is the mass density of aluminum. The buoyant force of the mercury on the cube is $\rho_M g A h$, where ρ_M is the mass density of mercury, A is the area of a cube face, and h is the extent to which the cube is submerged. Note that Ah is the volume of displaced mercury. The cube is in equilibrium, so the magnitudes of the two forces are the same: $\rho_A g V = \rho_M g A h$ and $h = \rho_A L / \rho_M$. $V = L^3$ and $A = L^2$ were used to obtain the result.

When the temperature changes all three quantities in the expression for h change. The change in h is

$$\Delta h = \frac{\partial h}{\partial \rho_A} \Delta \rho_A + \frac{\partial h}{\partial \rho_M} \Delta \rho_M + \frac{\partial h}{\partial L} \Delta L = \frac{L}{\rho_M} \Delta \rho_A - \frac{\rho_A L}{\rho_M^2} \Delta \rho_M + \frac{\rho_A}{\rho_M} \Delta L.$$

First consider the change in the density of aluminum. Suppose mass M of aluminum occupies volume V_A. The density is $\rho_A = M/V_A$ and the change in the density is

$$\Delta \rho_A = \frac{\mathrm{d}\rho_A}{\mathrm{d}V_A} \Delta V_A = -\frac{M}{V_A^2} \Delta V_A = -\frac{\rho_A}{V_A} \Delta V_A.$$

Now $\Delta V_A = 3\alpha V_A \Delta T$, so

$$\Delta \rho_A = -3\alpha \rho_A \Delta T.$$

Here α is coefficient of linear expansion for aluminum. Similarly for mercury

$$\Delta \rho_M = -\frac{\rho_M}{V_M} \Delta V_M.$$

Now $\Delta V_M = \beta V_M \Delta T$, where β is the coefficient of volume expansion for mercury. Thus

$$\Delta \rho_M = -\beta \rho_M \Delta T.$$

Finally, $\Delta L = \alpha L \Delta T$.

The expression for Δh becomes

$$\Delta h = \frac{L}{\rho_M}(-3\alpha \rho_A \Delta T) - \frac{\rho_A L}{\rho_M^2}(-\beta \rho_M \Delta T) + \frac{\rho_A}{\rho_M}(\alpha L \Delta T) = \frac{\rho_A}{\rho_M} L(\beta - 2\alpha) \Delta T$$

$$= \left(\frac{2.7\,\mathrm{g/cm^3}}{13.6\,\mathrm{g/cm^3}} \right) (20.0\,\mathrm{cm}) \left[1.80 \times 10^{-4}\,/^\circ\mathrm{C} - 2(23 \times 10^{-6}\,/^\circ\mathrm{C}) \right] (50^\circ\,\mathrm{C})$$

$$= 2.66 \times 10^{-2}\,\mathrm{cm}.$$

Chapter 20

2E

(a) The specific heat is given by $c = Q/m(T_f - T_i)$, where Q is the heat added, m is the mass of the sample, T_i is the initial temperature, and T_f is the final temperature. Thus

$$c = \frac{314\,\text{J}}{(30.0 \times 10^{-3}\,\text{kg})(45.0°\,\text{C} - 25.0°\,\text{C})} = 523\,\text{J/kg} \cdot \text{K}.$$

(b) If N is the number of moles of the substance and M is the mass per mole, then $m = NM$, so

$$N = \frac{m}{M} = \frac{30.0 \times 10^{-3}\,\text{kg}}{50 \times 10^{-3}\,\text{kg/mol}} = 0.600\,\text{mol}.$$

(c) The molar specific heat is given by

$$c_m = \frac{Q}{N(T_f - T_i)} = \frac{314\,\text{J}}{(0.600\,\text{mol})(45.0°\,\text{C} - 25.0°\,\text{C})} = 26.2\,\text{J/mol} \cdot \text{K}.$$

7E

The melting point of silver is 1235 K, so the temperature of the silver must first be raised from $15.0 + 273 = 288\,\text{K}$ to 1235 K. This requires heat

$$Q = cm(T_f - T_i) = (236\,\text{J/kg} \cdot \text{K})(0.130\,\text{kg})(1235°\,\text{C} - 288°\,\text{C}) = 2.91 \times 10^4\,\text{J}.$$

Now the silver at its melting point must be melted. If L_F is the heat of fusion for silver this requires

$$Q = mL_F = (0.130\,\text{kg})(105 \times 10^3\,\text{J/kg}) = 1.36 \times 10^4\,\text{J}.$$

The total heat required is $2.91 \times 10^4\,\text{J} + 1.36 \times 10^4\,\text{J} = 4.27 \times 10^4\,\text{J}$.

12E

(a) The heat generated is the power output of the drill multiplied by the time: $Q = Pt$. Use $1\,\text{hp} = 2545\,\text{Btu/h}$ to convert the given value of the power to Btu and $1\,\text{min} = (1/60)\,\text{h}$ to convert the given value of the time to hours. Then

$$Q = \frac{(0.400\,\text{hp})(2545\,\text{Btu/h})(2.00\,\text{min})}{60\,\text{min/h}} = 33.9\,\text{Btu}.$$

(b) Use $0.750Q = cm\,\Delta T$ to compute the rise in temperature. Here c is the specific heat of copper and m is the mass of the copper block. Table 1 gives $c = 386\,\text{J/kg} \cdot \text{K}$. Use $1\,\text{J} = 9.481 \times 10^{-4}\,\text{Btu}$ and $1\,\text{kg} = 6.852 \times 10^{-2}\,\text{slug}$ to show that

$$c = \frac{(386\,\text{J/kg} \cdot \text{K})(9.481 \times 10^{-4}\,\text{Btu/J})}{6.852 \times 10^{-2}\,\text{slug/kg}} = 5.341\,\text{Btu/slug} \cdot \text{K}.$$

The mass of the block is its weight W divided by g: $m = W/g = (1.60\,\text{lb})/(32\,\text{ft/s}^2) = 0.0500$ slug. Thus

$$\Delta T = \frac{0.750Q}{cm} = \frac{(0.750)(33.9\,\text{Btu})}{(5.341\,\text{Btu/slug}\cdot\text{K})(0.0500\,\text{slug})} = 95.3\,\text{K}.$$

This is equivalent to $(9/5)(95.3\,\text{K}) = 172°\,\text{F}$.

15E

Mass $m\ (= 0.100\,\text{kg})$ of water, with specific heat $c\ (= 4190\,\text{J/kg}\cdot\text{K})$, is raised from an initial temperature $T_i\ (= 23°\,\text{C})$ to its boiling point $T_f\ (= 100°\,\text{C})$. The heat input is given by $Q = cm(T_f - T_i)$. This must be the power output of the heater P multiplied by the time t; $Q = Pt$. Thus

$$t = \frac{Q}{P} = \frac{cm(T_f - T_i)}{P} = \frac{(4190\,\text{J/kg}\cdot\text{K})(0.100\,\text{kg})(100°\,\text{C} - 23°\,\text{C})}{200\,\text{W}} = 160\,\text{s}.$$

23P

Mass m of water must be raised from an initial temperature $T_i\ (= 59°\,\text{F} = 15°\,\text{C})$ to a final temperature $T_f\ (= 100°\,\text{C})$. If c is the specific heat of water then the energy required is $Q = cm(T_f - T_i)$. Each shake supplies energy mgh, where h is the distance moved during the downward stroke of the shake. If N is the total number of shakes then $Nmgh = Q$. If t is the time taken to raise the water to its boiling point then $(N/t)mgh = Q/t$. Notice that N/t is the rate R of shaking (30 shakes/min). Thus $Rmgh = Q/t$. The distance h is $1.0\,\text{ft} = 0.3048\,\text{m}$. Hence

$$t = \frac{Q}{Rmgh} = \frac{cm(T_f - T_i)}{Rmgh} = \frac{c(T_f - T_i)}{Rmgh}$$
$$= \frac{(4190\,\text{J/kg}\cdot\text{K})(100°\,\text{C} - 15°\,\text{C})}{(30\,\text{shakes/min})(9.8\,\text{m/s}^2)(0.3048\,\text{m})} = 3.97 \times 10^3\,\text{min}.$$

This is 2.8 d.

30P

(*a*) There are three possibilities:

1. None of the ice melts and the water-ice system reaches thermal equilibrium at a temperature that is at or below the melting point of ice.

2. The system reaches thermal equilibrium at the melting point of ice, with some of the ice melted.

3. All of the ice melts and the system reaches thermal equilibrium at a temperature at or above the melting point of ice.

First suppose that no ice melts. The temperature of the water decreases from T_{Wi} (= 25° C) to some final temperature T_f and the temperature of the ice increases from T_{Ii} (= −15° C) to T_f. If m_W is the mass of the water and c_W is its specific heat then the water rejects heat

$$Q = c_W m_W (T_{Wi} - T_f).$$

If m_I is the mass of the ice and c_I is its specific heat then the ice absorbs heat

$$Q = c_I m_I (T_f - T_{Ii}).$$

Since no energy is lost these two heats must be the same and

$$c_W m_W (T_{Wi} - T_f) = c_I m_I (T_f - T_{Ii}).$$

The solution for the final temperature is

$$
\begin{aligned}
T_f &= \frac{c_W m_W T_{Wi} + c_I m_I T_{Ii}}{c_W m_W + C_I m_I} \\
&= \frac{(4190\,\text{J/kg} \cdot \text{K})(0.200\,\text{kg})(25°\,\text{C}) + (2220\,\text{J/kg} \cdot \text{K})(0.100\,\text{kg})(-15°\,\text{C})}{(4190\,\text{J/kg} \cdot \text{K})(0.200\,\text{kg}) + (2220\,\text{J/kg} \cdot \text{K})(0.100\,\text{kg})} \\
&= 16.6°\,\text{C}.
\end{aligned}
$$

This is above the melting point of ice, so at least some of the ice must have melted. The calculation just completed does not take into account the melting of the ice and is in error.

Now assume the water and ice reach thermal equilibrium at $T_f = 0°$ C, with mass m ($< m_I$) of the ice melted. The heat rejected by the water is

$$Q = c_W m_W T_{Wi}$$

and the heat absorbed by the ice is

$$-c_I m_I T_{Ii} + m L_F,$$

where L_F is the heat of fusion for water. The first term is the energy required to warm all the ice from its initial temperature to 0° C and the second term is the energy required to melt mass m of the ice. The two heats are equal, so

$$c_W m_W T_{Wi} = -c_I m_I T_{Ii} + m L_F.$$

This equation can be solved for the mass m of ice melted:

$$
\begin{aligned}
m &= \frac{c_W m_W T_{Wi} + c_I m_I T_{Ii}}{L_F} \\
&= \frac{(4190\,\text{J/kg} \cdot \text{K})(0.200\,\text{kg})(25°\,\text{C}) + (2220\,\text{J/kg} \cdot \text{K})(0.100\,\text{kg})(-15°\,\text{C})}{333 \times 10^3\,\text{J/kg}} \\
&= 5.3 \times 10^{-2}\,\text{kg} = 53\,\text{g}.
\end{aligned}
$$

Since the total mass of ice present initially was 100 g, there is enough ice to bring the water temperature down to 0° C. This is the solution: the ice and water reach thermal equilibrium at a temperature of 0° C with 53 g of ice melted.

(*b*) Now there is less than 53 g of ice present initially. All the ice melts and the final temperature is above the melting point of ice. The heat rejected by the water is

$$Q = c_W m_W (T_{Wi} - T_f)$$

and the heat absorbed by the ice and the water it becomes when it melts is

$$Q = c_I m_I (-T_{Ii}) + c_W m_I T_f + m_I L_F.$$

The first term is the energy required to raise the temperature of the ice to 0° C, the second term is the energy required to raise the temperature of the melted ice from 0° C to T_f, and the third term is the energy required to melt all the ice. Since the two heats are equal

$$c_W m_W (T_{Wi} - T_f) = c_I m_I (-T_{Ii}) + c_W m_I T_f + m_I L_F.$$

The solution for T_f is

$$T_f = \frac{c_W m_W T_{Wi} + c_I m_I T_{Ii} - m_I L_F}{c_W (m_W + m_I)}.$$

Substitute given values to obtain $T_f = 2.5°$ C.

31P

If the ring diameter at 0.000° C is D_{r0} then its diameter when the ring and sphere are in thermal equilibrium is

$$D_r = D_{r0}(1 + \alpha_c T_f),$$

where T_f is the final temperature and α_c is the coefficient of linear expansion for copper. Similarly, if the sphere diameter at T_i (= 100.0° C) is D_{s0} then its diameter at the final temperature is

$$D_s = D_{s0}[1 + \alpha_a(T_f - T_i)],$$

where α_a is the coefficient of linear expansion for aluminum. At equilibrium the two diameters are equal, so

$$D_{r0}(1 + \alpha_c T_f) = D_{s0}[1 + \alpha_a(T_f - T_i)].$$

The solution for the final temperature is

$$
\begin{aligned}
T_f &= \frac{D_{r0} - D_{s0} + D_{s0}\alpha_a T_i}{D_{s0}\alpha_a - D_{r0}\alpha_c} \\
&= \frac{1.00000\,\text{in.} - 1.00200\,\text{in.} + (1.00200\,\text{in.})(23 \times 10^{-6}/°\text{C})(100°\,\text{C})}{(1.00200\,\text{in.})(23 \times 10^{-6}/°\text{C}) - (1.00000\,\text{in.})(17 \times 10^{-6}/°\text{C})} \\
&= 50.38°\,\text{C}.
\end{aligned}
$$

The heat absorbed by the ring is

$$Q = c_c m_r T_f,$$

where c_c is the specific heat of copper and m_r is the mass of the ring. The heat rejected up by the sphere is

$$Q = c_a m_s(T_i - T_f),$$

where c_a is the specific heat of aluminum and m_s is the mass of the sphere. Since these two heats are equal,

$$c_c m_r T_f = c_a m_s (T_i - T_f)$$

and

$$m_s = \frac{c_c m_r T_f}{c_a (T_i - T_f)} = \frac{(386 \,\text{J/kg} \cdot \text{K})(0.0200 \,\text{kg})(50.38°\,\text{C})}{(900 \,\text{J/kg} \cdot \text{K})(100°\,\text{C} - 50.38°\,\text{C})} = 8.71 \times 10^{-3} \,\text{kg}.$$

35E

One part of path A represents a constant pressure process. The volume changes from $1.0 \,\text{m}^3$ to $4.0 \,\text{m}^3$ while the pressure remains at $40 \,\text{Pa}$. The work done is

$$W = p \,\Delta V = (40 \,\text{Pa})(4.0 \,\text{m}^3 - 1.0 \,\text{m}^3) = 120 \,\text{J}.$$

The other part of the path represents a constant volume process. No work is done during this process. The total work done over the entire path is $120 \,\text{J}$

To find the work done over path B we need to know the pressure as a function of volume. Then we can evaluate the integral $W = \int p \,dV$. According to the graph the pressure is a linear function of the volume, so we may write $p = a + bV$, where a and b are constants. In order for the pressure to be $40 \,\text{Pa}$ when the volume is $1.0 \,\text{m}^3$ and $10 \,\text{Pa}$ when the volume is $4.00 \,\text{m}^3$ the values of the constants must be $a = 50 \,\text{Pa}$ and $b = -10 \,\text{Pa/m}^3$. Thus $p = 50 \,\text{Pa} - (10 \,\text{Pa/m}^3)V$ and

$$W = \int_1^4 p \,dV = \int_1^4 (50 - 10V) \,dV$$
$$= 50V \big|_1^4 - 5V^2 \big|_1^4 = 200 \,\text{J} - 50 \,\text{J} - 80 \,\text{J} + 5 \,\text{J} = 75 \,\text{J}.$$

One part of path C represents a constant pressure process in which the volume change from $1.0 \,\text{m}^3$ to $4.0 \,\text{m}^3$ while p remains at $10 \,\text{Pa}$. The work done is

$$p \,\Delta V = (10 \,\text{Pa})(4.0 \,\text{m}^3 - 1.0 \,\text{m}^3) = 30 \,\text{J}.$$

The other part of the process is at constant volume and no work is done. The total work is $30 \,\text{J}$.

Notice that the work is different for different paths.

39E

Since the process is a complete cycle (beginning and ending in the same thermodynamic state) the change in the internal energy is zero and the heat absorbed by the gas is equal to the work done by the gas: $Q = W$. In terms of the contributions of the individual parts of the cycle $Q_{AB} + Q_{BC} + Q_{CA} = W$ and $Q_{CA} = W - Q_{AB} - Q_{BC} = +15.0 \,\text{J} - 20.0 \,\text{J} - 0 = -5.0 \,\text{J}$. This means $5.0 \,\text{J}$ of energy leaves the gas in the form of heat.

40E

The work W done over the cycle can easily be calculated. Since the internal energy is the same at the beginning and end the heat absorbed Q equals the work done: $Q = W$. Over the portion of the cycle from A to B the pressure p is a linear function of the volume V and we may write

$$p = \frac{10}{3}\,\text{Pa} + \left(\frac{20}{3}\,\text{Pa/m}^3\right) V\,,$$

where the coefficients were chosen so that $p = 10\,\text{Pa}$ when $V = 1.0\,\text{m}^3$ and $p = 30\,\text{Pa}$ when $V = 4.0\,\text{m}^3$. The work done by the gas during this portion of the cycle is

$$W_{AB} = \int_1^4 p\,dV = \int_1^4 (\frac{10}{3} + \frac{20}{3}V)\,dV$$

$$= \frac{10}{3}V\big|_1^4 + \frac{10}{3}V^2\big|_1^4 = \frac{40}{3} + \frac{160}{3} - \frac{10}{3} - \frac{10}{3} = 60\,\text{J}\,.$$

The BC portion of the cycle is at constant pressure and the work done by the gas is $W_{BC} = p\Delta V = (30\,\text{Pa})(1.0\,\text{m}^3 - 4.0\,\text{m}^3) = -90\,\text{J}$. The CA portion of the cycle is at constant volume, so no work is done. The total work done by the gas is $W = W_{AB} + W_{BC} + W_{CA} = 60\,\text{J} - 90\,\text{J} + 0 = -30\,\text{J}$ and the total heat absorbed is $Q = W = -30\,\text{J}$. This means the gas loses 30 J of energy in the form of heat.

42P

(a) The change in internal energy ΔE_{int} is the same for path iaf and path ibf. According to the first law of thermodynamics $\Delta E_{\text{int}} = Q - W$, where Q is the heat absorbed and W is the work done by the system. Along iaf $\Delta E_{\text{int}} = Q - W = 50\,\text{cal} - 20\,\text{cal} = 30\,\text{cal}$. Along ibf $W = Q - \Delta E_{\text{int}} = 36\,\text{cal} - 30\,\text{cal} = 6\,\text{cal}$.

(b) Since the curved path is traversed from f to i the change in internal energy is $-30\,\text{cal}$ and $Q = \Delta E_{\text{int}} + W = -30\,\text{cal} - 13\,\text{cal} = -43\,\text{cal}$.

(c) Let $\Delta E_{\text{int}} = E_{\text{int},\,f} - E_{\text{int},\,i}$. Then $E_{\text{int},\,f} = \Delta E_{\text{int}} + E_{\text{int},\,i} = 30\,\text{cal} + 10\,\text{cal} = 40\,\text{cal}$.

(d) The work W_{bf} for the path bf is zero, so $Q_{bf} = E_{\text{int},\,f} - E_{\text{int},\,b} = 40\,\text{cal} - 22\,\text{cal} = 18\,\text{cal}$. For the path ibf $Q = 36\,\text{cal}$ so $Q_{ib} = Q - Q_{bf} = 36\,\text{cal} - 18\,\text{cal} = 18\,\text{cal}$.

43P

Let m_s be the mass of steam in the cylinder. It occupies volume $V_s = m_s/\rho_s$, where ρ_s is the density of the steam. Let m_w be the mass of water in the cylinder. It occupies volume $V_w = m_w/\rho_w$, where ρ_w is the density of the water. The total volume of the cylinder is

$$V = V_s + V_w = \frac{m_s}{\rho_s} + \frac{m_w}{\rho_w}$$

and it is changing at the rate

$$\frac{dV}{dt} = \frac{1}{\rho_s}\frac{dm_s}{dt} + \frac{1}{\rho_w}\frac{dm_w}{dt}\,.$$

Now all of the steam that condenses turns to water, so $dm_w/dt = -dm_s/dt$ and

$$\frac{dV}{dt} = \left(\frac{1}{\rho_s} - \frac{1}{\rho_w}\right)\frac{dm_s}{dt} = \left(\frac{\rho_w - \rho_s}{\rho_w \rho_s}\right)\frac{dm_s}{dt}\,.$$

The volume is changing because the piston is moving in. Write $V = Ah$, where A is the cross-sectional area of the cylinder and h is its length. Then $dV/dt = A(dh/dt) = -Av$, where v is the speed of the piston. The minus sign appears since the volume is getting smaller. Equate the two expression for dV/dt to obtain

$$-Av = \frac{\rho_w - \rho_s}{\rho_w \rho_s} \frac{dm_s}{dt}.$$

Thus

$$\frac{dm_s}{dt} = -\frac{Av\rho_w\rho_s}{\rho_w - \rho_s}$$

$$= -\frac{(2.0 \times 10^{-4}\,\mathrm{m^2})(0.30 \times 10^{-2}\,\mathrm{m/s})(0.998 \times 10^3\,\mathrm{kg/m^3})(0.60\,\mathrm{kg/m^3})}{998\,\mathrm{kg/m^3} - 0.60\,\mathrm{kg/m^3}}$$

$$= -3.6 \times 10^{-7}\,\mathrm{kg/s}.$$

(b) Energy is ejected in the form of heat because the steam is condensing. Since $Q = -L_V m_s$, where L_V is the heat of vaporization for water,

$$\frac{dQ}{dt} = -L_V \frac{dm_s}{dt} = -(2256 \times 10^3\,\mathrm{J/kg})(3.6 \times 10^{-7}\,\mathrm{kg/s}) = -0.81\,\mathrm{J/s}.$$

(c) Differentiate the first law with respect to time to obtain

$$\frac{dE_{\text{int}}}{dt} = \frac{dQ}{dt} - \frac{dW}{dt}.$$

Now $dW/dt = p(dV/dt) = -pAv$, where p is the pressure inside the cylinder. The air outside the cylinder pushes down on the piston with a force of $p_0 A$ where p_0 is atmospheric pressure. The force of gravity pulls down with a force of mg, where m is the mass of the piston, and the steam in the cylinder pushes up with a force of pA. Since the cylinder moves with constant velocity these forces must sum to zero. Thus $p = p_0 + (mg/A)$ and

$$\frac{dW}{dt} = -\left(p_0 + \frac{mg}{A}\right) Av$$

$$= -\left[1.013 \times 10^5\,\mathrm{Pa} + \frac{(2.0\,\mathrm{kg})(9.8\,\mathrm{m/s^2})}{2.0 \times 10^{-4}\,\mathrm{m^2}}\right] (2.0 \times 10^{-4}\,\mathrm{m^2})(0.30 \times 10^{-2}\,\mathrm{m/s})$$

$$= -0.12\,\mathrm{J/s}.$$

The rate of change of the internal energy is

$$\frac{dE_{\text{int}}}{dt} = \frac{dQ}{dt} - \frac{dW}{dt} = -0.81\,\mathrm{J/s} + 0.12\,\mathrm{J/s} = -0.69\,\mathrm{J/s}.$$

47E

The rate of heat flow is given by

$$H = kA \frac{T_H - T_C}{L},$$

where k is the thermal conductivity of copper (401 W/m · K), A is the cross-sectional area (in a plane perpendicular to the flow), L is the distance along the direction of flow between the points where the temperature is T_H and T_C. Thus

$$H = \frac{(401 \, \text{W/m} \cdot \text{K})(90.0 \times 10^{-4} \, \text{m}^2)(125°\,\text{C} - 10.0°\,\text{C})}{0.250 \, \text{m}} = 1.66 \times 10^3 \, \text{J/s} \,.$$

56P

(a) Use

$$H = kA \frac{T_H - T_C}{L} \,,$$

where H is the heat flow through an area A, between points a distance L apart with a temperature difference of $T_H - T_C$. Since the conductivity of glass is given in Table 4 as 1.0 W/m · K we use the Celsius scale to give the values of the temperature. A temperature of $-20°$ F is equivalent to $-28.9°$ C and a temperature of $72°$ F is equivalent to $22.2°$ C. Thus

$$\frac{H}{A} = k \frac{T_H - T_C}{L} = (1.0 \, \text{W/m} \cdot \text{K}) \left(\frac{22.2°\,\text{C} + 28.9°\,\text{C}}{3.0 \times 10^{-3} \, \text{m}} \right) = 1.7 \times 10^4 \, \text{W/m}^2 \,.$$

(b) The energy now passes in succession through 3 layers, one of air and two of glass. The heat flow H is the same in each layer and is given by

$$H = \frac{A(T_H - T_C)}{\sum L/k} \,,$$

where the sum in the denominator is over the layers. If L_g is the thickness of a glass layer, L_a is the thickness of the air layer, k_g is the thermal conductivity of glass, and k_a is the thermal conductivity of air, then the denominator is

$$\sum \frac{L}{k} = \frac{2L_g}{k_g} + \frac{L_a}{k_a} = \frac{2L_g k_a + L_a k_g}{k_a k_g}$$

and

$$\begin{aligned} \frac{H}{A} &= \frac{(T_H - T_C)k_a k_g}{2L_g k_a + L_a k_g} \\ &= \frac{(22.2°\,\text{C} + 28.9°\,\text{C})(0.026 \, \text{W/m} \cdot \text{K})(1.0 \, \text{W/m} \cdot \text{K})}{2(3.0 \times 10^{-3} \, \text{m})(0.026 \, \text{W/m} \cdot \text{K}) + (0.075 \, \text{m})(1.0 \, \text{W/m} \cdot \text{K})} \\ &= 18 \, \text{W/m}^2 \,. \end{aligned}$$

57P

Let h be the thickness of the slab and A be it area. Then the rate of heat flow through the slab is

$$H = \frac{kA(T_H - T_C)}{h} \,,$$

where k is the thermal conductivity of ice, T_H is the temperature of the water ($0°$ C), and T_C is the temperature of the air above the ice ($-10°$ C). The heat leaving the water freezes it, the heat required to freeze mass m of water being $Q = L_F m$, where L_F is the heat of fusion for water. Differentiate with respect to time and recognize that $dQ/dt = H$ to obtain

$$H = L_F \frac{dm}{dt}.$$

Now the mass of the ice is given by $m = \rho A h$, where ρ is the density of ice, so $dm/dt = \rho A(dh/dt)$ and

$$H = L_F \rho A \frac{dh}{dt}.$$

Equate the two expressions for H and solve for dh/dt:

$$\frac{dh}{dt} = \frac{k(T_H - T_C)}{L_F \rho h}.$$

Since $1\,cal = 4.186\,J$ and $1\,cm = 1 \times 10^{-2}\,m$, the thermal conductivity of ice has the SI value $k = (0.0040\,cal/s \cdot cm \cdot K)(4.186\,J/cal)/(1 \times 10^{-2}\,m/cm) = 1.674\,W/m \cdot K$. The SI value for the density of ice is $\rho = 0.92\,g/cm^3 = 0.92 \times 10^3\,kg/m^3$. Thus

$$\frac{dh}{dt} = \frac{(1.674\,W/m \cdot K)(0°\,C + 10°\,C)}{(333 \times 10^3\,J/kg)(0.92 \times 10^3\,kg/m^3)(0.050\,m)} = 1.1 \times 10^{-6}\,m/s = 0.40\,cm/h.$$

60P

(a) The situation as viewed from the ends of the studs is shown on the diagram (not to scale). Heat flows from left to right. We must consider two different types of tubes of flow. An example of the first type is indicated by the bottom two dotted lines on the diagram. Heat flows through the plaster board, then through a stud, then through a portion of the plywood, and finally through a portion of the pine. There are 16 of these tubes of flow present in the wall assembly, one for each stud. An example of the second type tube is indicated by the second and third dotted lines from the bottom. Heat flows through a portion of the paster board, the foam insulation, a portion of the plywood, and a portion of the pine. There are 15 tubes of flow of this type, one for each section of foam. For each tube the rate of heat flow is given by

$$H = \frac{A\,\Delta T}{\sum R}.$$

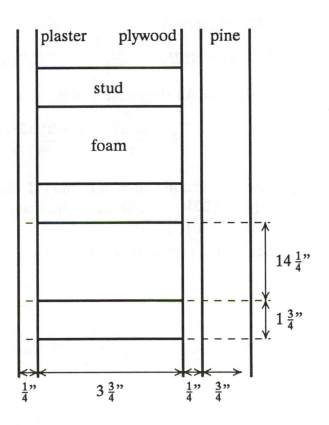

Here A is the cross-sectional area of the tube, ΔT is the temperature difference, and the sum in the denominator is over the R values of all the materials in the tube. The units of the R values are $ft^2 \cdot {}^\circ F \cdot h/Btu$, so we want to substitute an area in ft^2 and a temperature difference in ${}^\circ F$. H will then be in Btu/h.

Consider first the rate of flow H_s through a tube that contains a stud. The cross-sectional area is $A = [(1.75\,in.)/(12\,in./ft)](12\,ft) = 1.75\,ft^2$ where the thickness of the stud (1.75 in.) was converted to feet. The R value is proportional to the width, so the R value for a stud that is 3.75 in. wide and has an R value of $1.3\,ft^2 \cdot {}^\circ F \cdot h/Btu$ for a 1 in. width is $(1.3\,ft^2 \cdot {}^\circ F \cdot h/Btu \cdot in.)(3.75\,in.) = 4.875\,ft^2 \cdot {}^\circ F \cdot h/Btu$. The R value of the plaster board is $0.47\,ft^2 \cdot {}^\circ F \cdot h/Btu$, the R value of the plywood is $0.30\,ft^2 \cdot {}^\circ F \cdot h/Btu$ and the R value of the pine is $0.98\,ft^2 \cdot {}^\circ F \cdot h/Btu$, so

$$H_s = \frac{(1.75\,ft^2)(30^\circ F)}{(4.875 + 0.47 + 0.30 + 0.98)\,ft^2 \cdot {}^\circ F \cdot h/Btu} = 7.92\,Btu/h\,.$$

Now consider the rate of flow H_f through a tube that contains foam insulation. The cross-sectional area is $A = [(14.25\,in.)/(12\,in./ft)](12\,ft) = 14.25\,ft^2$ and the R value of the insulation is $(5.9\,ft^2 \cdot {}^\circ F \cdot h/Btu \cdot in.)(3.75\,in.) = 22.125\,ft^2 \cdot {}^\circ F \cdot h/Btu$. Thus

$$H_f = \frac{(14.25\,ft^2)(30^\circ F)}{(22.125 + 0.47 + 0.30 + 0.98)\,ft^2 \cdot {}^\circ F \cdot h/Btu} = 17.91\,Btu/h\,.$$

The total heat flow is

$$H = 16H_s + 15H_f = 16(7.92\,Btu/h) + 15(17.91\,Btu/h) = 395\,Btu/h\,.$$

(b) The R value for the entire wall is

$$R = \frac{A\,\Delta T}{H} = \frac{(20\,ft)(12\,ft)(30^\circ\,F)}{395\,Btu/h} = 18\,ft^2 \cdot {}^\circ F \cdot h/Btu\,.$$

(c) The cross-sectional area of a stud is $1.75\,ft^2$ and there are 16 studs. The total area of the wall is $(12\,ft)(20\,ft) = 240\,ft^2$, so the fraction of the total area that is occupied by studs is $16(1.75\,ft^2)/(240\,ft^2) = 0.12$.

(d) The fraction of heat flow through the studs is $16(7.92\,Btu/h)/(395\,Btu/h) = 0.32$.

Chapter 21

2E

Each atom has a mass of $m = M/N_A$, where M is the molar mass and N_A is the Avogadro constant. The molar mass of arsenic is $74.9\,\text{g/mol}$ or $74.9 \times 10^{-3}\,\text{kg/mol}$. 7.50×10^{24} arsenic atoms have a total mass of $(7.50 \times 10^{24})(74.9 \times 10^{-3}\,\text{kg/mol})/(6.02 \times 10^{23}\,\text{mol}^{-1}) = 0.933\,\text{kg}$.

7E

(a) Solve $pV = nRT$ for n:

$$n = \frac{pV}{RT} = \frac{(100\,\text{Pa})(1.0 \times 10^{-6}\,\text{m}^3)}{(8.31\,\text{J/mol} \cdot \text{K})(220\,\text{K})} = 5.47 \times 10^{-8}\,\text{mol}.$$

(b) The number of molecules N is the product of the number of moles n and the number of molecules in a mole N_A (the Avogadro constant). Thus $N = nN_A = (5.47 \times 10^{-6}\,\text{mol})(6.02 \times 10^{23}\,\text{mol}^{-1}) = 3.29 \times 10^{16}$ molecules.

10E

(a) Solve $pV = nRT$ for n. First convert the temperature to the Kelvin scale: $T = 40.0 + 273.15 = 313.15\,\text{K}$. Also convert the volume to m^3: $1000\,\text{cm}^3 = 1000 \times 10^{-6}\,\text{m}^3$. Then

$$n = \frac{pV}{RT} = \frac{(1.013 \times 10^5\,\text{Pa})(1000 \times 10^{-6}\,\text{m}^3)}{(8.31\,\text{J/mol} \cdot \text{K})(313.15\,\text{K})} = 3.88 \times 10^{-2}\,\text{mol}.$$

(b) Solve $pV = nRT$ for T:

$$T = \frac{pV}{nR} = \frac{(1.06 \times 10^5\,\text{Pa})(1500 \times 10^{-6}\,\text{m}^3)}{(3.88 \times 10^{-2}\,\text{mol})(8.31\,\text{J/mol} \cdot \text{K})} = 493\,\text{K} = 220°\,\text{C}.$$

14P

Since the pressure is constant the work is given by $W = p(V_2 - V_1)$. The initial volume is $V_1 = (AT_1 - BT_1^2)/p$, where T_1 is the initial temperature. The final volume is $V_2 = (AT_2 - BT_2^2)/p$. Thus $W = A(T_2 - T_1) - B(T_2^2 - T_1^2)$.

15P

Suppose the gas expands from volume V_i to volume V_f during the isothermal portion of the process. The work it does is

$$W = \int_{V_i}^{V_f} p\,dV = nRT \int_{V_i}^{V_f} \frac{dV}{V} = nRT \ln\frac{V_f}{V_i},$$

where $pV = nRT$ was used to replace p with nRT/V. Now $V_i = nRT/p_i$ and $V_f = nRT/p_f$, so $V_f/V_i = p_i/p_f$. Also replace nRT with p_iV_i to obtain

$$W = p_iV_i \ln \frac{p_i}{p_f}.$$

Since the initial gauge pressure is 1.03×10^5 Pa, $p_i = 1.03 \times 10^5$ Pa $+ 1.013 \times 10^5$ Pa $= 2.04 \times 10^5$ Pa. The final pressure is atmospheric pressure: $p_f = 1.013 \times 10^5$ Pa. Thus

$$W = (2.04 \times 10^5 \text{ Pa})(0.14 \text{ m}^3) \ln \frac{2.04 \text{ Pa}}{1.013 \text{ Pa}} = 2.00 \times 10^4 \text{ J}.$$

During the constant pressure portion of the process the work done by the gas is $W = p_f(V_i - V_f)$. Notice that the gas starts in a state with pressure p_f, so this is the pressure throughout this portion of the process. Also note that the volume decreases from V_f to V_i. Now $V_f = p_iV_i/p_f$, so

$$W = p_f \left(V_i - \frac{p_iV_i}{p_f}\right) = (p_f - p_i)V_i$$
$$= (1.013 \times 10^5 \text{ Pa} - 2.04 \times 10^5 \text{ Pa})(0.14 \text{ m}^3) = -1.44 \times 10^4 \text{ J}.$$

The total work done by the gas over the entire process is $W = 2.00 \times 10^4 \text{ J} - 1.44 \times 10^4 \text{ J} = 5.6 \times 10^3 \text{ J}$.

19P

Assume that the pressure of the air in the bubble is essentially the same as the pressure in the surrounding water. If d is the depth of the lake and ρ is the density of water, then the pressure at the bottom of the lake is $p_1 = p_0 + \rho g d$, where p_0 is atmospheric pressure. Since $p_1 V_1 = nRT_1$, the number of moles of gas in the bubble is $n = p_1 V_1/RT_1 = (p_0 + \rho g d)V_1/RT_1$, where V_1 is the volume of the bubble at the bottom of the lake and T_1 is the temperature there. At the surface of the lake the pressure is p_0 and the volume of the bubble is $V_2 = nRT_2/p_0$. Substitute for n to obtain

$$V_2 = \frac{T_2}{T_1} \frac{p_0 + \rho g d}{p_0} V_1$$
$$= \left(\frac{293 \text{ K}}{277 \text{ K}}\right) \left(\frac{1.013 \times 10^5 \text{ Pa} + (0.998 \times 10^3 \text{ kg/m}^3)(9.8 \text{ m/s}^2)(40 \text{ m})}{1.013 \times 10^5 \text{ Pa}}\right) (20 \text{ cm}^3)$$
$$= 100 \text{ cm}^3.$$

25E

According to kinetic theory the rms speed is

$$v_{\text{rms}} = \sqrt{\frac{3RT}{M}},$$

where T is the temperature and M is the molar mass. According to Table 1 the molar mass of molecular hydrogen is $2.02\,\text{g/mol} = 2.02 \times 10^{-3}\,\text{kg/mol}$, so

$$v_{\text{rms}} = \sqrt{\frac{3(8.31\,\text{J/mol}\cdot\text{K})(2.7\,\text{K})}{2.02 \times 10^{-3}\,\text{kg/mol}}} = 180\,\text{m/s}.$$

31P

On reflection only the normal component of the momentum changes, so for one molecule the change in momentum is $2mv\cos\theta$, where m is the mass of the molecule, v is its speed, and θ is the angle between its velocity and the normal to the wall. If N molecules collide with the wall the change in their total momentum is $\Delta P = 2Nmv\cos\theta$ and if the total time taken for the collisions is Δt then the average rate of change of the total momentum is $\Delta P/\Delta t = 2(N/\Delta t)mv\cos\theta$. This is the average force exerted by the N molecules on the wall and the pressure is the average force per unit area:

$$\begin{aligned}
p &= \frac{2}{A}\frac{N}{\Delta t}mv\cos\theta \\
&= \left(\frac{2}{2.0 \times 10^{-4}\,\text{m}^2}\right)(10^{23}\,\text{s}^{-1})(3.3 \times 10^{-27}\,\text{kg})(1.0 \times 10^3\,\text{m/s})\cos 55^\circ \\
&= 1.9 \times 10^3\,\text{Pa}.
\end{aligned}$$

Notice that the value given for the mass was converted to kg and the value given for the area was converted to m^2.

32E

(a) The average translational kinetic energy is given by $\overline{K} = \frac{3}{2}kT$, where k is the Boltzmann constant ($1.38 \times 10^{-23}\,\text{J/K}$) and T is the temperature on the Kelvin scale. Thus

$$\overline{K} = \frac{3}{2}(1.38 \times 10^{-23}\,\text{J/K})(1600\,\text{K}) = 3.31 \times 10^{-20}\,\text{J}.$$

(b) Since $1\,\text{eV} = 1.60 \times 10^{-19}\,\text{J}$, $\overline{K} = (3.31 \times 10^{-20}\,\text{J})/(1.60 \times 10^{-19}\,\text{J/eV}) = 0.21\,\text{eV}$.

37P

(a) Use $\epsilon = L_V/N$, where L_V is the heat of vaporization and N is the number of molecules per gram. The molar mass of atomic hydrogen is $1\,\text{g/mol}$ and the molar mass of atomic oxygen is $16\,\text{g/mol}$ so the molar mass of H_2O is $1 + 1 + 16 = 18\,\text{g/mol}$. There are $N_A = 6.02 \times 10^{23}$ molecules in a mole so the number of molecules in a gram of water is $(6.02 \times 10^{23}\,\text{mol}^{-1})/(18\,\text{g/mol}) = 3.34 \times 10^{22}$ molecules/g. Thus $\epsilon = (539\,\text{cal/g})/(3.34 \times 10^{22}\,\text{g}^{-1}) = 1.61 \times 10^{-20}\,\text{cal}$. This is $(1.61 \times 10^{-20}\,\text{cal})(4.186\,\text{J/cal}) = 6.76 \times 10^{-20}\,\text{J}$.

(b) The average translational kinetic energy is

$$\overline{K} = \frac{3}{2}kT = \frac{3}{2}(1.38 \times 10^{-23}\,\text{J/K})[(32.0 + 273.15)\,\text{K}] = 6.32 \times 10^{-21}\,\text{J}.$$

The ratio $\epsilon/\overline{K}$ is $(6.76 \times 10^{-20}\,\text{J})/(6.32 \times 10^{-21}\,\text{J}) = 10.7$.

38P

They are not equivalent. Avogadro's law does not tell how the pressure, volume, and temperature are related, so you cannot use it, for example, to calculate the change in volume when the pressure increases at constant temperature. The ideal gas law, however, implies Avogadro's law. It yields $N = nN_A = (pV/RT)N_A = pV/kT$, where $k = R/N_A$ was used. If the two gases have the same volume, the same pressure, and the same temperature, then pV/kT is the same for them. This implies that N is also the same.

40E

(a) According to Eq. 21–18 the mean free path for molecules in a gas is given by

$$\lambda = \frac{1}{\sqrt{2}\pi d^2 N/V},$$

where d is the diameter of a molecule and N is the number of molecules in volume V. Substitute $d = 2.0 \times 10^{-10}\,\text{m}$ and $N/V = 1 \times 10^6$ molecules/m^3 to obtain

$$\lambda = \frac{1}{\sqrt{2}\pi(2.0 \times 10^{-10}\,\text{m})^2(1 \times 10^6\,\text{m}^{-3})} = 6 \times 10^{12}\,\text{m}.$$

(b) At this altitude most of the gas particles are in orbit around the Earth and do not suffer randomizing collisions. The mean free path has little physical significance.

41E

Substitute $d = 1.0 \times 10^{-2}\,\text{m}$ and $N/V = 15/(1.0 \times 10^{-3}\,\text{m}^3) = 15 \times 10^3$ beans/m^3 into

$$\lambda = \frac{1}{\sqrt{2}\pi d^2 N/V}$$

to obtain

$$\lambda = \frac{1}{\sqrt{2}\pi(1.0 \times 10^{-2}\,\text{m})^2(15 \times 10^3\,\text{m}^{-3})} = 0.15\,\text{m}.$$

The conversion $1.00\,\text{L} = 1.00 \times 10^{-3}\,\text{m}^3$ was used.

43P

(a) Use $pV = nRT = NkT$, where p is the pressure, V is the volume, T is the temperature, n is the number of moles, and N is the number of molecules. The substitutions $N = nN_A$ and $k = R/N_A$ were made. Since 1 cm of mercury $= 1333\,\text{Pa}$, the pressure is $p = (10^{-7})(1333) = 1.333 \times 10^{-4}\,\text{Pa}$. Thus

$$\frac{N}{V} = \frac{p}{kT} = \frac{1.333 \times 10^{-4}\,\text{Pa}}{(1.38 \times 10^{-23}\,\text{J/K})(295\,\text{K})}$$
$$= 3.27 \times 10^{16}\,\text{molecules/m}^3 = 3.27 \times 10^{10}\,\text{molecules/cm}^3.$$

(b) The molecular diameter is $d = 2.00 \times 10^{-10}$ m, so the mean free path is

$$\lambda = \frac{1}{\sqrt{2}\pi d^2 N/V} = \frac{1}{\sqrt{2}\pi(2.00 \times 10^{-10}\,\text{m})^2(3.27 \times 10^{16}\,\text{m}^{-3})} = 172\,\text{m}\,.$$

48E

(a) The average speed is

$$\bar{v} = \frac{\sum v}{N}\,,$$

where the sum is over the speeds of the particles and N is the number of particles. Thus

$$\bar{v} = \frac{(2.0 + 3.0 + 4.0 + 5.0 + 6.0 + 7.0 + 8.0 + 9.0 + 10.0 + 11.0)\,\text{km/s}}{10} = 6.5\,\text{km/s}\,.$$

(b) The rms speed is given by

$$v_{\text{rms}} = \sqrt{\frac{\sum v^2}{N}}\,.$$

Now

$$\sum v^2 = \Big[(2.0)^2 + (3.0)^2 + (4.0)^2 + (5.0)^2 + (6.0)^2$$
$$+ (7.0)^2 + (8.0)^2 + (9.0)^2 + (10.0)^2 + (11.0)^2\Big]\,\text{km}^2/\text{s}^2 = 505\,\text{km}^2/\text{s}^2\,,$$

so

$$v_{\text{rms}} = \sqrt{\frac{505\,\text{km}^2/\text{s}^2}{10}} = 7.1\,\text{km/s}\,.$$

53P

(a) The rms speed of molecules in a gas is given by $v_{\text{rms}} = \sqrt{3RT/M}$, where T is the temperature and M is the molar mass of the gas. The speed required for escape from the Earth's gravitational pull is $v = \sqrt{2gr_e}$, where g is the acceleration due to gravity at the Earth's surface and $r_e\ (= 6.37 \times 10^6$ m) is the radius of the Earth. This expression can be derived easily. Take the zero of gravitational potential energy to be at infinity. Then the energy of a particle with speed v and mass m at the Earth's surface is $E = -mgr_e + \frac{1}{2}mv^2$. If the particle is just able to travel far away its kinetic energy must tend toward zero as its distance from the Earth becomes large without bound. This means $E = 0$ and $v = \sqrt{2gr_e}$.

Equate the expressions for the speeds to obtain $\sqrt{3RT/M} = \sqrt{2gr_e}$. The solution for T is $T = 2gr_e M/3R$. The molar mass of hydrogen is 2.02×10^{-3} kg/mol, so for that gas

$$T = \frac{2(9.8\,\text{m/s}^2)(6.37 \times 10^6\,\text{m})(2.02 \times 10^{-3}\,\text{kg/mol})}{3(8.31\,\text{J/mol} \cdot \text{K})} = 1.0 \times 10^4\,\text{K}\,.$$

The molar mass of oxygen is 32.0×10^{-3} kg/mol, so for that gas

$$T = \frac{2(9.8\,\text{m/s}^2)(6.37 \times 10^6\,\text{m})(32.0 \times 10^{-3}\,\text{kg/mol})}{3(8.31\,\text{J/mol} \cdot \text{K})} = 1.6 \times 10^5\,\text{K}\,.$$

(b) Now $T = 2g_m r_m M/3R$, where r_m ($= 1.74 \times 10^6$ m) is the radius of the moon and g_m ($= 0.16g$) is the acceleration due to gravity at the moon's surface. For hydrogen

$$T = \frac{2(0.16)(9.8\,\text{m/s}^2)(1.74 \times 10^6\,\text{m})(2.02 \times 10^{-3}\,\text{kg/mol})}{3(8.31\,\text{J/mol}\cdot\text{K})} = 4.4 \times 10^2\,\text{K}.$$

For oxygen

$$T = \frac{2(0.16)(9.8\,\text{m/s}^2)(1.74 \times 10^6\,\text{m})(32.0 \times 10^{-3}\,\text{kg/mol})}{3(8.31\,\text{J/mol}\cdot\text{K})} = 7.0 \times 10^3\,\text{K}.$$

(c) The temperature high in the Earth's atmosphere is great enough for a significant number of hydrogen atoms in the tail of the Maxwellian distribution to escape. As a result the atmosphere is depleted of hydrogen. On the other hand, very few oxygen atoms escape.

54P

(a) The root-mean-square speed is given by $v_{\text{rms}} = \sqrt{3RT/M}$. The molar mass of hydrogen is 2.02×10^{-3} kg/mol, so

$$v_{\text{rms}} = \sqrt{\frac{3(8.31\,\text{J/mol}\cdot\text{K})(4000\,\text{K})}{2.02 \times 10^{-3}\,\text{kg/mol}}} = 7.0 \times 10^3\,\text{m/s}.$$

(b) When the surfaces of the spheres that represent an H_2 molecule and an Ar atom are touching the distance between their centers is the sum of their radii: $d = r_1 + r_2 = 0.5 \times 10^{-8}$ cm $+ 1.5 \times 10^{-8}$ cm $= 2.0 \times 10^{-8}$ cm.

(c) The argon atoms are essentially at rest so in time t the hydrogen atom collides with all the argon atoms in a cylinder of radius d and length vt, where v is its speed. That is, the number of collisions is $\pi d^2 vt N/V$, where N/V is the concentration of argon atoms. The number of collisions per unit time is

$$\frac{\pi d^2 v N}{V} = \pi(2.0 \times 10^{-10}\,\text{m})^2(7.0 \times 10^3\,\text{m/s})(4.0 \times 10^{25}\,\text{m}^{-3}) = 3.5 \times 10^{10}\,\text{collisions/s}.$$

57P

(a) The distribution function gives the fraction of particles with speeds between v and $v + dv$, so its integral over all speeds is unity: $\int P(v)\,dv = 1$. Evaluate the integral by calculating the area under the curve in Fig. 21–20. The area of the triangular portion is half the product of the base and altitude, or $\frac{1}{2}av_0$. The area of the rectangular portion is the product of the sides, or av_0. Thus $\int P(v)\,dv = \frac{1}{2}av_0 + av_0 = \frac{3}{2}av_0$, so $\frac{3}{2}av_0 = 1$ and $a = 2/3v_0$.

(b) The number of particles with speeds between $1.5v_0$ and $2v_0$ is given by $N \int_{1.5v_0}^{2v_0} P(v)\,dv$. The integral is easy to evaluate since $P(v) = a$ throughout the range of integration. Thus the number of particles with speeds in the given range is $Na(2.0v_0 - 1.5v_0) = 0.5Nav_0 = N/3$, where $2/3v_0$ was substituted for a.

(c) The average speed is given by

$$\bar{v} = \int v P(v)\, dv.$$

For the triangular portion of the distribution $P(v) = av/v_0$ and the contribution of this portion is

$$\frac{a}{v_0}\int_0^{v_0} v^2\, dv = \frac{a}{3v_0}v_0^3 = \frac{av_0^2}{3} = \frac{2}{9}v_0,$$

where $2/3v_0$ was substituted for a. $P(v) = a$ in the rectangular portion and the contribution of this portion is

$$a\int_{v_0}^{2v_0} v\, dv = \frac{a}{2}(4v_0^2 - v_0^2) = \frac{3a}{2}v_0^2 = v_0.$$

Thus $\bar{v} = \frac{2}{9}v_0 + v_0 = 1.22v_0$.

(d) The mean-square speed is given by

$$v_{\text{rms}}^2 = \int v^2 P(v)\, dv.$$

The contribution of the triangular section is

$$\frac{a}{v_0}\int_0^{v_0} v^3\, dv = \frac{a}{4v_0}v_0^4 = \frac{1}{6}v_0^2.$$

The contribution of the rectangular portion is

$$a\int_{v_0}^{2v_0} v^2\, dv = \frac{a}{3}(8v_0^3 - v_0^3) = \frac{7a}{3}v_0^3 = \frac{14}{9}v_0^2.$$

Thus $v_{\text{rms}} = \sqrt{\frac{1}{6}v_0^2 + \frac{14}{9}v_0^2} = 1.3v_0$.

59E

According to the first law of thermodynamics $\Delta E_{\text{int}} = Q - W$. Since the process is isothermal $\Delta E_{\text{int}} = 0$ (the internal energy of an ideal gas depends only on the temperature) and $Q = W$. The work done by the gas as its volume expands from V_i to V_f at temperature T is

$$W = \int_{V_i}^{V_f} p\, dV = nRT\int_{V_i}^{V_f} \frac{dV}{V} = nRT\ln\frac{V_f}{V_i}.$$

For 1 mole $Q = W = RT\ln(V_f/V_i)$.

63P

When the temperature changes by ΔT the internal energy of the first gas changes by $n_1 C_1\,\Delta T$, the internal energy of the second gas changes by $n_2 C_2\,\Delta T$, and the internal energy of the third gas changes by $n_3 C_3\,\Delta T$. The change in the internal energy of the composite gas is

$\Delta E_{int} = (n_1 C_1 + n_2 C_2 + n_3 C_3) \Delta T$. This must be $(n_1 + n_2 + n_3) C \Delta T$, where C is the molar specific heat of the mixture. Thus

$$C = \frac{n_1 C_1 + n_2 C_2 + n_3 C_3}{n_1 + n_2 + n_3}.$$

68P

(a) Since the process is at constant pressure the heat added to the gas is given by $Q = nC_p \Delta T$, where n is the number of moles in the gas, C_p is the molar specific heat at constant pressure, and ΔT is the increase in temperature. For a diatomic ideal gas $C_p = \frac{7}{2}R$. Thus

$$Q = \frac{7}{2}nR\Delta T = \frac{7}{2}(4.00\,\text{mol})(8.314\,\text{J/mol} \cdot \text{K})(60.0\,\text{K}) = 6.98 \times 10^3\,\text{J}.$$

(b) The change in the internal energy is given by $\Delta E_{int} = nC_V \Delta T$. For a diatomic ideal gas $C_V = \frac{5}{2}R$, so

$$\Delta E_{int} = \frac{5}{2}nR\Delta T = \frac{5}{2}(4.00\,\text{mol})(8.314\,\text{J/mol} \cdot \text{K})(60.0\,\text{K}) = 4.99 \times 10^3\,\text{J}.$$

(c) According to the first law of thermodynamics $\Delta E_{int} = Q - W$, so

$$W = Q - \Delta E_{int} = 6.98 \times 10^3\,\text{J} - 4.99 \times 10^3\,\text{J} = 1.99 \times 10^3\,\text{J}.$$

(d) The change in the total translational kinetic energy is

$$\Delta K = \frac{3}{2}nR\Delta T = \frac{3}{2}(4.00\,\text{mol})(8.314\,\text{J/mol} \cdot \text{K})(60.0\,\text{K}) = 2.99 \times 10^3\,\text{J}.$$

70E

(a) Let p_i, V_i, and T_i represent the pressure, volume, and temperature of the initial state of the gas. Let p_f, V_f, and T_f represent the pressure, volume, and temperature of the final state. Since the process is adiabatic $p_i V_i^\gamma = p_f V_f^\gamma$, so

$$p_f = \left(\frac{V_i}{V_f}\right)^\gamma p_i = \left(\frac{4.3\,\text{L}}{0.76\,\text{L}}\right)^{1.4} (1.2\,\text{atm}) = 13.6\,\text{atm}.$$

Notice that since V_i and V_f have the same units, their units cancel and p_f has the same units as p_i.

(b) The gas obeys the ideal gas law $pV = nRT$, so $p_i V_i / p_f V_f = T_i / T_f$ and

$$T_f = \frac{p_f V_f}{p_i V_i} T_i = \left[\frac{(13.6\,\text{atm})(0.76\,\text{L})}{(1.2\,\text{atm})(4.3\,\text{L})}\right] (310\,\text{K}) = 620\,\text{K}.$$

Note that the units of $p_i V_i$ and $p_f V_f$ cancel since they are the same.

72E

Use the first law of thermodynamics: $\Delta E_{int} = Q - W$. The change in internal energy is $\Delta E_{int} = nC_V(T_2 - T_1)$, where C_V is the molar heat capacity for a constant volume process. Since the process is adiabatic $Q = 0$. Thus $W = -\Delta E_{int} = nC_V(T_1 - T_2)$.

83P

In the following C_V $(= \frac{3}{2}R)$ is the molar specific heat at constant volume, C_p $(= \frac{5}{2}R)$ is the molar specific heat at constant pressure, ΔT is the temperature change, and n is the number of moles.

(a) The process $1 \rightarrow 2$ takes place at constant volume. The heat added is

$$Q = nC_V \, \Delta T = \frac{3}{2}nR \, \Delta T$$
$$= \frac{3}{2}(1.00\,\text{mol})(8.314\,\text{J/mol} \cdot \text{K})(600\,\text{K} - 300\,\text{K}) = 3.74 \times 10^3 \, \text{J}.$$

The change in the internal energy is

$$\Delta E_{int} = nC_V \, \Delta T = 3.74 \times 10^3 \, \text{J}.$$

The work done by the gas is

$$W = Q - \Delta E_{int} = 0.$$

The process $2 \rightarrow 3$ is adiabatic. The heat added is zero. The change in the internal energy is

$$\Delta E_{int} = nC_V \, \Delta T = \frac{3}{2}nR \, \Delta T$$
$$= \frac{3}{2}(1.00\,\text{mol})(8.314\,\text{J/mol} \cdot \text{K})(455\,\text{K} - 600\,\text{K}) = -1.81 \times 10^3 \, \text{J}.$$

The work done by the gas is

$$W = Q - \Delta E_{int} = +1.81 \times 10^3 \, \text{J}.$$

The process $3 \rightarrow 1$ takes place at constant pressure. The heat added is

$$Q = nC_p \, \Delta T = \frac{5}{2}nR \, \Delta T$$
$$= \frac{5}{2}(1.00\,\text{mol})(8.314\,\text{J/mol} \cdot \text{K})(300\,\text{K} - 455\,\text{K}) = -3.22 \times 10^3 \, \text{J}.$$

The change in the internal energy is

$$\Delta E_{int} = nC_V \, \Delta T = \frac{3}{2}nR \, \Delta T$$
$$= \frac{3}{2}(1.00\,\text{mol})(8.314\,\text{J/mol} \cdot \text{K})(300\,\text{K} - 455\,\text{K}) = -1.29 \times 10^3 \, \text{J}.$$

The work done by the gas is

$$W = Q - \Delta E_{int} = -3.22 \times 10^3\,\mathrm{J} + 1.93 \times 10^3\,\mathrm{J} = -1.29 \times 10^3\,\mathrm{J}.$$

For the entire process the heat added is

$$Q = 3.74 \times 10^3\,\mathrm{J} + 0 - 3.22 \times 10^3\,\mathrm{J} = 520\,\mathrm{J},$$

the change in the internal energy is

$$\Delta E_{int} = 3.74 \times 10^3\,\mathrm{J} - 1.81 \times 10^3\,\mathrm{J} - 1.93 \times 10^3\,\mathrm{J} = 0,$$

and the work done by the gas is

$$W = 0 + 1.81 \times 10^3\,\mathrm{J} - 1.93 \times 10^3\,\mathrm{J} = 520\,\mathrm{J}.$$

(b) First find the initial volume. Since $p_1 V_1 = nRT_1$,

$$V_1 = \frac{nRT_1}{p_1} = \frac{(1.00\,\mathrm{mol})(8.314\,\mathrm{J/mol \cdot K})(300\,\mathrm{K})}{(1.013 \times 10^5\,\mathrm{Pa})} = 2.46 \times 10^{-2}\,\mathrm{m}^3.$$

Since $1 \rightarrow 2$ is a constant volume process $V_2 = V_1 = 2.46 \times 10^{-2}\,\mathrm{m}^3$.
The pressure for state 2 is

$$p_2 = \frac{nRT_2}{V_2} = \frac{(1.00\,\mathrm{mol})(8.314\,\mathrm{J/mol \cdot K})(600\,\mathrm{K})}{2.46 \times 10^{-2}\,\mathrm{m}^3} = 2.02 \times 10^5\,\mathrm{Pa}.$$

This is equivalent to 1.99 atm.

Since $3 \rightarrow 1$ is a constant pressure process, the pressure for state 3 is the same as the pressure for state 1: $p_3 = p_1 = 1.013 \times 10^5\,\mathrm{Pa}$ (1.00 atm). The volume for state 3 is

$$V_3 = \frac{nRT_3}{p_3} = \frac{(1.00\,\mathrm{mol})(8.314\,\mathrm{J/mol \cdot K})(455\,\mathrm{K})}{1.013 \times 10^5\,\mathrm{Pa}} = 3.73 \times 10^{-2}\,\mathrm{m}^3.$$

Chapter 22

4P

(a) Heat is added during the portion of the process from a to b. This portion occurs at constant volume (V_b), so $Q_{in} = nC_V \Delta T$. The gas is a monatomic ideal gas, so $C_V = \frac{3}{2}R$ and the ideal gas law gives $\Delta T = (1/nR)(p_bV_b - p_aV_a) = (1/nR)(p_b - p_a)V_b$. Thus $Q_{in} = \frac{3}{2}(p_b - p_a)V_b$. V_b and p_b are given. We need to find p_a. Now p_a is the same as p_c and points c and b are connected by an adiabatic process. Thus $p_cV_c^\gamma = p_bV_b^\gamma$ and

$$p_a = p_c = \left(\frac{V_b}{V_c}\right)^\gamma p_b = \left(\frac{1}{8.00}\right)^{5/3}(1.013 \times 10^6\,\text{Pa}) = 3.167 \times 10^4\,\text{Pa}.$$

The heat added is

$$Q_{in} = \frac{3}{2}(1.013 \times 10^6\,\text{Pa} - 3.167 \times 10^4\,\text{Pa})(1.00 \times 10^{-3}\,\text{m}^3) = 1.47 \times 10^3\,\text{J}.$$

(b) Heat leaves the gas during the portion of the process from c to a. This is a constant pressure process, so

$$Q_{out} = nC_p \Delta T = \frac{5}{2}(p_aV_a - p_cV_c) = \frac{5}{2}p_a(V_a - V_c)$$

$$= \frac{5}{2}(3.167 \times 10^4\,\text{Pa})(-7.00)(1.00 \times 10^{-3}\,\text{m}^3) = -5.54 \times 10^2\,\text{J}.$$

The substitution $V_a - V_c = V_a - 8.00V_a = -7.00V_a$ was made.

(c) For a complete cycle the change in the internal energy is zero and $W = Q = 1.47 \times 10^3\,\text{J} - 5.54 \times 10^2\,\text{J} = 9.18 \times 10^2\,\text{J}$.

(d) The efficiency is $e = W/Q_{in} = (9.18 \times 10^2\,\text{J})/(1.47 \times 10^3\,\text{J}) = 0.624$.

8E

(a) The efficiency is

$$e = \frac{T_H - T_C}{T_H} = \frac{(235 - 115)\,\text{K}}{(235 + 273)\,\text{K}} = 0.236.$$

Note that a temperature difference has the same value on the Kelvin and Celsius scales. Since the temperatures in the equation must be in kelvins, the temperature in the denominator was converted to the Kelvin scale.

(b) Since the efficiency is given by $e = |W|/|Q_H|$, $|W| = e|Q_H| = 0.236(6.30 \times 10^4\,\text{cal}) = 1.49 \times 10^4\,\text{cal}$. This is equivalent to $6.22 \times 10^4\,\text{J}$.

12E

For a Carnot cycle working as an engine the efficiency is related to the reservoir temperatures by $e = (T_H - T_C)/T_H$. Thus $T_H = (T_H - T_C)/e = (75\,\text{K})/(0.22) = 341\,\text{K}\,(= 68°\,\text{C})$. The temperature of the cold reservoir is $T_C = T_H - 75 = 341\,\text{K} - 75\,\text{K} = 266\,\text{K}\,(= -7°\,\text{C})$.

14E

An ideal refrigerator working between a hot reservoir at temperature T_H and a cold reservoir at temperature T_C has a coefficient of performance K that is given by $K = T_C/(T_H - T_C)$. For the refrigerator of this problem $T_H = 96°\,\text{F} = 309\,\text{K}$ and $T_C = 70°\,\text{F} = 294\,\text{K}$, so $K = (294\,\text{K})/(309\,\text{K} - 294\,\text{K}) = 19.6$. The coefficient of performance is the heat Q_C drawn from the cold reservoir divided by the work done: $K = Q_C/W$. Thus $Q_c = KW = (19.6)(1.0\,\text{J}) = 19.6\,\text{J}$.

17E

(a) In general, the efficiency of an engine operating between two reservoirs is given by $e = |W|/|Q_H|$, where W is the work done by the engine and Q_H is the heat taken in at the high temperature reservoir. For a Carnot engine $e = (T_H - T_C)/T_H$, where T_H is the temperature of the high temperature reservoir and T_C is the temperature of the low temperature reservoir. Thus

$$|W| = e|Q_H| = \left(\frac{T_H - T_C}{T_H}\right)|Q_H| = \left(\frac{320\,\text{K} - 260\,\text{K}}{320\,\text{K}}\right)(500\,\text{J}) = 94\,\text{J}\,.$$

(b) In general, the coefficient of performance for a refrigerator operating between two reservoirs is given by $K = |Q_C|/|W|$, where Q_C is the heat absorbed at the low temperature reservoir and W is the work done on the system. For a Carnot refrigerator $K = T_C/(T_H - T_C)$, where T_H and T_C have the same meanings as in part (a). Thus

$$|W| = \frac{|Q_C|}{K} = \left(\frac{T_H - T_C}{T_C}\right)|Q_C| = \left(\frac{320\,\text{K} - 260\,\text{K}}{260\,\text{K}}\right)(1000\,\text{J}) = 230\,\text{J}\,.$$

20P

The coefficient of performance for a refrigerator is given by $K = |Q_C|/|W|$, where Q_C is the heat absorbed from the cold reservoir and W is the work done by the refrigerator, a negative value. The first law of thermodynamics yields $Q_H + Q_C - W = 0$ for an integer number of cycles. Here Q_H is the heat ejected to the hot reservoir. Thus $Q_C = W - Q_H$. Q_H is negative and greater in magnitude than W, so $|Q_C| = Q_H - W$. Thus

$$K = \frac{Q_H - W}{W}\,.$$

The solution for W is $W = Q_H/(K + 1)$. In one hour

$$W = \frac{-1.8\,\text{Mcal}}{3.8 + 1} = -0.375\,\text{Mcal}\,.$$

The work done on the refrigerator is $-W = (0.375 \times 10^6\,\text{cal})(4.186\,\text{J/cal}) = 1.57 \times 10^6\,\text{J}$ and the rate at which work is done is $(1.57 \times 10^6\,\text{J})/(3600\,\text{s}) = 440\,\text{W}$.

24P

(a) On a T-S diagram an isothermal process is represented by a horizontal straight line and an adiabatic process is represented by a vertical straight line. Since a Carnot cycle consists of two isothermal and two adiabatic processes, it is represented by a rectangle.

(b) Heat is absorbed during the isothermal process at the high temperature. It is $Q_H = T_H \Delta S = (400\,\text{K})(0.60\,\text{J/K} - 0.10\,\text{J/K}) = 200\,\text{J}$. Heat is ejected during the isothermal process at the low temperature. It is $Q_C = T_C \Delta S = (250\,\text{K})(0.10\,\text{J/K} - 0.60\,\text{J/K}) = -125\,\text{J}$. The net heat absorbed over a cycle is $Q = Q_H + Q_C = 200\,\text{J} - 125\,\text{J} = 75\,\text{J}$.

(c) Since the internal energy does not change over a cycle the work done by the system equals the net heat absorbed, or 75 J.

27P

The efficiency of the engine is defined by $e = W/Q_1$ and is shown in the text to be $e = (T_1 - T_2)/T_1$, so $W/Q_1 = (T_1 - T_2)/T_1$. The coefficient of performance of the refrigerator is defined by $K = Q_4/W$ and is shown in the text to be $K = T_4/(T_3 - T_4)$, so $Q_4/W = T_4/(T_3 - T_4)$. Now $Q_4 = Q_3 - W$ so $(Q_3 - W)/W = T_4/(T_3 - T_4)$. The work done by the engine is used to drive the refrigerator so W is the same for the two. Solve the engine equation for W and substitute the resulting expression into the refrigerator equation. The engine equation yields $W = (T_1 - T_2)Q_1/T_1$ and the substitution yields

$$\frac{T_4}{T_3 - T_4} = \frac{Q_3}{W} - 1 = \frac{Q_3 T_1}{Q_1(T_1 - T_2)} - 1.$$

Solve for Q_3/Q_1:

$$\frac{Q_3}{Q_1} = \left(\frac{T_4}{T_3 - T_4} + 1\right)\left(\frac{T_1 - T_2}{T_1}\right) = \left(\frac{T_3}{T_3 - T_4}\right)\left(\frac{T_1 - T_2}{T_1}\right) = \frac{1 - (T_2/T_1)}{1 - (T_4/T_3)}.$$

29P

(a) If T_H is the temperature of the high temperature reservoir and T_C is the temperature of the low temperature reservoir, then the maximum efficiency of the engine is

$$e = \frac{T_H - T_C}{T_H} = \frac{(800 + 40)\,\text{K}}{(800 + 273)\,\text{K}} = 0.78.$$

(b) The efficiency is defined by $e = |W|/|Q_H|$, where W is the work done by the engine and Q_H is the heat input. W is positive. Over a complete cycle $Q_H = W + |Q_C|$, where Q_C is the heat output, so $e = W/(W + |Q_C|)$ and $|Q_C| = W[(1/e) - 1]$. Now $e = (T_H - T_C)/T_H$, where T_H is the temperature of the high temperature heat reservoir and T_C is the temperature of the low temperature reservoir. Thus $(1/e) - 1 = T_C/(T_H - T_C)$ and $|Q_C| = WT_C/(T_H - T_C)$. The heat output is used to melt ice at temperature T_i ($= -40°\,\text{C}$). The ice must be brought to $0°\,\text{C}$, then melted, so $|Q_C| = mc(T_f - T_i) + mL_F$, where m is the mass of ice melted, T_f is the melting temperature ($0°\,\text{C}$), c is the specific heat of ice, and L_F is the heat of fusion of ice. Thus $WT_C/(T_H - T_C) = mc(T_f - T_i) + mL_F$. Differentiate with respect to time and replace dW/dt with P, the power output of the engine. You should obtain $PT_C/(T_H - T_C) = (dm/dt)[c(T_f - T_i) + L_F]$. Thus

$$\frac{dm}{dt} = \left(\frac{PT_C}{T_H - T_C}\right)\left(\frac{1}{c(T_f - T_i) + L_F}\right).$$

Now $P = 100 \times 10^6 \, \text{W} = 2.39 \times 10^7 \, \text{cal/s}$, $T_C = 0 + 273 = 273 \, \text{K}$, $T_H = 800 + 273 = 1073 \, \text{K}$, $T_i = -40 + 273 = 233 \, \text{K}$, $T_f = 0 + 273 = 273 \, \text{K}$, $c = 0.50 \, \text{cal/g} \cdot \text{K}$ and $L_F = 80 \, \text{cal/g}$, so

$$\frac{dm}{dt} = \left[\frac{(2.39 \times 10^7 \, \text{cal/s})(273 \, \text{K})}{1073 \, \text{K} - 237 \, \text{K}} \right] \left[\frac{1}{(0.50 \, \text{cal/g} \cdot \text{K})(273 \, \text{K} - 233 \, \text{K}) + 80 \, \text{cal/g}} \right]$$

$$= 8.2 \times 10^4 \, \text{g/s} \,.$$

This is equivalent to 82 kg/s. Notice that the engine is now operated between 0° C and 800° C.

32P

(a) The pressure at 2 is $p_2 = 3p_1$, as given in the problem statement. The volume is $V_2 = V_1 = nRT_1/p_1$.

The process $4 \to 1$ is adiabatic so $p_4 V_4^\gamma = p_1 V_1^\gamma$ and

$$p_4 = \left(\frac{V_1}{V_4} \right)^\gamma p_1 = \frac{p_1}{4^\gamma}$$

since $V_4 = 4V_1$. The temperature at 4 is

$$T_4 = \frac{p_4 V_4}{nR} = \left(\frac{p_1}{4^\gamma} \right) \left(\frac{4nRT_1}{p_1} \right) \left(\frac{1}{nR} \right) = \frac{T_1}{4^{\gamma - 1}} \,.$$

The process $2 \to 3$ is adiabatic so $p_2 V_2^\gamma = p_3 V_3^\gamma$ and $p_3 = (V_2/V_3)^\gamma p_2$. Substitute $V_3 = 4V_1$, $V_2 = V_1$, and $p_2 = 3p_1$ to obtain

$$p_3 = \frac{3p_1}{4^\gamma} \,.$$

The temperature is

$$T_3 = \frac{p_3 V_3}{nR} = \left(\frac{1}{nR} \right) \left(\frac{3p_1}{4^\gamma} \right) \left(\frac{4nRT_1}{p_1} \right) = \frac{3T_1}{4^{\gamma - 1}} \,,$$

where $V_3 = V_4 = 4V_1 = 4nRT/p_1$ was used.

(b) The efficiency of the cycle is $e = W/Q_{12}$, where W is the total work done by the gas during the cycle and Q_{12} is the heat added during the $1 \to 2$ portion of the cycle, the only portion in which heat is added.

The work done during the portion of the cycle from 2 to 3 is $W_{23} = \int p \, dV$. Substitute $p = p_2 V_2^\gamma / V^\gamma$ to obtain

$$W_{23} = p_2 V_2^\gamma \int_{V_2}^{V_3} V^{-\gamma} \, dV = \left(\frac{p_2 V_2^\gamma}{\gamma - 1} \right) \left(V_2^{1-\gamma} - V_3^{1-\gamma} \right) \,.$$

Substitute $V_2 = V_1$, $V_3 = 4V_1$, and $p_3 = 3p_1$ to obtain

$$W_{23} = \left(\frac{3p_1 V_1}{1 - \gamma} \right) \left(1 - \frac{1}{4^{\gamma - 1}} \right) = \left(\frac{3nRT_1}{\gamma - 1} \right) \left(1 - \frac{1}{4^{\gamma - 1}} \right) \,.$$

Similarly the work done during the portion of the cycle from 4 to 1 is

$$W_{41} = \left(\frac{p_1 V_1^{\gamma}}{\gamma - 1}\right)\left(V_4^{1-\gamma} - V_1^{\gamma-1}\right) = -\left(\frac{p_1 V_1}{\gamma - 1}\right)\left(1 - \frac{1}{4^{\gamma-1}}\right) = -\left(\frac{nRT_1}{\gamma - 1}\right)\left(1 - \frac{1}{4^{\gamma-1}}\right).$$

No work is done during the $1 \rightarrow 2$ and $3 \rightarrow 4$ portions so the total work done by the gas during the cycle is

$$W = W_{23} + W_{41} = \left(\frac{2nRT_1}{\gamma - 1}\right)\left(1 - \frac{1}{4^{\gamma-1}}\right).$$

The heat added is $Q_{12} = nC_V(T_2 - T_1) = nC_V(3T_1 - T_1) = 2nC_V T_1$, where C_V is the molar specific heat at constant volume. Now $\gamma = C_p/C_V = (C_V + R)/C_V = 1 + (R/C_V)$, so $C_V = R/(\gamma - 1)$. Here C_p is the molar specific heat at constant pressure, which for an ideal gas is $C_p = C_V + R$. Thus $Q_{12} = 2nRT_1/(\gamma - 1)$. The efficiency is

$$e = \frac{2nRT_1}{\gamma - 1}\left(1 - \frac{1}{4^{\gamma-1}}\right)\frac{\gamma - 1}{2nRT_1} = 1 - \frac{1}{4^{\gamma-1}}.$$

34E

(a) The heat added is $Q = mc\,\Delta T$, where m is the mass of the copper block, c is the specific heat of copper, and ΔT is the change in temperature. Since c is given as $9.2 \times 10^{-2}\,\text{cal/g}\cdot{}^\circ\text{C}$, use $m = 1000\,\text{g}$ for the mass of the block. Thus $Q = (1000\,\text{g})(9.2 \times 10^{-2}\,\text{cal/g}\cdot{}^\circ\text{C})(75^\circ\,\text{C}) = 6.9 \times 10^3\,\text{cal}$.

(b) The change in entropy is given by the integral $\Delta S = \int(1/T)\,dQ$. Replace dQ with $mc\,dT$ and let T_i be the initial temperature and T_f be the final temperature. Then

$$\Delta S = mc\int_{T_i}^{T_f}\frac{dT}{T} = mc\ln\frac{T_f}{T_i}$$

$$= (1000\,\text{g})(9.2 \times 10^{-2}\,\text{cal/g}\cdot{}^\circ\text{C})\ln\frac{(273 + 100)\,\text{K}}{(273 + 25)\,\text{K}} = 21\,\text{cal/K}.$$

39E

The change in entropy is $\Delta S = \int(1/T)\,dQ = Q/T$, since the process is isothermal. Thus $Q = T\,\Delta S = [(132 + 273)\,\text{K}](46.0\,\text{J/K}) = 1.86 \times 10^4\,\text{J} = 4.45 \times 10^3\,\text{cal}$.

43E

(a) The heat that leaves the aluminum has magnitude $Q = m_a c_a(T_{ai} - T_f)$, where m_a is the mass of the aluminum, c_a is the specific heat of aluminum, T_{ai} is the initial temperature of the aluminum, and T_f is the final temperature of the aluminum-water system. The heat that enters the water is $Q = m_w c_w(T_f - T_{wi})$, where m_w is the mass of the water, c_w is the specific heat of water, and T_{wi} is the initial temperature of the water. The two heats are the same since no heat is lost, so $m_a c_a(T_{ai} - T_f) = m_w c_w(T_f - T_{wi})$ and

$$T_f = \frac{m_a c_a T_{ai} + m_w c_w T_{wi}}{m_a c_a + m_w c_w}.$$

The specific heat of aluminum is 0.215 cal/g·°C and the specific heat of water is 1.00 cal/g·°C. Thus

$$T_f = \frac{(200\,\text{g})(0.215\,\text{cal/g}\cdot\text{°C})(100°\,\text{C}) + (50.0\,\text{g})(1.00\,\text{cal/g}\cdot\text{°C})(20°\,\text{C})}{(200\,\text{g})(0.215\,\text{cal/g}\cdot\text{°C}) + (50.0\,\text{g})(1.00\,\text{cal/g}\cdot\text{°C})}$$

$$= 57.0°\,\text{C}.$$

This is equivalent to 330 K.

(b) Now temperatures must be given in kelvins: $T_{ai} = 393\,\text{K}$, $T_{wi} = 293\,\text{K}$, and $T_f = 330\,\text{K}$. For the aluminum $dQ = m_a c_a\, dT$ and the change in entropy is

$$\Delta S_a = \int \frac{dQ}{T} = m_a c_a \int_{T_{ai}}^{T_f} \frac{dT}{T} = m_a c_a \ln \frac{T_f}{T_{ai}}$$

$$= (200\,\text{g})(0.215\,\text{cal/g}\cdot\text{°C}) \ln \frac{330\,\text{K}}{373\,\text{K}} = -5.27\,\text{cal/K}.$$

(c) The entropy change for the water is

$$\Delta S_w = \int \frac{dQ}{T} = m_w c_w \int_{T_{wi}}^{T_f} \frac{dT}{T} = m_w c_w \ln \frac{T_f}{T_{wi}}$$

$$= (50.0\,\text{g})(1.00\,\text{cal/g}\cdot\text{°C}) \ln \frac{330\,\text{K}}{293\,\text{K}} = +5.95\,\text{cal/K}.$$

(d) The change in the total entropy of the aluminum-water system is $\Delta S = \Delta S_a + \Delta S_w = -5.27\,\text{cal/K} + 5.95\,\text{cal/K} = +0.68\,\text{cal/K}$.

44P

The ice warms to 0° C, then melts, and the resulting water warms to the temperature of the lake water 15° C.

As the ice warms the heat it receives when the temperature changes by dT is $dQ = mc_I\,dT$, where m is the mass of the ice and c_I is the specific heat of ice. If T_i (= 263 K) is the initial temperature and T_f (= 273 K) is the final temperature, then the change in its entropy is

$$\Delta S = \int \frac{dQ}{T} = mc_I \int_{T_i}^{T_f} \frac{dT}{T} = mc_I \ln \frac{T_f}{T_i}$$

$$= (10\,\text{g})(0.50\,\text{cal/g}\cdot\text{°C}) \ln \frac{273\,\text{K}}{263\,\text{K}} = 0.19\,\text{cal/K}.$$

Melting is an isothermal process. The heat leaving the ice is mL_F, where L_F is the heat of fusion for ice. Thus $\Delta S = Q/T = mL_F/T = (10\,\text{g})(79.5\,\text{cal/g})/(273\,\text{K}) = 2.91\,\text{cal/K}$.

For the warming of the water from the ice, the change in entropy is

$$\Delta S = mc_w \ln \frac{T_f}{T_i},$$

where c_w is the specific heat of water (1.00 cal/g·°C). Thus

$$\Delta S = (10\,\text{g})(1.00\,\text{cal/g}\cdot\text{°C}) \ln \frac{(273 + 15)\,\text{K}}{273\,\text{K}} = 0.54\,\text{cal/K}.$$

The total change in entropy for the ice and the water it becomes is $\Delta S = 0.19\,\text{cal/K} + 2.91\,\text{cal/K} + 0.54\,\text{cal/K} = 3.64\,\text{cal/K}$.

Since the temperature of the lake does not change significantly when the ice melts, the change in its entropy is $\Delta S = Q/T$, where Q is the heat it receives (the negative of the heat it supplies the ice) and T is its temperature. When the ice warms to $0°$ C,

$$Q = -mc_I(T_f - T_i) = -(10\,\text{g})(0.50\,\text{cal/g}\cdot°\text{C})(10°\,\text{C}) = -50\,\text{cal}.$$

When the ice melts

$$Q = -mL = -(10\,\text{g})(79.55\,\text{cal/g}) = -796\,\text{cal}.$$

When the water from the ice warms

$$Q = -mc_w(T_f - T_i) = -(10\,\text{g})(1.00\,\text{cal/g}\cdot°\text{C})(15°\,\text{C}) = -150\,\text{cal}.$$

The total heat leaving the lake water is $Q = -50\,\text{cal} - 796\,\text{cal} - 150\,\text{cal} = -996\,\text{cal}$. The change in entropy is

$$\Delta S = \frac{-996\,\text{cal/K}}{(273+15)\,\text{K}} = -3.46\,\text{cal/K}.$$

The change in the entropy of the ice-lake system is $\Delta S = (3.64 - 3.46)\,\text{cal/K} = 0.18\,\text{cal/K}$.

46P

(a) Since the gas is ideal, its pressure p is given in terms of the number of moles n, the volume V and the temperature T by $p = nRT/V$. The work done by the gas during the isothermal expansion is

$$W = \int_{V_1}^{V_2} p\,dV = nRT \int_{V_1}^{V_2} \frac{dV}{V} = nRT \ln \frac{V_2}{V_1}.$$

Substitute $V_2 = 2V_1$ to obtain

$$W = nRT \ln 2 = (4\,\text{mol})(8.314\,\text{J/mol}\cdot\text{K})(400\,\text{K})\ln 2 = 9.22 \times 10^3\,\text{J}.$$

(b) Since the expansion is isothermal, the change in entropy is given by $\Delta S = \int(1/T)\,dQ = Q/T$, where Q is the heat absorbed. According to the first law of thermodynamics $\Delta E_{int} = Q - W$. Now the internal energy of an ideal gas depends only on the temperature and not on the pressure and volume. Since the expansion is isothermal $\Delta E_{int} = 0$ and $Q = W$. Thus

$$\Delta S = \frac{W}{T} = \frac{9.22 \times 10^3\,\text{J}}{400\,\text{K}} = 23.1\,\text{J/K}.$$

(c) $\Delta S = 0$ for all reversible adiabatic processes.

51P

(a) The final mass of ice is $(1773\,\text{g} + 227\,\text{g})/2 = 1000\,\text{g}$. This means $773\,\text{g}$ of water froze. Energy in the form of heat left the system in the amount mL_F, where m is the mass of the water that froze and L_F is the heat of fusion of water. The process is isothermal, so the change in entropy is $\Delta S = Q/T = -mL_F/T = -(773\,\text{g})(79.5\,\text{cal/g})/(273\,\text{K}) = -225\,\text{cal/K}$.

(b) Now 773 g of ice is melted. The change in entropy is

$$\Delta S = \frac{Q}{T} = \frac{mL_F}{T} = +225\,\text{cal/K}\,.$$

(c) Yes, it is consistent with the second law of thermodynamics. Over the entire cycle the change in entropy of the water-ice system is zero even though part of the cycle is irreversible. However, the system is not closed. To consider a closed system we must include whatever exchanges energy with the ice and water. Suppose it is a constant temperature heat reservoir during the freezing portion of the cycle and a Bunsen burner during the melting portion. During freezing the entropy of the reservoir increases by 225 cal/K. As far as the reservoir-water-ice system is concerned, the process is adiabatic and reversible so its total entropy does not change. The melting process is irreversible, so the total entropy of the burner-water-ice system increases. The entropy of the burner either increases or else decreases by less than 225 cal/K.

53P

(a) Work is done only for the ab portion of the process. This portion is at constant pressure so the work done by the gas is

$$W = \int_{V_0}^{4V_0} p_0\,dV = p_0(4V_0 - V_0) = 3p_0V_0\,.$$

(b) Use the first law: $\Delta E_{\text{int}} = Q - W$. Since the process is at constant volume the work done by the gas is zero and $E_{\text{int}} = Q$. The heat Q absorbed by the gas is $Q = nC_V\,\Delta T$, where C_V is the molar specific heat at constant volume and ΔT is the change in temperature. Since the gas is a monatomic ideal gas, $C_V = \frac{3}{2}R$. Use the ideal gas law to find that the initial temperature is $T_b = p_bV_b/nR = 4p_0V_0/nR$ and that the final temperature is $T_c = p_cV_c/nR = (2p_0)(4V_0)/nR = 8p_0V_0/nR$. Thus

$$Q = \frac{3}{2}nR\left(\frac{8p_0V_0}{nR} - \frac{4p_0V_0}{nR}\right) = 6p_0V_0\,.$$

The change in the internal energy is $\Delta E_{\text{int}} = 6p_0V_0$. Since $n = 1\,\text{mol}$ this can also be written $Q = 6RT_0$.

Since the process is at constant volume use $dQ = nC_V\,dT$ to obtain

$$\Delta S = \int \frac{dQ}{T} = nC_V \int_{T_b}^{T_c} \frac{dT}{T} = nC_V \ln\frac{T_c}{T_b}\,.$$

Substitute $C_V = \frac{3}{2}R$. Use the ideal gas law to write

$$\frac{T_c}{T_b} = \frac{p_cV_c}{p_bV_b} = \frac{(2p_0)(4V_0)}{p_0(4V_0)} = 2\,.$$

Thus $\Delta S = \frac{3}{2}nR\ln 2$. Since $n = 1$ this is $\Delta S = \frac{3}{2}R\ln 2$.

(c) For a complete cycle $\Delta E_{\text{int}} = 0$ and $\Delta S = 0$.

Chapter 23

2E

The magnitude of the force is given by

$$F = \frac{1}{4\pi\epsilon_0} \frac{q_1 q_2}{r^2},$$

where q_1 and q_2 are the magnitudes of the charges and r is the distance between them. Thus

$$F = \frac{(8.99 \times 10^9\,\text{N}\cdot\text{m}^2/\text{C}^2)(3.00 \times 10^{-6}\,\text{C})(1.50 \times 10^{-6}\,\text{C})}{(12.0 \times 10^{-2}\,\text{m})^2} = 2.81\,\text{N}.$$

3E

The magnitude of the force that either charge exerts on the other is given by

$$F = \frac{1}{4\pi\epsilon_0} \frac{q_1 q_2}{r^2},$$

where r is the distance between them. Thus

$$r = \sqrt{\frac{q_1 q_2}{4\pi\epsilon_0 F}}$$

$$= \sqrt{\frac{(8.99 \times 10^9\,\text{N}\cdot\text{m}^2/\text{C}^2)(26.0 \times 10^{-6}\,\text{C})(47.0 \times 10^{-6}\,\text{C})}{5.70\,\text{N}}} = 1.38\,\text{m}.$$

12P

Assume the charge distributions are spherically symmetric so Coulomb's law can be used. Let q_1 and q_2 be the original charges and choose the coordinate system so the force on q_2 is positive if it is repelled by q_1. Take the distance between the charges to be r. Then the force on q_2 is

$$F_a = -\frac{1}{4\pi\epsilon_0} \frac{q_1 q_2}{r^2}.$$

The negative sign indicates that the spheres attract each other.

After the wire is connected the spheres, being identical, have the same charge. Since charge is conserved the total charge is the same as it was originally. This means the charge on each sphere is $(q_1 + q_2)/2$. The force is now one of repulsion and is given by

$$F_b = \frac{1}{4\pi\epsilon_0} \frac{(q_1 + q_2)^2}{4r^2}.$$

Solve the two force equations simultaneously for q_1 and q_2. The first gives

$$q_1 q_2 = -4\pi\epsilon_0 r^2 F_a = \frac{(0.500\,\text{m})^2 (0.108\,\text{N})}{8.99 \times 10^9\,\text{N} \cdot \text{m}^2/\text{C}^2} = -3.00 \times 10^{-12}\,\text{C}^2$$

and the second gives

$$q_1 + q_2 = r\sqrt{4 \times 4\pi\epsilon_0 F_b} = (0.500\,\text{m})\sqrt{\frac{4(0.0360\,\text{N})}{8.99 \times 10^9\,\text{N} \cdot \text{m}^2/\text{C}^2}} = 2.00 \times 10^{-6}\,\text{C}.$$

Thus

$$q_2 = \frac{-(3.00 \times 10^{-12}\,\text{C}^2)}{q_1}$$

and

$$q_1 - \frac{3.00 \times 10^{-12}\,\text{C}^2}{q_1} = 2.00 \times 10^{-6}\,\text{C}.$$

Multiply by q_1 to obtain the quadratic equation

$$q_1^2 - (2.00 \times 10^{-6}\,\text{C})q_1 - 3.00 \times 10^{-12}\,\text{C}^2 = 0.$$

The solutions are

$$q_1 = \frac{-2.00 \times 10^{-6}\,\text{C} \pm \sqrt{(2.00 \times 10^{-6}\,\text{C})^2 + 4(3.00 \times 10^{-12}\,\text{C}^2)}}{2}.$$

If the plus sign is used $q_1 = 1.00 \times 10^{-6}\,\text{C}$ and if the minus sign is used $q_1 = -3.00 \times 10^{-6}\,\text{C}$. Use $q_2 = (-3.00 \times 10^{-12})/q_1$ to calculate q_2. If $q_1 = 1.00 \times 10^{-6}\,\text{C}$ then $q_2 = -3.00 \times 10^{-6}\,\text{C}$ and if $q_1 = -3.00 \times 10^{-6}\,\text{C}$ then $q_2 = 1.00 \times 10^{-6}\,\text{C}$. Since the spheres are identical the solutions are essentially the same: one sphere originally had charge $1.00 \times 10^{-6}\,\text{C}$ and the other had charge $-3.00 \times 10^{-6}\,\text{C}$.

Another solution exists. If the signs of the charges are reversed the forces remain the same, so a charge of $-1.00 \times 10^{-6}\,\text{C}$ on one sphere and a charge of $3.00 \times 10^{-6}\,\text{C}$ on the other also satisfies the conditions of the problem.

15P

(a) If the system of three charges is to be in equilibrium the force on each charge must be zero. Let the third charge be q_0. It must lie between the other two or else the forces acting on it due to the other charges would be in the same direction and q_0 could not be in equilibrium. Suppose q_0 is a distance x from q, as shown on the diagram to the right. The force acting on q_0 is then given by

$$F_0 = \frac{1}{4\pi\epsilon_0} \left[\frac{q q_0}{x^2} - \frac{4 q q_0}{(L-x)^2} \right] = 0,$$

where the positive direction was taken to be toward the right. Solve this equation for x. Canceling common factors yields $1/x^2 = 4/(L-x)^2$ and taking the square root yields $1/x = 2/(L-x)$. The solution is $x = L/3$.

The force on q is

$$F_q = \frac{1}{4\pi\epsilon_0} \left[\frac{qq_0}{x^2} + \frac{4q^2}{L^2} \right] = 0.$$

Solve for q_0: $q_0 = -4qx^2/L^2 = -(4/9)q$, where $x = L/3$ was used.

The force on $4q$ is

$$F_{4q} = \frac{1}{4\pi\epsilon_0} \left[\frac{4q^2}{L^2} + \frac{4qq_0}{(L-x)^2} \right] = \frac{1}{4\pi\epsilon_0} \left[\frac{4q^2}{L^2} + \frac{4(-4/9)q^2}{(4/9)L^2} \right]$$

$$= \frac{1}{4\pi\epsilon_0} \left[\frac{4q^2}{L^2} - \frac{4q^2}{L^2} \right] = 0.$$

With $q_0 = -(4/9)q$ and $x = L/3$ all three charges are in equilibrium.

(b) If q_0 moves toward q the force of attraction exerted by q is greater in magnitude than the force of attraction exerted by $4q$ and q_0 continues to move toward q and away from its initial position. The equilibrium is unstable.

16P

(a) The magnitudes of the gravitational and electrical forces must be the same:

$$\frac{1}{4\pi\epsilon_0} \frac{q^2}{r^2} = G \frac{mM}{r^2},$$

where q is the charge on either body, r is the center-to-center separation of the Earth and moon, G is the universal gravitational constant, M is the mass of the Earth, and m is the mass of the moon. Solve for q:

$$q = \sqrt{4\pi\epsilon_0 GmM}.$$

According to Appendix C of the text, $M = 5.98 \times 10^{24}$ kg, and $m = 7.36 \times 10^{22}$ kg, so

$$q = \sqrt{\frac{(6.67 \times 10^{-11}\,\text{N} \cdot \text{m}^2/\text{kg}^2)(7.36 \times 10^{22}\,\text{kg})(5.98 \times 10^{24}\,\text{kg})}{8.99 \times 10^9\,\text{N} \cdot \text{m}^2/\text{C}^2}}$$

$$= 5.7 \times 10^{13}\,\text{C}.$$

Notice that the distance r cancels because both the electric and gravitational forces are proportional to $1/r^2$.

(b) The charge on a hydrogen ion is $e = 1.60 \times 10^{-19}$ C so there must be

$$\frac{q}{e} = \frac{5.7 \times 10^{13}\,\text{C}}{1.6 \times 10^{-19}\,\text{C}} = 3.6 \times 10^{32}\,\text{ions}.$$

Each ion has a mass of 1.67×10^{-27} kg so the total mass needed is

$$(3.6 \times 10^{32})(1.67 \times 10^{-27}\,\text{kg}) = 6.0 \times 10^5\,\text{kg}.$$

The magnitude of the force of either of the charges on the other is given by

$$F = \frac{1}{4\pi\epsilon_0}\frac{q(Q-q)}{r^2},$$

where r is the distance between the charges. You want the value of q that maximizes the function $f(q) = q(Q-q)$. Set the derivative df/dq equal to zero. This yields $2q - Q = 0$, or $q = Q/2$.

19P

(a) A force diagram for one of the balls is shown to the right. The force of gravity mg acts downward, the electrical force F_e of the other ball acts to the left, and the tension in the thread acts along the thread, at the angle θ to the vertical. The ball is in equilibrium, so its acceleration is zero. The y component of Newtons second law yields $T\cos\theta - mg = 0$ and the x component yields $T\sin\theta - F_e = 0$. Solve the first equation for T ($T = mg/\cos\theta$) and substitute the result into the second to obtain $mg\tan\theta - F_e = 0$.

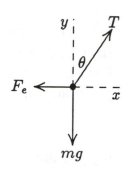

Now

$$\tan\theta = \frac{x/2}{\sqrt{L^2 - (x/2)^2}}.$$

If L is much larger than x we may neglect $x/2$ in the denominator and write $\tan\theta \approx x/2L$. This is equivalent to approximating $\tan\theta$ by $\sin\theta$. The magnitude of the electrical force of one ball on the other is

$$F_e = \frac{q^2}{4\pi\epsilon_0 x^2}.$$

When these two substitutions are made in the equation $mg\tan\theta = F_e$, it becomes

$$\frac{mgx}{2L} = \frac{1}{4\pi\epsilon_0}\frac{q^2}{x^2},$$

so

$$x^3 = \frac{1}{4\pi\epsilon_0}\frac{2q^2L}{mg},$$

and

$$x = \left(\frac{q^2L}{2\pi\epsilon_0 mg}\right)^{1/3}.$$

(b) Solve $x^3 = (1/4\pi\epsilon_0)(2q^2L/mg)$ for q:

$$q = \sqrt{\frac{4\pi\epsilon_0 mgx^3}{2L}} = \sqrt{\frac{(0.010\,\text{kg})(9.8\,\text{m/s}^2)(0.050\,\text{m})^3}{2(8.99\times10^9\,\text{N}\cdot\text{m}^2/\text{C}^2)(1.20\,\text{m})}} = \pm2.4\times10^{-8}\,\text{C}.$$

21P

(a) Since the rod is in equilibrium the net force acting on it is zero and the net torque about any point is also zero. Write an expression for the net torque about the bearing, equate it to zero, and solve for x. The charge Q on the left exerts an upward force of magnitude $(1/4\pi\epsilon_0)(qQ/h^2)$, at a distance $L/2$ from the bearing. Take the torque to be positive. The attached weight exerts a downward force of magnitude W, at a distance $L/2 - x$ from the bearing. This torque is positive. The charge Q on the right exerts an upward force of magnitude $(1/4\pi\epsilon_0)(2qQ/h^2)$, at a distance $L/2$ from the bearing. This torque is negative. The equation for rotational equilibrium is

$$\frac{1}{4\pi\epsilon_0}\frac{qQ}{h^2}\frac{L}{2} + W\left(\frac{L}{2} - x\right) - \frac{1}{4\pi\epsilon_0}\frac{2qQ}{h^2}\frac{L}{2} = 0.$$

The solution for x is

$$x = \frac{L}{2}\left(1 + \frac{1}{4\pi\epsilon_0}\frac{qQ}{h^2W}\right).$$

(b) The net force on the rod vanishes. If N is the magnitude of the upward force exerted by the bearing then

$$W - \frac{1}{4\pi\epsilon_0}\frac{qQ}{h^2} - \frac{1}{4\pi\epsilon_0}\frac{2qQ}{h^2} - N = 0.$$

Solve for h so that $N = 0$. The result is

$$h = \sqrt{\frac{1}{4\pi\epsilon_0}\frac{3qQ}{W}}.$$

24E

The mass of an electron is $m = 9.11 \times 10^{-31}$ kg so the number of electrons in a collection with total mass $M = 75.0$ kg is

$$N = \frac{M}{m} = \frac{75.0\,\text{kg}}{9.11 \times 10^{-31}\,\text{kg}} = 8.23 \times 10^{31}\text{ electrons}.$$

The total charge of the collection is

$$q = -Ne = -(8.23 \times 10^{31})(1.60 \times 10^{-19}\,\text{C}) = -1.32 \times 10^{13}\,\text{C}.$$

26E

(a) The magnitude of the force between the ions is given by

$$F = \frac{q^2}{4\pi\epsilon_0 r^2},$$

where q is the charge on either of them and r is the distance between them. Solve for the charge:

$$q = r\sqrt{4\pi\epsilon_0 F} = (5.0 \times 10^{-10})\sqrt{\frac{3.7 \times 10^{-9}\,\text{N}}{8.99 \times 10^9\,\text{N}\cdot\text{m}^2/\text{C}^2}} = 3.2 \times 10^{-19}\,\text{C}.$$

(b) Let N be the number of electrons missing from each ion. Then $Ne = q$, or

$$N = \frac{q}{e} = \frac{3.2 \times 10^{-19}\,\text{C}}{1.60 \times 10^{-19}\,\text{C}} = 2.$$

27E

(a) The magnitude of the force is given by

$$F = \frac{q^2}{4\pi\epsilon_0 r^2},$$

where q is the magnitude of the charge on each drop and r is the center-to-center separation of the drops. Thus

$$F = \frac{(8.99 \times 10^9\,\text{N} \cdot \text{m}^2/\text{C}^2)(1.00 \times 10^{-16}\,\text{C})^2}{(1.00 \times 10^{-2}\,\text{m})^2} = 8.99 \times 10^{-19}\,\text{N}.$$

(b) If N is the number of excess electrons on each drop then

$$N = -\frac{q}{e} = -\frac{-1.00 \times 10^{-16}\,\text{C}}{1.60 \times 10^{-19}\,\text{C}} = 625.$$

31P

If charge of magnitude q passes through the lamp in time t the current is $i = q/t$. The charge is the total charge on one mole of electrons, or $q = N_A e$, where N_A is the Avogadro constant. Thus $i = N_A e/t$, or

$$t = \frac{N_A e}{i} = \frac{(6.02 \times 10^{23})(1.60 \times 10^{-19}\,\text{C})}{0.83\,\text{A}} = 1.2 \times 10^5\,\text{s}.$$

This is equivalent to 1.3 d.

34P

(a) Every cesium ion at a corner of the cube exerts a force of the same magnitude on the chlorine ion at the cube center. Each force is a force of attraction and is directed toward the cesium ion that exerts it, along the body diagonal of the cube. We can pair every cesium ion with another, diametrically positioned at the opposite corner of the cube. Since the two ions in such a pair exerts forces that have the same magnitude but are oppositely directed, the two forces sum to zero and, since every cesium ion can be paired in this way, the total force on the chlorine ion is zero.

(b) Rather than remove a cesium ion, superpose charge $-e$ at the position of one cesium ion. This neutralizes the ion and, as far as the electrical force on the chlorine ion is concerned, it is equivalent to removing the ion. The forces of the 8 cesium ions at the cube corners sum to zero, so the only force on the chlorine ion is the force of the added charge.

The length of a body diagonal of a cube is $\sqrt{3}a$, where a is the length of a cube edge. Thus the distance from the center of the cube to a corner is $d = (\sqrt{3}/2)a$. The force has magnitude

$$F = \frac{1}{4\pi\epsilon_0}\frac{e^2}{d^2} = \frac{1}{4\pi\epsilon_0}\frac{e^2}{(3/4)a^2}$$
$$= \frac{(8.99 \times 10^9\,\text{N} \cdot \text{m}^2/\text{C}^2)(1.60 \times 10^{-19}\,\text{C})^2}{(3/4)(0.40 \times 10^{-9}\,\text{m})^2} = 1.9 \times 10^{-9}\,\text{N}.$$

Since both the added charge and the chlorine ion are negative the force is one of repulsion. The chlorine ion is pushed away from the site of the missing cesium ion.

38E

None of the reactions given include a beta decay, so the number of protons, the number of neutrons, and the number of electrons are each conserved. Atomic numbers (numbers of protons and numbers of electrons) and molar masses (combined numbers of protons and neutrons) can be found in Appendix D of the text.

(a) ^{1}H has 1 proton, 1 electron, and 0 neutrons and ^{9}Be has 4 protons, 4 electrons, and $9-4 = 5$ neutrons, so X has $1+4 = 5$ protons, $1+4 = 5$ electrons, and $0+5-1 = 4$ neutrons. One of the neutrons is freed in the reaction. X must be boron with a molar mass of $5 + 4 = 9\,\text{g/mol}$: ^{9}B.

(b) ^{12}C has 6 protons, 6 electrons, and $12 - 6 = 6$ neutrons and ^{1}H has 1 proton, 1 electron, and 0 neutrons, so X has $6 + 1 = 7$ protons, $6 + 1 = 7$ electrons, and $6 + 0 = 6$ neutrons. It must be nitrogen with a molar mass of $7 + 6 = 13\,\text{g/mol}$: ^{13}N.

(c) ^{15}N has 7 protons, 7 electrons, and $15-7 = 8$ neutrons; ^{1}H has 1 proton, 1 electron, and 0 neutrons; and ^{4}He has 2 protons, 2 electrons, and $4-2 = 2$ neutrons; so X has $7 + 1 - 2 = 6$ protons, 6 electrons, and $8 + 0 - 2 = 6$ neutrons. It must be carbon with a molar mass of $6 + 6 = 12$: ^{12}C.

41E

(a) The Planck time is

$$T_P = \sqrt{\frac{hG}{2\pi c^5}}.$$

The product of a speed and a time is a length, so we may take the Planck length to be

$$L_P = T_P c = \sqrt{\frac{hG}{2\pi c^5}}c^2 = \sqrt{\frac{hG}{2\pi c^3}}.$$

(b) Substitute values:

$$L_P = \sqrt{\frac{(6.63 \times 10^{-34}\,\text{J} \cdot \text{s})(6.67 \times 10^{-11}\,\text{N} \cdot \text{m}^2/\text{kg}^2)}{2\pi(3.00 \times 10^8\,\text{m/s})^3}} = 1.61 \times 10^{-35}\,\text{m}.$$

42P

(a) Suppose the Planck mass is the combination $h^n G^m c^p$, where n, m, p are exponents. Since the units of h are $kg \cdot m^2/s$, the units of G are $m^3/kg \cdot s^2$, and the units of c are m/s, the units of $h^n G^m c^p$ are

$$\left(\frac{kg^n m^{2m}}{s^n}\right) \left(\frac{m^{3m}}{kg^m s^{2m}}\right) \left(\frac{m^p}{s^p}\right) = kg^{(n-m)} m^{(2n+3m+p)} s^{(-n-2m-p)}.$$

For this combination of units to be kg, we want $n - m = 1$, $2n + 3m + p = 0$, and $-n - 2m - p = 0$. Solve these equations simultaneously for n, m, and p. Add the last two to get $n + m = 0$. This means $n = -m$ and the first two equations become $-2m = 1$ and $m + p = 0$. The first gives $m = -1/2$ and the second gives $p = -m = +1/2$. Finally $n = -m$ gives $n = +1/2$. Thus the Planck mass is

$$M_P = h^{1/2} G^{-1/2} c^{1/2} = \sqrt{hc/G}.$$

(b) Substitute the values:

$$M_P = \sqrt{\frac{(6.63 \times 10^{-34}\,J \cdot s)(3.00 \times 10^8\,m/s)}{6.67 \times 10^{-11}\,N \cdot m^2/kg^2}} = 5.46 \times 10^{-8}\,kg.$$

Chapter 24

5E

The diagram to the right is an edge view of the disk and shows the field lines above it. Near the disk the lines are perpendicular to the surface and since the disk is uniformly charged the lines are uniformly distributed over the surface. Far away from the disk the lines are like those of a single point charge (the charge on the disk). Extended back to the disk (along the dotted lines of the diagram) they intersect at the center of the disk.

If the disk is positively charged the lines are directed outward from the disk. If the disk is negatively charged they are directed inward toward the disk. Lines below the disk are exactly like those above.

9E

Since the magnitude of the electric field produced by a point charge q is given by $E = q/4\pi\epsilon_0 r^2$, where r is the distance from the charge to the point where the field has magnitude E, the charge is

$$q = 4\pi\epsilon_0 r^2 E = \frac{(0.50\,\text{m})^2(2.0\,\text{N/C})}{8.99 \times 10^9\,\text{N}\cdot\text{m}^2/\text{C}^2} = 5.6 \times 10^{-11}\,\text{C}.$$

12E

Since the charge is uniformly distributed throughout a sphere the electric field at the surface is exactly the same as it would be if the charge were all at the center. That is, the magnitude of the field is

$$E = \frac{q}{4\pi\epsilon_0 R^2},$$

where q is the magnitude of the total charge and R is the sphere radius. The magnitude of the total charge is Ze, so

$$E = \frac{Ze}{4\pi\epsilon_0 R^2} = \frac{(8.99 \times 10^9\,\text{N}\cdot\text{m}^2/\text{C}^2)(94)(1.60 \times 10^{-19}\,\text{C})}{(6.64 \times 10^{-15}\,\text{m})^2} = 3.07 \times 10^{21}\,\text{N/C}.$$

The field is normal to the surface and since the charge is positive it points outward from the surface.

17P

At points between the charges the individual electric fields are in the same direction and do not cancel. Charge q_2 has a greater magnitude than charge q_1 so a point of zero field must be closer to q_1 than to q_2. It must be to the right of q_1 on the diagram.

Put the origin at q_2 and let x be the coordinate of P, the point where the field vanishes. Then the total electric field at P is given by

$$E = \frac{1}{4\pi\epsilon_0}\left[\frac{q_2}{x^2} - \frac{q_1}{(x-d)^2}\right],$$

where q_1 and q_2 are the magnitudes of the charges. If the field is to vanish,

$$\frac{q_2}{x^2} = \frac{q_1}{(x-d)^2}.$$

Take the square root of both sides to obtain $\sqrt{q_1}/x = \sqrt{q_2}/(x-d)$. The solution for x is

$$x = \left(\frac{\sqrt{q_2}}{\sqrt{q_2}-\sqrt{q_1}}\right)d = \left(\frac{\sqrt{4.0q_1}}{\sqrt{4.0q_1}-\sqrt{q_1}}\right)d$$

$$= \left(\frac{2.0}{2.0-1.0}\right)d = 2.0d = (2.0)(50\,\text{cm}) = 100\,\text{cm}.$$

The point is 50 cm to the right of q_1.

22P

Choose the coordinate axes as shown on the diagram to the right. At the center of the square the electric fields produced by the charges at the lower left and upper right corners are both along the x axis and each points away from the center and toward the charge that produces it. Since each charge is a distance $d = \sqrt{2}a/2 = a/\sqrt{2}$ away from the center the net field due to these two charges is

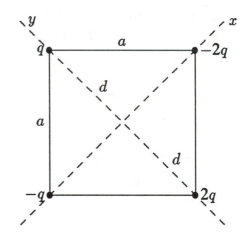

$$E_x = \frac{1}{4\pi\epsilon_0}\left[\frac{2q}{a^2/2} - \frac{q}{a^2/2}\right]$$

$$= \frac{1}{4\pi\epsilon_0}\frac{q}{a^2/2} = \frac{(8.99\times10^9\,\text{N}\cdot\text{m}^2/\text{C}^2)(1.0\times10^{-8}\,\text{C})}{(0.050\,\text{m})^2/2} = 7.19\times10^4\,\text{N/C}.$$

At the center of the square the field produced by the charges at the upper left and lower right corners are both along the y axis and each points toward the center, away from the charge that produces it. The net field produced at the center by these charges is

$$E_y = \frac{1}{4\pi\epsilon_0}\left[\frac{2q}{a^2/2} - \frac{q}{a^2/2}\right] = \frac{1}{4\pi\epsilon_0}\frac{q}{a^2/2} = 7.19\times10^4\,\text{N/C}.$$

The magnitude of the field is

$$E = \sqrt{E_x^2 + E_y^2} = \sqrt{2}(7.19\times10^4\,\text{N/C}) = 1.02\times10^5\,\text{N/C}$$

and the angle it makes with the x axis is

$$\theta = \tan^{-1}\frac{E_y}{E_x} = \tan^{-1}(1) = 45°.$$

It is upward in the diagram, from the center of the square toward the center of the upper side.

23E

The magnitude of the dipole moment is given by $p = qd$, where q is the positive charge in the dipole and d is the separation of the charges. For the dipole described in the problem $p = (1.60 \times 10^{-19}\,\text{C})(4.30 \times 10^{-9}\,\text{m}) = 6.88 \times 10^{-28}\,\text{C} \cdot \text{m}$. The dipole moment is a vector that points from the negative toward the positive charge.

27P

Think of the quadrupole as composed of two dipoles, each with dipole moment of magnitude $p = qd$. The moments point in opposite directions and produce fields in opposite directions at points on the dipole axis. Consider a point on the axis a distance z above the center of the quadrupole and take an upward pointing field to be positive. Then the field produced by the upper dipole of the pair is $qd/2\pi\epsilon_0(z - d/2)^3$ and the field produced by the lower is $-qd/2\pi\epsilon_0(z + d/2)^3$. Use the binomial expansions $(z - d/2)^{-3} \approx z^{-3} - 3z^{-4}(-d/2)$ and $(z + d/2)^{-3} \approx z^{-3} - 3z^{-4}(d/2)$ to obtain

$$E = \frac{qd}{2\pi\epsilon_0}\left[\frac{1}{z^3} + \frac{3d}{2z^4} - \frac{1}{z^3} + \frac{3d}{2z^4}\right] = \frac{6qd^2}{4\pi\epsilon_0 z^4}.$$

Let $Q = 2qd^2$. Then

$$E = \frac{3Q}{4\pi\epsilon_0 z^4}.$$

30P

The electric field at a point on the axis of a uniformly charged ring, a distance z from the ring center, is given by

$$E = \frac{qz}{4\pi\epsilon_0(z^2 + R^2)^{3/2}},$$

where q is the charge on the ring and R is the radius of the ring (see Eq. 24–19). For q positive the field points upward at points above the ring and downward at points below the ring. Take the positive direction to be upward. Then the force acting on an electron on the axis is

$$F = -\frac{eqz}{4\pi\epsilon_0(z^2 + R^2)^{3/2}}.$$

For small amplitude oscillations $z \ll R$ and z can be neglected in the denominator. Thus

$$F = -\frac{eqz}{4\pi\epsilon_0 R^3}.$$

The force is a restoring force: it pulls the electron toward the equilibrium point $z = 0$. Furthermore, the magnitude of the force is proportional to z, just as if the electron were attached to a spring with spring constant $k = eq/4\pi\epsilon_0 R^3$. The electron moves in simple harmonic motion with an angular frequency given by

$$\omega = \sqrt{\frac{k}{m}} = \sqrt{\frac{eq}{4\pi\epsilon_0 m R^3}} \,,$$

where m is the mass of the electron.

35P

Consider an infinitesimal section of the rod of length dx, a distance x from the left end, as shown in the diagram to the right. It contains charge $dq = \lambda\, dx$ and is a distance r from P. The magnitude of the field it produces at P is given by

$$dE = \frac{1}{4\pi\epsilon_0}\frac{\lambda\, dx}{r^2}.$$

The x component is

$$dE_x = -\frac{1}{4\pi\epsilon_0}\frac{\lambda\, dx}{r^2}\sin\theta$$

and the y component is

$$dE_y = -\frac{1}{4\pi\epsilon_0}\frac{\lambda\, dx}{r^2}\cos\theta.$$

Use θ as the variable of integration. Substitute $r = R/\cos\theta$, $x = R\tan\theta$, and $dx = (R/\cos^2\theta)\,d\theta$. The limits of integration are 0 and $\pi/2\,$rad. Thus

$$E_x = -\frac{\lambda}{4\pi\epsilon_0}\int_0^{\pi/2}\sin\theta\, d\theta = \frac{\lambda}{4\pi\epsilon_0}\cos\theta\Big|_0^{\pi/2} = -\frac{\lambda}{4\pi\epsilon_0 R}\,.$$

and

$$E_y = -\frac{\lambda}{4\pi\epsilon_0}\int_0^{\pi/2}\cos\theta\, d\theta = -\frac{\lambda}{4\pi\epsilon_0}\sin\theta\Big|_0^{\pi/2} = -\frac{\lambda}{4\pi\epsilon_0 R}\,.$$

Notice that $E_x = E_y$ no matter what the value of R. Thus **E** makes an angle of $45°$ with the negative x axis for all values of R.

38P

At a point on the axis of a uniformly charged disk a distance z above the center of the disk the magnitude of the electric field is

$$E = \frac{\sigma}{2\epsilon_0}\left[1 - \frac{z}{\sqrt{z^2 + R^2}}\right],$$

where R is the radius of the disk and σ is the surface charge density on the disk. See Eq. 24–27. The magnitude of the field at the center of the disk ($z = 0$) is $E_c = \sigma/2\epsilon_0$. You want to solve for the value of z such that $E/E_c = 1/2$. This means

$$1 - \frac{z}{\sqrt{z^2 + R^2}} = \frac{1}{2}$$

or

$$\frac{z}{\sqrt{z^2 + R^2}} = \frac{1}{2}.$$

Square both sides, then multiply them by $z^2 + R^2$ to obtain $z^2 = (z^2/4) + (R^2/4)$. Thus $z^2 = R^2/3$ and $z = R/\sqrt{3}$.

39E

The magnitude of the force acting on the electron is $F = eE$, where E is the magnitude of the electric field at its location. The acceleration of the electron is given by Newton's second law:

$$a = \frac{F}{m} = \frac{eE}{m} = \frac{(1.60 \times 10^{-19}\,\text{C})(2.00 \times 10^4\,\text{N/C})}{9.11 \times 10^{-31}\,\text{kg}} = 3.51 \times 10^{15}\,\text{m/s}^2.$$

43E

(a) The magnitude of the force on the particle is given by $F = qE$, where q is the magnitude of the charge carried by the particle and E is the magnitude of the electric field at the location of the particle. Thus

$$E = \frac{F}{q} = \frac{3.0 \times 10^{-6}\,\text{N}}{2.0 \times 10^{-9}\,\text{C}} = 1.5 \times 10^3\,\text{N/C}.$$

The force points downward and the charge is negative, so the field points upward.

(b) The magnitude of the electrostatic force on a proton is

$$F_e = eE = (1.60 \times 10^{-19}\,\text{C})(1.5 \times 10^3\,\text{N/C}) = 2.4 \times 10^{-16}\,\text{N}.$$

A proton is positively charged, so the force is in the same direction as the field, upward.

(c) The magnitude of the gravitational force on the proton is

$$F_g = mg = (1.67 \times 10^{-27}\,\text{kg})(9.8\,\text{m/s}^2) = 1.6 \times 10^{-26}\,\text{N}.$$

The force is downward.

(d) The ratio of the forces is

$$\frac{F_e}{F_g} = \frac{2.4 \times 10^{-16}\,\text{N}}{1.64 \times 10^{-26}\,\text{N}} = 1.5 \times 10^{10}.$$

46E

(a) The magnitude of the force acting on the proton is $F = eE$, where E is the magnitude of the electric field. According to Newton's second law the acceleration of the proton is $a = F/m = eE/m$, where m is the mass of the proton. Thus

$$a = \frac{(1.60 \times 10^{-19}\,\text{C})(2.00 \times 10^4\,\text{N/C})}{1.67 \times 10^{-27}\,\text{kg}} = 1.92 \times 10^{12}\,\text{m/s}^2\,.$$

(b) Assume the proton starts from rest and use the kinematic equations $x = \frac{1}{2}at^2$ and $v = at$ to show that $v = \sqrt{2ax} = \sqrt{2(1.92 \times 10^{12}\,\text{m/s}^2)(0.0100\,\text{m})} = 1.96 \times 10^5\,\text{m/s}$.

49E

When the drop is in equilibrium the force of gravity is balanced by the force of the electric field: $mg = qE$, where m is the mass of the drop, q is the charge on the drop, and E is the magnitude of the electric field. The mass of the drop is given by $m = (4\pi/3)r^3\rho$, where r is its radius and ρ is its mass density. Thus

$$\begin{aligned} q &= \frac{mg}{E} = \frac{4\pi r^3 \rho g}{3E} \\ &= \frac{4\pi(1.64 \times 10^{-6}\,\text{m})^3(851\,\text{kg/m}^3)(9.8\,\text{m/s}^2)}{3(1.92 \times 10^5\,\text{N/C})} = 8.0 \times 10^{-19}\,\text{C} \end{aligned}$$

and $q/e = (8.0 \times 10^{-19}\,\text{C})/(1.60 \times 10^{-19}\,\text{C}) = 5$.

54P

Take the positive direction to be to the right in the diagram. The acceleration of the proton is $a_p = eE/m_p$ and the acceleration of the electron is $a_e = -eE/m_e$, where E is the magnitude of the electric field, m_p is the mass of the proton, and m_e is the mass of the electron. Take the origin to be at the initial position of the proton. Then the coordinate of the proton at time t is $x = \frac{1}{2}a_p t^2$ and the coordinate of the electron is $x = L + \frac{1}{2}a_e t^2$. They pass each other when their coordinates are the same, or $\frac{1}{2}a_p t^2 = L + \frac{1}{2}a_e t^2$. This means $t^2 = 2L/(a_p - a_e)$ and

$$\begin{aligned} x &= \frac{a_p}{a_p - a_e}L = \frac{eE/m_p}{(eE/m_p) + (eE/m_e)}L = \frac{m_e}{m_e + m_p}L \\ &= \frac{9.11 \times 10^{-31}\,\text{kg}}{9.11 \times 10^{-31}\,\text{kg} + 1.67 \times 10^{-27}\,\text{kg}}(0.050\,\text{m}) = 2.7 \times 10^{-5}\,\text{m}\,. \end{aligned}$$

55P

(a) Suppose the pendulum is at the angle θ with the vertical. The force diagram is shown to the right. T is the tension in the thread, mg is the force of gravity, and qE is the force of the electric field. The field points upward and the charge is positive, so the force is upward. Take the angle shown to be positive. Then the torque on the sphere about the point where the thread is attached to the upper plate is $\tau = -(mg-qE)\ell\sin\theta$. If $mg > qE$ then the torque is a restoring torque; it tends to pull the pendulum back to its equilibrium position.

If the amplitude of the oscillation is small, $\sin\theta$ can be replaced by θ in radians and the torque is $\tau = -(mg - qE)\ell\theta$. The torque is proportional to the angular displacement and the pendulum moves in simple harmonic motion. Its angular frequency is $\omega = \sqrt{(mg - qE)\ell/I}$, where I is the rotational inertia of the pendulum. Since $I = m\ell^2$ for a simple pendulum,

$$\omega = \sqrt{\frac{(mg - qE)\ell}{m\ell^2}} = \sqrt{\frac{g - qE/m}{\ell}}$$

and the period is

$$T = \frac{2\pi}{\omega} = 2\pi\sqrt{\frac{\ell}{g - qE/m}}\,.$$

If $qE > mg$ the torque is not a restoring torque and the pendulum does not oscillate.

(b) The force of the electric field is now downward and the torque on the pendulum is $\tau = -(mg + qE)\ell\theta$ if the angular displacement is small. The period of oscillation is

$$T = 2\pi\sqrt{\frac{\ell}{g + qE/m}}\,.$$

56P

The electric field is upward in the diagram and the charge is negative, so the force of the field on it is downward. The magnitude of the acceleration is $a = eE/m$, where E is the magnitude of the field and m is the mass of the electron. Its numerical value is

$$a = \frac{(1.60 \times 10^{-19}\,\text{C})(2.00 \times 10^3\,\text{N/C})}{9.11 \times 10^{-31}\,\text{kg}} = 3.51 \times 10^{14}\,\text{m/s}^2\,.$$

Put the origin of a coordinate system at the initial position of the electron. Take the x axis to be horizontal and positive to the right; take the y axis to be vertical and positive toward the top of the page. The kinematic equations are $x = v_0 t \cos\theta$, $y = v_0 t \sin\theta - \frac{1}{2}at^2$, and $v_y = v_0 \sin\theta - at$.

First find the greatest y coordinate attained by the electron. If it is less than d the electron does not hit the upper plate. If it is greater than d it will hit the upper plate if the corresponding x coordinate is less than L. The greatest y coordinate occurs when $v_y = 0$. This means $v_0 \sin\theta - at = 0$ or $t = (v_0/a)\sin\theta$ and

$$y_{max} = \frac{v_0^2 \sin^2\theta}{a} - \frac{1}{2}a\frac{v_0^2 \sin^2\theta}{a^2} = \frac{1}{2}\frac{v_0^2 \sin^2\theta}{a}$$
$$= \frac{(6.00 \times 10^6\,\text{m/s})^2 \sin^2 45°}{2(3.51 \times 10^{14}\,\text{m/s}^2)} = 2.56 \times 10^{-2}\,\text{m}\,.$$

Since this is greater than $d\,(= 2.00\,\text{cm})$ the electron might hit the upper plate.

Now find the x coordinate of the position of the electron when $y = d$. Since

$$v_0 \sin \theta = (6.00 \times 10^6 \, \text{m/s}) \sin 45° = 4.24 \times 10^6 \, \text{m/s}$$

and

$$2ad = 2(3.51 \times 10^{14} \, \text{m/s}^2)(0.0200 \, \text{m}) = 1.40 \times 10^{13} \, \text{m}^2/\text{s}^2 \,,$$

the solution to $d = v_0 t \sin \theta - \frac{1}{2}at^2$ is

$$t = \frac{v_0 \sin \theta - \sqrt{v_0^2 \sin^2 \theta - 2ad}}{a}$$

$$= \frac{4.24 \times 10^6 \, \text{m/s} - \sqrt{(4.24 \times 10^6 \, \text{m/s})^2 - 1.40 \times 10^{13} \, \text{m}^2/\text{s}^2}}{3.51 \times 10^{14} \, \text{m/s}^2} = 6.43 \times 10^{-9} \, \text{s} \,.$$

The negative root was used because we want the *earliest* time for which $y = d$.
The x coordinate is

$$x = v_0 t \cos \theta$$

$$= (6.00 \times 10^6 \, \text{m/s})(6.43 \times 10^{-9} \, \text{s}) \cos 45° = 2.72 \times 10^{-2} \, \text{m} \,.$$

This is less than L so the electron hits the upper plate at $x = 2.72 \, \text{cm}$.

60P

The magnitude of the torque acting on the dipole is given by $\tau = pE \sin \theta$, where p is the magnitude of the dipole moment, E is the magnitude of the electric field, and θ is the angle between the dipole moment and the field. It is a restoring torque: it always tends to rotate the dipole moment toward the direction of the electric field. If θ is positive the torque is negative and vice-versa. Write $\tau = -pE \sin \theta$. If the amplitude of the motion is small we may replace $\sin \theta$ with θ in radians. Thus $\tau = -pE\theta$. Since the magnitude of the torque is proportional to the angle of rotation, the dipole oscillates in simple harmonic motion, just like a torsional pendulum with torsion constant $\kappa = pE$. The angular frequency ω is given by

$$\omega^2 = \frac{\kappa}{I} = \frac{pE}{I} \,,$$

where I is the rotational inertia of the dipole. The frequency of oscillation is

$$f = \frac{\omega}{2\pi} = \frac{1}{2\pi}\sqrt{\frac{pE}{I}} \,.$$

Chapter 25

2E

The vector area **A** and the electric field **E** are shown on the diagram to the right. The angle θ between them is $180° - 35° = 145°$ so the electric flux through the area is $\Phi = \mathbf{E} \cdot \mathbf{A} = EA \cos\theta = (1800\,\text{N/C})(3.2 \times 10^{-3}\,\text{m})^2 \cos 145° = -1.5 \times 10^{-2}\,\text{N} \cdot \text{m}^2/\text{C}$.

7E

Use Gauss' law: $\epsilon_0 \Phi = q$, where Φ is the total flux through the cube surface and q is the net charge inside the cube. Thus

$$\Phi = \frac{q}{\epsilon_0} = \frac{1.8 \times 10^{-6}\,\text{C}}{8.85 \times 10^{-12}\,\text{C}^2/\text{N} \cdot \text{m}^2} = 2.0 \times 10^5\,\text{N} \cdot \text{m}^2/\text{C}.$$

11P

Let A be the area of one face of the cube, E_u be the magnitude of the electric field at the upper face, and E_ℓ be the magnitude of the field at the lower face. Since the field is downward the flux through the upper face is negative and the flux through the lower face is positive. The flux through the other faces is zero, so the total flux through the cube surface is $\Phi = A(E_\ell - E_u)$. The net charge inside the cube is given by Gauss' law:

$$q = \epsilon_0 \Phi = \epsilon_0 A(E_\ell - E_u) = (8.85 \times 10^{-12}\,\text{C}^2/\text{N} \cdot \text{m}^2)(100\,\text{m})^2(100\,\text{N/C} - 60.0\,\text{N/C})$$
$$= 3.54 \times 10^{-6}\,\text{C} = 3.54\,\mu\text{C}.$$

13P

The total flux through any surface that completely surrounds the point charge is q/ϵ_0. If you stack identical cubes side by side and directly on top of each other, you will find that 8 cubes meet at any corner. Thus one-eighth of the field lines emanating from the point charge pass through a cube with a corner at the charge and the total flux through the surface of such a cube is $q/8\epsilon_0$. Now the field lines are radial, so at each of the 3 cube faces that meet at the charge the lines are parallel to the face and the flux through the face is zero. The fluxes through each of the other 3 faces are the same so the flux through each of them is one-third the total. That is, the flux through each of these faces is $(1/3)(q/8\epsilon_0) = q/24\epsilon_0$.

16E

(a) The charge on the surface of the sphere is the product of the surface charge density σ and the surface area of the sphere ($4\pi r^2$, where r is the radius). Thus

$$q = 4\pi r^2 \sigma = 4\pi \left(\frac{1.2\,\text{m}}{2}\right)^2 (8.1 \times 10^{-6}\,\text{C/m}^2) = 3.66 \times 10^{-5}\,\text{C}.$$

(b) Choose a Gaussian surface in the form a sphere, concentric with the conducting sphere and with a slightly larger radius. The flux is given by Gauss' law:

$$\Phi = \frac{q}{\epsilon_0} = \frac{3.66 \times 10^{-5}\,\text{C}}{8.85 \times 10^{-12}\,\text{C}^2/\text{N}\cdot\text{m}^2} = 4.1 \times 10^6\,\text{N}\cdot\text{m}^2/\text{C}.$$

19P

(a) Consider a Gaussian surface that is completely within the conductor and surrounds the cavity. Since the electric field is zero everywhere on the surface, the net charge it encloses is zero. The net charge is the sum of the charge q in the cavity and the charge q_w on the cavity wall, so $q + q_w = 0$ and $q_w = -q = -3.0 \times 10^{-6}\,\text{C}$.

(b) The net charge Q of the conductor is the sum of the charge on the cavity wall and the charge q_s on the outer surface of the conductor, so $Q = q_w + q_s$ and $q_s = Q - q_w = (10 \times 10^{-6}\,\text{C}) - (-3.0 \times 10^{-6}\,\text{C}) = +1.3 \times 10^{-5}\,\text{C}$.

21E

The magnitude of the electric field produced by a uniformly charged infinite line is $E = \lambda/2\pi\epsilon_0 r$, where λ is the linear charge density and r is the distance from the line to the point where the field is measured. See Eq. 24–14. Thus

$$\lambda = 2\pi\epsilon_0 E r = 2\pi(8.85 \times 10^{12}\,\text{C}^2/\text{N}\cdot\text{m}^2)(4.5 \times 10^4\,\text{N/C})(2.0\,\text{m}) = 5.0 \times 10^{-6}\,\text{C/m}.$$

26P

The electric field is radially outward from the central wire. You want to find its magnitude in the region between the wire and the cylinder as a function of the distance r from the wire. Use a Gaussian surface in the form of a cylinder with radius r and length ℓ, concentric with the wire. The radius is greater than the radius of the wire and less than in the radius of the inner cylinder wall. Only the charge on the wire is enclosed by the Gaussian surface; denote it by q. The area of the rounded surface of the Gaussian cylinder is $2\pi r\ell$ and the flux through it is $\Phi = 2\pi r\ell E$. If we neglect fringing there is no flux through the ends of the cylinder, so Φ is the total flux. Gauss' law yields $q = 2\pi\epsilon_0 r\ell E$. Since the magnitude of the field at the cylinder wall is known take the Gaussian surface to coincide with the wall. Then r is the radius of the wall and

$$q = 2\pi(8.85 \times 10^{-12}\,\text{C}^2/\text{N}\cdot\text{m}^2)(0.014\,\text{m})(0.16\,\text{m})(2.9 \times 10^4\,\text{N/C})$$
$$= 3.6 \times 10^{-9}\,\text{C}.$$

30P

(a) The diagram to the right shows a cross section of the charged cylinder (solid circle). Consider a Gaussian surface in the form of a cylinder with radius r and length ℓ, concentric with the cylinder of charge The cross section is shown as a dotted circle. Use Gauss' law to find an expression for the magnitude of the electric field at the Gaussian surface.

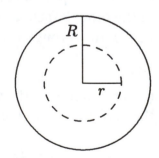

The charge within the Gaussian cylinder is $q = \rho V = \pi r^2 \ell \rho$, where $V\ (= \pi r^2 \ell)$ is the volume of the cylinder.

If ρ is positive the electric field lines are radially outward and are therefore normal to the rounded portion of the Gaussian cylinder and are distributed uniformly over it. None pass through the ends of the cylinder. Thus the total flux through the Gaussian cylinder is $\Phi = EA = 2\pi r\ell E$, where A $(= 2\pi r\ell)$ is the area of rounded portion of the cylinder.

Gauss' law $(\epsilon_0\Phi = q)$ yields $2\pi\epsilon_0 r\ell E = \pi r^2\ell\rho$, so

$$E = \frac{\rho r}{2\epsilon_0}.$$

(b) Take the Gaussian surface to be a cylinder with length ℓ and radius r (greater than R). The flux is again $\Phi = 2\pi r\ell E$. The charge enclosed is the total charge in a section of the charged cylinder with length ℓ. That is, $q = \pi R^2\ell\rho$. Gauss' law yields $2\pi\epsilon_0 r\ell E = \pi R^2\ell\rho$, so

$$E = \frac{R^2\rho}{2\epsilon_0 r}.$$

32E

(a) To calculate the electric field at a point very close to the center of a large, uniformly charged, conducting plate, we may replace the finite plate with an infinite plate with the same area charge density and take the magnitude of the field to be $E = \sigma/\epsilon_0$, where σ is the area charge density for the surface just under the point. The charge is distributed uniformly over both sides of the original plate, with half being on the side near the field point. Thus

$$\sigma = \frac{q}{2A} = \frac{6.0 \times 10^{-6}\,\text{C}}{2(0.080\,\text{m})^2} = 4.69 \times 10^{-4}\,\text{C/m}^2.$$

The magnitude of the field is

$$E = \frac{4.69 \times 10^{-4}\,\text{C/m}^2}{8.85 \times 10^{-12}\,\text{C}^2/\text{N}\cdot\text{m}^2} = 5.3 \times 10^7\,\text{N/C}.$$

The field is normal to the plate and since the charge on the plate is positive it points away from the plate.

(b) At a point far away from the plate the electric field is nearly that of a point particle with charge equal to the total charge on the plate. The magnitude of the field is $E = q/4\pi\epsilon_0 r^2$, where r is the distance from the plate. Thus

$$E = \frac{(8.99 \times 10^9\,\text{N}\cdot\text{m}^2/\text{C}^2)(6.0 \times 10^{-6}\,\text{C})}{(30\,\text{m})^2} = 60\,\text{N/C}.$$

34P

The forces acting on the ball are shown in the diagram to the right. The gravitational force has magnitude mg, where m is the mass of the ball; the electrical force has magnitude qE, where q is the charge on the ball and E is the electric field at the position of the ball; and the tension in the thread is denoted by T. The electric field produced by the plate is normal to the plate and points to the right. Since the ball is positively charged the electric force on it also points to the right. The tension in the thread makes the angle θ $(= 30°)$ with the vertical.

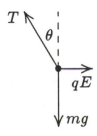

Since the ball is in equilibrium the net force on it vanishes. The sum of the horizontal components yields $qE - T\sin\theta = 0$ and the sum of the vertical components yields $T\cos\theta - mg = 0$. The expression $T = qE/\sin\theta$, from the first equation, is substituted into the second to obtain $qE = mg\tan\theta$.

The electric field produced by a large uniform plane of charge is given by $E = \sigma/2\epsilon_0$, where σ is the surface charge density. Thus

$$\frac{q\sigma}{2\epsilon_0} = mg\tan\theta$$

and

$$\sigma = \frac{2\epsilon_0 mg\tan\theta}{q}$$

$$= \frac{2(8.85 \times 10^{-12}\,\text{C}^2/\text{N}\cdot\text{m}^2)(1.0 \times 10^{-6}\,\text{kg})(9.8\,\text{m/s}^2)\tan 30°}{2.0 \times 10^{-8}\,\text{C}}$$

$$= 5.0 \times 10^{-9}\,\text{C/m}^2\,.$$

35P

The charge on the metal plate, which is negative, exerts a force of repulsion on the electron and stops it. First find an expression for the acceleration of the electron, then use kinematics to find the stopping distance. Take the initial direction of motion of the electron to be positive. Then the electric field is given by $E = \sigma/\epsilon_0$, where σ is the surface charge density on the plate. The force on the electron is $F = -eE = -e\sigma/\epsilon_0$ and the acceleration is

$$a = \frac{F}{m} = -\frac{e\sigma}{\epsilon_0 m}\,,$$

where m is the mass of the electron.

The force is constant, so we use constant acceleration kinematics. If v_0 is the initial velocity of the electron, v is the final velocity, and x is the distance traveled between the initial and final positions, then $v^2 - v_0^2 = 2ax$. Set $v = 0$ and replace a with $-e\sigma/\epsilon_0 m$, then solve for x. You should get

$$x = -\frac{v_0^2}{2a} = \frac{\epsilon_0 mv_0^2}{2e\sigma}\,.$$

Now $\frac{1}{2}mv_0^2$ is the initial kinetic energy K_0, so

$$x = \frac{\epsilon_0 K_0}{e\sigma}\,.$$

You must convert the given value of K_0 to joules. Since $1.00\,\text{eV} = 1.60 \times 10^{-19}\,\text{J}$, $100\,\text{eV} = 1.60 \times 10^{-17}\,\text{J}$. Thus

$$x = \frac{(8.85 \times 10^{-12}\,\text{C}^2/\text{N}\cdot\text{m}^2)(1.60 \times 10^{-17}\,\text{J})}{(1.60 \times 10^{-19}\,\text{C})(2.0 \times 10^{-6}\,\text{C/m}^2)} = 4.4 \times 10^{-4}\,\text{m}\,.$$

39P

(a) Use a Gaussian surface in the form of a box with rectangular sides. The cross section is shown with dotted lines in the diagram to the right. It is centered at the central plane of the slab so the left and right faces are each a distance x from the central plane. Take the thickness of the rectangular solid to be a, the same as its length, so the left and right faces are squares.

The electric field is normal to the left and right faces and is uniform over them. If ρ is positive it points outward at both faces: toward the left at the left face and toward the right at the right face. Furthermore, the magnitude is the same at both faces. The electric flux through each of these faces is Ea^2. The field is parallel to the other faces of the Gaussian surface and the flux through them is zero. The total flux through the Gaussian surface is $\Phi = 2Ea^2$.

The volume enclosed by the Gaussian surface is $2a^2x$ and the charge contained within it is $q = 2a^2x\rho$. Gauss' law yields $2\epsilon_0 Ea^2 = 2a^2x\rho$. Solve for E:

$$E = \frac{\rho x}{\epsilon_0}.$$

(b) Take a Gaussian surface of the same shape and orientation, but with $x > d/2$, so the left and right faces are outside the slab. The total flux through the surface is again $\Phi = 2Ea^2$ but the charge enclosed is now $q = a^2d\rho$. Gauss's law yields $2\epsilon_0 Ea^2 = a^2d\rho$, so

$$E = \frac{\rho d}{2\epsilon_0}.$$

40E

Charge is distributed uniformly over the surface of the sphere and the electric field it produces at points outside the sphere is like the field of a point particle with charge equal to the net charge on the sphere. That is, the magnitude of the field is given by $E = q/4\pi\epsilon_0 r^2$, where q is the magnitude of the charge on the sphere and r is the distance from the center of the sphere to the point where the field is measured. Thus

$$q = 4\pi\epsilon_0 r^2 E = \frac{(0.15\,\text{m})^2(3.0 \times 10^3\,\text{N/C})}{8.99 \times 10^9\,\text{N} \cdot \text{m}^2/\text{C}^2} = 7.5 \times 10^{-9}\,\text{C}.$$

The field points inward, toward the sphere center, so the charge is negative: -7.5×10^{-9} C.

45E

Use Gauss' law to find an expression for the magnitude of the electric field a distance r from the center of the atom. The field is radially outward and is uniform over any sphere centered at the atom's center. Take the Gaussian surface to be a sphere of radius r with its center at the center of the atom. If E is the magnitude of the field then the total flux through the Gaussian sphere is $\Phi = 4\pi r^2 E$. The charge enclosed by the Gaussian surface is the positive charge at the center of the atom and that portion of the negative charge within the surface. Since the negative charge is uniformly distributed throughout a sphere of radius R we can compute the charge inside the Gaussian sphere using a ratio of volumes. That is, the negative charge inside is $-Zer^3/R^3$. Thus the total charge enclosed is $Ze - Zer^3/R^3$. Gauss' law yields

$$4\pi\epsilon_0 r^2 E = Ze\left(1 - \frac{r^3}{R^3}\right).$$

Solve for E:

$$E = \frac{Ze}{4\pi\epsilon_0}\left(\frac{1}{r^2} - \frac{r}{R^3}\right).$$

48P

At all points where there is an electric field it is radially outward. For each part of the problem use a Gaussian surface in the form of a sphere that is concentric with the sphere of charge and passes through the point where the electric field is to be found. The field is uniform on the surface, so

$$\oint \mathbf{E} \cdot d\mathbf{A} = 4\pi r^2 E,$$

where r is the radius of the Gaussian surface.

(a) Here r is less than a and the charge enclosed by the Gaussian surface is $q(r/a)^3$. Gauss' law yields

$$4\pi r^2 E = \left(\frac{q}{\epsilon_0}\right)\left(\frac{r}{a}\right)^3,$$

so

$$E = \frac{qr}{4\pi\epsilon_0 a^3}.$$

(b) Here r is greater than a but less than b. The charge enclosed by the Gaussian surface is q, so Gauss' law becomes

$$4\pi r^2 E = \frac{q}{\epsilon_0}$$

and

$$E = \frac{q}{4\pi\epsilon_0 r^2}.$$

(c) The shell is conducting so the electric field inside it is zero.

(d) For $r > c$ the charge enclosed by the Gaussian surface is zero (charge q is inside the shell cavity and charge $-q$ is on the shell). Gauss' law yields

$$4\pi r^2 E = 0,$$

so $E = 0$.

(e) Consider a Gaussian surface that lies completely within the conducting shell. Since the electric field is everywhere zero on the surface $\oint \mathbf{E} \cdot d\mathbf{A} = 0$ and, according to Gauss' law, the net charge enclosed by the surface is zero. If Q_i is the charge on the inner surface of the shell, then $q + Q_i = 0$ and $Q_i = -q$. Let Q_o be the charge on the outer surface of the shell. Since the net charge on the shell is $-q$, $Q_i + Q_o = -q$. This means $Q_o = -q - Q_i = -q - (-q) = 0$.

51P

The proton is in uniform circular motion, with the electrical force of the charge on the sphere providing the centripetal force. According to Newton's second law $F = mv^2/r$, where F is the magnitude of the force, v is the speed of the proton, and r is the radius of its orbit, essentially the same as the radius of the sphere.

The magnitude of the force on the proton is $F = eq/4\pi\epsilon_0 r^2$, where q is the magnitude of the charge on the sphere. Thus

$$\frac{1}{4\pi\epsilon_0} \frac{eq}{r^2} = \frac{mv^2}{r},$$

so

$$q = \frac{4\pi\epsilon_0 mv^2 r}{e} = \frac{(1.67 \times 10^{-27}\,\text{kg})(3.00 \times 10^5\,\text{m/s})^2(0.0100\,\text{m})}{(8.99 \times 10^9\,\text{N} \cdot \text{m}^2/\text{C}^2)(1.60 \times 10^{-19}\,\text{C})}$$

$$= 1.04 \times 10^{-9}\,\text{C}.$$

The force must be inward, toward the center of the sphere, and since the proton is positively charged the electric field must also be inward. The charge on the sphere is negative: $q = -1.04 \times 10^{-9}\,\text{C}$.

53P

You wish to find an expression for the electric field inside the shell in terms of A and the distance from the center of the shell, then select A so the field does not depend on the distance.

Use a Gaussian surface in the form of a sphere with radius r_g, concentric with the spherical shell and within it ($a < r_g < b$). Gauss' law will be used to find the magnitude of the electric field a distance r_g from the shell center.

The charge that is both in the shell and within the Gaussian sphere is given by the integral $q_s = \int \rho\, dV$ over the portion of the shell within the Gaussian surface. Since the charge distribution has spherical symmetry, we may take dV to be the volume of a spherical shell with radius r and infinitesimal thickness dr: $dV = 4\pi r^2\, dr$. Thus

$$q_s = 4\pi \int_a^{r_g} \rho r^2\, dr = 4\pi \int_a^{r_g} \frac{A}{r} r^2\, dr = 4\pi A \int_a^{r_g} r\, dr = 2\pi A(r_g^2 - a^2).$$

The total charge inside the Gaussian surface is $q + q_s = q + 2\pi A(r_g^2 - a^2)$.

The electric field is radial, so the flux through the Gaussian surface is $\Phi = 4\pi r_g^2 E$, where E is the magnitude of the field. Gauss' law yields

$$4\pi\epsilon_0 E r_g^2 = q + 2\pi A(r_g^2 - a^2).$$

Solve for E:

$$E = \frac{1}{4\pi\epsilon_0}\left[\frac{q}{r_g^2} + 2\pi A - \frac{2\pi A a^2}{r_g^2}\right].$$

For the field to be uniform the first and last terms in the brackets must cancel. They do if $q - 2\pi A a^2 = 0$ or $A = q/2\pi a^2$.

55P

Suppose there is no charge in the immediate vicinity of q but q is in equilibrium under the combined forces of charges at other places. The net electric field at the position P of q is zero but q experiences an electric force if it is moved away from P. You want to show that it is not possible to set up an electric field that pushes q back toward P for every direction of its displacement.

Assume q is at P and surround it with an extremely small Gaussian surface in the form of a sphere centered at P. Now displace q from P to some point on the Gaussian sphere. If an electric force pushes q back there must be an inward pointing electric field at that point on the surface. If an electric field pushes q toward P no matter where it is placed on the surface, then there is an inward pointing electric field at every point on the surface. The net flux through the surface is not zero and, according to Gauss' law there must be charge within the Gaussian surface. This is a contradiction. We conclude that the field acting on a charge cannot push the charge back toward P for every possible displacement and that therefore the charge cannot be in stable equilibrium.

If there are places on the Gaussian surface where the electric field points inward and pushes q back toward its original position, then there must be other places where the field points outward and pushes q away from its original position.

Chapter 26

2E

(a) An ampere is a coulomb per second, so

$$84\,\text{A} \cdot \text{h} = \left(84\,\frac{\text{C} \cdot \text{h}}{\text{s}}\right)\left(3600\,\frac{\text{s}}{\text{h}}\right) = 3.0 \times 10^5\,\text{C}\,.$$

(b) The change in potential energy is $\Delta U = q\,\Delta V = (3.0 \times 10^5\,\text{C})(12\,\text{V}) = 3.6 \times 10^6\,\text{J}$.

3P

(a) When charge q moves through a potential difference ΔV its potential energy changes by $\Delta U = q\,\Delta V$. In this case, $\Delta U = (30\,\text{C})(1.0 \times 10^9\,\text{V}) = 3.0 \times 10^{10}\,\text{J}$.

(b) Equate the final kinetic energy of the automobile to the energy released by the lightning: $\Delta U = \frac{1}{2}mv^2$, where m is the mass of the automobile and v is its final speed. Thus

$$v = \sqrt{\frac{2\,\Delta U}{m}} = \sqrt{\frac{2(3.0 \times 10^{10}\,\text{J})}{1000\,\text{kg}}} = 7.7 \times 10^3\,\text{m/s}\,.$$

(c) Equate the energy required to melt mass m of ice to the energy released by the lightning: $\Delta U = mL_F$, where L_F is the heat of fusion for water. Thus

$$m = \frac{\Delta U}{L_F} = \frac{3.0 \times 10^{10}\,\text{J}}{3.3 \times 10^5\,\text{J/kg}} = 9.0 \times 10^4\,\text{kg}\,.$$

9E

The electric field produced by an infinite sheet of charge has magnitude $E = \sigma/2\epsilon_0$, where σ is the surface charge density. The field is normal to the sheet and is uniform. Place the origin of a coordinate system at the sheet and take the x axis to be parallel to the field and positive in the direction of the field. Then the electric potential is

$$V = V_s - \int_0^x E\,dx = V_s - Ex\,,$$

where V_s is the potential at the sheet. The equipotential surfaces are surfaces of constant x; that is, they are planes that are parallel to the plane of charge. If two surfaces are separated by Δx then their potentials differ by $\Delta V = -E\Delta x = (\sigma/2\epsilon_0)\Delta x$. Thus

$$\Delta x = \frac{2\epsilon_0\,\Delta V}{\sigma} = \frac{2(8.85 \times 10^{-12}\,\text{C}^2/\text{N} \cdot \text{m}^2)(50\,\text{V})}{0.10 \times 10^{-6}\,\text{C/m}^2} = 8.8 \times 10^{-3}\,\text{m}\,.$$

The potential difference between the wire and cylinder is given, not the linear charge density on the wire. Use Gauss' law to find an expression for the electric field a distance r from the center of the wire, between the wire and the cylinder, in terms of the linear charge density. Then integrate with respect to r to find an expression for the potential difference between the wire and cylinder in terms of the linear charge density. Use this result to obtain an expression for the linear charge density in terms of the potential difference and substitute the result into the equation for the electric field. This will give the electric field in terms of the potential difference and will allow you to compute numerical values for the field at the wire and at the cylinder.

For the Gaussian surface use a cylinder of radius r and length ℓ, concentric with the wire and cylinder. The electric field is normal to the rounded portion of the cylinder's surface and is uniform over that surface. This means the electric flux through the Gaussian surface is given by $2\pi r \ell E$, where E is the magnitude of the electric field. The charge enclosed by the Gaussian surface is $q = \lambda \ell$, where λ is the linear charge density on the wire. Gauss' law yields $2\pi \epsilon_0 r \ell E = \lambda \ell$. Thus

$$E = \frac{\lambda}{2\pi \epsilon_0 r} .$$

Since the field is radial, the difference in the potential V_c of the cylinder and the potential V_w of the wire is

$$\Delta V = V_w - V_c = -\int_{r_c}^{r_w} E\, dr = \int_{r_w}^{r_c} \frac{\lambda}{2\pi \epsilon_0 r}\, dr = \frac{\lambda}{2\pi \epsilon_0} \ln \frac{r_c}{r_w} ,$$

where r_w is the radius of the wire and r_c is the radius of the cylinder. This means that

$$\lambda = \frac{2\pi \epsilon_0 \, \Delta V}{\ln(r_c/r_w)}$$

and

$$E = \frac{\lambda}{2\pi \epsilon_0 r} = \frac{\Delta V}{r \ln(r_c/r_w)} .$$

(a) Substitute r_w for r to obtain the field at the surface of the wire:

$$E = \frac{\Delta V}{r_w \ln(r_c/r_w)} = \frac{850\,\text{V}}{(0.65 \times 10^{-6}\,\text{m}) \ln\left[(1.0 \times 10^{-2}\,\text{m})/(0.65 \times 10^{-6}\,\text{m})\right]}$$
$$= 1.36 \times 10^8\,\text{V/m} .$$

(b) Substitute r_c for r to find the field at the surface of the cylinder:

$$E = \frac{\Delta V}{r_w \ln(r_c/r_w)} = \frac{850\,\text{V}}{(1.0 \times 10^{-2}\,\text{m}) \ln\left[(1.0 \times 10^{-2}\,\text{m})/(0.65 \times 10^{-6}\,\text{m})\right]}$$
$$= 8.82 \times 10^3\,\text{V/m} .$$

13P

(a) Use Gauss' law to find expressions for the electric field inside and outside the spherical charge distribution. Since the field is radial the electric potential can be written as an integral of the field along a sphere radius, extended to infinity. Since different expressions for the field apply in different regions the integral must be split into two parts, one from infinity to the surface of the distribution and one from the surface to a point inside.

Outside the charge distribution the magnitude of the field is $E = q/4\pi\epsilon_0 r^2$ and the potential is $V = q/4\pi\epsilon_0 r$, where r is the distance from the center of the distribution.

To find an expression for the magnitude of the field inside the charge distribution use a Gaussian surface in the form of a sphere with radius r, concentric with the distribution. The field is normal to the Gaussian surface and its magnitude is uniform over it, so the electric flux through the surface is $4\pi r^2 E$. The charge enclosed is qr^3/R^3. Gauss' law becomes

$$4\pi\epsilon_0 r^2 E = \frac{qr^3}{R^3},$$

so

$$E = \frac{qr}{4\pi\epsilon_0 R^3}.$$

If V_s is the potential at the surface of the distribution ($r = R$) then the potential at a point inside, a distance r from the center, is

$$V = V_s - \int_R^r E\,dr = V_s - \frac{q}{4\pi\epsilon_0 R^3}\int_R^r r\,dr = V_s - \frac{qr^2}{8\pi\epsilon_0 R^3} + \frac{q}{8\pi\epsilon_0 R}.$$

The potential at the surface can be found by replacing r with R in the expression for the potential at points outside the distribution. It is $V_s = q/4\pi\epsilon_0 R$. Thus

$$V = \frac{q}{4\pi\epsilon_0}\left[\frac{1}{R} - \frac{r^2}{2R^3} + \frac{1}{2R}\right] = \frac{q}{8\pi\epsilon_0 R^3}(3R^2 - r^2).$$

(b) In Problem 11 the electric potential was taken to be zero at the center of the sphere. In this problem it is zero at infinity. According to the expression derived in part (a) the potential at the center of the sphere is $V_c = 3q/8\pi\epsilon_0 R$. Thus $V - V_c = -qr^2/8\pi\epsilon_0 R^3$. This is the result of Problem 11.

(c) The potential difference is

$$\Delta V = V_s - V_c = \frac{2q}{8\pi\epsilon_0 R} - \frac{3q}{8\pi\epsilon_0 R} = -\frac{q}{8\pi\epsilon_0 R}.$$

The same value as is given by the expression obtained in Problem 11.

(d) Only potential differences have physical significance, not the value of the potential at any particular point. The same value can be added to the potential at every point without changing the electric field, for example. Changing the reference point from the center of the distribution to infinity changes the value of the potential at every point but it does not change any potential differences.

14P

(*a*) For $r > r_2$ the field is like that of a point charge and

$$V = \frac{1}{4\pi\epsilon_0}\frac{Q}{r},$$

where the zero of potential was taken to be at infinity.

(*b*) To find the potential in the region $r_1 < r < r_2$, first use Gauss's law to find an expression for the electric field, then integrate along a radial path from r_2 to r. The Gaussian surface is a sphere of radius r, concentric with the shell. The field is radial and therefore normal to the surface. Its magnitude is uniform over the surface, so the flux through the surface is $\Phi = 4\pi r^2 E$. The volume of the shell is $(4\pi/3)(r_2^3 - r_1^3)$, so the charge density is

$$\rho = \frac{3Q}{4\pi(r_2^3 - r_1^3)}$$

and the charge enclosed by the Gaussian surface is

$$q = \left(\frac{4\pi}{3}\right)(r^3 - r_1^3)\rho = Q\left(\frac{r^3 - r_1^3}{r_2^3 - r_1^3}\right).$$

Gauss' law yields

$$4\pi\epsilon_0 r^2 E = Q\left(\frac{r^3 - r_1^3}{r_2^3 - r_1^3}\right)$$

and the magnitude of the electric field is

$$E = \frac{Q}{4\pi\epsilon_0}\frac{r^3 - r_1^3}{r^2(r_2^3 - r_1^3)}.$$

If V_s is the electric potential at the outer surface of the shell ($r = r_2$) then the potential a distance r from the center is given by

$$V = V_s - \int_{r_2}^r E\,dr = V_s - \frac{Q}{4\pi\epsilon_0}\frac{1}{r_2^3 - r_1^3}\int_{r_2}^r\left(r - \frac{r_1^3}{r^2}\right)dr$$

$$= V_s - \frac{Q}{4\pi\epsilon_0}\frac{1}{r_2^3 - r_1^3}\left(\frac{r^2}{2} - \frac{r_2^2}{2} + \frac{r_1^3}{r} - \frac{r_1^3}{r_2}\right).$$

The potential at the outer surface is found by placing $r = r_2$ in the expression found in part (*a*). It is $V_s = Q/4\pi\epsilon_0 r_2$. Make this substitution and collect terms to find

$$V = \frac{Q}{4\pi\epsilon_0}\frac{1}{r_2^3 - r_1^3}\left(\frac{3r_2^2}{2} - \frac{r^2}{2} - \frac{r_1^3}{r}\right).$$

Since $\rho = 3Q/4\pi(r_2^3 - r_1^3)$ this can also be written

$$V = \frac{\rho}{3\epsilon_0}\left(\frac{3r_2^2}{2} - \frac{r^2}{2} - \frac{r_1^3}{r}\right).$$

(c) The electric field vanishes in the cavity, so the potential is everywhere the same inside and has the same value as at a point on the inside surface of the shell. Put $r = r_1$ in the result of part (b). After collecting terms the result is

$$V = \frac{Q}{4\pi\epsilon_0} \frac{3(r_2^2 - r_1^2)}{2(r_2^3 - r_1^3)},$$

or in terms of the charge density

$$V = \frac{\rho}{2\epsilon_0}(r_2^2 - r_1^2).$$

(d) The solutions agree at $r = r_1$ and at $r = r_2$.

19E

If the electric potential is zero at infinity then at the surface of a uniformly charged sphere it is $V = q/4\pi\epsilon_0 R$, where q is the charge on the sphere and R is the sphere radius. Thus $q = 4\pi\epsilon_0 RV$ and the number of electrons is

$$N = \frac{|q|}{e} = \frac{4\pi\epsilon_0 R|V|}{e} = \frac{(1.0 \times 10^{-6}\,\text{m})(400\,\text{V})}{(8.99 \times 10^9\,\text{N}\cdot\text{m}^2/\text{C}^2)(1.60 \times 10^{-19}\,\text{C})} = 2.8 \times 10^5.$$

24P

Assume the charge on the Earth is distributed with spherical symmetry. If the electric potential is zero at infinity then at the surface of the Earth it is $V = q/4\pi\epsilon_0 R$, where q is the charge on the Earth and $R\,(= 6.37 \times 10^6\,\text{m})$ is the radius of the Earth. The magnitude of the electric field at the surface is $E = q/4\pi\epsilon_0 R^2$, so $V = ER = (100\,\text{V/m})(6.37 \times 10^6\,\text{m}) = 6.4 \times 10^8\,\text{V}$.

26P

(a) The electric potential V at the surface of the drop, the charge q on the drop, and the radius R of the drop are related by $V = q/4\pi\epsilon_0 R$. Thus

$$R = \frac{q}{4\pi\epsilon_0 V} = \frac{(8.99 \times 10^9\,\text{N}\cdot\text{m}^2/\text{C}^2)(30 \times 10^{-12}\,\text{C})}{500\,\text{V}} = 5.4 \times 10^{-4}\,\text{m}.$$

(b) After the drops combine the total volume is twice the volume of an original drop, so the radius R' of the combined drop is given by $(R')^3 = 2R^3$ and $R' = 2^{1/3}R$. The charge is twice the charge of original drop: $q' = 2q$. Thus

$$V' = \frac{1}{4\pi\epsilon_0}\frac{q'}{R'} = \frac{1}{4\pi\epsilon_0}\frac{2q}{2^{1/3}R} = \frac{2V}{2^{1/3}} = \frac{2(500\,\text{V})}{2^{1/3}} = 790\,\text{V}.$$

33P

A positive charge q is a distance $r - d$ from P, another positive charge q is a distance r from P, and a negative charge $-q$ is a distance $r + d$ from P. Sum the individual electric potentials created at P to find the total:

$$V = \frac{q}{4\pi\epsilon_0}\left[\frac{1}{r - d} + \frac{1}{r} - \frac{1}{r + d}\right].$$

Use the binomial theorem to approximate $1/(r - d)$ for r much larger than d:

$$\frac{1}{r - d} = (r - d)^{-1} \approx (r)^{-1} - (r)^{-2}(-d) = \frac{1}{r} + \frac{d}{r^2}.$$

Similarly,

$$\frac{1}{r + d} \approx \frac{1}{r} - \frac{d}{r^2}.$$

Only the first two terms of each expansion were retained. Thus

$$V \approx \frac{q}{4\pi\epsilon_0}\left[\frac{1}{r} + \frac{d}{r^2} + \frac{1}{r} - \frac{1}{r} + \frac{d}{r^2}\right] = \frac{q}{4\pi\epsilon_0}\left[\frac{1}{r} + \frac{2d}{r^2}\right] = \frac{q}{4\pi\epsilon_0 r}\left[1 + \frac{2d}{r}\right].$$

34P

A charge $-5q$ is a distance $2d$ from P, a charge $-5q$ is a distance d from P, and two charges $+5q$ are each a distance d from P, so the electric potential at P is

$$V = \frac{q}{4\pi\epsilon_0}\left[-\frac{5}{2d} - \frac{5}{d} + \frac{5}{d} + \frac{5}{d}\right] = -\frac{5q}{8\pi\epsilon_0}.$$

The zero of the electric potential was taken to be at infinity.

39P

(a) All the charge is the same distance R from C, so the electric potential at C is

$$V = \frac{1}{4\pi\epsilon_0}\left[\frac{Q}{R} - \frac{6Q}{R}\right] = -\frac{5Q}{4\pi\epsilon_0 R},$$

where the zero was taken to be at infinity.

(b) All the charge is the same distance from P. That distance is $\sqrt{R^2 + z^2}$, so the electric potential at P is

$$V = \frac{1}{4\pi\epsilon_0}\left[\frac{Q}{\sqrt{R^2 + z^2}} - \frac{6Q}{\sqrt{R^2 + z^2}}\right] = -\frac{5Q}{4\pi\epsilon_0\sqrt{R^2 + z^2}}.$$

40P

The disk is uniformly charged. This means that when the full disk is present each quadrant contributes equally to the electric potential at P, so the potential at P due to a single quadrant is one-fourth the potential due to the entire disk. First find an expression for the potential at P due to the entire disk.

Consider a ring of charge with radius r and width dr. Its area is $2\pi r\,dr$ and it contains charge $dq = 2\pi\sigma r\,dr$. All the charge in it is a distance $\sqrt{r^2 + z^2}$ from P, so the potential it produces at P is

$$dV = \frac{1}{4\pi\epsilon_0}\frac{2\pi\sigma r\,dr}{\sqrt{r^2 + z^2}} = \frac{\sigma r\,dr}{2\epsilon_0\sqrt{r^2 + z^2}}.$$

The total potential at P is

$$V = \frac{\sigma}{2\epsilon_0} \int_0^R \frac{r\,dr}{\sqrt{r^2 + z^2}} = \frac{\sigma}{2\epsilon_0} \sqrt{r^2 + z^2}\bigg|_0^R = \frac{\sigma}{2\epsilon_0}\left[\sqrt{R^2 + z^2} - z\right] .$$

The potential V_{sq} at P due to a single quadrant is

$$V_{sq} = \frac{V}{4} = \frac{\sigma}{8\epsilon_0}\left[\sqrt{R^2 + z^2} - z\right] .$$

48P

(a) The charge on every part of the ring is the same distance from any point P on the axis. This distance is $r = \sqrt{z^2 + R^2}$, where R is the radius of the ring and z is the distance from the center of the ring to P. The electric potential at P is

$$V = \frac{1}{4\pi\epsilon_0}\int \frac{dq}{r} = \frac{1}{4\pi\epsilon_0}\int \frac{dq}{\sqrt{z^2 + R^2}} = \frac{1}{4\pi\epsilon_0}\frac{1}{\sqrt{z^2 + R^2}}\int dq = \frac{1}{4\pi\epsilon_0}\frac{q}{\sqrt{z^2 + R^2}} .$$

(b) The electric field is along the axis and its component is given by

$$E = -\frac{\partial V}{\partial z} = -\frac{q}{4\pi\epsilon_0}\frac{d}{dz}(z^2 + R^2)^{-1/2}$$

$$= -\frac{q}{4\pi\epsilon_0}\left(\frac{1}{2}\right)(z^2 + R^2)^{-3/2}(2z) = \frac{q}{4\pi\epsilon_0}\frac{z}{(z^2 + R^2)^{3/2}} .$$

This agrees with the result of Section 24–6.

49P

(a) Put the origin at the right end of the rod and consider an infinitesimal segment of the rod, located at $x = x'$ (a negative coordinate). It has length dx' and contains charge $dq = \lambda\,dx'$. Its distance from P is $d = x - x'$ and the potential it creates at P is

$$dV = \frac{1}{4\pi\epsilon_0}\frac{dq}{d} = \frac{1}{4\pi\epsilon_0}\frac{\lambda\,dx'}{x - x'} .$$

To find the total potential at P, integrate over the rod:

$$V = \frac{\lambda}{4\pi\epsilon_0}\int_{-L}^0 \frac{dx'}{x - x'} = -\frac{\lambda}{4\pi\epsilon_0}\ln(x - x')\bigg|_{-L}^0 = \frac{\lambda}{4\pi\epsilon_0}\ln\frac{x + L}{x} .$$

(b) Differentiate the electric potential with respect to x to find the x component of the electric field:

$$E_x = -\frac{\partial V}{\partial x} = -\frac{\lambda}{4\pi\epsilon_0}\frac{\partial}{\partial x}\ln\frac{x + L}{x} = -\frac{\lambda}{4\pi\epsilon_0}\frac{x}{x + L}\left(\frac{1}{x} - \frac{x + L}{x^2}\right) = \frac{\lambda}{4\pi\epsilon_0}\frac{L}{x(x + L)} .$$

(c) Consider two points an equal distance on either side of P, along a line that is perpendicular to the x axis. The difference in the electric potential divided by their separation gives the transverse component of the electric field. Since the two points are situated symmetrically with respect to the rod, their potentials are the same and the potential difference is zero. Thus the transverse component of the electric field is zero.

52E

(a) The charges are equal and are the same distance from C. Use the Pythagorean theorem to find the distance $r = \sqrt{(d/2)^2 + (d/2)^2} = d/\sqrt{2}$. The electric potential at C is the sum of the potential due to the individual charges but since they produce the same potential, it is twice that of either one:

$$V = \frac{2q}{4\pi\epsilon_0}\frac{\sqrt{2}}{d} = \frac{2\sqrt{2}q}{4\pi\epsilon_0 d}$$

$$= \frac{(8.99 \times 10^9\,\text{N}\cdot\text{m}^2/\text{C}^2)(2)\sqrt{2}(2.0 \times 10^{-6}\,\text{C})}{0.020\,\text{m}} = 2.5 \times 10^6\,\text{V}\,.$$

(b) As you move the charge into position from far away the potential energy changes from zero to qV, where V is the electric potential at the final location of the charge. The change in the potential energy equals the work you must do to bring the charge in: $W = qV = (2.0 \times 10^{-6}\,\text{C})(2.5 \times 10^6\,\text{V}) = 5.1\,\text{J}$.

(c) The work calculated in part (b) represents the potential energy of the interactions between the charge brought in from infinity and the other two charges. To find the total potential energy of the three-charge system you must add the potential energy of the interaction between the fixed charges. Their separation is d so this potential energy is $q^2/4\pi\epsilon_0 d$. The total potential energy is

$$U = W + \frac{q^2}{4\pi\epsilon_0 d}$$

$$= 5.1\,\text{J} + \frac{(8.99 \times 10^9\,\text{N}\cdot\text{m}^2/\text{C}^2)(2.0 \times 10^{-6}\,\text{C})^2}{0.020\,\text{m}} = 6.9\,\text{J}\,.$$

59P

(a) Let $\ell\,(= 0.15\,\text{m})$ be the length of the rectangle and $w\,(= 0.050\,\text{m})$ be its width. Charge q_1 is a distance ℓ from point A and charge q_2 is a distance w, so the electric potential at A is

$$V_A = \frac{1}{4\pi\epsilon_0}\left[\frac{q_1}{\ell} + \frac{q_2}{w}\right]$$

$$= (8.99 \times 10^9\,\text{N}\cdot\text{m}^2/\text{C}^2)\left[\frac{-5.0 \times 10^{-6}\,\text{C}}{0.15\,\text{m}} + \frac{2.0 \times 10^{-6}\,\text{C}}{0.050\,\text{m}}\right]$$

$$= 6.0 \times 10^4\,\text{V}\,.$$

(b) Charge q_1 is a distance w from point b and charge q_2 is a distance ℓ, so the electric potential at B is

$$V_B = \frac{1}{4\pi\epsilon_0}\left[\frac{q_1}{w} + \frac{q_2}{\ell}\right]$$

$$= (8.99 \times 10^9\,\text{N}\cdot\text{m}^2/\text{C}^2)\left[\frac{-5.0 \times 10^{-6}\,\text{C}}{0.050\,\text{m}} + \frac{2.0 \times 10^{-6}\,\text{C}}{0.15\,\text{m}}\right]$$

$$= -7.8 \times 10^5\,\text{V}\,.$$

(c) Since the kinetic energy is zero at the beginning and end of the trip, the work done by an external agent equals the change in the potential energy of the system. The potential energy is the product of the charge q_3 and the electric potential. If U_A is the potential energy when q_3 is at A and U_B is the potential energy when q_3 is at B, then the work done in moving the charge from B to A is $W = U_A - U_B = q_3(V_A - V_B) = (3.0 \times 10^{-6}\,\text{C})(6.0 \times 10^4\,\text{V} + 7.8 \times 10^5\,\text{V}) = 2.5\,\text{J}$.

(d) The work done by the external agent is positive, so the energy of the three-charge system increases.

(e) and (f) The electrostatic force is conservative, so the work is the same no matter what the path.

61P

The particle with charge $-q$ has both potential and kinetic energy and both these change when the radius of the orbit is changed. Find an expression for the total energy in terms of the orbit radius. Q provides the centripetal force required for $-q$ to move in uniform circular motion. The magnitude of the force is $F = Qq/4\pi\epsilon_0 r^2$, where r is the orbit radius. The acceleration of $-q$ is v^2/r, where v is its speed. Newton's second law yields $Qq/4\pi\epsilon_0 r^2 = mv^2/r$, so $mv^2 = Qq/4\pi\epsilon_0 r$ and the kinetic energy is $K = \frac{1}{2}mv^2 = Qq/8\pi\epsilon_0 r$. The potential energy is $U = -Qq/4\pi\epsilon_0 r$ and the total energy is

$$E = K + U = \frac{Qq}{8\pi\epsilon_0 r} - \frac{Qq}{4\pi\epsilon_0 r} = -\frac{Qq}{8\pi\epsilon_0 r}.$$

When the orbit radius is r_1 the energy is $E_1 = -Qq/8\pi\epsilon_0 r_1$ and when it is r_2 the energy is $E_2 = -Qq/8\pi\epsilon_0 r_2$. The difference $E_2 - E_1$ is the work W done by an external agent to change the radius:

$$W = E_2 - E_1 = -\frac{Qq}{8\pi\epsilon_0}\left(\frac{1}{r_2} - \frac{1}{r_1}\right) = \frac{Qq}{8\pi\epsilon_0}\left(\frac{1}{r_1} - \frac{1}{r_2}\right).$$

64P

Use the conservation of energy principle. The initial potential energy is $U_i = q^2/4\pi\epsilon_0 r_1$, the initial kinetic energy is $K_i = 0$, the final potential energy is $U_f = q^2/4\pi\epsilon_0 r_2$, and the final kinetic energy is $K_f = \frac{1}{2}mv^2$, where v is the final speed of the particle. Conservation of energy yields

$$\frac{q^2}{4\pi\epsilon_0 r_1} = \frac{q^2}{4\pi\epsilon_0 r_2} + \frac{1}{2}mv^2.$$

The solution for v is

$$v = \sqrt{\frac{2q^2}{4\pi\epsilon_0 m}\left(\frac{1}{r_1} - \frac{1}{r_2}\right)}$$

$$= \sqrt{\frac{(8.99 \times 10^9\,\text{N}\cdot\text{m}^2/\text{C}^2)(2)(3.1 \times 10^{-6}\,\text{C})^2}{20 \times 10^{-6}\,\text{kg}}\left(\frac{1}{0.90 \times 10^{-3}\,\text{m}} - \frac{1}{2.5 \times 10^{-3}\,\text{m}}\right)}$$

$$= 2.5 \times 10^3\,\text{m/s}.$$

(a) The potential energy is

$$U = \frac{q^2}{4\pi\epsilon_0 d} = \frac{(8.99 \times 10^9 \, \text{N} \cdot \text{m}^2/\text{C}^2)(5.0 \times 10^{-6} \, \text{C})^2}{1.00 \, \text{m}} = 0.225 \, \text{J},$$

relative to the potential energy at infinite separation.

(b) Each sphere repels the other with a force that has magnitude

$$F = \frac{q^2}{4\pi\epsilon_0 d^2} = \frac{(8.99 \times 10^9 \, \text{N} \cdot \text{m}^2/\text{C}^2)(5.0 \times 10^{-6} \, \text{C})^2}{(1.00 \, \text{m})^2} = 0.225 \, \text{N}.$$

According to Newton's second law the acceleration of each sphere is the force divided by the mass of the sphere. Let m_A and m_B be the masses of the spheres. The acceleration of sphere A is

$$a_A = \frac{F}{m_A} = \frac{0.225 \, \text{N}}{5.0 \times 10^{-3} \, \text{kg}} = 45.0 \, \text{m/s}^2$$

and the acceleration of sphere B is

$$a_B = \frac{F}{m_B} = \frac{0.225 \, \text{N})}{10 \times 10^{-3} \, \text{kg}} = 22.5 \, \text{m/s}^2.$$

(c) Energy is conserved. The initial potential energy is $U = 0.225 \, \text{J}$, as calculated in part (a). The initial kinetic energy is zero since the spheres start from rest. The final potential energy is zero since the spheres are then far apart. The final kinetic energy is $\frac{1}{2}m_A v_A^2 + \frac{1}{2}m_B v_B^2$, where v_A and v_B are the final velocities. Thus

$$U = \frac{1}{2}m_A v_A^2 + \frac{1}{2}m_B v_B^2.$$

Momentum is also conserved, so

$$0 = m_A v_A + m_B v_B.$$

Solve these equations simultaneously for v_A and v_B.

Substitute $v_B = -(m_A/m_B)v_A$, from the momentum equation, into the energy equation and collect terms. You should obtain $U = \frac{1}{2}(m_A/m_B)(m_A + m_B)v_A^2$. Thus

$$v_A = \sqrt{\frac{2U m_B}{m_A(m_A + m_B)}}$$

$$= \sqrt{\frac{2(0.225 \, \text{J})(10 \times 10^{-3} \, \text{kg})}{(5.0 \times 10^{-3} \, \text{kg})(5.0 \times 10^{-3} \, \text{kg} + 10 \times 10^{-3} \, \text{kg})}} = 7.75 \, \text{m/s}.$$

Now calculate v_B:

$$v_B = -\frac{m_A}{m_B}v_A = -\left(\frac{5.0 \times 10^{-3} \, \text{kg}}{10 \times 10^{-3} \, \text{kg}}\right)(7.75 \, \text{m/s}) = -3.87 \, \text{m/s}.$$

70P

Use the conservation of energy principle. Take the potential energy to be zero when the moving electron is far away from the fixed electrons. The final potential energy is then $U_f = 2e^2/4\pi\epsilon_0 d$, where d is the half the distance between the fixed electrons. The initial kinetic energy is $K_i = \frac{1}{2}mv^2$, where m is the mass of an electron and v is the initial speed of the moving electron. The final kinetic energy is zero. Thus $K_i = U_f$ or $\frac{1}{2}mv^2 = 2e^2/4\pi\epsilon_0 d$. Hence

$$v = \sqrt{\frac{4e^2}{4\pi\epsilon_0 dm}} = \sqrt{\frac{(8.99 \times 10^9\,\text{N}\cdot\text{m}^2/\text{C}^2)(4)(1.60 \times 10^{-19}\,\text{C})^2}{(0.010\,\text{m})(9.11 \times 10^{-31}\,\text{kg})}} = 3.2 \times 10^2\,\text{m/s}.$$

75E

If the electric potential is zero at infinity, then the potential at the surface of the sphere is given by $V = q/4\pi\epsilon_0 r$, where q is the charge on the sphere and r is its radius. Thus

$$q = 4\pi\epsilon_0 rV = \frac{(0.15\,\text{m})(1500\,\text{V})}{8.99 \times 10^9\,\text{N}\cdot\text{m}^2/\text{C}^2} = 2.5 \times 10^{-8}\,\text{C}.$$

79P

(a) The electric potential is the sum of the contributions of the individual spheres. Let q_1 be the charge on one and q_2 be the charge on the other. The point halfway between them is the same distance $d\ (= 1.0\,\text{m})$ from the center of each sphere, so the potential at the halfway point is

$$V = \frac{q_1 + q_2}{4\pi\epsilon_0 d} = \frac{(8.99 \times 10^9\,\text{N}\cdot\text{m}^2/\text{C}^2)(1.0 \times 10^{-8}\,\text{C} - 3.0 \times 10^{-8}\,\text{C})}{1.0\,\text{m}} = -1.80 \times 10^2\,\text{V}.$$

(b) The distance from the center of one sphere to the surface of the other is $d - R$, where R is the radius of either sphere. The potential of either one of the spheres is due to the charge on that sphere and the charge on the other sphere. The potential at the surface of sphere 1 is

$$V_1 = \frac{1}{4\pi\epsilon_0}\left[\frac{q_1}{R} + \frac{q_2}{d-R}\right]$$

$$= (8.99 \times 10^9\,\text{N}\cdot\text{m}^2/\text{C}^2)\left[\frac{1.0 \times 10^{-8}\,\text{C}}{0.030\,\text{m}} - \frac{3.0 \times 10^{-8}\,\text{C}}{1.0\,\text{m} - 0.030\,\text{m}}\right]$$

$$= 2.7 \times 10^3\,\text{V}.$$

The potential at the surface of sphere 2 is

$$V_2 = \frac{1}{4\pi\epsilon_0}\left[\frac{q_1}{d-R} + \frac{q_2}{R}\right]$$

$$= (8.99 \times 10^9\,\text{N}\cdot\text{m}^2/\text{C}^2)\left[\frac{1.0 \times 10^{-8}\,\text{C}}{1.0\,\text{m} - 0.030\,\text{m}} - \frac{3.0 \times 10^{-8}\,\text{C}}{0.030\,\text{m}}\right]$$

$$= -8.9 \times 10^3\,\text{V}.$$

84E

(a) Since energy is conserved and the α particle starts from rest its final kinetic energy is the change in its potential energy: $K = \Delta U$. The change in its potential energy is $\Delta U = q\,\Delta V$, where q is its charge ($2e$) and ΔV is the difference in the electric potential along the accelerator. Thus $K = 2e\,\Delta V = 2(1.60 \times 10^{-19}\,\text{C})(1.0 \times 10^6\,\text{V}) = 3.2 \times 10^{-13}\,\text{J}$.

(b) The charge on the proton is e, so

$$K = e\,\Delta V = (1.60 \times 10^{-19}\,\text{C})(1.0 \times 10^6\,\text{V}) = 1.6 \times 10^{-13}\,\text{J}\,.$$

(c) Since $K = \frac{1}{2}mv^2$, where m is the mass of a particle and v is its speed,

$$v = \sqrt{\frac{2K}{m}} = \sqrt{\frac{2q\,\Delta V}{m}}\,,$$

Now an α particle has twice the charge of a proton and 4 times the mass so the ratio of the final speeds is $v_p/v_\alpha = \sqrt{2}$. For $\Delta V = 1.0 \times 10^6\,\text{V}$, $v_p = 1.4 \times 10^7\,\text{m/s}$ and $v_\alpha = 9.8 \times 10^6\,\text{m/s}$.

86P

If Q is the charge on the shell and R is its radius, then the electric potential of the shell is given by $V = Q/4\pi\epsilon_0 R$ and the magnitude of the electric field just outside the shell is given by $E = Q/4\pi\epsilon_0 R^2$, so $V = RE$. Calculate R for the operating potential of the generator and the breakdown electric field:

$$R = \frac{V}{E} = \frac{9.0\,\text{MV}}{100\,\text{MV/m}} = 9.0 \times 10^{-2}\,\text{m}\,.$$

For larger values of R the electric field is smaller in magnitude so R must be larger than 9.0 cm to avoid breakdown.

(b) For each infinitesimal amount of charge dq that is transferred, work $dW = V\,dq$ must be done. The power required is the rate of doing work: $P = dW/dt = V\,dq/dt$, where dq/dt is the rate at which charge is transferred. Thus $P = (9.0 \times 10^6\,\text{V})(300 \times 10^{-6}\,\text{C/s}) = 2.7 \times 10^3\,\text{W}$.

(c) As the belt moves a distance $\Delta \ell$ it delivers charge $\Delta q = \sigma w \Delta \ell$, so the charge delivered per unit time is $\Delta q/\Delta t = \sigma w \Delta \ell/\Delta t$, where Δt is the time required for the belt to move the distance $\Delta \ell$. Now $\Delta \ell/\Delta t = v$, where v is the speed of the belt, so $\Delta q/\Delta t = \sigma w v$ and

$$\sigma = \frac{1}{wv}\frac{\Delta q}{\Delta t} = \frac{300 \times 10^{-6}\,\text{C/s}}{(0.50\,\text{m})(30\,\text{m/s})} = 2.0 \times 10^{-5}\,\text{C/m}^2\,.$$

Chapter 27

3E

Charge flows until the potential difference across the capacitor is the same as the emf of the battery. The charge on the capacitor is then $q = C\mathcal{E}$ and this is the same as the total charge that has passed through the battery. Thus $q = (25 \times 10^{-6}\,\text{F})(120\,\text{V}) = 3.0 \times 10^{-3}\,\text{C}$.

5E

(a) The capacitance of a parallel plate capacitor is given by $C = \epsilon_0 A/d$, where A is the area of each plate and d is the plate separation. Since the plates are circular the plate area is $A = \pi R^2$, where R is the radius of a plate. Thus

$$C = \frac{\epsilon_0 \pi R^2}{d} = \frac{(8.85 \times 10^{-12}\,\text{F/m})\pi(8.2 \times 10^{-2}\,\text{m})^2}{1.3 \times 10^{-3}\,\text{m}} = 1.4 \times 10^{-10}\,\text{F} = 140\,\text{pF}\,.$$

(b) The charge on the positive plate is given by $q = CV$, where V is the potential difference across the plates. Thus $q = (1.4 \times 10^{-10}\,\text{F})(120\,\text{V}) = 1.7 \times 10^{-8}\,\text{C} = 17\,\text{nC}$.

7E

The capacitance of a cylindrical capacitor is given by $C = 2\pi\epsilon_0 L/\ln(b/a)$, where L is the length, a is the inner radius, and b is the outer radius. Thus

$$C = \frac{2\pi(8.85 \times 10^{-12}\,\text{F/m})(2.4 \times 10^{-2}\,\text{m})}{\ln\left[(18\,\text{mm})/(1.6\,\text{mm})\right]} = 5.5 \times 10^{-13}\,\text{F} = 0.55\,\text{pF}\,.$$

The diameters can be used instead of the radii since the radii enter the equation as a ratio.

11E

You want to find the radius of the combined spheres, then use $C = 4\pi\epsilon_0 R$ to find the capacitance. When the drops combine the volume is doubled. It is then $V = 2(4\pi/3)R^3$. The new radius R' is given by

$$\frac{4\pi}{3}(R')^3 = 2\frac{4\pi}{3}R^3\,,$$

so

$$R' = 2^{1/3}R\,.$$

The new capacitance is $C' = 4\pi\epsilon_0 R' = 4\pi\epsilon_0 2^{1/3}R = 2^{1/3}C = 1.26C$.

15E

The equivalent capacitance is given by $C_{\text{eq}} = q/V$, where q is the total charge on all the capacitors and V is the potential difference across any one of them. For N identical capacitors in parallel $C_{\text{eq}} = NC$, where C is the capacitance of one of them. Thus $NC = q/V$ and

$$N = \frac{q}{VC} = \frac{1.00\,\text{C}}{(110\,\text{V})(1.00 \times 10^{-6}\,\text{F})} = 9090\,.$$

26P

Let x be the separation of the plates in the lower capacitor. Then the plate separation in the upper capacitor is $a - b - x$. The capacitance of the lower capacitor is $C_\ell = \epsilon_0 A/x$ and the capacitance of the upper capacitor is $C_u = \epsilon_0 A/(a - b - x)$, where A is the plate area. Since the two capacitors are in series the equivalent capacitance is determined from

$$\frac{1}{C_{eq}} = \frac{1}{C_\ell} + \frac{1}{C_u} = \frac{x}{\epsilon_0 A} + \frac{a - b - x}{\epsilon_0 A} = \frac{a - b}{\epsilon_0 A}.$$

Thus the equivalent capacitance is given by $C_{eq} = \epsilon_0 A/(a - b)$ and is independent of x.

28P

(a) After the switches are closed the potential differences across the capacitors are the same and the two capacitors are in parallel. The potential difference from a to b is given by $V_{ab} = Q/C_{eq}$, where Q is the net charge on the combination and C_{eq} is the equivalent capacitance. The equivalent capacitance is $C_{eq} = C_1 + C_2 = 4.0 \times 10^{-6}\,\text{F}$. The total charge on the combination is the net charge on either pair of connected plates. The charge on capacitor 1 is

$$q_1 = C_1 V = (1.0 \times 10^{-6}\,\text{F})(100\,\text{V}) = 1.0 \times 10^{-4}\,\text{C}$$

and the charge on capacitor 2 is

$$q_2 = C_2 V = (3.0 \times 10^{-6}\,\text{F})(100\,\text{V}) = 3.0 \times 10^{-4}\,\text{C},$$

so the net charge on the combination is $3.0 \times 10^{-4}\,\text{C} - 1.0 \times 10^{-4}\,\text{C} = 2.0 \times 10^{-4}\,\text{C}$. The potential difference is

$$V_{ab} = \frac{2.0 \times 10^{-4}\,\text{C}}{4.0 \times 10^{-6}\,\text{F}} = 50\,\text{V}.$$

(b) The charge on capacitor 1 is now $q_1 = C_1 V_{ab} = (1.0 \times 10^{-6}\,\text{F})(50\,\text{V}) = 5.0 \times 10^{-5}\,\text{C}$.

(c) The charge on capacitor 2 is now $q_2 = C_2 V_{ab} = (3.0 \times 10^{-6}\,\text{F})(50\,\text{V}) = 1.5 \times 10^{-4}\,\text{C}$.

29P

The charges on capacitors 2 and 3 are the same, so these capacitors may be replaced by an equivalent capacitance determined from

$$\frac{1}{C_{eq}} = \frac{1}{C_2} + \frac{1}{C_3} = \frac{C_2 + C_3}{C_2 C_3}.$$

Thus $C_{eq} = C_2 C_3/(C_2 + C_3)$. The charge on the equivalent capacitor is the same as the charge on either of the two capacitors in the combination and the potential difference across the equivalent capacitor is given by q_2/C_{eq}. The potential difference across capacitor 1 is q_1/C_1, where q_1 is the charge on this capacitor.

The potential difference across the combination of capacitors 2 and 3 must be the same as the potential difference across capacitor 1, so $q_1/C_1 = q_2/C_{eq}$. Now some of the charge originally on capacitor 1 flows to the combination of 2 and 3. If q_0 is the original charge, conservation of charge yields $q_1 + q_2 = q_0 = C_1V_0$, where V_0 is the original potential difference across capacitor 1.

Solve the two equations $q_1/C_1 = q_2/C_{eq}$ and $q_1 + q_2 = C_1V_0$ for q_1 and q_2. The second equation yields

$$q_2 = C_1V_0 - q_1$$

and, when this is substituted into the first, the result is

$$\frac{q_1}{C_1} = \frac{C_1V_0 - q_1}{C_{eq}}.$$

Solve for q_1. You should get

$$q_1 = \frac{C_1^2V_0}{C_{eq} + C_1} = \frac{C_1^2V_0}{\dfrac{C_2C_3}{C_2 + C_3} + C_1} = \frac{C_1^2(C_2 + C_3)V_0}{C_1C_2 + C_1C_3 + C_2C_3}.$$

The charges on capacitors 2 and 3 are

$$q_2 = q_3 = C_1V_0 - q_1 = C_1V_0 - \frac{C_1^2(C_2 + C_3)V_0}{C_1C_2 + C_1C_3 + C_2C_3} = \frac{C_1C_2C_3V_0}{C_1C_2 + C_1C_3 + C_2C_3}.$$

34E

The energy stored by a capacitor is given by $U = \frac{1}{2}CV^2$, where V is the potential difference across its plates. You must convert the given value of the energy to joules. Since a joule is a watt·second, simply multiply by $(10^3\,\text{W/kW})(3600\,\text{s/h})$ to obtain $10\,\text{kW}\cdot\text{h} = 3.6 \times 10^7\,\text{J}$. Thus

$$C = \frac{2U}{V^2} = \frac{2(3.6 \times 10^7\,\text{J})}{(1000\,\text{V})^2} = 72\,\text{F}.$$

37E

The total energy is the sum of the energies stored in the individual capacitors. Since they are connected in parallel the potential difference V across the capacitors is the same and the total energy is $U = \frac{1}{2}(C_1 + C_2)V^2 = \frac{1}{2}(2.0 \times 10^{-6}\,\text{F} + 4.0 \times 10^{-6}\,\text{F})(300\,\text{V})^2 = 0.27\,\text{J}$.

46P

(a) Let q be the charge on the positive plate. Since the capacitance of a parallel-plate capacitor is given by $\epsilon_0 A/d$, the charge is $q = CV = \epsilon_0 AV/d$. After the plates are pulled apart their separation is $2d$ and the potential difference is V'. Then $q = \epsilon_0 AV'/2d$ and

$$V' = \frac{2d}{\epsilon_0 A}q = \frac{2d}{\epsilon_0 A}\frac{\epsilon_0 A}{d}V = 2V.$$

(b) The initial energy stored in the capacitor is

$$U_i = \frac{1}{2}CV^2 = \frac{\epsilon_0 A V^2}{2d}$$

and the final energy stored is

$$U_f = \frac{1}{2}\frac{\epsilon_0 A}{2d}(V')^2 = \frac{1}{2}\frac{\epsilon_0 A}{2d}4V^2 = \frac{\epsilon_0 A V^2}{d}.$$

This is twice the initial energy.

(c) The work done to pull the plates apart is the difference in the energy: $W = U_f - U_i = \epsilon_0 A V^2/2d$.

47P

You first need to find an expression for the energy stored in a cylinder of radius R and length L, whose surface lies between the inner and outer cylinders of the capacitor ($a < R < b$). The energy density at any point is given by $u = \frac{1}{2}\epsilon_0 E^2$, where E is the magnitude of the electric field at that point. If q is the charge on the surface of the inner cylinder then the magnitude of the electric field at a point a distance r from the cylinder axis is given by

$$E = \frac{q}{2\pi\epsilon_0 L r}$$

(see Eq. 27–12) and the energy density at that point is given by

$$u = \frac{1}{2}\epsilon_0 E^2 = \frac{q^2}{8\pi^2\epsilon_0 L^2 r^2}.$$

The energy in the cylinder is the volume integral

$$U_R = \int u\,dV.$$

Now $dV = 2\pi r L\,dr$, so

$$U_R = \int_a^R \frac{q^2}{8\pi^2\epsilon_0 L^2 r^2}2\pi r\,dr = \frac{q^2}{4\pi\epsilon_0 L^2}\int_a^R \frac{dr}{r} = \frac{q^2}{4\pi\epsilon_0 L^2}\ln\frac{R}{a}.$$

To find an expression for the total energy stored in the capacitor, replace R with b:

$$U_b = \frac{q^2}{4\pi\epsilon_0 L^2}\ln\frac{b}{a}.$$

You want the ratio U_R/U_b to be 1/2, so

$$\ln\frac{R}{a} = \frac{1}{2}\ln\frac{b}{a}$$

or, since $\frac{1}{2}\ln(b/a) = \ln(\sqrt{b/a})$, $\ln(R/a) = \ln(\sqrt{b/a})$. This means $R/a = \sqrt{b/a}$ or $R = \sqrt{ab}$.

49P

The charge is held constant while the plates are being separated, so write the expression for the stored energy as $U = q^2/2C$, where q is the charge and C is the capacitance. The capacitance of a parallel-plate capacitor is given by $C = \epsilon_0 A/x$, where A is the plate area and x is the plate separation, so

$$ U = \frac{q^2 x}{2\epsilon_0 A} . $$

If the plate separation increases by dx the energy increases by $dU = (q^2/2\epsilon_0 A)\, dx$. Suppose the agent pulling the plate apart exerts force F. Then the agent does work $F\, dx$ and if the plates begin and end at rest this must equal the increase in stored energy. Thus

$$ F\, dx = \left(\frac{q^2}{2\epsilon_0 A} \right) dx $$

and

$$ F = \frac{q^2}{2\epsilon_0 A} . $$

The net force on a plate is zero so this must also be the magnitude of the force one plate exerts on the other.

The force can also be computed as the product of the charge q on one plate and the electric field E_1 due to the charge on the other plate. Recall that the field produced by a uniform plane surface of charge is $E_1 = q/2\epsilon_0 A$. Thus $F = q^2/2\epsilon_0 A$.

50P

According to the result of Problem 27–49 the force on either capacitor plate is given by $F = q^2/2\epsilon_0 A$, where q is the charge on one plate and A is the area of a plate. The electric field between the plates is $E = q/\epsilon_0 A$, so $q = \epsilon_0 AE$ and

$$ F = \frac{\epsilon_0^2 A^2 E^2}{2\epsilon_0 A} = \frac{1}{2}\epsilon_0 AE^2 . $$

The force per unit area of plate is

$$ \frac{F}{A} = \frac{1}{2}\epsilon_0 E^2 . $$

Note that the field E that enters this equation is the total field, due to charges on both plates.

51P

According to the result of Problem 27–50 the electrostatic force acting on a small area ΔA is $F_e = \frac{1}{2}\epsilon_0 E^2\, \Delta A$. The electric field at the surface is $E = q/4\pi\epsilon_0 R^2$, where q is the charge on the bubble. Thus

$$ F_e = \frac{1}{2}\epsilon_0 \frac{q^2\, \Delta A}{16\pi^2\epsilon_0^2 R^4} = \frac{q^2\, \Delta A}{32\pi^2\epsilon_0 R^4} . $$

This force is outward. The force of the gas inside is the product of the pressure inside and the area: $F_g = p(V_0/V)\Delta A$. Since $V_0 = (4\pi/3)R_0^3$ and $V = (4\pi/3)R^3$,

$$F_g = p\left(\frac{R_0^3}{R^3}\right)\Delta A.$$

This force is outward. The force of the air outside is $F_a = p\Delta A$. This force is inward. Since the bubble surface is in equilibrium, the sum of the forces must vanish: $F_e + F_g - F_a = 0$. This means

$$\frac{q^2}{32\pi^2\epsilon_0 R^4} + p\frac{R_0^3}{R^3} - p = 0.$$

Solve for q^2. You should get

$$q^2 = 32\pi^2\epsilon_0 R^4 p\left(1 - \frac{R_0^3}{R^3}\right) = 32\pi^2\epsilon_0 pR(R^3 - R_0^3).$$

53E

The capacitance with the dielectric in place is given by $C = \kappa C_0$, where C_0 is the capacitance before the dielectric is inserted. The energy stored is given by $U = \frac{1}{2}CV^2 = \frac{1}{2}\kappa C_0 V^2$, so

$$\kappa = \frac{2U}{C_0 V^2} = \frac{2(7.4 \times 10^{-6}\,\text{J})}{(7.4 \times 10^{-12}\,\text{F})(652\,\text{V})^2} = 4.7.$$

According to Table 27–2 you should use pyrex.

56E

The capacitance of a cylindrical capacitor is given by

$$C = \kappa C_0 = \frac{2\pi\kappa\epsilon_0 L}{\ln(b/a)},$$

where C_0 is the capacitance without the dielectric, κ is the dielectric constant, L is the length, a is the inner radius, and b is the outer radius. The capacitance per unit length of the cable is

$$\frac{C}{L} = \frac{2\pi\kappa\epsilon_0}{\ln(b/a)} = \frac{2\pi(2.6)(8.85 \times 10^{-12}\,\text{F/m})}{\ln\left[(0.60\,\text{mm})/(0.10\,\text{mm})\right]} = 8.1 \times 10^{-11}\,\text{F/m} = 81\,\text{pF/m}.$$

57P

The capacitance is given by $C = \kappa C_0 = \kappa\epsilon_0 A/d$, where C_0 is the capacitance without the dielectric, κ is the dielectric constant, A is the plate area, and d is the plate separation. The electric field between the plates is given by $E = V/d$, where V is the potential difference between the plates. Thus $d = V/E$ and $C = \kappa\epsilon_0 AE/V$. Solve for A:

$$A = \frac{CV}{\kappa\epsilon_0 E}.$$

For the area to be a minimum, the electric field must be the greatest it can be without breakdown occurring. That is,

$$A = \frac{(7.0 \times 10^{-8} \, \text{F})(4.0 \times 10^3 \, \text{V})}{2.8(8.85 \times 10^{-12} \, \text{F/m})(18 \times 10^6 \, \text{V/m})} = 0.63 \, \text{m}^2 .$$

64P

Assume there is charge q on one plate and charge $-q$ on the other. Calculate the electric field at points between the plates and use the result to find an expression for the potential difference V between the plates, in terms of q. The capacitance is $C = q/V$.

The electric field in the upper half of the region between the plates is

$$E_1 = \frac{q}{\kappa_1 \epsilon_0 A} ,$$

where A is the plate area. The electric field in the lower half is

$$E_2 = \frac{q}{\kappa_2 \epsilon_0 A} .$$

Take $d/2$ to be the thickness of each dielectric. Since the field is uniform in each region the potential difference between the plates is

$$V = \frac{E_1 d}{2} + \frac{E_2 d}{2} = \frac{qd}{2\epsilon_0 A} \left[\frac{1}{\kappa_1} + \frac{1}{\kappa_2} \right] = \frac{qd}{2\epsilon_0 A} \frac{\kappa_1 + \kappa_2}{\kappa_1 \kappa_2} ,$$

so

$$C = \frac{q}{V} = \frac{2\epsilon_0 A}{d} \frac{\kappa_1 \kappa_2}{\kappa_1 + \kappa_2} .$$

Notice that this expression is exactly the same as the expression for the equivalent capacitance of two capacitors in series, one with dielectric constant κ_1 and the other with dielectric constant κ_2. Each has plate area A and plate separation $d/2$. Also notice that if $\kappa_1 = \kappa_2$ the expression reduces to $C = \kappa_1 \epsilon_0 A/d$, the correct result for a parallel plate capacitor with plate area A, plate separation d, and dielectric constant κ_1.

66E

(a) The electric field in the region between the plates is given by $E = V/d$, where V is the potential difference between the plates and d is the plate separation. The capacitance is given by $C = \kappa \epsilon_0 A/d$, where A is the plate area and κ is the dielectric constant, so $d = \kappa \epsilon_0 A/C$ and

$$E = \frac{VC}{\kappa \epsilon_0 A} = \frac{(50 \, \text{V})(100 \times 10^{-12} \, \text{F})}{5.4(8.85 \times 10^{-12} \, \text{F/m})(100 \times 10^{-4} \, \text{m}^2)} = 1.0 \times 10^4 \, \text{V/m} .$$

(b) The free charge on the plates is $q_f = CV = (100 \times 10^{-12} \, \text{F})(50 \, \text{V}) = 5.0 \times 10^{-9} \, \text{C}.$

(c) The electric field is produced by both the free and induced charge. Since the field of a large uniform layer of charge is $q/2\epsilon_0 A$, the field between the plates is

$$E = \frac{q_f}{2\epsilon_0 A} + \frac{q_f}{2\epsilon_0 A} - \frac{q_i}{2\epsilon_0 A} - \frac{q_i}{2\epsilon_0 A},$$

where the first term is due to the positive free charge on one plate, the second is due to the negative free charge on the other plate, the third is due to the positive induced charge on one dielectric surface, and the fourth is due to the negative induced charge on the other dielectric surface. Note that the field due to the induced charge is opposite the field due to the free charge, so they tend to cancel. The induced charge is therefore

$$q_i = q_f - \epsilon_0 A E$$
$$= 5.0 \times 10^{-9}\,\mathrm{C} - (8.85 \times 10^{-12}\,\mathrm{F/m})(100 \times 10^{-4}\,\mathrm{m}^2)(1.0 \times 10^4\,\mathrm{V/m})$$
$$= 4.1 \times 10^{-9}\,\mathrm{C} = 4.1\,\mathrm{nC}.$$

71P

Assume the charge on one plate is $+q$ and the charge on the other plate is $-q$. Find an expression for the electric field in each region, in terms of q, then use the result to find an expression for the potential difference V between the plates. The capacitance is $C = q/V$.

The electric field in the dielectric is $E_d = q/\kappa\epsilon_0 A$, where κ is the dielectric constant and A is the plate area. Outside the dielectric (but still between the capacitor plates) the field is $E = q/\epsilon_0 A$. The field is uniform in each region so the potential difference across the plates is

$$V = E_d b + E(d - b) = \frac{qb}{\kappa\epsilon_0 A} + \frac{q(d-b)}{\epsilon_0 A} = \frac{q}{\epsilon_0 A}\frac{b + \kappa(d-b)}{\kappa}.$$

The capacitance is

$$C = \frac{q}{V} = \frac{\kappa\epsilon_0 A}{\kappa(d-b) + b} = \frac{\kappa\epsilon_0 A}{\kappa d - b(\kappa - 1)}.$$

The result does not depend on where the dielectric is located between the plates; it might be touching one plate or it might have a vacuum gap on each side.

For the capacitor of Sample Problem 27–10, $\kappa = 2.61$, $A = 115\,\mathrm{cm}^2 = 115 \times 10^{-4}\,\mathrm{m}^2$, $d = 1.24\,\mathrm{cm} = 1.24 \times 10^{-2}\,\mathrm{m}$, and $b = 0.78\,\mathrm{cm} = 0.78 \times 10^{-2}\,\mathrm{m}$, so

$$C = \frac{2.61(8.85 \times 10^{-12}\,\mathrm{F/m})(115 \times 10^{-4}\,\mathrm{m}^2)}{2.61(1.24 \times 10^{-2}\,\mathrm{m} - 0.78 \times 10^{-2}\,\mathrm{m}) + 0.78 \times 10^{-2}\,\mathrm{m}}$$
$$= 1.34 \times 10^{-11}\,\mathrm{F} = 13.4\,\mathrm{pF},$$

in agreement with the result found in the sample problem.

If $b = 0$ and $\kappa = 1$, then the expression derived above yields $C = \epsilon_0 A/d$, the correct expression for a parallel-plate capacitor with no dielectric. If $b = d$ then the derived expression yields $C = \kappa\epsilon_0 A/d$, the correct expression for a parallel-plate capacitor completely filled with a dielectric.

Chapter 28

1E

(a) The charge that passes through any cross section is the product of the current and time. Since $4.0\,\text{min} = (4.0\,\text{min})(60\,\text{s/min}) = 240\,\text{s}$, $q = it = (5.0\,\text{A})(240\,\text{s}) = 1200\,\text{C}$.

(b) The number of electrons N is given by $q = Ne$, where e is the magnitude of the charge on an electron. Thus $N = q/e = (1200\,\text{C})/(1.60 \times 10^{-19}\,\text{C}) = 7.5 \times 10^{21}$.

3P

Suppose the charge on the sphere increases by Δq in time Δt. Then in that time its potential increases by

$$\Delta V = \frac{\Delta q}{4\pi\epsilon_0 r},$$

where r is the radius of the sphere. This means

$$\Delta q = 4\pi\epsilon_0 r\, \Delta V.$$

Now $\Delta q = (i_{\text{in}} - i_{\text{out}})\Delta t$, where i_{in} is the current entering the sphere and i_{out} is the current leaving. Thus

$$\Delta t = \frac{\Delta q}{i_{\text{in}} - i_{\text{out}}} = \frac{4\pi\epsilon_0 r\, \Delta V}{i_{\text{in}} - i_{\text{out}}}$$

$$= \frac{(0.10\,\text{m})(1000\,\text{V})}{(8.99 \times 10^9\,\text{F/m})(1.0000020\,\text{A} - 1.0000000\,\text{A})} = 5.6 \times 10^{-3}\,\text{s}.$$

5E

(a) The magnitude of the current density is given by $J = nqv_d$, where n is the number of particles per unit volume, q is the charge on each particle, and v_d is the drift speed of the particles. The particle concentration is $n = 2.0 \times 10^8\,\text{cm}^{-3} = 2.0 \times 10^{14}\,\text{m}^{-3}$, the charge is $q = 2e = 2(1.60 \times 10^{-19}\,\text{C}) = 3.20 \times 10^{-19}\,\text{C}$, and the drift speed is $1.0 \times 10^5\,\text{m/s}$. Thus

$$J = (2 \times 10^{14}\,\text{m}^{-3})(3.2 \times 10^{-19}\,\text{C})(1.0 \times 10^5\,\text{m/s}) = 6.4\,\text{A/m}^2.$$

Since the particles are positively charged the current density is in the same direction as their motion, to the north.

(b) The current cannot be calculated unless the cross-sectional area of the beam is known. Then $i = JA$ can be used.

7E

The cross-sectional area of wire is given by $A = \pi r^2$, where r is its radius. The magnitude of the current density is $J = i/A = i/\pi r^2$, so

$$r = \sqrt{\frac{i}{\pi J}} = \sqrt{\frac{0.50\,\text{A}}{\pi(440 \times 10^4\,\text{A/m}^2)}} = 1.9 \times 10^{-4}\,\text{m}.$$

The diameter is $D = 2r = 2(1.9 \times 10^{-4}\,\text{m}) = 3.8 \times 10^{-4}\,\text{m}$.

14P

(a) The charge that strikes the surface in time Δt is given by $\Delta q = i\,\Delta t$, where i is the current. Since each particle carries charge $2e$, the number of particles that strike the surface is

$$N = \frac{\Delta q}{2e} = \frac{i\,\Delta t}{2e} = \frac{(0.25 \times 10^{-6}\,\text{A})(3.0\,\text{s})}{2(1.6 \times 10^{-19}\,\text{C})} = 2.3 \times 10^{12}\,.$$

(b) Now let N be the number of particles in a length L of the beam. They will all pass through the beam cross section at one end in time $t = L/v$, where v is the particle speed. The current is the charge that moves through the cross section per unit time. That is, $i = 2eN/t = 2eNv/L$. Thus $N = iL/2ev$.

Now find the particle speed. The kinetic energy of a particle is

$$K = 20\,\text{MeV} = (20 \times 10^6\,\text{eV})(1.60 \times 10^{-19}\,\text{J/eV}) = 3.2 \times 10^{-12}\,\text{J}\,.$$

Since $K = \frac{1}{2}mv^2$, $v = \sqrt{2K/m}$. The mass of an alpha particle is 4 times the mass of a proton or $m = 4(1.67 \times 10^{-27}\,\text{kg}) = 6.68 \times 10^{-27}\,\text{kg}$, so

$$v = \sqrt{\frac{2(3.2 \times 10^{-12}\,\text{J})}{6.68 \times 10^{-27}\,\text{kg}}} = 3.1 \times 10^7\,\text{m/s}$$

and

$$N = \frac{iL}{2ev} = \frac{(0.25 \times 10^{-6})(20 \times 10^{-2}\,\text{m})}{2(1.60 \times 10^{-19}\,\text{C})(3.1 \times 10^7\,\text{m/s})} = 5.0 \times 10^3\,.$$

(c) Use conservation of energy. The initial kinetic energy is zero, the final kinetic energy is $20\,\text{MeV} = 3.2 \times 10^{-12}\,\text{J}$, the initial potential energy is $qV = 2eV$, and the final potential energy is zero. Here V is the electric potential through which the particles are accelerated. Conservation of energy leads to $K_f = U_i = 2eV$, so

$$V = \frac{K_f}{2e} = \frac{3.2 \times 10^{-12}\,\text{J}}{2(1.60 \times 10^{-19}\,\text{C})} = 10 \times 10^6\,\text{V}\,.$$

17E

The resistance of the wire is given by $R = \rho L/A$, where ρ is the resistivity of the material, L is the length of the wire, and A is the cross-sectional area of the wire. The cross-sectional area is $A = \pi r^2 = \pi(0.50 \times 10^{-3}\,\text{m})^2 = 7.85 \times 10^{-7}\,\text{m}^2$. Here $r = 0.50\,\text{mm} = 0.50 \times 10^{-3}\,\text{m}$ is the radius of the wire. Thus

$$\rho = \frac{RA}{L} = \frac{(50 \times 10^{-3}\,\Omega)(7.85 \times 10^{-7}\,\text{m}^2)}{2.0\,\text{m}} = 2.0 \times 10^{-8}\,\Omega \cdot \text{m}\,.$$

18E

Since the potential difference V and current i are related by $V = iR$, where R is the resistance of the electrician, the fatal voltage is $V = (50 \times 10^{-3}\,\text{A})(2000\,\Omega) = 100\,\text{V}$.

19E

The resistance of the coil is given by $R = \rho L/A$, where L is the length of the wire, ρ is the resistivity of copper, and A is the cross-sectional area of the wire. Since each turn of wire has length $2\pi r$, where r is the radius of the coil, $L = (250)2\pi r = (250)(2\pi)(0.12\,\text{m}) = 188.5\,\text{m}$. If r_w is the radius of the wire, its cross-sectional area is $A = \pi r_w^2 = \pi(0.65 \times 10^{-3}\,\text{m})^2 = 1.33 \times 10^{-6}\,\text{m}^2$. According to Table 28–1, the resistivity of copper is $1.69 \times 10^{-8}\,\Omega \cdot \text{m}$. Thus

$$R = \frac{\rho L}{A} = \frac{(1.69 \times 10^{-8}\,\Omega \cdot \text{m})(188.5\,\text{m})}{1.33 \times 10^{-6}\,\text{m}^2} = 2.4\,\Omega.$$

27E

Since the mass and density of the material do not change, the volume remains the same. If L_0 is the original length, L is the new length, A_0 is the original cross-sectional area, and A is the new cross-sectional area, then $L_0 A_0 = LA$ and $A = L_0 A_0/L = L_0 A_0/3L_0 = A_0/3$. The new resistance is

$$R = \frac{\rho L}{A} = \frac{\rho 3 L_0}{A_0/3} = 9\frac{\rho L_0}{A_0} = 9R_0,$$

where R_0 is the original resistance. Thus $R = 9(6.0\,\Omega) = 54\,\Omega$.

30P

The resistance of conductor A is given by

$$R_A = \frac{\rho L}{\pi r_A^2},$$

where r_A is the radius of the conductor. If r_o is the outside diameter of conductor B and r_i is its inside diameter, then its cross-sectional area is $\pi(r_o^2 - r_i^2)$ and its resistance is

$$R_B = \frac{\rho L}{\pi(r_o^2 - r_i^2)}.$$

The ratio is

$$\frac{R_A}{R_B} = \frac{r_o^2 - r_i^2}{r_A^2} = \frac{(1.0\,\text{mm})^2 - (0.50\,\text{mm})^2}{(0.50\,\text{mm})^2} = 3.0.$$

36P

Use $J = E/\rho$, where E is the magnitude of the electric field in the wire, J is the magnitude of the current density, and ρ is the resistivity of the material. The electric field is given by $E = V/L$, where V is the potential difference along the wire and L is the length of the wire. Thus $J = V/L\rho$ and

$$\rho = \frac{V}{LJ} = \frac{115\,\text{V}}{(10\,\text{m})(1.4 \times 10^4\,\text{A/m}^2)} = 8.2 \times 10^{-4}\,\Omega \cdot \text{m}.$$

41P

(a) Let ΔT be the change in temperature and β be the coefficient of linear expansion for copper. Then $\Delta L = \beta L \Delta T$ and

$$\frac{\Delta L}{L} = \beta \Delta T = (1.7 \times 10^{-5} /°C)(1.0° \, C) = 1.7 \times 10^{-5} .$$

This is 0.0017%.

The fractional change in area is

$$\frac{\Delta A}{A} = 2\beta \Delta T = 2(1.7 \times 10^{-5} /°C)(1.0° \, C) = 3.4 \times 10^{-5} .$$

This is 0.0034%.

For small changes in the resistivity ρ, length L, and area A of a wire, the change in the resistance is given by

$$\Delta R = \frac{\partial R}{\partial \rho} \Delta\rho + \frac{\partial R}{\partial L} \Delta L + \frac{\partial R}{\partial A} \Delta A .$$

Since $R = \rho L/A$, $\partial R/\partial\rho = L/A = R/\rho$, $\partial R/\partial L = \rho/A = R/L$, and $\partial R/\partial A = -\rho L/A^2 = -R/A$. Furthermore, $\Delta\rho/\rho = \alpha \Delta T$, where α is the temperature coefficient of resistivity for copper ($4.3 \times 10^{-3} /°C$, according to Table 28–1). Thus

$$\frac{\Delta R}{R} = \frac{\Delta\rho}{\rho} + \frac{\Delta L}{L} - \frac{\Delta A}{A} = (\alpha + \beta - 2\beta)\Delta T = (\alpha - \beta)\Delta T$$
$$= (4.3 \times 10^{-3} /°C - 1.7 \times 10^{-5} /°C)(1.0° \, C) = 4.3 \times 10^{-3} .$$

This is 0.43%.

(b) The fractional change in resistivity is much larger than the fractional change in length and area. Changes in length and area affect the resistance much less than changes in resistivity.

42P

(a) Assume, as in Fig. 28–20, that the cone has current i, from bottom to top. The current is the same through every cross-section. We can find an expression for the electric field at every cross section, in terms of the current and then use this expression to find the potential difference V from bottom to top of the cone. The resistance of the cone is given by $R = V/i$.

Consider any cross section of the cone. Let J denote the current density at that cross section and assume it is uniform over the cross section. Then the current through the cross section is given by $i = \int J \, dA = \pi r^2 J$, where r is the radius of the cross section. Now $J = E/\rho$, where ρ is the resistivity and E is the magnitude of the electric field at the cross section. Thus $i = \pi r^2 E/\rho$ and $E = i\rho/\pi r^2$. The current density and electric field have different values on different cross sections because different cross sections have different radii.

Let x measure distance downward from the upper surface of the cone. The radius increases linearly with x, so we may write

$$r = a + \frac{b-a}{L} x .$$

The coefficients in this function have been chosen so $r = a$ when $x = 0$ and $r = b$ when $x = L$. Thus

$$E = \frac{i\rho}{\pi} \left[a + \frac{b-a}{L} x \right]^{-2}.$$

The potential difference between the upper and lower surfaces of the cone is given by

$$V = -\int_0^L E\, dx = -\frac{i\rho}{\pi} \int_0^L \left[a + \frac{b-a}{L} x \right]^{-2} dx$$

$$= \frac{i\rho}{\pi} \frac{L}{b-a} \left[a + \frac{b-a}{L} x \right]^{-1} \Bigg|_0^L = \frac{i\rho}{\pi} \frac{L}{b-a} \left[\frac{1}{a} - \frac{1}{b} \right]$$

$$= \frac{i\rho}{\pi} \frac{L}{b-a} \frac{b-a}{ab} = \frac{i\rho L}{\pi ab}.$$

The resistance is

$$R = \frac{V}{i} = \frac{\rho L}{\pi ab}.$$

(b) If $b = a$ then $R = \rho L / \pi a^2 = \rho L / A$, where $A = \pi a^2$ is the cross-sectional area of the cylinder.

45E

The power dissipated is given by the product of the current and the potential difference:

$$P = iV = (7.0 \times 10^{-3}\,\text{A})(80 \times 10^3\,\text{V}) = 560\,\text{W}.$$

48E

(a) Electrical energy is transferred to heat at a rate given by

$$P = \frac{V^2}{R},$$

where V is the potential difference across the heater and R is the resistance of the heater. Thus

$$P = \frac{(120\,\text{V})^2}{14\,\Omega} = 1.0 \times 10^3\,\text{W} = 1.0\,\text{kW}.$$

(b) The cost is given by

$$C = (1.0\,\text{kW})(5.0\,\text{h})(5.0\,\textit{\textcent}/\text{kW} \cdot \text{h}) = 25\,\textit{\textcent}.$$

56P

(a) Let P be the power dissipated, i be the current in the heater, and V be the potential difference across the heater. They are related by $P = iV$. Solve for i:

$$i = \frac{P}{V} = \frac{1250\,\text{W}}{115\,\text{V}} = 10.9\,\text{A}.$$

(b) According to Ohm's law $V = iR$, where R is the resistance of the heater. Solve for R:

$$R = \frac{V}{i} = \frac{115\,\text{V}}{10.9\,\text{A}} = 10.6\,\Omega\,.$$

(c) The thermal energy E generated by the heater in time t $(= 1.0\,\text{h} = 3600\,\text{s})$ is

$$E = Pt = (1250\,\text{W})(3600\,\text{s}) = 4.5 \times 10^6\,\text{J}\,.$$

58P

Let R_H be the resistance at the higher temperature $(800°\,\text{C})$ and let R_L be the resistance at the lower temperature $(200°\,\text{C})$. Since the potential difference is the same for the two temperatures, the power dissipated at the lower temperature is $P_L = V^2/R_L$, and the power dissipated at the higher temperature is $P_H = V^2/R_H$, so $P_L = (R_H/R_L)P_H$. Now $R_L = R_H + \alpha R_H\,\Delta T$, where ΔT is the temperature difference $T_L - T_H = -600°\,\text{C}$. Thus

$$P_L = \frac{R_H}{R_H + \alpha R_H\,\Delta T}\,P_H = \frac{P_H}{1 + \alpha\,\Delta T} = \frac{500\,\text{W}}{1 + (4.0 \times 10^{-4}/°\text{C})(-600°\,\text{C})} = 660\,\text{W}\,.$$

60P

(a) The charge q that flows past any cross section of the beam in time Δt is given by $q = i\,\Delta t$ and the number of electrons is $N = q/e = (i/e)\,\Delta t$. This is the number of electrons that are accelerated. Thus

$$N = \frac{(0.50\,\text{A})(0.10 \times 10^{-6}\,\text{s})}{1.60 \times 10^{-19}\,\text{C}} = 3.1 \times 10^{11}\,.$$

(b) Over a long time t the total charge is $Q = nqt$, where n is the number of pulses per unit time and q is the charge in one pulse. The average current is given by $\bar{\imath} = Q/t = nq$. Now $q = i\,\Delta t = (0.50\,\text{A})(0.10 \times 10^{-6}\,\text{s}) = 5.0 \times 10^{-8}\,\text{C}$, so

$$\bar{\imath} = (500\,\text{s}^{-1})(5.0 \times 10^{-8}\,\text{C}) = 2.5 \times 10^{-5}\,\text{A}\,.$$

(c) The accelerating potential difference is $V = K/e$, where K is the final kinetic energy of an electron. Since $K = 50\,\text{MeV}$, the accelerating potential is $V = 50\,\text{kV} = 5.0 \times 10^7\,\text{V}$. During a pulse the power output is

$$P = iV = (0.50\,\text{A})(5.0 \times 10^7\,\text{V}) = 2.5 \times 10^7\,\text{W}\,.$$

This is the peak power. The average power is

$$\bar{P} = \bar{\imath}V = (2.5 \times 10^{-5}\,\text{A})(5.0 \times 10^7\,\text{V}) = 1.3 \times 10^3\,\text{W}\,.$$

Chapter 29

2E

The chemical energy of the battery is reduced by $\Delta E = q\mathcal{E}$, where q is the charge that passes through in time $\Delta t = 6.0\,\text{min}$ and $\mathcal{E}$ is the emf of the battery. If i is the current, then $q = i\,\Delta t$ and

$$\Delta E = i\mathcal{E}\,\Delta t = (5.0\,\text{A})(6.0\,\text{V})(6.0\,\text{min})(60\,\text{s/min}) = 1.1 \times 10^4\,\text{J}\,.$$

Notice the conversion of time from minutes to seconds.

4P

If P is the rate at which the battery delivers energy and Δt is the time, the $\Delta E = P\,\Delta t$ is the energy delivered in time Δt. If q is the charge that passes through the battery in time Δt and $\mathcal{E}$ is the emf of the battery, then $\Delta E = q\mathcal{E}$. Equate the two expressions for ΔE and solve for Δt:

$$\Delta t = \frac{q\mathcal{E}}{P} = \frac{(120\,\text{A}\cdot\text{h})(12\,\text{V})}{100\,\text{W}} = 14.4\,\text{h}\,.$$

8E

(a) Let i be the current in the circuit and take it to be positive if it is to the left in R_1. Use Kirchhoff's loop rule: $\mathcal{E}_1 - iR_2 - iR_1 - \mathcal{E}_2 = 0$. Solve for i:

$$i = \frac{\mathcal{E}_1 - \mathcal{E}_2}{R_1 + R_2} = \frac{12\,\text{V} - 6.0\,\text{V}}{4.0\,\Omega + 8.0\,\Omega} = 0.50\,\text{A}\,.$$

A positive value was obtained, so the current is counterclockwise around the circuit.

(b) If i is the current in a resistor R, then the power dissipated by that resistor is given by $P = i^2 R$. For R_1 the power dissipated is

$$P_1 = (0.50\,\text{A})^2(4.0\,\Omega) = 1.0\,\text{W}$$

and for R_2 the power dissipated is

$$P_2 = (0.50\,\text{A})^2(8.0\,\Omega) = 2.0\,\text{W}\,.$$

(c) If i is the current in a battery with emf $\mathcal{E}$, then the battery supplies energy at the rate $P = i\mathcal{E}$ provided the current and emf are in the same direction. The battery absorbs energy at the rate $P = i\mathcal{E}$ if the current and emf are in opposite directions. For $\mathcal{E}_1$ the power is

$$P_1 = (0.50\,\text{A})(12\,\text{V}) = 6.0\,\text{W}$$

and for $\mathcal{E}_2$ it is

$$P_2 = (0.50\,\text{A})(6.0\,\text{V}) = 3.0\,\text{W}\,.$$

In battery 1 the current is in the same direction as the emf so this battery supplies energy to the circuit. The battery is discharging. The current in battery 2 is opposite the direction of the emf, so this battery absorbs energy from the circuit. It is charging.

11E

(a) If i is the current and ΔV is the potential difference then the power absorbed is given by $P = i\,\Delta V$. Thus

$$\Delta V = \frac{P}{i} = \frac{50\,\text{W}}{1.0\,\text{A}} = 50\,\text{V}.$$

Since energy is absorbed, point A is at a higher potential than point B; that is $V_A - V_B = 50\,\text{V}$.

(b) The end-to-end potential difference is given by $V_A - V_B = +iR + \mathcal{E}$, where $\mathcal{E}$ is the emf of element C and is taken to be positive if it is to the left in the diagram. Thus $\mathcal{E} = V_A - V_B - iR = 50\,\text{V} - (1.0\,\text{A})(2.0\,\Omega) = 48\,\text{V}$.

(c) A positive value was obtained for $\mathcal{E}$, so it is toward the left. The negative terminal is at B.

20P

(a) The circuit is shown in the diagram to the right. The current is taken to be positive if it is clockwise. The potential difference across battery 1 is given by $V_1 = \mathcal{E} - ir_1$ and for this to be zero the current must be $i = \mathcal{E}/r_1$. Kirchhoff's loop rule gives $2\mathcal{E} - ir_1 - ir_2 - iR = 0$. Substitute $i = \mathcal{E}/r_1$ and solve for R. You should get $R = r_1 - r_2$.

(b) Since R must be positive, r_1 must be greater than r_2. The potential difference across the battery with the larger internal resistance can be made to vanish with the proper choice of R; the potential difference across the battery with the smaller potential difference cannot be made to vanish.

22P

(a) The current in the circuit is

$$i = \frac{\mathcal{E}}{r + R},$$

so the rate of energy dissipation in R is

$$P = i^2 R = \frac{\mathcal{E}^2 R}{(r + R)^2}.$$

You want to find the value of R that maximizes P. To do this find an expression for the derivative with respect to R, set it equal to zero, and solve for R. The derivative is

$$\frac{dP}{dR} = \frac{\mathcal{E}^2}{(r + R)^2} - \frac{2\mathcal{E}^2 R}{(r + R)^3} = \frac{\mathcal{E}^2(r - R)}{(r + R)^3}.$$

Thus $r - R = 0$ and $R = r$. For R very small P increases with R. For R large P decreases with increasing R. So $R = r$ is a maximum, not a minimum.

(b) Substitute $R = r$ into $P = \mathcal{E}^2 R/(r + R)^2$ to obtain $P = \mathcal{E}^2/4r$.

26E

The potential difference across each resistor is $V = 25.0\,\text{V}$. Since the resistors are identical, the current in each is

$$i = \frac{V}{R} = \frac{25.0\,\text{V}}{18.0\,\Omega} = 1.39\,\text{A}.$$

The total current through the battery is $i_{\text{total}} = 4(1.39\,\text{A}) = 5.56\,\text{A}$.

You might use the idea of equivalent resistance. For 4 identical resistors in parallel the equivalent resistance is determined by

$$\frac{1}{R_{\text{eq}}} = \frac{4}{R},$$

so that $R_{\text{eq}} = R/4$. When a potential difference of 25.0 V is applied to the equivalent resistor the current through it is the same as the total current through the 4 resistors in parallel. Thus

$$i_{\text{total}} = \frac{V}{R_{\text{eq}}} = \frac{4V}{R} = \frac{4(25.0\,\text{V})}{18.0\,\Omega} = 5.56\,\text{A}.$$

29E

Let i_1 be the current in R_1 and take it to be positive if it is to the right. Let i_2 be the current in R_2 and take it to be positive if it is upward. When the loop rule is applied to the lower loop, the result is

$$\mathcal{E}_2 - i_1 R_1 = 0.$$

and when it is applied to the upper loop, the result is

$$\mathcal{E}_1 - \mathcal{E}_2 - \mathcal{E}_3 - i_2 R_2 = 0.$$

The first equation yields

$$i_1 = \frac{\mathcal{E}_2}{R_1} = \frac{5.0\,\text{V}}{100\,\Omega} = 0.050\,\text{A}.$$

The second yields

$$i_2 = \frac{\mathcal{E}_1 - \mathcal{E}_2 - \mathcal{E}_3}{R_2} = \frac{6.0\,\text{V} - 5.0\,\text{V} - 4.0\,\text{V}}{50\,\Omega} = -0.060\,\text{A}.$$

The negative sign indicates that the current in R_2 is actually downward.

If V_b is the potential at point b, then the potential at point a is $V_a = V_b + \mathcal{E}_3 + \mathcal{E}_2$, so $V_a - V_b = \mathcal{E}_3 + \mathcal{E}_2 = 4.0\,\text{V} + 5.0\,\text{V} = 9.0\,\text{V}$.

33E

(a) Let $\mathcal{E}$ be the emf of the battery. When the bulbs are connected in parallel the potential difference across them is the same and is the same as the emf of the battery. The power dissipated by bulb 1 is $P_1 = \mathcal{E}^2/R_1$ and the power dissipated by bulb 2 is $P_2 = \mathcal{E}^2/R_2$. Since R_1 is greater than R_2, bulb 2 dissipates energy at a greater rate than bulb 1 and is the brighter of the two.

(b) When the bulbs are connected in series the current in them is the same. The power dissipate by bulb 1 is now $P_1 = i^2 R_1$ and the power dissipated by bulb 2 is $P_2 = i^2 R_2$. Since R_1 is greater than R_2 greater power is dissipated by bulb 1 than by bulb 2 and bulb 1 is the brighter of the two.

35E

Let r be the resistance of each of the thin wires. Since they are in parallel the resistance R of the composite can be determined from

$$\frac{1}{R} = \frac{9}{r},$$

or $R = r/9$. Now

$$r = \frac{4\rho\ell}{\pi d^2}$$

and

$$R = \frac{4\rho\ell}{\pi D^2},$$

where ρ is the resistivity of copper. $A = \pi d^2/4$ was used for the cross-sectional area of a single wire and a similar expression was used for the cross-sectional area of the thick wire. Since the single thick wire is to have the same resistance as the composite,

$$\frac{4\rho\ell}{\pi D^2} = \frac{4\rho\ell}{9\pi d^2}.$$

Solve for D and obtain $D = 3d$.

39P

Divide the resistors into groups of n resistors each, with all the resistors of a group connected in series. Suppose there are m such groups, with the groups connected in parallel. The scheme is shown in the diagram on the right for $n = 3$ and $m = 2$. Let R be the resistance of any one the resistors. Then the resistance of a series group is nR and the resistance of the total array can be determined from

$$\frac{1}{R_{\text{total}}} = \frac{m}{nR}.$$

Since we want $R_{\text{total}} = R$, we must select $n = m$.

The current is the same in every resistor and there are n^2 resistors, so the maximum total power that can be dissipated is $P_{\text{total}} = n^2 P$, where P is the maximum power that can be dissipated by any one of the resistors.

You want $P_{\text{total}} > 5.0P$. Since $P = 1.0\,\text{W}$, n^2 must be larger than 5.0. Since n must be an integer, the smallest it can be is 3. The least number of resistors is $n^2 = 9$.

40P

(a) The batteries are identical and, because they are connected in parallel, the potential differences across them are the same. This means the currents in them are the same. Let i be the current in either battery and take it to be positive to the left. According to the junction rule the current in R is $2i$ and it is positive to the right. The loop rule applied to either loop containing a battery and R yields $\mathcal{E} - ir - 2iR = 0$, so

$$i = \frac{\mathcal{E}}{r + 2R}.$$

The power dissipated in R is

$$P = i^2 R = \frac{\mathcal{E}^2 R}{(r + 2R)^2}.$$

Find the maximum by setting the derivative with respect to R equal to zero. The derivative is

$$\frac{dP}{dR} = \frac{\mathcal{E}^2}{(r + 2R)^2} - \frac{4\mathcal{E}^2 R}{(r + 2R)^3} = \frac{\mathcal{E}^2(r - 2R)}{(r + 2R)^3}.$$

It vanishes and P is a maximum if $R = r/2$.

(b) Substitute $R = r/2$ into $P = \mathcal{E}^2 R/(r + 2R)^2$ to obtain

$$P_{max} = \frac{\mathcal{E}^2 r/2}{(2r)^2} = \frac{\mathcal{E}^2}{8r}.$$

46P

(a) First find the currents. Let i_1 be the current in R_1 and take it to be positive if it is upward. Let i_2 be the current in R_2 and take it to be positive if it is to the left. Let i_3 be the current in R_3 and take it to be positive if it is to the right. The junction rule produces

$$i_1 + i_2 + i_3 = 0.$$

The loop rule applied to the left-hand loop produces

$$\mathcal{E}_1 - i_3 R_3 + i_1 R_1 = 0$$

and applied to the right-hand loop produces

$$\mathcal{E}_2 - i_2 R_2 + i_1 R_1 = 0.$$

Substitute $i_1 = -i_2 - i_3$, from the first equation, into the other two to obtain

$$\mathcal{E}_1 - i_3 R_3 - i_2 R_1 - i_3 R_1 = 0$$

and

$$\mathcal{E}_2 - i_2 R_2 - i_2 R_1 - i_3 R_1 = 0.$$

The first of these yields

$$i_3 = \frac{\mathcal{E}_1 - i_2 R_1}{R_1 + R_3}.$$

Substitute this into the second equation and solve for i_2. You should obtain

$$i_2 = \frac{\mathcal{E}_2(R_1 + R_3) - \mathcal{E}_1 R_1}{R_1 R_2 + R_1 R_3 + R_2 R_3}$$
$$= \frac{(1.00\,\text{V})(5.00\,\Omega + 4.00\,\Omega) - (3.00\,\text{V})(5.00\,\Omega)}{(5.00\,\Omega)(2.00\,\Omega) + (5.00\,\Omega)(4.00\,\Omega) + (2.00\,\Omega)(4.00\,\Omega)} = -0.158\,\text{A}.$$

Substitute into the expression for i_3 to obtain

$$i_3 = \frac{\mathcal{E}_1 - i_2 R_1}{R_1 + R_3} = \frac{3.00\,\text{V} - (-0.158\,\text{A})(5.00\,\Omega)}{5.00\,\Omega + 4.00\,\Omega} = 0.421\,\text{A}.$$

Finally,

$$i_1 = -i_2 - i_3 = -(-0.158\,\text{A}) - (0.421\,\text{A}) = -0.263\,\text{A}.$$

Note that the current in R_1 is actually downward and the current in R_2 is to the right. The current in R_3 is also to the right.

The power dissipated in R_1 is

$$P_1 = i_1^2 R_1 = (-0.263\,\text{A})^2(5.00\,\Omega) = 0.346\,\text{W},$$

the power dissipated in R_2 is

$$P_2 = i_2^2 R_2 = (-0.158\,\text{A})^2(2.00\,\Omega) = 0.0499\,\text{W},$$

and the power dissipated in R_3 is

$$P_3 = i_3^2 R_3 = (0.421\,\text{A})^2(4.00\,\Omega) = 0.709\,\text{W}.$$

(b) The power supplied by $\mathcal{E}_1$ is

$$i_3 \mathcal{E}_1 = (0.421\,\text{A})(3.00\,\text{V}) = 1.26\,\text{W}$$

and the power supplied by $\mathcal{E}_2$ is

$$i_2 \mathcal{E}_2 = (-0.158\,\text{A})(1.00\,\text{V}) = -0.158\,\text{W}.$$

The negative sign indicates that $\mathcal{E}_2$ is actually absorbing energy from the circuit.

50P

(a) The copper wire and the aluminum jacket are connected in parallel, so the potential difference is the same for them. Since the potential difference is the product of the current and the resistance, $i_C R_C = i_A R_A$, where i_C is the current in the copper, i_A is the current in the aluminum, R_C is the resistance of the copper, and R_A is the resistance of the aluminum. The resistance of either component is given by $R = \rho L/A$, where ρ is the resistivity, L is the length, and A is the cross-sectional area. The resistance of the copper wire is

$$R_C = \frac{\rho_C L}{\pi a^2}$$

and the resistance of the aluminum jacket is

$$R_A = \frac{\rho_A L}{\pi(b^2 - a^2)}.$$

Substitute these expressions into $i_C R_C = i_A R_A$, and cancel the common factors L and π to obtain

$$\frac{i_C \rho_C}{a^2} = \frac{i_A \rho_A}{b^2 - a^2}.$$

Solve this equation simultaneously with $i = i_C + i_A$, where i is the total current. You should get

$$i_C = \frac{a^2 \rho_C i}{(b^2 - a^2)\rho_C + a^2 \rho_A}$$

and

$$i_A = \frac{(b^2 - a^2)\rho_C i}{(b^2 - a^2)\rho_C + a^2 \rho_A}.$$

The denominators are the same and each has the value

$$(b^2 - a^2)\rho_C + a^2 \rho_A = \left[(0.380 \times 10^{-3}\,\mathrm{m})^2 - (0.250 \times 10^{-3}\,\mathrm{m})^2\right] (1.69 \times 10^{-8}\,\Omega \cdot \mathrm{m})$$
$$+ (0.250 \times 10^{-3}\,\mathrm{m})^2 (2.75 \times 10^{-8}\,\Omega \cdot \mathrm{m})$$
$$= 3.10 \times 10^{-15}\,\Omega \cdot \mathrm{m}^3.$$

Thus

$$i_C = \frac{(0.250 \times 10^{-3}\,\mathrm{m})^2 (2.75 \times 10^{-8}\,\Omega \cdot \mathrm{m})(2.00\,\mathrm{A})}{3.10 \times 10^{-15}\,\Omega \cdot \mathrm{m}^3} = 1.11\,\mathrm{A}$$

and

$$i_A = \frac{\left[(0.380 \times 10^{-3}\,\mathrm{m})^2 - (0.250 \times 10^{-3}\,\mathrm{m})^2\right] (1.69 \times 10^{-8}\,\Omega \cdot \mathrm{m})(2.00\,\mathrm{A})}{3.10 \times 10^{-15}\,\Omega \cdot \mathrm{m}^3}$$
$$= 0.893\,\mathrm{A}.$$

(b) Consider the copper wire. If V is the potential difference, then the current is given by $V = i_C R_C = i_C \rho_C L/\pi a^2$, so

$$L = \frac{\pi a^2 V}{i_C \rho_C} = \frac{\pi(0.250 \times 10^{-3}\,\mathrm{m})^2 (12.0\,\mathrm{V})}{(1.11\,\mathrm{A})(1.69 \times 10^{-8}\,\Omega \cdot \mathrm{m})} = 126\,\mathrm{m}.$$

(a) A potential difference is applied across points 1 and 3 in the diagram of the text. Notice that the point labelled 2 is one cube edge from 1 and one cube edge from 3. So is the point labelled 4. These points must be at the same potential. Similarly the points labelled 6 and 8 must be at the same potential. The cube can be distorted so that points 2 and 4 come together and points 6 and 8 come together, without changing the current in any branch. When this is done the circuit can be drawn as in the figure shown to the right. The numbered points correspond to points on the original diagram of the text. Each of the resistors has the same value, R.

Notice that the two resistors in the upper left of the diagram are in parallel and can be replaced by a resistor with their equivalent resistance, $R/2$. So can the pair in the upper right, the pair in the lower left, the pair in the lower right, and the vertical pair in the center. When this is done the circuit looks like that shown on the left below. Now the resistor on the left and the one on the bottom left are in series and have an equivalent resistance of $3R/2$. Similarly, the resistor on the right and the one on the bottom right have the same equivalent resistance. When these pairs of resistors are replaced with their equivalents the circuit looks like that shown on the right below. Be sure you understand why neither of the two resistors along the top of the circuit are in series with any of the others.

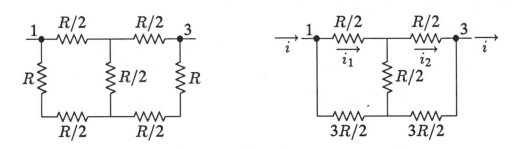

This is as far as we can go by combining resistors in parallel and series. Find the equivalent resistance by assuming a current i enters the circuit at point 1 and leaves at point 3. Find the current in resistors, then use the results to calculate the potential difference V between points 1 and 3. The equivalent resistance of the circuit is $R_{13} = V/i$.

Let i_1 be the current in the upper left resistor and take it to be positive to the right. Let i_2 be the current in the upper right resistor and take it to be positive to the right. Then the current in the bottom left resistor is $i - i_1$ and it is positive if it is to the right. The current in the bottom right resistor is $i - i_2$ and it is positive if it is to the right. Finally the current in the central resistor is $i_2 - i_1$ and it is positive if it is upward. These currents automatically satisfy the junction rule. When the loop rule is applied to the left-hand loop the result is

$$-i_1 \frac{R}{2} + (i_2 - i_1)\frac{R}{2} + (i - i_1)\frac{3R}{2} = 0$$

and when it is applied to the right-hand loop the result is

$$-i_2 \frac{R}{2} + (i - i_2) \frac{3R}{2} - (i_2 - i_1) \frac{R}{2} = 0.$$

The solution to these equation is $i_1 = i_2 = 3i/4$. The potential difference between points 1 and 3 is $V = i_1(R/2) + i_2(R/2) = 3iR/4$ and the equivalent resistance is $R_{13} = V/i = 3R/4$.
(b) Now a potential difference is applied across the points 1 and 7 on the diagram of the text. Points 3 and 6 must be at the same potential and points 4 and 5 must be at the same potential. Once the points in each of these pairs are placed together, the circuit can be drawn as in the diagram on the right.

Again 5 pairs of parallel resistors can be identified and each pair can be replaced by an equivalent resistor. When this is done the circuit on the left below is obtained. There are now two series connections, each with an equivalent resistance of $3R/2$. When these replacements are made the circuit on the right below is obtained.

Now find the equivalent resistance of this circuit. Suppose current i enters at 1 and leaves at 7. Let i_1 be the current in the upper left resistor and take it to be positive if it is to the right. Let i_2 be the current in the upper right resistor and take it to be positive if it is to the right. The current in the lower left resistor is $i - i_1$ and it is positive if it is to the right. The current in the lower right resistor is $i - i_2$ and it is positive if it is to the right. The current in the central resistor is $i_2 - i_1$ and it is positive if it is upward. The loop rule applied to the left-hand loop yields

$$-i_1 \frac{R}{2} + (i_2 - i_1) \frac{R}{2} + (i - i_1) \frac{3R}{2} = 0$$

and applied to the right-hand loop yields

$$-i_2 \frac{3R}{2} + (i - i_2) \frac{R}{2} - (i_2 - i_1) \frac{R}{2} = 0.$$

The solution to these equations is $i_1 = 2i/3$ and $i_2 = i/3$.

The potential difference between points 1 and 7 is $V = i_1(R/2) + i_2(3R/2) = i(R/3) + i(R/2) = 5iR/6$ and the equivalent resistance is $R_{17} = V/i = 5R/6$.

58P

The current in R_2 is i. Let i_1 be the current in R_1 and take it to be downward. According to the junction rule the current in the voltmeter is $i - i_1$ and it is downward. Apply the loop rule to the left-hand loop to obtain

$$\mathcal{E} - iR_2 - i_1 R_1 - ir = 0.$$

Apply the loop rule to the right-hand loop to obtain

$$i_1 R_1 - (i - i_1)R_V = 0.$$

Solve these equations for i_1. The second equation yields

$$i = \frac{R_1 + R_V}{R_V} i_1.$$

Substitute this into the first equation to obtain

$$\mathcal{E} - \frac{(R_2 + r)(R_1 + R_V)}{R_V} i_1 + R_1 i_1 = 0.$$

This has the solution

$$i_1 = \frac{\mathcal{E} R_V}{(R_2 + r)(R_1 + R_V) + R_1 R_V}.$$

The reading on the voltmeter is

$$
\begin{aligned}
i_1 R_1 &= \frac{\mathcal{E} R_V R_1}{(R_2 + r)(R_1 + R_V) + R_1 R_V} \\
&= \frac{(3.0\,\text{V})(5.0 \times 10^3\,\Omega)(250\,\Omega)}{(300\,\Omega + 100\,\Omega)(250\,\Omega + 5.0 \times 10^3\,\Omega) + (250\,\Omega)(5.0 \times 10^3\,\Omega)} \\
&= 1.12\,\text{V}.
\end{aligned}
$$

The current in the absence of the voltmeter can be obtained by taking the limit as R_V becomes infinitely large. Then

$$i_1 R_1 = \frac{\mathcal{E} R_1}{R_1 + R_2 + r} = \frac{(3.0\,\text{V})(250\,\Omega)}{250\,\Omega + 300\,\Omega + 100\,\Omega} = 1.15\,\text{V}.$$

The fractional error is

$$\frac{1.15 - 1.12}{1.15} = 0.030.$$

This is 3.0%.

63P

Let i_1 be the current in R_1 and R_2 and take it to be positive if it is toward point a in R_1. Let i_2 be the current in R_s and R_x and take it to be positive if it is toward b in R_s. The loop rule yields $(R_1 + R_2)i_1 - (R_x + R_s)i_2 = 0$. Since points a and b are at the same potential

$i_1 R_1 = i_2 R_s$. The second equation gives $i_2 = i_1 R_1 / R_s$. This expression is substituted into the first equation to obtain

$$(R_1 + R_2)i_1 = (R_x + R_s) \frac{R_1}{R_s} i_1 .$$

Solve for R_x. You should get $R_x = R_2 R_s / R_1$.

66E

During charging the charge on the positive plate of the capacitor is given by Eq. 29–29, with $RC = \tau$. That is,

$$q = C\mathcal{E} \left[1 - e^{-t/\tau} \right] ,$$

where C is the capacitance, $\mathcal{E}$ is applied emf, and τ is the time constant. You want the time for which $q = 0.99 C\mathcal{E}$, so

$$0.99 = 1 - e^{-t/tau} .$$

Thus

$$e^{-t/\tau} = 0.01 .$$

Take the natural logarithm of both sides to obtain $t/\tau = -\ln 0.01 = 4.6$ and $t = 4.6\tau$.

69P

(a) The potential difference V across the plates of a capacitor is related to the charge q on the positive plate by $V = q/C$, where C is capacitance. Since the charge on a discharging capacitor is given by $q = q_0 e^{-t/\tau}$, where q_0 is the charge at time $t = 0$ and τ is the time constant, this means

$$V = V_0 e^{-t/\tau} ,$$

where q_0/C was replaced by V_0, the initial potential difference. Solve for τ by dividing the equation by V_0 and taking the natural logarithm of both sides. The result is:

$$\tau = -\frac{t}{\ln(V/V_0)} = -\frac{10.0\,\text{s}}{\ln[(1.00\,\text{V})/(100\,\text{V})]} = 2.17\,\text{s} .$$

(b) At $t = 17.0\,\text{s}$, $t/\tau = (17.0\,\text{s})/(2.17\,\text{s}) = 7.83$ and

$$V = V_0 e^{-t/\tau} = (100\,\text{V}) e^{-7.83} = 3.96 \times 10^{-2}\,\text{V} .$$

71P

(a) The initial energy stored in a capacitor is given by

$$U_C = \frac{q_0^2}{2C} ,$$

where C is the capacitance and q_0 is the initial charge on one plate. Thus

$$q_0 = \sqrt{2CU_C} = \sqrt{2(1.0 \times 10^{-6}\,\text{F})(0.50\,\text{J})} = 1.0 \times 10^{-3}\,\text{C} .$$

(b) The charge as a function of time is given by $q = q_0 e^{-t/\tau}$, where τ is the time constant. The current is the derivative with respect to time of the charge:

$$i = -\frac{dq}{dt} = \frac{q_0}{\tau} e^{-t/\tau}.$$

The initial current is given by this function evaluated for $t = 0$: $i_0 = q_0/\tau$. The time constant is

$$\tau = RC = (1.0 \times 10^{-6}\,\text{F})(1.0 \times 10^6\,\Omega) = 1.0\,\text{s}.$$

Thus

$$i_0 = \frac{1.0 \times 10^{-3}\,\text{C}}{1.0\,\text{s}} = 1.0 \times 10^{-3}\,\text{A}.$$

(c) Substitute $q = q_0 e^{-t/\tau}$ into $V_C = q/C$ to obtain

$$V_C = \frac{q_0}{C} e^{-t/\tau} = \frac{1.0 \times 10^{-3}\,\text{C}}{1.0 \times 10^{-6}\,\text{F}} e^{-t/1.0\,\text{s}} = (1.0 \times 10^3\,\text{V})e^{-t},$$

where t is in seconds.

Substitute $i = (q_0/\tau) e^{-t/\tau}$ into $V_R = iR$ to obtain

$$V_R = \frac{q_0 R}{\tau} e^{-t/\tau} = \frac{(1.0 \times 10^{-3}\,\text{C})(1.0 \times 10^6\,\Omega)}{1.0\,\text{s}} e^{-t/1.0\,\text{s}} = (1.0 \times 10^3\,\text{V})e^{-t},$$

where t is in seconds.

(d) Substitute $i = (q_0/\tau) e^{-t/\tau}$ into $P = i^2 R$ to obtain

$$P = \frac{q_0^2 R}{\tau^2} e^{-2t/\tau} = \frac{(1.0 \times 10^{-3}\,\text{C})^2(1.0 \times 10^6\,\Omega)}{(1.0\,\text{s})^2} e^{-2t/1.0\,\text{s}} = (1.0\,\text{W})e^{-2t},$$

where t is in seconds.

72P

(a) The charge on the positive plate of the capacitor is given by

$$q = C\mathcal{E}\left[1 - e^{-t/\tau}\right],$$

where $\mathcal{E}$ is the emf of the battery, C is the capacitance, and τ is the time constant. The value of τ is

$$\tau = RC = (3.00 \times 10^6\,\Omega)(1.00 \times 10^{-6}\,\text{F}) = 3.00\,\text{s}.$$

At $t = 1.00\,\text{s}$,

$$\frac{t}{\tau} = \frac{1.00\,\text{s}}{3.00\,\text{s}} = 0.333$$

and the rate at which the charge is increasing is

$$\frac{dq}{dt} = \frac{C\mathcal{E}}{\tau} e^{-t/\tau} = \frac{(1.00 \times 10^{-6}\,\text{F})(4.00\,\text{V})}{3.00\,\text{s}} e^{-0.333} = 9.55 \times 10^{-7}\,\text{C/s}.$$

(b) The energy stored in the capacitor is given by

$$U_C = \frac{q^2}{2C}.$$

and its rate of change is

$$\frac{dU_c}{dt} = \frac{q}{C}\frac{dq}{dt}.$$

Now, when $t = 1.00\,\text{s}$,

$$q = C\mathcal{E}\left[1 - e^{-t/\tau}\right] = (1.00 \times 10^{-6})(4.00\,\text{V})\left[1 - e^{-0.333}\right] = 1.13 \times 10^{-6}\,\text{C},$$

so

$$\frac{dU_C}{dt} = \left(\frac{1.13 \times 10^{-6}\,\text{C}}{1.00 \times 10^{-6}\,\text{F}}\right)(9.55 \times 10^{-7}\,\text{C/s}) = 1.08 \times 10^{-6}\,\text{W}.$$

(c) The rate at which energy is being dissipated in the resistor is given by $P = i^2R$. The current is $9.55 \times 10^{-7}\,\text{A}$, so

$$P = (9.55 \times 10^{-7}\,\text{A})^2(3.00 \times 10^6\,\Omega) = 2.74 \times 10^{-6}\,\text{W}.$$

(d) The rate at which energy is delivered by the battery is

$$i\mathcal{E} = (9.55 \times 10^{-7}\,\text{A})(4.00\,\text{V}) = 3.82 \times 10^{-6}\,\text{W}.$$

The energy delivered by the battery is either stored in the capacitor or dissipated in the resistor. Conservation of energy requires that $i\mathcal{E} = (q/C)(dq/dt) + i^2R$. The numerical results support the conservation principle.

78P

(a) At $t = 0$ the capacitor is completely uncharged and the current in the capacitor branch is as it would be if the capacitor were replaced by a wire. Let i_1 be the current in R_1 and take it to be positive if it is to the right. Let i_2 be the current in R_2 and take it to be positive if it is downward. Let i_3 be the current in R_3 and take it to be positive if it is downward. The junction rule produces $i_1 = i_2 + i_3$, the loop rule applied to the left-hand loop produces $\mathcal{E} - i_1R_1 - i_2R_2 = 0$, and the loop rule applied to the right-hand loop produces $i_2R_2 - i_3R_3 = 0$. Since the resistances are all the same you can simplify the mathematics by replacing R_1, R_2, and R_3 with R. The solution to the three simultaneous equations is

$$i_1 = \frac{2\mathcal{E}}{3R} = \frac{2(1.2 \times 10^3\,\text{V})}{3(0.73 \times 10^6\,\Omega)} = 1.1 \times 10^{-3}\,\text{A}$$

and

$$i_2 = i_3 = \frac{\mathcal{E}}{3R} = \frac{1.2 \times 10^3\,\text{V}}{3(0.73 \times 10^6\,\Omega)} = 5.5 \times 10^{-4}\,\text{A}.$$

At $t = \infty$ the capacitor is fully charged and the current in the capacitor branch is zero. Then $i_1 = i_2$ and the loop rule yields

$$\mathcal{E} - i_1 R_1 - i_1 R_2 = 0.$$

The solution is

$$i_1 = i_2 = \frac{\mathcal{E}}{2R} = \frac{1.2 \times 10^3\,\text{V}}{2(0.73 \times 10^6\,\Omega)} = 8.2 \times 10^{-4}\,\text{A}.$$

(b) Take the upper plate of the capacitor to be positive. This is consistent with current into that plate. The junction and loop equations are $i_1 = i_2 + i_3$, $\mathcal{E} - i_1 R - i_2 R = 0$, and $-(q/C) - i_3 R + i_2 R = 0$. Use the first equation to substitute for i_1 in the second and obtain $\mathcal{E} - 2i_2 R - i_3 R = 0$. Thus $i_2 = (\mathcal{E} - i_3 R)/2R$. Substitute this expression into the third equation above to obtain $-(q/C) - (i_3 R) + (\mathcal{E}/2) - (i_3 R/2) = 0$. Now replace i_3 with dq/dt and obtain

$$\frac{3R}{2}\frac{dq}{dt} + \frac{q}{C} = \frac{\mathcal{E}}{2}.$$

This is just like the equation for an RC series circuit, except that the time constant is $\tau = 3RC/2$ and the impressed potential difference is $\mathcal{E}/2$. The solution is

$$q = \frac{C\mathcal{E}}{2}\left[1 - e^{-2t/3RC}\right].$$

The current in the capacitor branch is

$$i_3 = \frac{dq}{dt} = \frac{\mathcal{E}}{3R}e^{-2t/3RC}.$$

The current in the center branch is

$$i_2 = \frac{\mathcal{E}}{2R} - \frac{i_3}{2} = \frac{\mathcal{E}}{2R} - \frac{\mathcal{E}}{6R}e^{-2t/3RC}$$

$$= \frac{\mathcal{E}}{6R}\left[3 - e^{-2t/3RC}\right]$$

and the potential difference across R_2 is

$$V_2 = i_2 R = \frac{\mathcal{E}}{6}\left[3 - e^{-2t/3RC}\right].$$

The graph of V_2 vs. t is shown to the right.

(c) For $t = 0$, $e^{-2t/3RC}$ is 1 and $V_2 = \mathcal{E}/3 = (1.2 \times 10^3\,\text{V})/3 = 400\,\text{V}$. For $t = \infty$, $e^{-2t/3RC}$ is zero and $V_2 = \mathcal{E}/2 = (1.2 \times 10^3\,\text{V})/2 = 600\,\text{V}$.

(d) "A long time" means after many time constants. Then the current in the capacitor branch is very small and can be approximated by zero.

Chapter 30

4E

(a) The magnitude of the magnetic force on the proton is given by $F_B = evB \sin \phi$, where v is the speed of the proton, B is the magnitude of the magnetic field, and ϕ is the angle between the particle velocity and the field when they are drawn with their tails at the same point. Thus

$$v = \frac{F_B}{eB \sin \phi} = \frac{6.50 \times 10^{-17}\,\text{N}}{(1.60 \times 10^{-19}\,\text{C})(2.60 \times 10^{-3}\,\text{T}) \sin 23.0°} = 4.00 \times 10^5\,\text{m/s}.$$

(b) The kinetic energy of the proton is

$$K = \frac{1}{2}mv^2 = \frac{1}{2}(1.67 \times 10^{-27}\,\text{kg})(4.00 \times 10^5\,\text{m/s})^2 = 1.34 \times 10^{-16}\,\text{J}.$$

This is $(1.34 \times 10^{-16}\,\text{J})/(1.60 \times 10^{-19}\,\text{J/eV}) = 835\,\text{eV}$.

7P

(a) The diagram shows the electron traveling to the north. The magnetic field is into the page. The right-hand rule tells us that $\mathbf{v} \times \mathbf{B}$ is to the west, but since the electron is negatively charged the magnetic force on it is to the east.

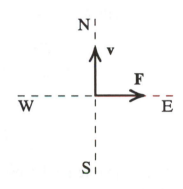

(b) Use $F = ma$, with $F = evB \sin \phi$. Here v is the speed of the electron, B is the magnitude of the magnetic field, and ϕ is the angle between the electron velocity and the magnetic field. The velocity and field are perpendicular to each other so $\phi = 90°$ and $\sin \phi = 1$. Thus $a = evB/m$.

The electron speed can be found from its kinetic energy. Since $K = \frac{1}{2}mv^2$,

$$v = \sqrt{\frac{2K}{m}} = \sqrt{\frac{2(12.0 \times 10^3\,\text{eV})(1.60 \times 10^{-19}\,\text{J/eV})}{9.11 \times 10^{-31}\,\text{kg}}} = 6.49 \times 10^7\,\text{m/s}.$$

Thus

$$a = \frac{evB}{m} = \frac{(1.60 \times 10^{-19}\,\text{C})(6.49 \times 10^7\,\text{m/s})(55.0 \times 10^{-6}\,\text{T})}{9.11 \times 10^{-31}\,\text{kg}} = 6.27 \times 10^{14}\,\text{m/s}^2.$$

(c) The electron follows a circular path. Its acceleration is given by v^2/R, where R is the radius of the path. Thus

$$R = \frac{v^2}{a} = \frac{(6.49 \times 10^7\,\text{m/s})^2}{6.27 \times 10^{14}\,\text{m/s}^2} = 6.72\,\text{m}.$$

The solid curve on the diagram is the path. Suppose it subtends the angle θ at the center. ℓ (= 0.200 m) is the length of the tube and d is the deflection. The right triangle yields $d = R - R\cos\theta$ and $\ell = R\sin\theta$. Thus $R\cos\theta = R - d$ and $R\sin\theta = \ell$. Square both these equations and add the results to obtain $R^2 = (R-d)^2 + \ell^2$, or $d^2 - 2Rd + \ell^2 = 0$. The solution is $d = R \pm \sqrt{R^2 - \ell^2}$. The plus sign corresponds to an angle of $180° - \theta$; the minus sign corresponds to the correct solution.

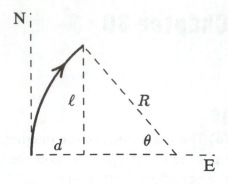

Since ℓ is much less than R, use the binomial theorem to expand $\sqrt{R^2 - \ell^2}$. The first two terms are $R - \frac{1}{2}\ell^2/R$, so $d = \frac{1}{2}\ell^2/R$ and when numerical values are substituted, $d = 0.00298$ m is obtained.

10E

(a) The total force on the electron is $\mathbf{F} = -e(\mathbf{E} + \mathbf{v} \times \mathbf{B})$, where $\mathbf{E}$ is the electric field, $\mathbf{B}$ is the magnetic field and $\mathbf{v}$ is the electron velocity. The magnitude of the magnetic force is $evB\sin\phi$, where ϕ is the angle between the velocity and the field. Since the total force must vanish, $B = E/v\sin\phi$. The force is the smallest it can be when the field is perpendicular to the velocity and $\phi = 90°$. Then $B = E/v$. Use $K = \frac{1}{2}mv^2$ to find the speed:

$$v = \sqrt{\frac{2K}{m}} = \sqrt{\frac{2(2.5 \times 10^3\,\text{eV})(1.60 \times 10^{-19}\,\text{J/eV})}{9.11 \times 10^{-31}\,\text{kg}}} = 2.96 \times 10^7\,\text{m/s}\,.$$

Thus

$$B = \frac{E}{v} = \frac{10 \times 10^3\,\text{V/m}}{2.96 \times 10^7\,\text{m/s}} = 3.37 \times 10^{-4}\,\text{T}\,.$$

The magnetic field must be perpendicular to both the electric field and the velocity of the electron.

(b) A proton will pass undeflected if its velocity is the same as that of the electron. Both the electric and magnetic forces reverse direction, but they still cancel.

13P

Since the total force, given by $\mathbf{F} = e(\mathbf{E} + \mathbf{v} \times \mathbf{B})$, vanishes, the electric field $\mathbf{E}$ must be perpendicular to both the particle velocity $\mathbf{v}$ and the magnetic field $\mathbf{B}$. The magnetic field is perpendicular to the velocity so $\mathbf{v} \times \mathbf{B}$ has magnitude vB and the magnitude of the electric field is given by $E = vB$. Since the particle has charge e and is accelerated through a potential difference V, $\frac{1}{2}mv^2 = eV$ and $v = \sqrt{2eV/m}$. Thus

$$E = B\sqrt{\frac{2eV}{m}} = (1.2\,\text{T})\sqrt{\frac{2(1.60 \times 10^{-19}\,\text{C})(10 \times 10^3\,\text{V})}{(6.0\,\text{u})(1.661 \times 10^{-27}\,\text{kg/u})}} = 6.8 \times 10^5\,\text{V/m}\,.$$

Notice that the mass, given in u, must be converted to kg.

19E

Since the magnetic field is perpendicular to the particle velocity the magnitude of the magnetic force is given by $F_B = evB$ and the acceleration of the electron has magnitude $a = F_B/m$, where v is the speed of the electron and m is its mass. B is the magnitude of the magnetic field. Since the electron is traveling with uniform speed in a circle, its acceleration is $a = v^2/r$, where r is the radius of the circle. Thus $evB/m = v^2/r$ and

$$B = \frac{mv}{er} = \frac{(9.11 \times 10^{-31}\,\text{kg})(1.3 \times 10^6\,\text{m/s})}{(1.60 \times 10^{-19}\,\text{C})(0.35\,\text{m})} = 2.1 \times 10^{-5}\,\text{T}.$$

23E

(a) In the accelerating process the electron loses potential energy eV and gains the same amount of kinetic energy. Since it starts from rest, $\frac{1}{2}mv^2 = eV$ and

$$v = \sqrt{\frac{2eV}{m}} = \sqrt{\frac{2(1.60 \times 10^{-19}\,\text{C})(350\,\text{V})}{9.11 \times 10^{-31}\,\text{kg}}} = 1.11 \times 10^7\,\text{m/s}.$$

(b) The electron travels with constant speed around a circle. The magnetic force on it has magnitude $F_B = evB$ and its acceleration is v^2/R, where R is the radius of the circle. Newton's second law yields $evB = mv^2/R$, so

$$R = \frac{mv}{eB} = \frac{(9.11 \times 10^{-31}\,\text{kg})(1.11 \times 10^7\,\text{m/s})}{(1.60 \times 10^{-19}\,\text{C})(200 \times 10^{-3}\,\text{T})} = 3.16 \times 10^{-4}\,\text{m}.$$

27E

Orient the magnetic field so it is perpendicular to the plane of page. Then the electron will travel with constant speed around a circle in the plane of the page. The magnetic force on an electron has magnitude $F_B = evB$, where v is the speed of the electron and B is the magnitude of the magnetic field. If r is the radius of the circle, the acceleration of the electron has magnitude $a = v^2/r$. Newton's second law yields $evB = mv^2/r$, so the radius of the circle is given by $r = mv/eB$. The kinetic energy of the electron is $K = \frac{1}{2}mv^2$, so $v = \sqrt{2K/m}$. Thus

$$r = \frac{m}{eB}\sqrt{\frac{2K}{m}} = \sqrt{\frac{2mK}{e^2B^2}}.$$

This must be less than d, so

$$\sqrt{\frac{2mK}{e^2B^2}} \leq d$$

or

$$B \geq \sqrt{\frac{2mK}{e^2d^2}}.$$

If the electrons are to travel as shown in Fig. 30-32 the magnetic field must be out of the page. Then the magnetic force is toward the center of the circular path.

31P

During acceleration by the electric field, outside the magnetic field, the ion loses potential energy qV, where V is the accelerating potential difference and q is the charge on the ion. Since it starts from rest it gains kinetic energy $\frac{1}{2}mv^2$, where v is its final speed and m is its mass. Energy is conserved, so $qV = \frac{1}{2}mv^2$ and $v = \sqrt{2qV/m}$. While in the magnetic field it travels with constant speed around a circle of radius $x/2$. The magnetic force is qvB and its acceleration is $v^2/(x/2)$, so $qvB = 2mv^2/x$ or $qB = 2mv/x$. Substitute $v = \sqrt{2qV/m}$ and solve for m. You should get $m = qB^2x^2/8V$.

33P

(a) Solve the result of Problem 31 for B. You should get

$$B = \sqrt{\frac{8Vm}{qx^2}}.$$

Evaluate this expression using $x = 2.00\,\text{m}$:

$$B = \sqrt{\frac{8(100 \times 10^3\,\text{V})(3.92 \times 10^{-25}\,\text{kg})}{(3.20 \times 10^{-19}\,\text{C})(2.00\,\text{m})^2}} = 0.495\,\text{T}.$$

(b) Let N be the number of ions that are separated by the machine per unit time. The current is $i = qN$ and the mass that is separated per unit time is $M = mN$, where m is the mass of a single ion. M has the value

$$M = 100\,\text{mg/h} = \frac{100 \times 10^{-6}\,\text{kg}}{3600\,\text{s}} = 2.78 \times 10^{-8}\,\text{kg/s}.$$

Since $N = M/m$ we have

$$i = \frac{qM}{m} = \frac{(3.20 \times 10^{-19}\,\text{C})(2.78 \times 10^{-8}\,\text{kg/s})}{3.92 \times 10^{-25}\,\text{kg}} = 2.27 \times 10^{-2}\,\text{A}.$$

(c) Each ion deposits an energy of qV in the cup, so the energy deposited in time Δt is given by

$$E = NqV\,\Delta t = \frac{iqV}{q}\,\Delta t = iV\,\Delta t.$$

For $\Delta t = 1.0\,\text{h}$,

$$E = (2.27 \times 10^{-2}\,\text{A})(100 \times 10^3\,\text{V})(3600\,\text{s}) = 8.17 \times 10^8\,\text{J}.$$

To obtain the second expression, i/q was substituted for N.

35P

(a) If v is the speed of the positron then $v \sin \phi$ is the component of its velocity in the plane that is perpendicular to the magnetic field. Here ϕ is the angle between the velocity and the field (89°). Newton's second law yields $eBv\sin\phi = m(v\sin\phi)^2/r$, where r is the radius of the orbit. Thus $r = (mv/eB)\sin\phi$. The period is given by

$$T = \frac{2\pi r}{v\sin\phi} = \frac{2\pi m}{eB} = \frac{2\pi(9.11\times 10^{-31}\,\text{kg})}{(1.60\times 10^{-19}\,\text{C})(0.10\,\text{T})} = 3.58\times 10^{-10}\,\text{s}\,.$$

The equation for r was substituted to obtain the second expression for T.

(b) The pitch is the distance traveled along the line of the magnetic field in a time interval of one period. Thus $p = vT\cos\phi$. Use the kinetic energy to find the speed: $K = \frac{1}{2}mv^2$ means

$$v = \sqrt{\frac{2K}{m}} = \sqrt{\frac{2(2.0\times 10^3\,\text{eV})(1.60\times 10^{-19}\,\text{J/eV})}{9.11\times 10^{-31}\,\text{kg}}} = 2.651\times 10^7\,\text{m/s}\,.$$

Thus

$$p = (2.651\times 10^7\,\text{m/s})(3.58\times 10^{-10}\,\text{s})\cos 89° = 1.66\times 10^{-4}\,\text{m}\,.$$

(c) The orbit radius is

$$R = \frac{mv\sin\phi}{eB} = \frac{(9.11\times 10^{-31}\,\text{kg})(2.651\times 10^7\,\text{m/s})\sin 89°}{(1.60\times 10^{-19}\,\text{C})(0.10\,\text{T})} = 1.51\times 10^{-3}\,\text{m}\,.$$

37P

(a) The radius r of the circular orbit is given by $r = p/eB$, where B is the magnitude of the magnetic field. The relativisitic expression $p = mv/\sqrt{1 - v^2/c^2}$ must be used for the magnitude of the momentum. Here v is the speed of the proton, m is its mass, and c is the speed of light. Hence

(b)

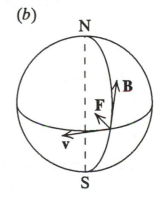

$$r = \frac{mv}{eB\sqrt{1 - v^2/c^2}}\,.$$

Square both sides and solve for v. The result is

$$v = \frac{reBc}{\sqrt{m^2c^2 + r^2e^2B^2}}\,.$$

Substitute $r = 6.37\times 10^6\,\text{m}$ (the radius of the Earth), $e = 1.6022\times 10^{-19}\,\text{C}$ (the charge on a proton), $B = 41\times 10^{-6}\,\text{T}$, $m = 1.6726\times 10^{-27}\,\text{kg}$ (the mass of a proton), and $c = 2.9979\times 10^8\,\text{m/s}$ to obtain $v = 2.9977\times 10^8\,\text{m/s}$.

42P

Approximate the total distance by the number of revolutions times the circumference of the orbit corresponding to the average energy. This should be a good approximation since the deuteron receives the same energy each revolution and its period does not depend on its energy. The deuteron accelerates twice in each cycle and each time it receives an energy of $qV = 80 \times 10^3$ eV. Since its final energy is 16.6 MeV the number of revolutions it makes is

$$n = \frac{16.6 \times 10^6 \text{ eV}}{2(80 \times 10^3 \text{ eV})} = 104 .$$

Its average energy during the accelerating process is 8.3 MeV. The radius of the orbit is given by $r = mv/qB$, where v is the deuteron's speed. Since this is given by $v = \sqrt{2K/m}$, the radius is

$$r = \frac{m}{qB} \sqrt{\frac{2K}{m}} = \frac{1}{qB} \sqrt{2Km} .$$

For the average energy

$$r = \frac{\sqrt{2(8.3 \times 10^6 \text{ eV})(1.60 \times 10^{-19} \text{ J/eV})(3.34 \times 10^{-27} \text{ kg})}}{(1.60 \times 10^{-19} \text{ C})(1.57 \text{ T})} = 0.375 \text{ m} .$$

The total distance traveled is about $n2\pi r = (104)(2\pi)(0.375) = 245$ m.

44E

The magnitude of the magnetic force on the wire is given by $F_B = iLB \sin \phi$, where i is the current in the wire, L is the length of the wire, B is the magnitude of the magnetic field, and ϕ is the angle between the current and the field. In this case $\phi = 70°$. Thus

$$F_B = (5000 \text{ A})(100 \text{ m})(60.0 \times 10^{-6} \text{ T}) \sin 70° = 28.2 \text{ N} .$$

Apply the right-hand rule to the vector product $\mathbf{F}_B = i\mathbf{L} \times \mathbf{B}$ to show that the force is to the west.

46P

The magnetic force on the wire must be upward and have a magnitude equal to the gravitational force mg on the wire. Apply the right-hand rule to show that the current must be from left to right. Since the field and the current are perpendicular to each other the magnitude of the magnetic force is given by $F_B = iLB$, where L is the length of the wire. The condition that the tension in the supports vanish is $iLB = mg$, which yields

$$i = \frac{mg}{LB} = \frac{(0.0130 \text{ kg})(9.8 \text{ m/s}^2)}{(0.620 \text{ m})(0.440 \text{ T})} = 0.467 \text{ A} .$$

52P

The magnetic force must push horizontally on the rod to overcome the force of friction. But it can be oriented so it also pulls up on the rod and thereby reduces both the normal force and the force of friction.

Suppose the magnetic field makes the angle θ with the vertical. The diagram to the right shows the view from the end of the sliding rod. The forces are also shown: F_B is the force of the magnetic field if the current is out of the page, mg is the force of gravity, N is the normal force of the stationary rails on the rod, and f is the force of friction. Notice that the magnetic force makes the angle θ with the *horizontal*. When the rod is on the verge of sliding, the net force acting on it is zero and the magnitude of the frictional force is given by $f = \mu_s N$, where μ_s is the coefficient of static friction. The magnetic field is perpendicular to the wire so the magnitude of the magnetic force is given by $F_B = iLB$, where i is the current in the rod and L is the length of the rod.

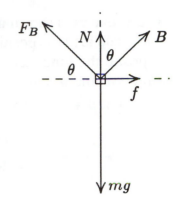

The vertical component of Newton's second law yields

$$N + iLB \sin\theta - mg = 0$$

and the horizontal component yields

$$iLB \cos\theta - \mu_s N = 0.$$

Solve the second equation for N and substitute the resulting expression into the first equation, then solve for B. You should get

$$B = \frac{\mu_s mg}{iL(\cos\theta + \mu_s \sin\theta)}.$$

The minimum value of B occurs when $\cos\theta + \mu_s \sin\theta$ is a maximum. Set the derivative of $\cos\theta + \mu_s \sin\theta$ equal to zero and solve for θ. You should get $\theta = \tan^{-1}\mu_s = \tan^{-1}(0.60) = 31°$. Now evaluate the expression for the minimum value of B:

$$B_{min} = \frac{0.60(1.0\,\text{kg})(9.8\,\text{m/s}^2)}{(50\,\text{A})(1.0\,\text{m})(\cos 31° + 0.60 \sin 31°)} = 0.10\,\text{T}.$$

54P

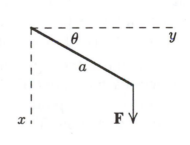

The situation is shown in the left diagram above. The z axis is along the hinge and the magnetic field is in the positive y direction. A torque around the hinge is associated with the wire opposite the hinge and not with the other wires. The force on this wire is in the positive x direction and has magnitude $F = NibB$, where N is the number of turns.

The right diagram shows the view from above. The magnitude of the torque is given by

$$\tau = Fa\cos\theta = NibBa\cos\theta$$
$$= 20(0.10\,\text{A})(0.10\,\text{m})(0.50 \times 10^{-3}\,\text{T})(0.050\,\text{m})\cos 30°$$
$$= 4.33 \times 10^{-3}\,\text{N}\cdot\text{m}.$$

Use the right-hand rule to show that the torque is directed downward, in the negative z direction.

56P

If N closed loops are formed from the wire of length L, the circumference of each loop is L/N, the radius of each loop is $R = L/2\pi N$, and the area of each loop is $A = \pi R^2 = \pi(L/2\pi N)^2 = L^2/4\pi N^2$. For maximum torque, orient the plane of the loops parallel to the magnetic field, so the dipole moment is perpendicular to the field. The magnitude of the torque is then

$$\tau = NiAB = (Ni)\left(\frac{L^2}{4\pi N^2}\right)B = \frac{iL^2B}{4\pi N}.$$

To maximize the torque take N to have the smallest possible value, 1. Then

$$\tau = \frac{iL^2B}{4\pi}.$$

59P

Consider an infinitesimal segment of the loop, of length ds. The magnetic field is perpendicular to the segment so the magnetic force on it is has magnitude $dF = iB\,ds$. The diagram shows the direction of the force for the segment on the far right of the loop. The horizontal component of the force has magnitude $dF_h = (iB\cos\theta)\,ds$ and points inward toward the center of the loop. The vertical component has magnitude $dF_v = (iB\sin\theta)\,ds$ and points upward.

Now sum the forces on all the segments of the loop. The horizontal component of the total force vanishes since each segment of wire can be paired with another, diametrically opposite, segment. The horizontal components of these forces are both toward the center of the loop and thus in opposite directions. The vertical component of the total force is

$$F_v = iB\sin\theta \int ds = (iB\sin\theta)2\pi a.$$

Notice the i, B, and θ have the same value for every segment and so can be factored from the integral.

60P

(a) The current in the galvanometer should be 1.62 mA when the potential difference across the resistor-galvanometer combination is 1.00 V. The potential difference across the galvanometer alone is $iR_g = (1.62 \times 10^{-3}\,\text{A})(75.3\,\Omega) = 0.122$ V so the resistor must be in series with the galvanometer and the potential difference across it must be $1.00\,\text{V} - 0.122\,\text{V} = 0.878$ V. The resistance should be $R = (0.878\,\text{V})/(1.62 \times 10^{-3}\,\text{A}) = 542\,\Omega$.

(b) The current in the galvanometer should be 1.62 mA when the total current in the resistor-galvanometer combination is 50.0 mA. The resistor should be in parallel with the galvanometer and the current through it should be $50\,\text{mA} - 1.62\,\text{mA} = 48.38$ mA. The potential difference across the resistor is the same as that across the galvanometer, 0.122 V, so the resistance should be $R = (0.122\,\text{V})/(48.38 \times 10^{-3}\,\text{A}) = 2.52\,\Omega$.

61P

In the diagram to the right μ is the magnetic dipole moment of the wire loop and **B** is the magnetic field. Since the plane of the loop is parallel to the incline the dipole moment is normal to the incline. The forces acting on the cylinder are the force of gravity mg, acting downward from the center of mass, the normal force of the incline N, acting perpendicularly to the incline through the center of mass, and the force of friction f, acting up the incline at the point of contact.

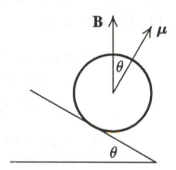

Take the x axis to be positive down the incline. Then the x component of Newton's second law for the center of mass yields

$$mg\sin\theta - f = ma.$$

For purposes of calculating the torque take the axis of the cylinder to be the axis of rotation. The magnetic field produces a torque with magnitude $\mu B \sin\theta$ and the force of friction produces a torque with magnitude fr, where r is the radius of the cylinder. The first tends to produce an angular acceleration in the counterclockwise direction and the second tends to produce an angular acceleration in the clockwise direction. Newton's second law for rotation about the center of the cylinder, $\tau = I\alpha$, gives

$$fr - \mu B \sin\theta = I\alpha.$$

Since you want the current that holds the cylinder in place, set $a = 0$ and $\alpha = 0$, then use one equation to eliminate f from the other. You should obtain $mgr = \mu B$. The loop is rectangular with two sides of length L and two of length $2r$, so its area is $A = 2rL$ and the dipole moment is $\mu = NiA = 2NirL$. Thus $mgr = 2NirLB$ and

$$i = \frac{mg}{2NLB} = \frac{(0.250\,\text{kg})(9.8\,\text{m/s}^2)}{2(10.0)(0.100\,\text{m})(0.500\,\text{T})} = 2.45\,\text{A}.$$

62E

(a) The magnitude of the magnetic dipole moment is given by $\mu = NiA$, where N is the number of turns, i is the current in each turn, and A is the area of a loop. In this case the loops are circular, so $A = \pi r^2$, where r is the radius of a turn. Thus

$$i = \frac{\mu}{N\pi r^2} = \frac{2.30 \, \text{A} \cdot \text{m}^2}{(160)(\pi)(0.0190 \, \text{m})^2} = 12.7 \, \text{A}.$$

(b) The maximum torque occurs when the dipole moment is perpendicular to the field (or the plane of the loop is parallel to the field). It is given by $\tau = \mu B = (2.30 \, \text{A·m}^2)(35.0 \times 10^{-3} \, \text{T}) = 8.05 \times 10^{-2} \, \text{N} \cdot \text{m}$.

67P

The magnetic dipole moment is $\boldsymbol{\mu} = \mu(0.60\,\mathbf{i} - 0.80\,\mathbf{j})$, where $\mu = NiA = Ni\pi r^2 = 1(0.20 \, \text{A})\pi(0.080 \, \text{m})^2 = 4.02 \times 10^{-4} \, \text{A} \cdot \text{m}^2$. Here i is the current in the loop, N is the number of turns, A is the area of the loop, and r is its radius.

(a) The torque is

$$\begin{aligned}
\boldsymbol{\tau} = \boldsymbol{\mu} \times \mathbf{B} &= \mu(0.60\,\mathbf{i} - 0.80\,\mathbf{j}) \times (0.25\,\mathbf{i} + 0.30\,\mathbf{k}) \\
&= \mu\,[(0.60)(0.30)(\mathbf{i} \times \mathbf{k}) - (0.80)(0.25)(\mathbf{j} \times \mathbf{i}) - (0.80)(0.30)(\mathbf{j} \times \mathbf{k})] \\
&= \mu[-0.18\,\mathbf{j} + 0.20\,\mathbf{k} - 0.24\,\mathbf{i}].
\end{aligned}$$

Here $\mathbf{i} \times \mathbf{k} = -\mathbf{j}$, $\mathbf{j} \times \mathbf{i} = -\mathbf{k}$, and $\mathbf{j} \times \mathbf{k} = \mathbf{i}$ were used. We also used $\mathbf{i} \times \mathbf{i} = 0$. Substitute the value for μ to obtain

$$\boldsymbol{\tau} = [-0.965 \times 10^{-4}\,\mathbf{i} - 7.23 \times 10^{-4}\,\mathbf{j} + 8.04 \times 10^{-4}\,\mathbf{k}]\,\text{N} \cdot \text{m}.$$

(b) The potential energy of the dipole is given by

$$\begin{aligned}
U = -\boldsymbol{\mu} \cdot \mathbf{B} &= -\mu(0.60\,\mathbf{i} - 0.80\,\mathbf{j}) \cdot (0.25\,\mathbf{i} + 0.30\,\mathbf{k}) \\
&= -\mu(0.60)(0.25) = -0.15\mu = -6.0 \times 10^{-4} \, \text{J}.
\end{aligned}$$

Here $\mathbf{i} \cdot \mathbf{i} = 1$, $\mathbf{i} \cdot \mathbf{k} = 0$, $\mathbf{j} \cdot \mathbf{i} = 0$, and $\mathbf{j} \cdot \mathbf{k} = 0$ were used.

Chapter 31

3E

(a) The magnitude of the magnetic field due to the current in the wire, at a point a distance r from the wire, is given by

$$B = \frac{\mu_0 i}{2\pi r}.$$

Put $r = 20\,\text{ft} = 6.10\,\text{m}$. Then

$$B = \frac{(4\pi \times 10^{-7}\,\text{T} \cdot \text{m/A})(100\,\text{A})}{2\pi(6.10\,\text{m})} = 3.3 \times 10^{-6}\,\text{T} = 3.3\,\mu\text{T}.$$

(b) This is about one-sixth the magnitude of the Earth's field. It will affect the compass reading.

7E

(a) The field due to the wire, at a point 8.0 cm from the wire, must be $39\,\mu\text{T}$ and must be directed toward due south. Since $B = \mu_0 i / 2\pi r$,

$$i = \frac{2\pi r B}{\mu_0} = \frac{2\pi(0.080\,\text{m})(39 \times 10^{-6}\,\text{T})}{4\pi \times 10^{-7}\,\text{T} \cdot \text{m/A}} = 16\,\text{A}.$$

(b) The current must be from west to east to produce a field to the south at points below it.

15P

Sum the fields of the two straight wires and the circular arc. Look at the derivation of the expression for the field of a long straight wire, leading to Eq. 31–10. Since the wires we are considering are infinite in only one direction the field of either of them is half the field of an infinite wire. That is, the magnitude is $\mu_0 i / 4\pi r$, where r is the distance from the end of the wire to the center of the arc. It is the radius of the arc. The fields of both wires are out of the page at the center of the arc.

Now find an expression for the field of the arc, at its center. Divide the arc into infinitesimal segments. Each segment produces a field in the same direction. If ds is the length of a segment the magnitude of the field it produces at the arc center is $(\mu_0 i / 4\pi r^2)\,ds$. If θ is the angle subtended by the arc in radians, then $r\theta$ is the length of the arc and the total field of the arc is $\mu_0 i \theta / 4\pi r$. For the arc of the diagram the field is into the page. The total field at the center, due to the wires and arc together, is

$$B = \frac{\mu_0 i}{4\pi r} + \frac{\mu_0 i}{4\pi r} - \frac{\mu_0 i \theta}{4\pi r} = \frac{\mu_0 i}{4\pi r}(2 - \theta).$$

For this to vanish θ must be 2 radians.

16P

First find the magnetic field of a circular arc at its center. Let ds be an infinitesimal segment of the arc and **r** be the vector from the segment to the arc center. ds and **r** are perpendicular to each other, so the contribution of the segment to the field at the center has magnitude

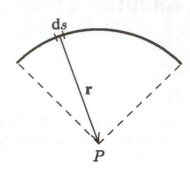

$$dB = \frac{\mu_0 i \, ds}{4\pi r^2}.$$

The field is into the page if the current is from left to right in the diagram and out of the page if the current is from right to left. All segments contribute magnetic fields in the same direction. Furthermore r is the same for all of them. Thus the magnitude of the total field at the center is given by

$$B = \frac{\mu_0 i s}{4\pi r^2} = \frac{\mu_0 i \theta}{4\pi r}.$$

Here s is the arc length and θ is the angle (in radians) subtended by the arc at its center. The second expression was obtained by replacing s with $r\theta$. θ must be in radians for this expression to be valid.

Now consider the circuit of Fig. 31-36. The magnetic field produced by the inner arc has magnitude $\mu_0 i \theta / 4\pi b$ and is out of the page. The field produced by the outer arc has magnitude $\mu_0 i \theta / 4\pi a$ and is into the page. The two straight segments of the circuit do not produce fields at the center of the arcs because the vector **r** from any point on them to the center is parallel or antiparallel to the current at that point. If the positive direction is out of the page then the total magnetic field at the center is

$$B = \frac{\mu_0 i \theta}{4\pi} \left[\frac{1}{b} - \frac{1}{a} \right].$$

Since $b < a$ the total field is out of the page.

17P

Put the x axis along the wire with the origin at the midpoint and the current in the positive x direction. All segments of the wire produce magnetic fields at P that are into the page so we simply divide the wire into infinitesimal segments and sum the fields due to all the segments. The diagram shows one infinitesimal segment, with width dx. According to the Biot-Savart law the magnitude of the field it produces at P is given by

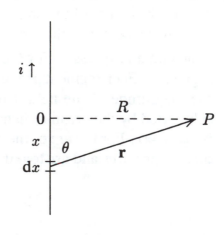

$$dB = \frac{\mu_0 i}{4\pi} \frac{\sin \theta}{r^2} \, dx.$$

θ and r are functions of x. Replace r with $\sqrt{x^2 + R^2}$ and $\sin\theta$ with $R/r = R/\sqrt{x^2 + R^2}$, then integrate from $x = -L/2$ to $x = L/2$. The total field is

$$B = \frac{\mu_0 i R}{4\pi} \int_{-L/2}^{L/2} \frac{dx}{(x^2 + R^2)^{3/2}} = \frac{\mu_0 i R}{4\pi} \frac{1}{R^2} \frac{x}{(x^2 + R^2)^{1/2}}\bigg|_{-L/2}^{L/2} = \frac{\mu_0 i}{2\pi R} \frac{L}{\sqrt{L^2 + 4R^2}}.$$

If $L \gg R$ then R^2 in the denominator can be ignored and

$$B = \frac{\mu_0 i}{2\pi R}$$

is obtained. This is the field of a long straight wire. For points close to a finite wire the field is quite similar to that of an infinitely long wire.

20P

The result of Problem 17 is used four times, once for each of the sides of the square loop. A point on the axis of the loop is also on a perpendicular bisector of each of the loop sides. The diagram shows the field due to one of the loop sides, the one on the left. In the expression found in Problem 17 replace L with a and R with $\sqrt{x^2 + a^2/4} = \frac{1}{2}\sqrt{4x^2 + a^2}$. The field due to the side is therefore

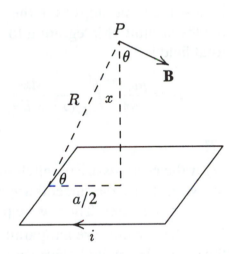

$$B = \frac{\mu_0 i a}{\pi\sqrt{4x^2 + a^2}\sqrt{4x^2 + 2a^2}}.$$

The field is in the plane of the dotted triangle shown and is perpendicular to the line from the midpoint of the loop side to the point P. It therefore makes the angle θ with the vertical.

When the fields of the four sides are summed vectorially the horizontal components add to zero. The vertical components are all the same, so the total field is given by

$$B_{\text{total}} = 4B\cos\theta = \frac{4Ba}{2R} = \frac{4Ba}{\sqrt{4x^2 + a^2}}.$$

Thus

$$B_{\text{total}} = \frac{4\mu_0 i a^2}{\pi(4x^2 + a^2)\sqrt{4x^2 + 2a^2}}.$$

For $x = 0$ the expression reduces to

$$B_{\text{total}} = \frac{4\mu_0 i a^2}{\pi a^2 \sqrt{2}a} = \frac{2\sqrt{2}\mu_0 i}{\pi a},$$

in agreement with the result of Problem 18.

21P

The square has sides of length $L/4$. The magnetic field at the center of the square is given by the result of Problem 20, with $a = L/4$ and $x = 0$. It is

$$B_{sq} = \frac{8\sqrt{2}\mu_0 i}{\pi L} = 11.31 \frac{\mu_0 i}{\pi L}.$$

The radius of the circle is $R = L/2\pi$. Use Eq. 31–24 of the text, with $R = L/2\pi$ and $z = 0$. The field is

$$B_{circ} = \frac{\pi^2 \mu_0 i}{\pi L} = 9.87 \frac{\mu_0 i}{\pi L}.$$

The square produces the larger magnetic field.

22P

Follow the same steps as in the solution of Problem 17 above but replace R with D, change the lower limit of integration to $-L$, and change the upper limit to 0. The magnitude of the total field is

$$B = \frac{\mu_0 i D}{4\pi} \int_{-L}^{0} \frac{dx}{(x^2 + D^2)^{3/2}} = \frac{\mu_0 i D}{4\pi} \frac{1}{D^2} \frac{x}{(x^2 + D^2)^{1/2}} \bigg|_{-L}^{0} = \frac{\mu_0 i}{4\pi D} \frac{L}{\sqrt{L^2 + D^2}}.$$

27E

(a) If the currents are parallel the two fields are in opposite directions in the region between the wires. Since the currents are the same the total field is zero along the line that runs halfway between the wires. There is no possible current for which the field does not vanish.

(b) If the currents are antiparallel the fields are in the same direction in the region between the wires. At a point halfway between they have the same magnitude, $\mu_0 i/2\pi r$. Thus the total field at the midpoint has magnitude $B = \mu_0 i/\pi r$ and

$$i = \frac{\pi r B}{\mu_0} = \frac{\pi (0.040\,\text{m})(300 \times 10^{-6}\,\text{T})}{4\pi \times 10^{-7}\,\text{T} \cdot \text{m/A}} = 30\,\text{A}.$$

34P

Each wire produces a field with magnitude given by $B = \mu_0 i/2\pi r$, where r is the distance from the corner of the square to the center. According to the Pythagorean theorem the diagonal of the square has length $\sqrt{2}a$, so $r = a/\sqrt{2}$ and $B = \mu_0 i/\sqrt{2}\pi a$. The fields due to the wires at the upper left and lower right corners both point toward the upper right corner of the square. The fields due to the wires at the upper right and lower left corners both point toward the upper left corner. The horizontal components cancel and the vertical components sum to

$$B_{total} = 4\frac{\mu_0 i}{\sqrt{2}\pi a} \cos 45° = \frac{2\mu_0 i}{\pi a}$$

$$= \frac{2(4\pi \times 10^{-7}\,\text{T} \cdot \text{m/A})(20\,\text{A})}{\pi (0.20\,\text{m})} = 8.0 \times 10^{-5}\,\text{T}.$$

In the calculation $\cos 45°$ was replaced with $1/\sqrt{2}$. The total field points upward.

40E

(a) Two of the currents are out of the page and one is into the page, so the net current enclosed by the path is 2.0 A, out of the page. Since the path is traversed in the clockwise sense a current into the page is positive and a current out of the page is negative, as indicated by the right-hand rule associated with Ampere's law. Thus

$$\oint \mathbf{B} \cdot \mathbf{ds} = -\mu_0 i = -(2.0\,\text{A})(4\pi \times 10^{-7}\,\text{T}\cdot\text{m/A}) = -2.5 \times 10^{-6}\,\text{T}\cdot\text{m}.$$

(b) The net current enclosed by the path is zero (two currents are out of the page and two are into the page), so $\oint \mathbf{B} \cdot \mathbf{ds} = \mu_0 i = 0$.

45P

Use Ampere's law: $\oint \mathbf{B} \cdot \mathbf{ds} = \mu_0 i$, where the integral is around a closed loop and i is the net current through the loop. For the dotted loop shown on the diagram $i = 0$. The integral $\int \mathbf{B} \cdot \mathbf{ds}$ is zero along the bottom, right, and top sides of the loop. Along the right side the field is zero, along the top and bottom sides the field is perpendicular to ds. If ℓ is the length of the left edge then direct integration yields $\oint \mathbf{B} \cdot \mathbf{ds} = B\ell$, where B is the magnitude of the field at the left side of the loop. Since neither B nor ℓ is zero, Ampere's law is contradicted. We conclude that the geometry shown for the magnetic field lines is in error. The lines actually bulge outward and their density decreases gradually, not precipitously as shown.

50P

(a) Take the magnetic field at a point within the hole to be the sum of the fields due to two current distributions. The first is the solid cylinder obtained by filling the hole and has a current density that is the same as that in the original cylinder with the hole. The second is the solid cylinder that fills the hole. It has a current density with the same magnitude as that of the original cylinder but it is in the opposite direction. Notice that if these two situations are superposed the total current in the region of the hole is zero.

Recall that a solid cylinder carrying current i, uniformly distributed over a cross section, produces a magnetic field with magnitude $B = \mu_0 i r/2\pi R^2$ a distance r from its axis, inside the cylinder. Here R is the radius of the cylinder.

For the cylinder of this problem the current density is

$$J = \frac{i}{A} = \frac{i}{\pi(a^2 - b^2)},$$

where $A\ (= \pi(a^2 - b^2))$ is the cross-sectional area of the cylinder with the hole. The current in the cylinder without the hole is

$$I_1 = JA = \pi J a^2 = \frac{ia^2}{a^2 - b^2}$$

and the magnetic field it produces at a point inside, a distance r_1 from its axis, has magnitude

$$B_1 = \frac{\mu_0 I_1 r_1}{2\pi a^2} = \frac{\mu_0 i r_1 a^2}{2\pi a^2 (a^2 - b^2)} = \frac{\mu_0 i r_1}{2\pi (a^2 - b^2)}.$$

The current in the cylinder that fills the hole is

$$I_2 = \pi J b^2 = \frac{i b^2}{a^2 - b^2}$$

and the field it produces at a point inside, a distance r_2 from the its axis, has magnitude

$$B_2 = \frac{\mu_0 I_2 r_2}{2\pi b^2} = \frac{\mu_0 i r_2 b^2}{2\pi b^2 (a^2 - b^2)} = \frac{\mu_0 i r_2}{2\pi (a^2 - b^2)}.$$

At the center of the hole this field is zero and the field there is exactly the same as it would be if the hole were filled. Place $r_1 = d$ in the expression for B_1 and obtain

$$B = \frac{\mu_0 i d}{2\pi (a^2 - b^2)}.$$

for the field at the center of the hole. The field points upward in the diagram if the current is out of the page.

(b) If $b = 0$ the formula for the field becomes

$$B = \frac{\mu_0 i d}{2\pi a^2}.$$

This correctly gives the field of a solid cylinder carrying a uniform current i, at a point inside the cylinder a distance d from the axis. If $d = 0$ the formula gives $B = 0$. This is correct for the field on the axis of a cylindrical shell carrying a uniform current.

(c)The diagram shows the situation in a cross-sectional plane of the cylinder. P is a point within the hole, A is on the axis of the cylinder, and C is on the axis of the hole. The magnetic field due to the cylinder without the hole, carrying a uniform current out of the page, is labeled $\mathbf{B_1}$ and the magnetic field of the cylinder that fills the hole, carrying a uniform current into the page, is labeled $\mathbf{B_2}$. The line from A to P makes the angle θ_1 with the line that joins the centers of the cylinders and the line from C to P makes the angle θ_2 with that line, as shown. $\mathbf{B_1}$ is perpendicular to the line from A to P and so makes the angle θ_1 with the vertical. Similarly, $\mathbf{B_2}$ is perpendicular to the line from C to P and so makes the angle θ_2 with the vertical.

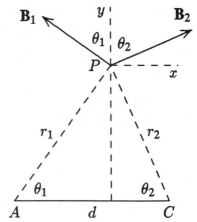

The x component of the total field is

$$B_x = B_2 \sin\theta_2 - B_1 \sin\theta_1 = \frac{\mu_0 i r_2}{2\pi(a^2 - b^2)} \sin\theta_2 - \frac{\mu_0 i r_1}{2\pi(a^2 - b^2)} \sin\theta_1$$

$$= \frac{\mu_0 i}{2\pi(a^2 - b^2)} [r_2 \sin\theta_2 - r_1 \sin\theta_1] \, .$$

As the diagram shows $r_2 \sin\theta_2 = r_1 \sin\theta_1$, so $B_x = 0$. The y component is given by

$$B_y = B_2 \cos\theta_2 + B_1 \cos\theta_1 = \frac{\mu_0 i r_2}{2\pi(a^2 - b^2)} \cos\theta_2 + \frac{\mu_0 i r_1}{2\pi(a^2 - b^2)} \cos\theta_1$$

$$= \frac{\mu_0 i}{2\pi(a^2 - b^2)} [r_2 \cos\theta_2 + r_1 \cos\theta_1] \, .$$

The diagram shows that $r_2 \cos\theta_2 + r_1 \cos\theta_1 = d$, so

$$B_y = \frac{\mu_0 i d}{2\pi(a^2 - b^2)} \, .$$

This is identical to the result found in part (a) for the field on the axis of the hole. It is independent of r_1, r_2, θ_1, and θ_2, showing that the field is uniform in the hole.

51P

(a) Suppose the field is not parallel to the sheet, as shown in the upper diagram. Reverse the direction of the current. According to the Biot-Savart law the field reverses, so it will be as in the second diagram. Now rotate the sheet by 180° about a line that perpendicular to the sheet. The field, of course, will rotate with it and end up in the direction shown in the bottom diagram. The current distribution is now exactly as it was originally, so the field must also be as it was originally. But it is not. Only if the field is parallel to the sheet will be final direction of the field be the same as the original direction. If the current is out of the page any infinitesimal portion of the sheet, in the form of a long straight wire, produces a field that is to the left above the sheet and to the right below the sheet. The field must be as drawn in Fig. 31–59.

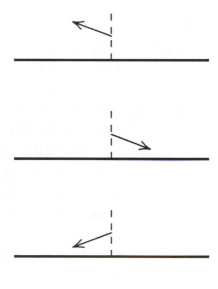

Integrate the tangential component of the magnetic field around the rectangular loop shown with dotted lines. The upper and lower edges are the same distance from the current sheet and each has length L. This means the field has the same magnitude along these edges. It points to the left along the upper edge and to the right along the lower.

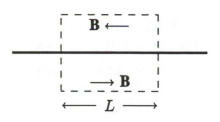

If the integration is carried out in the counterclockwise sense the contribution of the upper edge is BL, the contribution of the lower edge is also BL, and the contribution of each of the sides is zero because the field is perpendicular to the sides. Thus $\oint \mathbf{B} \cdot \mathbf{ds} = 2BL$. The total current through the loop is λL. Ampere's law yields $2BL = \mu_0 \lambda L$, so $B = \mu_0 \lambda / 2$.

54E

The magnetic field inside an ideal solenoid is given by Eq. 31–21. The number of turns per unit length is $n = (200\,\text{turns})/(0.25\,\text{m}) = 800\,\text{turns/m}$. Thus

$$B = \mu_0 n i_0 = (4\pi \times 10^{-7}\,\text{T} \cdot \text{m/A})(800\,\text{m}^{-1})(0.30\,\text{A}) = 3.0 \times 10^{-4}\,\text{T}.$$

56E

(a) Use Eq. 31–22. The inner radius is $r = 15.0\,\text{cm}$ so the field there is

$$B = \frac{\mu_0 i_0 N}{2\pi r} = \frac{(4\pi \times 10^{-7}\,\text{T} \cdot \text{m/A})(0.800\,\text{A})(500)}{2\pi(0.150\,\text{m})} = 5.33 \times 10^{-4}\,\text{T}.$$

(b) The outer radius is $r = 20.0\,\text{cm}$. The field there is

$$B = \frac{\mu_0 i_0 N}{2\pi r} = \frac{(4\pi \times 10^{-7}\,\text{T} \cdot \text{m/A})(0.800\,\text{A})(500)}{2\pi(0.200\,\text{m})} = 4.00 \times 10^{-4}\,\text{T}.$$

59P

Consider a circle of radius r, inside the toroid and concentric with it. The current that passes through the region between the circle and the outer rim of the toroid is $N i_0$, where N is the number of turns and i_0 is the current. The current per unit length of circle is $\lambda = N i_0 / 2\pi r$ and $\mu_0 \lambda$ is $\mu_0 N i_0 / 2\pi r$, the magnitude of the magnetic field at the circle. Since the field is zero outside a toroid, this is also the change in the magnitude of the field encountered as you move from the circle to the outside.

The equality is not really surprising in light of Ampere's law. You are moving perpendicularly to the magnetic field lines. Consider an extremely narrow loop, with the narrow sides along field lines and the two long sides perpendicular to the field lines. If B_1 is the field at one end and B_2 is the field at the other end then $\oint \mathbf{B} \cdot \mathbf{ds} = (B_2 - B_1)w$, where w is the width of the loop. The current through the loop is $w\lambda$, so Ampere's law yields $(B_2 - B_1)w = \mu_0 w \lambda$ and $B_2 - B_1 = \mu_0 \lambda$.

62P

(a) The magnetic force must be directed toward the center of the orbit. For the particle orbit shown $\mathbf{v} \times \mathbf{B}$ is away from the orbit center, so the particle must be negative.

(b) The radius of curvature is given by

$$R = \frac{mv}{qB},$$

where B is the local magnitude of the magnetic field, v is the particle speed, m is the particle mass, and q is the particle charge. The magnetic field does no work on the particle and so does not change its kinetic energy or speed. This means for two different points in the toroid $R_1 B_1 = R_2 B_2$. The magnetic field is given by

$$B = \frac{\mu_0 i_0 N}{2\pi r},$$

where r is the distance from the toroid center. This means $r_1 B_1 = r_2 B_2$. Divide the first equation by the second to obtain $R_1/r_1 = R_2/r_2$. Thus

$$R_2 = \frac{r_2}{r_1} R_1 = \frac{110\,\text{cm}}{125\,\text{cm}}(11.0\,\text{cm}) = 9.68\,\text{cm}.$$

63E

The magnitude of the dipole moment is given by $\mu = NiA$, where N is the number of turns, i is the current, and A is the area. Use $A = \pi R^2$, where R is the radius. Thus

$$\mu = (200)(0.30\,\text{A})\pi(0.050\,\text{m})^2 = 0.47\,\text{A} \cdot \text{m}^2.$$

66E

(a) As for Exercise 63, the magnitude of the dipole moment is given by $\mu = NiA$, where N is the number of turns, i is the current, and A is the area. Use $A = \pi R^2$, where R is the radius. Thus

$$\mu = Ni\pi R^2 = (300)(4.0\,\text{A})\pi(0.025\,\text{m})^2 = 2.36\,\text{A} \cdot \text{m}^2.$$

(b) The magnetic field on the axis of a magnetic dipole, a distance z away, is given by Eq. 31–25:

$$B = \frac{\mu_0}{2\pi} \frac{\mu}{z^3}.$$

Solve for z:

$$z = \left[\frac{\mu_0}{2\pi} \frac{\mu}{B} \right]^{1/3} = \left[\frac{4\pi \times 10^{-7}\,\text{T} \cdot \text{m/A}}{2\pi} \frac{2.36\,\text{A} \cdot \text{m}^2}{5.0 \times 10^{-6}\,\text{T}} \right]^{1/3} = 46\,\text{cm}.$$

70P

(a) The magnitude of the magnetic field on the axis of a circular loop, a distance z from the loop center, is given by Eq. 31–24:

$$B = \frac{N\mu_0 i R^2}{2(R^2 + z^2)^{3/2}},$$

where R is the radius of the loop, N is the number of turns, and i is the current. Both of the loops in the problem have the same radius, the same number of turns, and carry the same current. The currents are in the same sense and the fields they produce are in the same direction in the region between them. Place the origin at the center of the left-hand loop and let x be the coordinate of a point on the axis between the loops. To calculate the field of the left-hand loop set $z = x$ in the equation above. The chosen point on the axis is a distance $s - x$ from the center of the right-hand loop. To calculate the field it produces put $z = s - x$ in the equation above. The total field at the point is therefore

$$B = \frac{N\mu_0 i R^2}{2}\left[\frac{1}{(R^2 + x^2)^{3/2}} + \frac{1}{(R^2 + x^2 - 2sx + s^2)^{3/2}}\right].$$

Its derivative with respect to x is

$$\frac{dB}{dx} = -\frac{N\mu_0 i R^2}{2}\left[\frac{3x}{(R^2 + x^2)^{5/2}} + \frac{3(x - s)}{(R^2 + x^2 - 2sx + s^2)^{5/2}}\right].$$

When this is evaluated for $x = s/2$ (the midpoint between the loops) the result is

$$\frac{dB}{dx}\bigg|_{s/2} = -\frac{N\mu_0 i R^2}{2}\left[\frac{3s/2}{(R^2 + s^2/4)^{5/2}} - \frac{3s/2}{(R^2 + s^2/4 - s^2 + s^2)^{5/2}}\right] = 0,$$

independently of the value of s.

(b) The second derivative is

$$\frac{d^2 B}{dx^2} = \frac{N\mu_0 i R^2}{2}\left[-\frac{3}{(R^2 + x^2)^{5/2}} + \frac{15x^2}{(R^2 + x^2)^{7/2}}\right.$$
$$\left. - \frac{3}{(R^2 + x^2 - 2sx + s^2)^{5/2}} + \frac{15(x - s)^2}{(R^2 + x^2 - 2sx + s^2)^{7/2}}\right].$$

At $x = s/2$,

$$\frac{d^2 B}{dx^2}\bigg|_{s/2} = \frac{N\mu_0 i R^2}{2}\left[-\frac{6}{(R^2 + s^2/4)^{5/2}} + \frac{30s^2/4}{(R^2 + s^2/4)^{7/2}}\right]$$
$$= \frac{N\mu_0 R^2}{2}\frac{-6(R^2 + s^2/4) + 30s^2/4}{(R^2 + s^2/4)^{7/2}} = 3N\mu_0 i R^2 \frac{s^2 - R^2}{(R^2 + s^2/4)^{7/2}}.$$

Clearly, this is zero if $s = R$.

Chapter 32

3E

The magnetic field is normal to the plane of the loop and is uniform over the loop. Thus at any instant the magnetic flux through the loop is given by $\Phi_B = AB = \pi r^2 B$, where A ($= \pi r^2$) is the area of the loop. According to Faraday's law the magnitude of the emf in the loop is

$$\mathcal{E} = \frac{d\Phi_B}{dt} = \pi r^2 \frac{dB}{dt} = \pi(0.055\,\text{m})^2(0.16\,\text{T/s}) = 1.52 \times 10^{-3}\,\text{V}\,.$$

10P

(a) The magnitude of the magnetic field inside the solenoid is $B = \mu_0 n i_s$, where n is the number of turns per unit length and i_s is the current. The field is parallel to the solenoid axis, so the flux through a cross section of the solenoid is $\Phi_B = A_s B = \mu_0 \pi r_s^2 n i_s$, where A_s ($= \pi r_s^2$) is the cross-sectional area of the solenoid. Since the magnetic field is zero outside the solenoid this is also the flux through the coil. The emf in the coil has magnitude

$$\mathcal{E} = \frac{Nd\Phi}{dt} = \mu_0 \pi r_s^2 N n \frac{di_s}{dt}$$

and the current in the coil is

$$i_c = \frac{\mathcal{E}}{R} = \frac{\mu_0 \pi r_s^2 N n}{R} \frac{di_s}{dt}\,,$$

where N is the number of turns in the coil and R is the resistance of the coil. According to Sample Problem 32–1 the current changes linearly by 3.0 A in 50 ms, so $di_s/dt = (3.0\,\text{A})/(50 \times 10^{-3}\,\text{s}) = 60\,\text{A/s}$. Thus

$$i_c = \frac{(4\pi \times 10^{-7}\,\text{T} \cdot \text{m/A})\pi(0.016\,\text{m})^2(120)(220 \times 10^2\,\text{m}^{-1})}{5.3\,\Omega} = 30 \times 10^{-2}\,\text{A}\,.$$

(b) As the magnetic field changes, an electric field appears and affects the motions of the conduction electrons in the coil. See Chapter 32.

14P

Thermal energy is generated at the rate $\mathcal{E}^2/R$, where $\mathcal{E}$ is the emf in the wire and R is the resistance of the wire. The resistance is given by $R = \rho L/A$, where ρ is the resistivity of copper, L is the length of the wire, and A is the cross-sectional area of the wire. The resistivity can be found in Table 28–1. Thus

$$R = \frac{(1.69 \times 10^{-8}\,\Omega \cdot \text{m})(0.500\,\text{m})}{\pi(0.500 \times 10^{-3}\,\text{m})^2} = 1.076 \times 10^{-2}\,\Omega\,.$$

Faraday's law is used to find the emf. If B is the magnitude of the magnetic field through the loop, then $\mathcal{E} = A \, dB/dt$, where A is the area of the loop. The radius r of the loop is $r = L/2\pi$ and its area is $\pi r^2 = \pi L^2/4\pi^2 = L^2/4\pi$. Thus

$$\mathcal{E} = \frac{L^2}{4\pi} \frac{dB}{dt} = \frac{(0.500 \, \text{m})^2}{4\pi}(10.0 \times 10^{-3} \, \text{T/s}) = 1.989 \times 10^{-4} \, \text{V} .$$

The rate of thermal energy generation is

$$P = \frac{\mathcal{E}^2}{R} = \frac{(1.989 \times 10^{-4} \, \text{V})^2}{1.076 \times 10^{-2} \, \Omega} = 3.68 \times 10^{-6} \, \text{W} .$$

16P

(a) In the region of the smaller loop the magnetic field produced by the larger loop may be taken to be uniform and equal to its value at the center of the smaller loop, on the axis. Eq. 31–24, with $z = x$ and much greater than R, gives

$$B = \frac{\mu_0 i R^2}{2x^3}$$

for the magnitude. The field is upward in the diagram. The magnetic flux through the smaller loop is the product of this field and the area (πr^2) of the smaller loop:

$$\Phi_B = \frac{\pi \mu_0 i r^2 R^2}{2x^3} .$$

(b) The emf is given by Faraday's law:

$$\mathcal{E} = -\frac{d\Phi_B}{dt} = -\left(\frac{\pi \mu_0 i r^2 R^2}{2}\right) \frac{d}{dt}\left(\frac{1}{x^3}\right) = -\left(\frac{\pi \mu_0 i r^2 R^2}{2}\right)\left(-\frac{3}{x^4}\frac{dx}{dt}\right) = \frac{3\pi \mu_0 i r^2 R^2 v}{2x^4} .$$

(c) The field of the larger loop is upward and decreases with distance away from the loop. As the smaller loop moves away the flux through it decreases. The induced current will be directed so as to produce a magnetic field that is upward through the smaller loop, in the same direction as the field of the larger loop. It will be counterclockwise as viewed from above, in the same direction as the current in the larger loop.

17P

(a) The emf induced around the loop is given by Faraday's law: $\mathcal{E} = -d\Phi_B/dt$ and the current in the loop is given by $i = \mathcal{E}/R = -(1/R)(d\Phi_B/dt)$. The charge that passes through the resistor from time zero to time t is given by the integral

$$q = \int_0^t i \, dt = -\frac{1}{R}\int_0^t \frac{d\Phi_B}{dt}\, dt = -\frac{1}{R}\int_{\Phi_B(0)}^{\Phi_B(t)} d\Phi_B = \frac{1}{R}[\Phi_B(0) - \Phi_B(t)] .$$

All that matters is the change in the flux, not how it was changed.

(b) If $\Phi_B(t) = \Phi_B(0)$ then $q = 0$. This does not mean that the current was zero for any extended time during the interval. If Φ_B increases and then decreases back to its original value there is current in the resistor while Φ_B is changing. It is in one direction at first, then in the opposite direction. When equal charge has passed through the resistor in opposite directions the net charge is zero.

19P

(a) Let L be the length of a side of the square circuit. Then the magnetic flux through the circuit is $\Phi_B = L^2 B/2$ and the induced emf has magnitude

$$\mathcal{E}_i = -\frac{d\Phi_B}{dt} = -\frac{L^2}{2}\frac{dB}{dt}.$$

Now $B = 0.042 - 0.870t$ and $dB/dt = -0.870\,\text{T/s}$. Thus

$$\mathcal{E}_i = \frac{(2.00\,\text{m})^2}{2}(0.870\,\text{T/s}) = 1.74\,\text{V}.$$

The magnetic field is out of the page and decreasing so the induced emf is counterclockwise around the circuit, in the same direction as the emf of the battery. The total emf is $\mathcal{E} + \mathcal{E}_i = 20.0\,\text{V} + 1.74\,\text{V} = 21.7\,\text{V}$.

(b) The current is in the sense of the total emf, counterclockwise.

21P*

(a) Suppose each wire has radius R and the distance between their axes is a. Consider a single wire and calculate the flux through a rectangular area with the axis of the wire along one side. Take this side to have length L and the other dimension of the rectangle to be a. The magnetic field is everywhere perpendicular to the rectangle. First consider the part of the rectangle that is inside the wire. The field a distance r from the axis is given by $B = \mu_0 i r/2\pi R^2$ and the flux through the strip of length L and width dr at that distance is $(\mu_0 i r/2\pi R^2)L\,dr$. Thus the flux through the area inside the wire is

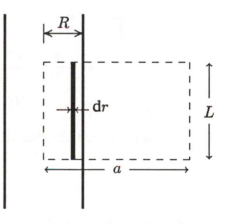

$$\Phi_{\text{in}} = \int_0^R \frac{\mu_0 i L}{2\pi R^2}r\,dr = \frac{\mu_0 i L}{4\pi}.$$

Now consider the region outside the wire. There the field is given by $B = \mu_0 i/2\pi r$ and the flux through an infinitesimally thin strip is $(\mu_0 i/2\pi r)L\,dr$. The flux through the whole region is

$$\Phi_{\text{out}} = \int_R^a \frac{\mu_0 i L}{2\pi}\frac{dr}{r} = \frac{\mu_0 i L}{2\pi}\ln\left(\frac{a}{R}\right).$$

The total flux through the area bounded by the dotted lines is the sum of the two contributions:

$$\Phi = \frac{\mu_0 i L}{4\pi}\left[1 + 2\ln\left(\frac{a}{R}\right)\right].$$

Now include the contribution of the other wire. Since the currents are in the same direction the two contributions have the same sign. They also have the same magnitude, so

$$\Phi_{\text{total}} = \frac{\mu_0 i L}{2\pi}\left[1 + 2\ln\left(\frac{a}{R}\right)\right].$$

The total flux per unit length is

$$\frac{\Phi_{\text{total}}}{L} = \frac{\mu_0 i}{2\pi}\left[1 + 2\ln\left(\frac{a}{R}\right)\right] = \frac{(4\pi \times 10^{-7}\,\text{T}\cdot\text{m/A})(10\,\text{A})}{2\pi}\left[1 + 2\ln\left(\frac{20\,\text{mm}}{1.25\,\text{mm}}\right)\right]$$
$$= 1.31 \times 10^{-5}\,\text{Wb/m}.$$

(b) Again consider the flux of a single wire. The flux inside the wire itself is again $\Phi_{\text{in}} = \mu_0 i L/4\pi$. The flux inside the region of the other wire is

$$\Phi_{\text{out}} = \int_{a-R}^{a}\frac{\mu_0 i L}{2\pi}\frac{dr}{r} = \frac{\mu_0 i L}{2\pi}\ln\left(\frac{a}{a-R}\right).$$

Double this to include the flux of the other wire (inside the first wire) and divide by L to obtain the flux per unit length. The total flux per unit length that is inside the wires is

$$\frac{\Phi_{\text{wires}}}{L} = \frac{\mu_0 i}{2\pi}\left[1 + 2\ln\left(\frac{a}{a-R}\right)\right]$$
$$= \frac{(4\pi \times 10^{-7}\,\text{T}\cdot\text{m/A})(10\,\text{A})}{2\pi}\left[1 + 2\ln\left(\frac{20\,\text{mm}}{20\,\text{mm} - 1.25\,\text{mm}}\right)\right]$$
$$= 2.26 \times 10^{-6}\,\text{Wb/m}.$$

The fraction of the total flux that is inside the wires is

$$\frac{2.26 \times 10^{-6}\,\text{Wb/m}}{1.31 \times 10^{-5}\,\text{Wb/m}} = 0.17.$$

(c) The contributions of the two wires to the total flux have the same magnitudes but now the currents are in opposite directions, so the contributions have opposite signs. This means $\Phi_{\text{total}} = 0$.

23E

(a) The flux changes because the area bounded by the rod and rails increases as the rod moves. Suppose that at some instant the rod is a distance x from the right-hand end of the rails and has speed v. Then the flux through the area is $\Phi_B = BA = BLx$, where L is the distance between the rails. According to Faraday's law the magnitude of the emf induced is $\mathcal{E} = d\Phi_B/dt = BL\,(dx/dt) = BLv = (0.350\,\text{T})(0.250\,\text{m})(0.550\,\text{m/s}) = 4.81 \times 10^{-2}\,\text{V}$.

(b) Use Ohm's law. If R is the resistance of the rod then the current in the rod is $i = \mathcal{E}/R = (4.81 \times 10^{-2}\,\text{V})/(18.0\,\Omega) = 2.67 \times 10^{-3}\,\text{A}$.

24E

(a) Let x be the distance from the right end of the rails to the rod. The area enclosed by the rod and rails is Lx and the magnetic flux through the area is $\Phi_B = BLx$. The emf induced is $\mathcal{E} = d\Phi_B/dt = BL\,dx/dt = BLv$, where v is the speed of the rod. Thus $\mathcal{E} = (1.2\,\text{T})(0.10\,\text{m})(5.0\,\text{m/s}) = 0.60\,\text{V}$.

(b) If R is the resistance of the rod, the current in the loop is $i = \mathcal{E}/R = (0.60\,\text{V})/(0.40\,\Omega) = 1.5\,\text{A}$. Since the rod moves to the left in the diagram, the flux increases. The induced current must produce a magnetic field that is into the page in the region bounded by the rod and rails. To do this the current must be clockwise.

(c) The rate of generation of thermal energy by the resistance of the rod is $P = \mathcal{E}^2/R = (0.60\,\text{V})^2/(0.40\,\Omega) = 0.90\,\text{W}$.

(d) Since the rod moves with constant velocity the net force on it must be zero. This means the force of the external agent has the same magnitude as the magnetic force but is in the opposite direction. The magnitude of the magnetic force is $F_B = iLB = (1.5\,\text{A})(0.10\,\text{m})(1.2\,\text{T}) = 0.18\,\text{N}$. Since the field is out of the page and the current is upward through the rod the magnetic force is to the right. The force of the external agent must be 0.18 N, to the left.

(e) As the rod moves an infinitesimal distance dx the external agent does work $dW = F\,dx$, where F is the force of the agent. The force is in the direction of motion, so the work done by the agent is positive. The rate at which the agent does work is $dW/dt = F\,dx/dt = Fv = (0.18\,\text{N})(5.0\,\text{m/s}) = 0.90\,\text{W}$, the same as the rate at which thermal energy is generated. The energy supplied by the external agent is converted completely to thermal energy.

29P

(a) The area of the loop is $A = ab$. Suppose that at some instant of time the normal to the loop makes the angle θ with the magnetic field. The magnetic flux through the loop is then $\Phi_B = NabB\cos\theta$ and the emf induced around the loop is

$$\mathcal{E} = -\frac{d\Phi_B}{dt} = -\frac{d}{dt}[NabB\cos\theta] = [NabB\sin\theta]\frac{d\theta}{dt}.$$

In terms of the frequency of rotation f and the time t, θ is given by $\theta = 2\pi ft$ and $d\theta/dt = 2\pi f$. The emf is therefore

$$\mathcal{E} = 2\pi fNabB\sin(2\pi ft).$$

This can be written $\mathcal{E} = \mathcal{E}_0 \sin(2\pi ft)$, where $\mathcal{E}_0 = 2\pi fNabB$.

(b) You want $2\pi fNabB = 150\,\text{V}$. This means

$$Nab = \frac{\mathcal{E}_0}{2\pi fB} = \frac{150\,\text{V}}{2\pi(60.0\,\text{rev/s})(0.500\,\text{T})} = 0.796\,\text{m}^2.$$

Any loop for which $Nab = 0.796\,\text{m}^2$ will do the job. An example is $N = 100$ turns, $a = b = 8.92$ cm.

34P

(a) Let x be the distance from the right end of the rails to the rod and find an expression for the magnetic flux through the area enclosed by the rod and rails. The magnetic field is not uniform but varies with distance from the long straight wire. The field is normal to the area and has magnitude $B = \mu_0 i/2\pi r$, where r is the distance from the wire and i is the current in the wire. Consider an infinitesimal strip of length x and width dr, parallel to the wire and a distance r from it. The area of this strip is $A = x\,dr$ and the flux through it is $d\Phi_B = (\mu_0 i x/2\pi r)\,dr$. The total flux through the area enclosed by the rod and rails is

$$\Phi_B = \frac{\mu_0 i x}{2\pi} \int_a^{a+L} \frac{dr}{r} = \frac{\mu_0 i x}{2\pi} \ln\left(\frac{a+L}{a}\right).$$

According to Faraday's law the emf induced in the loop is

$$\mathcal{E} = \frac{d\Phi}{dt} = \frac{\mu_0 i}{2\pi}\frac{dx}{dt}\ln\left(\frac{a+L}{a}\right) = \frac{\mu_0 i v}{2\pi}\ln\left(\frac{a+L}{a}\right)$$
$$= \frac{(4\pi \times 10^{-7}\,\text{T}\cdot\text{m/A})(100\,\text{A})(5.00\,\text{m/s})}{2\pi}\ln\left(\frac{1.00\,\text{cm} + 10.0\,\text{cm}}{1.00\,\text{cm}}\right)$$
$$= 2.40 \times 10^{-4}\,\text{V}.$$

(b) If R is the resistance of the rod then the current in the conducting loop is $i_\ell = \mathcal{E}/R = (2.40 \times 10^{-4}\,\text{V})/(0.400\,\Omega) = 6.00 \times 10^{-4}\,\text{A}$. Since the flux is increasing the magnetic field produced by the induced current must be into the page in the region enclosed by the rod and rails. This means the current is clockwise.

(c) Thermal energy is be generated at the rate $P = i_\ell^2 R = (6.00 \times 10^{-4}\,\text{A})^2(0.400\,\Omega) = 1.44 \times 10^{-7}\,\text{W}$.

(d) Since the rod moves with constant velocity the net force on it is zero. The force of the external agent must have the same magnitude as the magnetic force and must be in the opposite direction. The magnitude of the magnetic force on an infinitesimal segment of the rod, with length dr and a distance r from the long straight wire, is $dF_B = i_\ell B\,dr = (\mu_0 i_\ell i/2\pi r)\,dr$. The total magnetic force on the rod has magnitude

$$F_B = \frac{\mu_0 i_\ell i}{2\pi} \int_a^{a+L} \frac{dr}{r} = \frac{\mu_0 i_\ell i}{2\pi}\ln\left(\frac{a+L}{a}\right)$$
$$= \frac{(4\pi \times 10^{-7}\,\text{T}\cdot\text{m/A})(6.00 \times 10^{-4}\,\text{A})(100\,\text{A})}{2\pi}\ln\left(\frac{1.00\,\text{cm} + 10.0\,\text{cm}}{1.00\,\text{cm}}\right)$$
$$= 2.87 \times 10^{-8}\,\text{N}.$$

Since the field is out of the page and the current in the rod is upward in the diagram, this force is toward the right. The external agent must apply a force of $2.87 \times 10^{-8}\,\text{N}$, to the left.

(e) The external agent does work at the rate $P = Fv = (2.87 \times 10^{-8}\,\text{N})(5.00\,\text{m/s}) = 1.44 \times 10^{-7}\,\text{W}$. This is the same as the rate at which thermal energy is generated in the rod. All the energy supplied by the agent is converted to thermal energy.

36P

Use Faraday's law to find an expression for the emf induced by the changing magnetic field. First find an expression for the magnetic flux through the loop. Since the field depends on y but not on x, divide the area into strips of length L and width dy, parallel to the x axis. Here L is the length of one side of the square. At time t the flux through a strip with coordinate y is $d\Phi_B = BL\,dy = 4.0Lt^2y\,dy$ and the total flux through the square is

$$\Phi_B = \int_0^L 4.0Lt^2y\,dy = 2.0L^3t^2\,.$$

According to Faraday's law the magnitude of the emf around the square is

$$\mathcal{E} = \frac{d\Phi_B}{dt} = \frac{d}{dt}\left(2.0L^3t^2\right) = 4.0L^3t\,.$$

At $t = 2.5\,\text{s}$ this is $4.0(0.020\,\text{m})^3(2.5\,\text{s}) = 8.0 \times 10^{-5}\,\text{V}$.

The externally-produced magnetic field is out of the page and is increasing with time. The induced current produces a field that is into the page, so it must be clockwise. The induced emf is also clockwise.

38P*

(a) Since the changing magnetic flux through the area enclosed by the rod and rails induces a current, the magnetic field exerts a force on the rod. The magnetic force is horizontal and to the left. It tends to slow the rod, while the gravitational force on the rod tends to speed it up. Since the magnetic force is zero when the rod is at rest and increases with the speed of the rod, a terminal speed is reached when the net force acting on the rod is zero.

First assume the rod has speed v and calculate the magnetic force on it. Let x represent the distance from the rod to the fixed cross rail at the bottom of the incline. The area of the loop formed by the rod and rails is $A = \ell x$ and, since the normal to the area makes the angle θ with the magnetic field, the magnetic flux through the loop is $\Phi = B\ell x \cos\theta$. According to Faraday's law the emf induced around the loop is $\mathcal{E} = B\ell v \cos\theta$. If R is the resistance in the rod then the induced current is $i = \mathcal{E}/R = (B\ell v/R)\cos\theta$ and the magnitude of the magnetic force is $F_B = i\ell B = (B^2\ell^2v/R)\cos\theta$. This force is perpendicular to both the field and the current. It is horizontal and to the left.

The diagram shows the forces on the rod: mg is the force of gravity, F_B is the magnetic force, and N is the normal force of the incline. Take the x axis to be down the incline. The magnetic force makes the angle θ with the incline and the force of gravity makes the angle θ with the normal to the incline. The x component of Newton's second law yields $mg\sin\theta - F_B\cos\theta = ma$, where a is the acceleration of the rod. At terminal velocity $a = 0$ and $F_B\cos\theta = mg\sin\theta$.

Substitute the expression for F_B found above and solve for v. You should get

$$v = \frac{mgR\sin\theta}{B^2\ell^2\cos^2\theta}.$$

(b) Thermal energy is generated in the rod at the rate $P_T = i^2 R$ and since $i = (B\ell v/R)\cos\theta$, this is $P_T = (B^2\ell^2 v^2/R)\cos^2\theta$. At terminal velocity $v = (mgR\sin\theta)/(B^2\ell^2\cos^2\theta)$, so

$$P_T = \frac{m^2 g^2 R\sin^2\theta}{B^2\ell^2\cos^2\theta}.$$

Suppose the rod is a vertical distance h above the base of the incline. Its potential energy is then $U = mgh = mgx\sin\theta$ where $h\sin\theta$ was substituted for x. The rate at which gravitational potential energy is lost is given by $P_g = dU/dt = mg(dx/dt)\sin\theta = mgv\sin\theta$. Substitute the expression for the terminal velocity for v to obtain

$$P_g = \frac{m^2 g^2 R\sin^2\theta}{B^2\ell^2\cos^2\theta},$$

the same as the expression for the rate of thermal energy generation. Note that the expression for the terminal velocity must be used. Until terminal velocity is reached some gravitational potential energy is converted to kinetic energy as the rod goes faster.

(c) If the magnetic field is downward the direction of the current will be reversed but the magnetic force will be in the same direction and the motion of the rod will be unaffected.

40E

(a) The field point is inside the solenoid, so Eq. 32–24 applies. The magnitude of the induced electric field is

$$E = \frac{1}{2}\frac{dB}{dt}\,r = \frac{1}{2}(6.5 \times 10^{-3}\,\text{T/s})(0.0220\,\text{m}) = 7.15 \times 10^{-5}\,\text{V/m}.$$

(b) Now the field point is outside the solenoid and Eq. 32–25 applies. The magnitude of the induced field is

$$E = \frac{1}{2}\frac{dB}{dt}\frac{R^2}{r} = \frac{1}{2}(6.5 \times 10^{-3}\,\text{T/s})\frac{(0.0600\,\text{m})^2}{(0.0820\,\text{m})} = 1.43 \times 10^{-4}\,\text{V/m}.$$

44P

Use Faraday's law in the form $\oint \mathbf{E} \cdot d\mathbf{s} = -(d\Phi_B/dt)$. Integrate around the dotted path shown in Fig. 32–53. At all points on the upper and lower sides the electric field is either perpendicular to the side or else it vanishes. Assume it vanishes at all points on the right side (outside the capacitor). On the left side it is parallel to the side and has constant magnitude. Thus direct integration yields $\oint \mathbf{E} \cdot d\mathbf{s} = EL$, where L is the length of the left side of the rectangle. The magnetic field is zero and remains zero, so $d\Phi_B/dt = 0$. Faraday's law leads to a contradiction: $EL = 0$, but neither E nor L is zero. There must be an electric field along the right side of the rectangle.

Chapter 33

1E

Since $N\Phi = Li$, where N is the number of turns, L is the inductance, and i is the current,

$$\Phi = \frac{Li}{N} = \frac{(8.0 \times 10^{-3}\,\text{H})(5.0 \times 10^{-3}\,\text{A})}{400} = 1.0 \times 10^{-7}\,\text{Wb}.$$

8P

The area of integration for the calculation of the magnetic flux is bounded by the two dotted lines and the boundaries of the wires. If the origin is taken to be on the axis of the right-hand wire and r measures distance from that axis, it extends from $r = a$ to $r = d - a$. Consider the right-hand wire first. In the region of integration the field it produces is into the page and has magnitude $B = \mu_0 i / 2\pi r$. Divide the region into strips of length ℓ and width dr, as shown. The flux through the strip a distance r from the axis of the wire is $d\Phi = B\ell\,dr$ and the flux through the entire region is

$$\Phi = \frac{\mu_0 i \ell}{2\pi} \int_a^{d-a} \frac{dr}{r} = \frac{\mu_0 i \ell}{2\pi} \ln\left(\frac{d-a}{a}\right).$$

The other wire produces the same result, so the total flux through the dotted rectangle is

$$\Phi_{\text{total}} = \frac{\mu_0 i \ell}{\pi} \ln\left(\frac{d-a}{a}\right).$$

The inductance is Φ_{total} divided by i:

$$L = \frac{\Phi_{\text{total}}}{i} = \frac{\mu_0 \ell}{\pi} \ln\left(\frac{d-a}{a}\right).$$

10E

Since $\mathcal{E} = -L(di/dt)$, cause the current to change at the rate

$$\frac{di}{dt} = \frac{\mathcal{E}}{L} = \frac{60\,\text{V}}{12\,\text{H}} = 5.0\,\text{A/s}.$$

You might, for example, uniformly reduce the current to zero in 33 ms.

15E

Starting with zero current at time $t = 0$, when the switch is closed, the current in an RL series circuit at a later time t is given by

$$i = \frac{\mathcal{E}}{R}\left(1 - e^{-t/\tau_L}\right),$$

where τ_L is the inductive time constant, $\mathcal{E}$ is the emf, and R is the resistance. You want to calculate the time t for which $i = 0.9990\mathcal{E}/R$. This means

$$0.9990\frac{\mathcal{E}}{R} = \frac{\mathcal{E}}{R}\left(1 - e^{-t/\tau_L}\right),$$

so

$$0.9990 = 1 - e^{-t/\tau_L}$$

or

$$e^{-t/\tau_L} = 0.0010.$$

Take the natural logarithm of both sides to obtain $-(t/\tau) = \ln(0.0010) = -6.91$. Thus $t = 6.91\tau_L$. That is, 6.91 inductive time constants must elapse.

16E

The current in the circuit is given by

$$i = i_0 e^{-t/\tau_L},$$

where i_0 is the initial current (at time $t = 0$) and τ_L $(= L/R)$ is the inductive time constant. Solve for τ_L. Divide by i_0 and take the natural logarithm of both sides of the resulting equation to obtain

$$\ln\left(\frac{i}{i_0}\right) = -\frac{t}{\tau_L}.$$

This yields

$$\tau_L = -\frac{t}{\ln[i/i_0]} = -\frac{1.0\,\text{s}}{\ln\left[(10 \times 10^{-3}\,\text{A})/(1.0\,\text{A})\right]} = 0.217\,\text{s}.$$

Thus $R = L/\tau_L = (10\,\text{H})/(0.217\,\text{s}) = 46\,\Omega$.

19E

(a) If the battery is switched into the circuit at time $t = 0$, then the current at a later time t is given by

$$i = \frac{\mathcal{E}}{R}\left(1 - e^{-t/\tau_L}\right),$$

where $\tau_L = L/R$. You want to find the time for which $i = 0.800\mathcal{E}/R$. This means

$$0.800 = 1 - e^{-t/\tau_L}$$

or
$$e^{-t/\tau_L} = 0.200.$$

Take the natural logarithm of both sides to obtain $-(t/\tau_L) = \ln(0.200) = -1.609$. Thus

$$t = 1.609\tau_L = \frac{1.609L}{R} = \frac{1.609(6.30 \times 10^{-6}\,\text{H})}{1.20 \times 10^3\,\Omega} = 8.45 \times 10^{-9}\,\text{s}.$$

(b) At $t = 1.0\tau_L$ the current in the circuit is

$$i = \frac{\mathcal{E}}{R}(1 - e^{-1.0}) = \left(\frac{14.0\,\text{V}}{1.20 \times 10^3\,\Omega}\right)(1 - e^{-1.0}) = 7.37 \times 10^{-3}\,\text{A}.$$

25P

(a) The inductor prevents a fast build-up of the current through it, so immediately after the switch is closed the current in the inductor is zero. This means

$$i_1 = i_2 = \frac{\mathcal{E}}{R_1 + R_2} = \frac{100\,\text{V})}{10.0\,\Omega + 20.0\,\Omega} = 3.33\,\text{A}.$$

(b) A long time later the current reaches steady state and no longer changes. The emf across the inductor is zero and the circuit behaves as if it were replaced by a wire. The current in R_3 is $i_1 - i_2$. Kirchhoff's loop rule gives

$$\mathcal{E} - i_1 R_1 - i_2 R_2 = 0$$

and

$$\mathcal{E} - i_1 R_1 - (i_1 - i_2)R_3 = 0.$$

Solve these simultaneously for i_1 and i_2. The results are

$$i_1 = \frac{\mathcal{E}(R_2 + R_3)}{R_1 R_2 + R_1 R_3 + R_2 R_3}$$
$$= \frac{(100\,\text{V})(20.0\,\Omega + 30.0\,\Omega)}{(10.0\,\Omega)(20.0\,\Omega) + (10.0\,\Omega)(30.0\,\Omega) + (20.0\,\Omega)(30.0\,\Omega)}$$
$$= 4.55\,\text{A}$$

and

$$i_2 = \frac{\mathcal{E}R_3}{R_1 R_2 + R_1 R_3 + R_2 R_3}$$
$$= \frac{(100\,\text{V})(30.0\,\Omega)}{(10.0\,\Omega)(20.0\,\Omega) + (10.0\,\Omega)(30.0\,\Omega) + (20.0\,\Omega)(30.0\,\Omega)}$$
$$= 2.73\,\text{A}.$$

(c) The left-hand branch is now broken. If its inductance is zero the current immediately drops to zero when the switch is opened. That is, $i_1 = 0$. The current in R_3 changes only slowly because there is an inductor in its branch. Immediately after the switch is opened it has the same value as it had just before the switch was opened. That value is $4.55\,\text{A} - 2.73\,\text{A} = 1.82\,\text{A}$. The current in R_2 is the same as that in R_3, 1.82 A.

(d) There are no longer any sources of emf in the circuit, so all currents eventually drop to zero.

28P*

(a) Assume i is from left to right through the closed switch. Let i_1 be the current in the resistor and take it to be downward. Let i_2 be the current in the inductor and also take it to be downward. The junction rule gives $i = i_1 + i_2$ and the loop rule gives $i_1 R - L(di_2/dt) = 0$. According to the junction rule, $(di_1/dt) = -(di_2/dt)$. Substitute into the loop equation to obtain

$$L\frac{di_1}{dt} + i_1 R = 0.$$

This equation is similar to the one given in Section 33-4, just before Eq. 33-20, and its solution is the function given as Eq. 33-20:

$$i_1 = i_0 e^{-Rt/L},$$

where i_0 is the current through the resistor at $t = 0$, just after the switch is closed. Now just after the switch is closed the inductor prevents the rapid build-up of current in its branch, so at that time $i_2 = 0$ and $i_1 = i$. Thus $i_0 = i$, so

$$i_1 = i e^{-Rt/L}$$

and

$$i_2 = i - i_1 = i \left[1 - e^{-Rt/L}\right].$$

(b) When $i_2 = i_1$,

$$e^{-Rt/L} = 1 - e^{-Rt/L},$$

so

$$e^{-Rt/L} = \frac{1}{2}.$$

Take the natural logarithm of both sides and use $\ln(1/2) = -\ln 2$ to obtain $(Rt/L) = \ln 2$ or

$$t = \frac{L}{R} \ln 2.$$

34P

(a) If the battery is applied at time $t = 0$ the current is given by

$$i = \frac{\mathcal{E}}{R}\left(1 - e^{-t/\tau_L}\right),$$

where $\mathcal{E}$ is the emf of the battery, R is the resistance, and τ_L is the inductive time constant. In terms of R and the inductance L, $\tau_L = L/R$. Solve the current equation for the time constant. First obtain

$$e^{-t/\tau_L} = 1 - \frac{iR}{\mathcal{E}},$$

then take the natural logarithm of both sides to obtain

$$-\frac{t}{\tau_L} = \ln\left[1 - \frac{iR}{\mathcal{E}}\right].$$

Since

$$\ln\left[1 - \frac{iR}{\mathcal{E}}\right] = \ln\left[1 - \frac{(2.00 \times 10^{-3}\,\text{A})(10.0 \times 10^3\,\Omega)}{50.0\,\text{V}}\right] = -.5108,$$

the inductive time constant is $\tau_L = t/0.5108 = (5.00 \times 10^{-3}\,\text{s})/(0.5108) = 9.79 \times 10^{-3}\,\text{s}$ and the inductance is

$$L = \tau_L R = (9.79 \times 10^{-3}\,\text{s})(10.0 \times 10^3\,\Omega) = 97.9\,\text{H}.$$

(b) The energy stored in the coil is

$$U_B = \frac{1}{2}Li^2 = \frac{1}{2}(97.9\,\text{H})(2.00 \times 10^{-3}\,\text{A})^2 = 1.96 \times 10^{-4}\,\text{J}.$$

37P

Suppose that the switch has been in position a for a long time, so the current has reached the steady-state value i_0. The energy stored in the inductor is $U_B = \frac{1}{2}Li_0^2$. Now the switch is thrown to position b at time $t = 0$. Thereafter the current is given by

$$i = i_0 e^{-t/\tau_L},$$

where τ_L is the inductive time constant, given by $\tau = L/R$. The rate at which thermal energy is generated in the resistor is given by

$$P = i^2 R = i_0^2 R e^{-2t/\tau_L}.$$

Over a long time period the energy dissipated is

$$E = \int_0^\infty P\,dt = i_0^2 R \int_0^\infty e^{-2t/\tau_L}\,dt = -\frac{1}{2}i_0^2 R\tau_L e^{-2t/\tau_L}\Big|_0^\infty = \frac{1}{2}i_0^2 R\tau_L.$$

Substitute $\tau_L = L/R$ to obtain

$$E = \frac{1}{2}Li_0^2,$$

the same as the total energy originally stored in the inductor.

38E

(a) At any point the magnetic energy density is given by $u_B = B^2/2\mu_0$, where B is the magnitude of the magnetic field at that point. Inside a solenoid $B = \mu_0 n i$, where n is the number of turns per unit length and i is the current. For the solenoid of this problem $n = (950)/(0.850\,\text{m}) = 1.118 \times 10^3\,\text{m}^{-1}$. The magnetic energy density is

$$u_B = \frac{1}{2}\mu_0 n^2 i^2 = \frac{1}{2}(4\pi \times 10^{-7}\,\text{T}\cdot\text{m/A})(1.118 \times 10^3\,\text{m}^{-1})^2(6.60\,\text{A})^2 = 34.2\,\text{J/m}^3.$$

(b) Since the magnetic field is uniform inside an ideal solenoid, the total energy stored in the field is $U_B = u_B V$, where V is the volume of the solenoid. V is calculated as the product of the cross-sectional area and the length. Thus

$$U_B = (34.2\,\text{J/m}^3)(17.0 \times 10^{-4}\,\text{m}^2)(0.850\,\text{m}) = 4.94 \times 10^{-2}\,\text{J}.$$

39E

The magnetic energy stored in the toroid is given by $U_B = \frac{1}{2}Li^2$, where L is its inductance and i is the current. The energy is also given by $U_B = u_B V$, where u_B is the average energy density and V is the volume. Thus

$$i = \sqrt{\frac{2u_B V}{L}} = \sqrt{\frac{2(70.0\,\text{J/m}^3)(0.0200\,\text{m}^3)}{90.0 \times 10^{-3}\,\text{H}}} = 5.58\,\text{A}.$$

44P

(a) The energy density at any point is given by $u_B = B^2/2\mu_0$, where B is the magnitude of the magnetic field. The magnitude of the field inside a toroid, a distance r from the center, is given by Eq. 31–22: $B = \mu_0 i N/2\pi r$, where N is the number of turns and i is the current. Thus

$$u_B = \frac{1}{2\mu_0}\left(\frac{\mu_0 i N}{2\pi r}\right)^2 = \frac{\mu_0 i^2 N^2}{8\pi^2 r^2}.$$

(b) Evaluate the integral $U_B = \int u_B\, dV$ over the volume of the toroid. A circular strip with radius r, height h, and thickness dr has volume $dV = 2\pi r h\, dr$, so

$$U_B = \frac{\mu_0 i^2 N^2}{8\pi^2} 2\pi h \int_a^b \frac{dr}{r} = \frac{\mu_0 i^2 N^2 h}{4\pi} \ln\left(\frac{b}{a}\right).$$

Numerically,

$$U_B = \frac{(4\pi \times 10^{-7}\,\text{T}\cdot\text{m/A})(0.500\,\text{A})^2(1250)^2(13 \times 10^{-3}\,\text{m})}{4\pi} \ln\left(\frac{95\,\text{mm}}{52\,\text{mm}}\right)$$

$$= 3.06 \times 10^{-4}\,\text{J}.$$

(c) The inductance is given by Eq. 33–7:

$$L = \frac{\mu_0 N^2 h}{2\pi} \ln\left(\frac{b}{a}\right)$$

so the energy is given by

$$U_B = \frac{1}{2}Li^2 = \frac{\mu_0 N^2 i^2 h}{4\pi} \ln\left(\frac{b}{a}\right).$$

This the exactly the same as the expression found in part (b) and yields the same numerical result.

47E

(a) The mutual inductance M is given by

$$\mathcal{E}_1 = M \frac{di_2}{dt},$$

where $\mathcal{E}_1$ is the emf in coil 1 due to the changing current i_2 in coil 2. Thus

$$M = \frac{\mathcal{E}_1}{di_2/dt} = \frac{25.0 \times 10^{-3}\,\text{V}}{15.0\,\text{A/s}} = 1.67 \times 10^{-3}\,\text{H}.$$

(b) The flux linkage in coil 2 is

$$N_2 \Phi_{21} = M i_1 = (1.67 \times 10^{-3}\,\text{H})(3.60\,\text{A}) = 6.01 \times 10^{-3}\,\text{Wb}.$$

49P

(a) Assume the current is changing at the rate di/dt and calculate the total emf across both coils. First consider the left-hand coil. The magnetic field due to the current in that coil points to the left. So does the magnetic field due to the current in coil 2. When the current increases both fields increase and both changes in flux contribute emf's in the same direction. Thus the emf in coil 1 is

$$\mathcal{E}_1 = -(L_1 + M) \frac{di}{dt}.$$

The magnetic field in coil 2 due to the current in that coil points to the left, as does the field in coil 2 due to the current in coil 1. The two sources of emf are again in the same direction and the emf in coil 2 is

$$\mathcal{E}_2 = -(L_2 + M) \frac{di}{dt}.$$

The total emf across both coils is

$$\mathcal{E} = \mathcal{E}_1 + \mathcal{E}_2 = -(L_1 + L_2 + 2M) \frac{di}{dt}.$$

This is exactly the emf that would be produced if the coils were replaced by a single coil with inductance $L_{\text{eq}} = L_1 + L_2 + 2M$.

(b) Reverse the leads of coil 2 so the current enters at the back of coil rather than the front as pictured in the diagram. Then the field produced by coil 2 at the site of coil 1 is opposite the field produced by coil 1 itself. The fluxes have opposite signs. An increasing current in coil 1 tends to increase the flux in that coil but an increasing current in coil 2 tends to decrease it. The emf across coil 1 is

$$\mathcal{E}_1 = -(L_1 - M) \frac{di}{dt}.$$

Similarly the emf across coil 2 is

$$\mathcal{E}_2 = -(L_2 - M) \frac{di}{dt}.$$

The total emf across both coils is

$$\mathcal{E} = -(L_1 + L_2 - 2M)\frac{di}{dt}.$$

This the same as the emf that would be produced by a single coil with inductance $L_{eq} = L_1 + L_2 - 2M$.

52P

Assume the current in solenoid 1 is i and calculate the flux linkage in solenoid 2. The mutual inductance is this flux linkage divided by i. The magnetic field inside solenoid 1 is parallel to the axis and has uniform magnitude $B = \mu_0 i n_1$, where n_1 is the number of turns per unit length of the solenoid. The cross-sectional area of the solenoid is πR_1^2 and since the field is normal to a cross section the flux through a cross section is

$$\Phi = AB = \pi R_1^2 \mu_0 n_1 i.$$

Since the magnetic field is zero outside the solenoid, this is also the flux through a cross section of solenoid 2. The number of turns in a length ℓ of solenoid 2 is $N_2 = n_2\ell$ and the flux linkage is

$$N_2\Phi = n_2\ell\pi R_1^2 \mu_0 n_1 i.$$

The mutual inductance is

$$M = \frac{N_2\Phi}{i} = \pi R_1^2 \ell \mu_0 n_1 n_2.$$

M does not depend on R_2 because there is no magnetic field in the region between the solenoids. Changing R_2 does not change the flux through solenoid 2, but changing R_1 does.

Chapter 34

5P

(a) The period of rotation is $T = 2\pi/\omega$ and in this time all the charge passes any fixed point near the ring. The average current is $i = q/T = q\omega/2\pi$ and the magnitude of the magnetic dipole moment is

$$\mu = iA = \frac{q\omega}{2\pi}\pi r^2 = \frac{1}{2}q\omega r^2.$$

(b) Curl the fingers of your right hand in the direction of rotation. Since the charge is positive your thumb points in the direction of the dipole moment. It is the same as the direction of the angular momentum of the ring.

8P

Use Gauss' law for magnetism: $\oint \mathbf{B} \cdot d\mathbf{A} = 0$. Write $\oint \mathbf{B} \cdot d\mathbf{A} = \Phi_1 + \Phi_2 + \Phi_C$, where Φ_1 is the magnetic flux through the first end mentioned, Φ_2 is the magnetic flux through the second end mentioned, and Φ_C is the magnetic flux through the curved surface. Over the first end the magnetic field is inward, so the flux is $\Phi_1 = -25.0\,\mu\text{Wb}$. Over the second end the magnetic field is uniform, normal to the surface, and outward, so the flux is $\Phi_2 = AB = \pi r^2 B$, where A is the area of the end and r is the radius of the cylinder. It value is

$$\Phi_2 = \pi(0.120\,\text{m})^2(1.60 \times 10^{-3}\,\text{T}) = +7.24 \times 10^{-5}\,\text{Wb} = +72.4\,\mu\text{Wb}.$$

Since the three fluxes must sum to zero,

$$\Phi_C = -\Phi_1 - \Phi_2 = 25.0\,\mu\text{Wb} - 72.4\,\mu\text{Wb} = -47.4\,\mu\text{Wb}.$$

The minus sign indicates that the flux is inward through the curved surface.

10E

The horizontal component of the Earth's magnetic field is given by $B_h = B\cos\phi_i$, where B is the magnitude of the field and ϕ_i is the inclination angle. Thus

$$B = \frac{B_h}{\cos\phi_i} = \frac{16\,\mu\text{T}}{\cos 73°} = 55\,\mu\text{T}.$$

15P

At a distance r from the center of the Earth, the magnitude of the magnetic field is given by

$$B = \frac{\mu_0\mu}{4\pi r^3}\sqrt{1 + 3\sin^2\lambda_m},$$

where μ is the Earth's dipole moment and λ_m is the magnetic latitude. The ratio of the field magnitudes for two different distances at the same latitude is

$$\frac{B_2}{B_1} = \frac{r_1^3}{r_2^3}.$$

Take B_1 to be at the surface and B_2 to be half of B_1. Set r_1 equal to the radius r_e of the Earth and r_2 equal to $r_e + h$, where h is altitude at which B is half its value at the surface. Thus

$$\frac{1}{2} = \frac{r_e^3}{(r_e + h)^3}.$$

Take the cube root of both sides and solve for h. You should get

$$h = \left(2^{1/3} - 1\right) r_e = \left(2^{1/3} - 1\right) (6370\,\text{km}) = 1660\,\text{km}.$$

19E

The magnetization is the dipole moment per unit volume, so the dipole moment is given by $\mu = MV$, where M is the magnetization and V is the volume of the cylinder. Use $V = \pi r^2 L$, where r is the radius of the cylinder and L is its length. Thus

$$\mu = M\pi r^2 L = (5.30 \times 10^3\,\text{A/m})\pi(0.500 \times 10^{-2}\,\text{m})^2(5.00 \times 10^{-2}\,\text{m}) = 2.08 \times 10^{-2}\,\text{J/T}.$$

21P

For the measurements carried out the largest ratio of the magnetic field to the temperature is $(0.50\,\text{T})/(10\,\text{K}) = 0.050\,\text{T/K}$. Look at Fig. 34-11 to see if this is in the region where the magnetization is a linear function of the ratio. It is quite close to the origin, so we conclude that the magnetization obeys Curie's law.

24P

(a) A charge e traveling with uniform speed v around a circular path of radius r takes time $T = 2\pi r/v$ to complete one orbit, so the average current is

$$i = \frac{e}{T} = \frac{ev}{2\pi r}.$$

The magnitude of the dipole moment is this times the area of the orbit:

$$\mu = \frac{ev}{2\pi r}\pi r^2 = \frac{evr}{2}.$$

Since the magnetic force, with magnitude evB, is centripetal, Newton's law yields $evB = mv^2/r$, so

$$r = \frac{mv}{eB}.$$

Thus
$$\mu = \frac{1}{2}(ev)\left(\frac{mv}{eB}\right) = \left(\frac{1}{B}\right)\left(\frac{1}{2}mv^2\right) = \frac{K_e}{B}.$$

The magnetic force $-e\mathbf{v} \times \mathbf{B}$ must point toward the center of the circular path. If the magnetic field is into the page, for example, the electron will travel clockwise around the circle. Since the electron is negative, the current is in the opposite direction, counterclockwise and, by the right-hand rule for dipole moments, the dipole moment is out of the page. That is, the dipole moment is directed opposite to the magnetic field vector.

(b) Notice that the charge canceled in the derivation of $\mu = K_e/B$. Thus the relation $\mu = K_i/B$ holds for a positive ion. If the magnetic field is into the page, the ion travels counterclockwise around a circular orbit and the current is in the same direction. Thus the dipole moment is again out of the page, opposite to the magnetic field.

(c) The magnetization is given by $M = \mu_e n_e + \mu_i n_i$, where μ_e is the dipole moment of an electron, n_e is the electron concentration, μ_i is the dipole moment of an ion, and n_i is the ion concentration. Since $n_e = n_i$, we may write n for both concentrations. Substitute $\mu_e = K_e/B$ and $\mu_i = K_i/B$ to obtain

$$M = \frac{n}{B}[K_e + K_i] = \frac{5.3 \times 10^{21}\,\text{m}^{-3}}{1.2\,\text{T}}\left[6.2 \times 10^{-20}\,\text{J} + 7.6 \times 10^{-21}\,\text{J}\right] = 310\,\text{A/m}.$$

27P

An electric field with circular field lines is induced as the magnetic field is turned on. Suppose the magnetic field increases linearly from zero to B in time t. According to Eq. 32–24 the magnitude of the electric field at the orbit is given by

$$E = \left(\frac{r}{2}\right)\frac{dB}{dt} = \left(\frac{r}{2}\right)\frac{B}{t},$$

where r is the radius of the orbit. The induced electric field is tangent to the orbit and changes the speed of the electron, the change in speed being given by

$$\Delta v = at = \frac{eE}{m}t = \left(\frac{e}{m}\right)\left(\frac{r}{2}\right)\left(\frac{B}{t}\right)t = \frac{erB}{2m}.$$

The average current associated with the circulating electron is $i = ev/2\pi r$ and the dipole moment is

$$\mu = Ai = \left(\pi r^2\right)\left(\frac{ev}{2\pi r}\right) = \frac{1}{2}evr.$$

The change in the dipole moment is

$$\Delta\mu = \frac{1}{2}er\,\Delta v = \frac{1}{2}er\frac{erB}{2m} = \frac{e^2r^2B}{4m}.$$

28E

The Curie temperature for iron is $770°$ C. If x is the depth at which the temperature has this value, then $10°\,\text{C} + (30°\,\text{C/km})x = 770°\,\text{C}$ or

$$x = \frac{770°\,\text{C} - 10°\,\text{C}}{30°\,\text{C/km}} = 25\,\text{km}.$$

29E

(a) The field of a dipole along its axis is given by Eq. 31–25:

$$B = \frac{\mu_0}{2\pi} \frac{\mu}{z^3},$$

where μ is the dipole moment and z is the distance from the dipole. Thus

$$B = \frac{(4\pi \times 10^{-7}\,\text{T}\cdot\text{m/A})(1.5 \times 10^{-23}\,\text{J/T})}{2\pi(10 \times 10^{-9}\,\text{m})} = 3.0 \times 10^{-6}\,\text{T}.$$

(b) The energy of a magnetic dipole μ in a magnetic field $\mathbf{B}$ is given by $U = \boldsymbol{\mu} \cdot \mathbf{B} = \mu B \cos\phi$, where ϕ is the angle between the dipole moment and the field. The energy required to turn it end-for-end (from $\phi = 0°$ to $\phi = 180°$) is

$$\Delta U = 2\mu B = 2(1.5 \times 10^{-23}\,\text{J/T})(3.0 \times 10^{-6}\,\text{T}) = 9.0 \times 10^{-29}\,\text{J} = 5.6 \times 10^{-10}\,\text{eV}.$$

The mean kinetic energy of translation at room temperature is about 0.04 eV (see Sample Problem 34–4). Thus if dipole-dipole interactions were responsible for aligning dipoles, collisions would easily randomize the directions of the moments and they would not remain aligned.

30E

The saturation magnetization corresponds to complete alignment of all atomic dipoles and is given by $M_{\text{sat}} = \mu n$, where n is the number of atoms per unit volume and μ is the magnetic dipole moment of an atom. The number of nickel atoms per unit volume is $n = \rho/m$, where ρ is the density of nickel and m is the mass of a single nickel atom, calculated using $m = M/N_A$, where M is the atomic mass of nickel and N_A is the Avogadro constant. Thus

$$n = \frac{\rho N_A}{M} = \frac{(8.90\,\text{g/cm}^3)(6.02 \times 10^{23}\,\text{atoms/mol})}{58.71\,\text{g/mol}}$$

$$= 9.126 \times 10^{22}\,\text{atoms/cm}^3 = 9.126 \times 10^{28}\,\text{atoms/m}^3.$$

The dipole moment of a single atom of nickel is

$$\mu = \frac{M_{\text{sat}}}{n} = \frac{4.70 \times 10^5\,\text{A/m}}{9.126 \times 10^{28}\,\text{m}^3} = 5.15 \times 10^{-24}\,\text{A}\cdot\text{m}^2.$$

32P

(a) If the magnetization of the sphere is saturated the total dipole moment is $\mu_{\text{total}} = N\mu$, where N is the number of iron atoms in the sphere and μ is the dipole moment of an iron atom. We wish to find the radius of an iron sphere with N iron atoms. The mass of such a sphere is Nm, where m is the mass of an iron atom. It is also given by $4\pi\rho R^3/3$, where ρ is the density of iron and R is the radius of the sphere. Thus $Nm = 4\pi\rho R^3/3$ and

$$N = \frac{4\pi\rho R^3}{3m}.$$

Substitute this into $\mu_{total} = N\mu$ to obtain

$$\mu_{total} = \frac{4\pi\rho R^3 \mu}{3m}.$$

Solve for R and obtain

$$R = \left[\frac{3m\mu_{total}}{4\pi\rho\mu}\right]^{1/3}.$$

The mass of an iron atom is

$$m = 56\,u = (56\,u)(1.66 \times 10^{-27}\,kg/u) = 9.30 \times 10^{-26}\,kg.$$

So

$$R = \left[\frac{3(9.30 \times 10^{-26}\,kg)(8.0 \times 10^{22}\,J/T)}{4\pi(14 \times 10^3\,kg/m^3)(2.1 \times 10^{-23}\,J/T)}\right]^{1/3} = 1.8 \times 10^5\,m.$$

(*b*) The volume of the sphere is

$$V_s = \frac{4\pi}{3}R^3 = \frac{4\pi}{3}(1.82 \times 10^5\,m)^3 = 2.53 \times 10^{16}\,m^3$$

and the volume of the Earth is

$$V_e = \frac{4\pi}{3}(6.37 \times 10^6\,m)^3 = 1.08 \times 10^{21}\,m^3,$$

so the fraction of the Earth's volume that is occupied by the sphere is

$$\frac{2.53 \times 10^{16}\,m^3}{1.08 \times 10^{21}\,m^3} = 2.3 \times 10^{-5}.$$

34P

(*a*) Use Eq. 34–16 for the magnitude of the toroidal field: $B_0 = \mu_0 n i_p$, where n is the number of turns per unit length of toroid and i_p is the current required to produce the field (in the absence of the ferromagnetic material). Use the average radius ($\bar{r} = 5.5$ cm) to calculate n:

$$n = \frac{N}{2\pi\bar{r}} = \frac{400\,turns}{2\pi(5.5 \times 10^{-2}\,m)} = 1.16 \times 10^3\,turns/m.$$

Thus

$$i_p = \frac{B_0}{\mu_0 n} = \frac{0.20 \times 10^{-3}\,T}{(4\pi \times 10^{-7}\,T \cdot m/A)(1.16 \times 10^3\,m^{-1})} = 0.14\,A.$$

(*b*) If Φ is the magnetic flux through the secondary coil then the magnitude of the emf induced in that coil is $\mathcal{E} = N(d\Phi/dt)$ and the current in the secondary is $i_s = \mathcal{E}/R$, where R is the resistance of the coil. Thus

$$i_s = \left(\frac{N}{R}\right)\frac{d\Phi}{dt}.$$

The charge that passes through the secondary when the primary current is turned on is

$$q = \int i_s \, dt = \frac{N}{R} \int \frac{d\Phi}{dt} \, dt = \frac{N}{R} \int_0^{\Phi} d\Phi = \frac{N\Phi}{R}.$$

The magnetic field through the secondary coil has magnitude $B = B_0 + B_M = 801B_0$, where B_M is the field of the magnetic dipoles in the magnetic material. The total field is perpendicular to the plane of the secondary coil, so the magnetic flux is $\Phi = AB$, where A is the area of the Rowland ring (the field is inside the ring, not in the region between the ring and coil). If r is the radius of the ring's cross section then $A = \pi r^2$. Thus

$$\Phi = 801\pi r^2 B_0.$$

The radius r is $(6.0 \, \text{cm} - 5.0 \, \text{cm})/2 = 0.50 \, \text{cm}$ and

$$\Phi = 801\pi (0.50 \times 10^{-2} \, \text{m})^2 (0.20 \times 10^{-3} \, \text{T}) = 1.26 \times 10^{-5} \, \text{Wb}.$$

Thus

$$q = \frac{50(1.26 \times 10^{-5} \, \text{Wb})}{8.0 \, \Omega} = 7.9 \times 10^{-5} \, \text{C}.$$

Chapter 35

4E

(a) All the energy in the circuit resides in the capacitor when it has its maximum charge. The current is then zero. If C is the capacitance and Q is the maximum charge on the capacitor then the total energy is

$$U = \frac{Q^2}{2C} = \frac{(2.90 \times 10^{-6}\,\text{C})^2}{2(3.60 \times 10^{-6}\,\text{F})} = 1.17 \times 10^{-6}\,\text{J}.$$

(b) When the capacitor is fully discharged the current is a maximum and all the energy resides in the inductor. If I is the maximum current, then $U = LI^2/2$ and

$$I = \sqrt{\frac{2U}{L}} = \sqrt{\frac{2(1.17 \times 10^{-6}\,\text{J})}{75 \times 10^{-3}\,\text{H}}} = 5.59 \times 10^{-3}\,\text{A}.$$

8P

(a) The mass m corresponds to the inductance, so $m = 1.25\,\text{kg}$.

(b) The spring constant k corresponds to the reciprocal of the capacitance. Since the total energy is given by $U = Q^2/2C$, where Q is the maximum charge on the capacitor and C is the capacitance,

$$C = \frac{Q^2}{2U} = \frac{(175 \times 10^{-6}\,\text{C})^2}{2(5.70 \times 10^{-6}\,\text{J})} = 2.69 \times 10^{-3}\,\text{F}$$

and

$$k = \frac{1}{2.69 \times 10^{-3}\,\text{m/N}} = 372\,\text{N/m}.$$

(c) The maximum displacement x_m corresponds to the maximum charge, so

$$x_m = 175 \times 10^{-6}\,\text{m}.$$

(d) The maximum speed v_m corresponds to the maximum current. The maximum current is

$$I = Q\omega = \frac{Q}{\sqrt{LC}} = \frac{175 \times 10^{-6}\,\text{C}}{\sqrt{(1.25\,\text{H})(2.69 \times 10^{-3}\,\text{F})}} = 3.02 \times 10^{-3}\,\text{A}.$$

Thus $v_m = 3.02 \times 10^{-3}\,\text{m/s}.$

11E

If T is the period of oscillation, the time required is $t = T/4$. The period is given by $T = 2\pi/\omega = 2\pi\sqrt{LC}$, where ω is the angular frequency of oscillation, L is the inductance, and C is the capacitance. Hence

$$t = \frac{T}{4} = \frac{2\pi\sqrt{LC}}{4} = \frac{2\pi\sqrt{(0.050\,\text{H})(4.0 \times 10^{-6}\,\text{F})}}{4} = 7.0 \times 10^{-4}\,\text{s}.$$

12E

When switch S_1 is closed and the others are open, the inductor is essentially out of the circuit and what remains is an RC circuit. The time constant is $\tau_C = RC$. When switch S_2 is closed and the others are open the capacitor is essentially out of the circuit and what remains is an LR circuit with time constant $\tau_L = L/R$. When switch S_3 is closed and the others are open the resistor is essentially out of the circuit and what remains is an LC circuit that oscillates with period $T = 2\pi\sqrt{LC}$. Substitute $L = R\tau_L$ and $C = \tau_C/R$ to obtain $T = 2\pi\sqrt{\tau_C\tau_L}$.

18P

(a) After the switch is thrown to position b the circuit is an LC circuit. The angular frequency of oscillation is $\omega = 1/\sqrt{LC}$ and the frequency is

$$f = \frac{\omega}{2\pi} = \frac{1}{2\pi\sqrt{LC}} = \frac{1}{2\pi\sqrt{(54.0 \times 10^{-3}\,\text{H})(6.20 \times 10^{-6}\,\text{F})}} = 275\,\text{Hz}.$$

(b) When the switch is thrown, the capacitor is charged to $V = 34.0\,\text{V}$ and the current is zero. Thus the maximum charge on the capacitor is $Q = VC = (34.0\,\text{V})(6.20 \times 10^{-6}\,\text{F}) = 2.11 \times 10^{-4}\,\text{C}$. The current amplitude is

$$I = \omega Q = 2\pi f Q = 2\pi(275\,\text{Hz})(2.11 \times 10^{-4}\,\text{C}) = 0.365\,\text{A}.$$

21P

(a) At any time the total energy U in the circuit is the sum of the energy U_E in the electric field of the capacitor and the energy U_B in the magnetic field of the inductor. When $U_E = 0.500 U_B$, then $U_B = 2.00 U_E$ and $U = U_E + U_B = 3.00 U_E$. U_E is given by $q^2/2C$, where q is the charge on the capacitor and C is the capacitance. The total energy U is given by $Q^2/2C$, where Q is the maximum charge on the capacitor, so $Q^2/2C = 3.00 q^2/2C$ or $q = Q/\sqrt{3.00} = 0.577Q$.

(b) If the capacitor is fully charged at time $t = 0$ then the charge on the capacitor is given by $q(t) = Q\cos\omega t$, where ω is the angular frequency of oscillation. The condition $q = 0.577Q$ is satisfied when $\cos\omega t = 0.557$ or $\omega t = 0.955\,\text{rad}$. Since $\omega = 2\pi/T$, where T is the period of oscillation, $t = 0.955T/2\pi = 0.152T$.

24P

(a) Since the frequency of oscillation f is related to the inductance L and capacitance C by $f = 1/2\pi\sqrt{LC}$, the smaller value of C gives the larger value of f. Hence, $f_{\text{max}} = 1/2\pi\sqrt{LC_{\text{min}}}$, $f_{\text{min}} = 1/2\pi\sqrt{LC_{\text{max}}}$, and

$$\frac{f_{\text{max}}}{f_{\text{min}}} = \frac{\sqrt{C_{\text{max}}}}{\sqrt{C_{\text{min}}}} = \frac{\sqrt{365\,\text{pF}}}{\sqrt{10\,\text{pF}}} = 6.0.$$

(b) You want to choose the additional capacitance C so the ratio of the frequencies is

$$r = \frac{1.60\,\text{MHz}}{0.54\,\text{MHz}} = 2.96.$$

Since the additional capacitor is in parallel with the tuning capacitor, its capacitance adds to that of the tuning capacitor. If C is in picofarads, then

$$\frac{\sqrt{C + 365\,\text{pF}}}{\sqrt{C + 10\,\text{pF}}} = 2.96.$$

The solution for C is

$$C = \frac{(365\,\text{pF}) - (2.96)^2(10\,\text{pF})}{(2.96)^2 - 1} = 36\,\text{pF}.$$

Solve $f = 1/2\pi\sqrt{LC}$ for L. For the minimum frequency $C = 365\,\text{pF} + 36\,\text{pF} = 401\,\text{pF}$ and $f = 0.54\,\text{MHz}$. Thus

$$L = \frac{1}{(2\pi)^2 C f^2} = \frac{1}{(2\pi)^2(401 \times 10^{-12}\,\text{F})(0.54 \times 10^6\,\text{Hz})^2} = 2.2 \times 10^{-4}\,\text{H}.$$

25P

(a) The total energy U is the sum of the energies in the inductor and capacitor. If q is the charge on the capacitor, C is the capacitance, i is the current, and L is the inductance, then

$$
\begin{aligned}
U = U_E + U_B &= \frac{q^2}{2C} + \frac{i^2 L}{2} \\
&= \frac{(3.80 \times 10^{-6}\,\text{C})^2}{2(7.80 \times 10^{-6}\,\text{F})} + \frac{(9.20 \times 10^{-3}\,\text{A})^2(25.0 \times 10^{-3}\,\text{H})}{2} = 1.98 \times 10^{-6}\,\text{J}.
\end{aligned}
$$

(b) Solve $U = Q^2/2C$ for the maximum charge Q:

$$Q = \sqrt{2CU} = \sqrt{2(7.80 \times 10^{-6}\,\text{F})(1.98 \times 10^{-6}\,\text{J})} = 5.56 \times 10^{-6}\,\text{C}.$$

(c) Solve $U = I^2 L/2$ for the maximum current I:

$$I = \sqrt{\frac{2U}{L}} = \sqrt{\frac{2(1.98 \times 10^{-6}\,\text{J})}{25.0 \times 10^{-3}\,\text{H}}} = 1.26 \times 10^{-2}\,\text{A}.$$

(d) If q_0 is the charge on the capacitor at time $t = 0$, then $q_0 = Q \cos \phi$ and

$$\phi = \cos^{-1}\left(\frac{q}{Q}\right) = \cos^{-1}\left(\frac{3.80 \times 10^{-6}\,\text{C}}{5.56 \times 10^{-6}\,\text{C}}\right) = \pm 46.9°.$$

For $\phi = +46.9°$ the charge on the capacitor is decreasing, for $\phi = -46.9°$ it is increasing. To check this calculate the derivative of q with respect to time, evaluated for $t = 0$. You should get $-\omega Q \sin \phi$. You want this to be positive. Since $\sin(+46.9°)$ is positive and $\sin(-46.9°)$ is negative the correct value for increasing charge is $\phi = -46.9°$.

(e) Now you want the derivative to be negative and $\sin \phi$ to be positive. Take $\phi = +46.9°$.

26P

(a) The charge is given by $q(t) = Q \sin \omega t$, where Q is the maximum charge on the capacitor and ω is the angular frequency of oscillation. A sine function was chosen so that $q = 0$ at time $t = 0$. The current is

$$i(t) = \frac{dq}{dt} = \omega Q \cos \omega t$$

and at $t = 0$ it is $I = \omega Q$. Since $\omega = 1/\sqrt{LC}$,

$$Q = I\sqrt{LC} = (2.00\,\text{A})\sqrt{(3.00 \times 10^{-3}\,\text{H})(2.70 \times 10^{-6}\,\text{F})} = 1.80 \times 10^{-4}\,\text{C}.$$

(b) The energy stored in the capacitor is given by

$$U_E = \frac{q^2}{2C} = \frac{Q^2 \sin^2 \omega t}{2C}$$

and its rate of change is

$$\frac{dU_E}{dt} = \frac{Q^2 \omega \sin \omega t \cos \omega t}{C}.$$

Use the trigonometric identity $\cos \omega t \sin \omega t = \frac{1}{2} \sin(2\omega t)$ to write this

$$\frac{dU_E}{dt} = \frac{\omega Q^2}{2C} \sin(2\omega t).$$

The greatest rate of change occurs when $\sin(2\omega t) = 1$ or $2\omega t = \pi/2\,\text{rad}$. This means

$$t = \frac{\pi}{4\omega} = \frac{\pi T}{4(2\pi)} = \frac{T}{8},$$

where T is the period of oscillation. The relationship $\omega = 2\pi/T$ was used.

(c) Substitute $\omega = 2\pi/T$ and $\sin(2\omega t) = 1$ into $dU_E/dt = (\omega Q^2/2C)\sin(2\omega t)$ to obtain

$$\left(\frac{dU_E}{dt}\right)_{\text{max}} = \frac{2\pi Q^2}{2TC} = \frac{\pi Q^2}{TC}.$$

Now $T = 2\pi\sqrt{LC} = 2\pi\sqrt{(3.00 \times 10^{-3}\,\text{H})(2.70 \times 10^{-6}\,\text{F})} = 5.655 \times 10^{-4}\,\text{s}$, so

$$\left(\frac{dU_E}{dt}\right)_{\text{max}} = \frac{\pi(1.80 \times 10^{-4}\,\text{C})^2}{(5.655 \times 10^{-4}\,\text{s})(2.70 \times 10^{-6}\,\text{F})} = 66.7\,\text{W}.$$

Notice that this is a positive result, indicating that the energy in the capacitor is indeed increasing at $t = T/8$.

30P*

The energy needed to charge the 100 μF capacitor to 300 V is

$$\frac{1}{2}C_{100}V^2 = \frac{1}{2}(100 \times 10^{-6}\,\text{F})(300\,\text{V})^2 = 4.50\,\text{J}.$$

The energy originally in the 900 μF capacitor is

$$\frac{1}{2}C_{900}V^2 = \frac{1}{2}(900 \times 10^{-6}\,\text{F})(100\,\text{V})^2 = 4.5\,\text{J}.$$

All the energy originally in the 900 μF capacitor must be transferred to the 100 μF capacitor. The plan is to store it temporarily in the inductor. To do this leave switch S_1 open and close switch S_2. Wait until the 900 μF capacitor is completely discharged and the current in the right-hand loop is a maximum. This is one quarter of the period of oscillation. Since

$$T_{900} = 2\pi\sqrt{LC_{900}} = 2\pi\sqrt{(10.0\,\text{H})(900 \times 10^{-6}\,\text{F})} = 0.596\,\text{s},$$

you should wait $(0.596\,\text{s})/4 = 0.149\,\text{s}$. At that instant close switch S_1 and open switch S_2 so the current is in the left-hand loop. Now wait one quarter of the period of oscillation of the left-hand LC circuit and open switch S_1. The 100 μF capacitor then has maximum charge and all the energy resides in it. The period of oscillation is

$$T_{100} = 2\pi\sqrt{LC_{100}} = 2\pi\sqrt{(10.0\,\text{H})(100 \times 10^{-6}\,\text{F})} = 0.199\,\text{s}$$

and you must keep S_1 closed for $(0.199\,\text{s})/4 = 0.0497\,\text{s}$ before opening it again.

31E

Let T be the period of oscillation and $\omega\ (= 1/\sqrt{LC})$ be the angular frequency. Here L is the inductance and C is the capacitance. The time required for 50.0 cycles is

$$t = 50.0T = (50.0)\left(\frac{2\pi}{\omega}\right) = (50.0)\left(2\pi\sqrt{LC}\right)$$

$$= (50.0)\left(2\pi\sqrt{(220 \times 10^{-3}\,\text{H})(12.0 \times 10^{-6}\,\text{F})}\right) = 0.5104\,\text{s}.$$

The maximum charge on the capacitor decays according to

$$q_{max} = Qe^{-Rt/2L},$$

where Q is the charge at time $t = 0$ and R is the resistance in the circuit. Divide by Q and take the natural logarithm of both sides to obtain

$$\ln\left(\frac{q_{max}}{Q}\right) = -\frac{Rt}{2L}$$

and

$$R = -\frac{2L}{t}\ln\left(\frac{q_{max}}{Q}\right) = -\frac{2(220 \times 10^{-3}\,\text{H})}{0.5104\,\text{s}}\ln(0.99) = 8.66 \times 10^{-3}\,\Omega.$$

33P

Since the maximum energy in the capacitor each cycle is given by $q_{max}^2/2C$, where q_{max} is the maximum charge and C is the capacitance, you want the time for which

$$\frac{q_{max}^2}{2C} = \frac{1}{2}\frac{Q^2}{2C}.$$

This means $q_{max} = Q/\sqrt{2}$. Now q_{max} is given by

$$q_{max} = Qe^{-Rt/2L},$$

where R is the resistance and L is the inductance in the circuit. Divide by Q and take the natural logarithm of both sides to obtain

$$\ln\left(\frac{q_{max}}{Q}\right) = -\frac{Rt}{2L}.$$

Solve for t:

$$t = -\frac{2L}{R}\ln\left(\frac{q_{max}}{Q}\right) = -\frac{2L}{R}\ln\left(\frac{1}{\sqrt{2}}\right) = \frac{L}{R}\ln 2.$$

The identities $\ln(1/\sqrt{2}) = -\ln\sqrt{2} = -\frac{1}{2}\ln 2$ were used to obtain the last form of the result.

36P*

Let t be a time at which the capacitor is fully charged in some cycle and let $q_{max\,1}$ be the charge on the capacitor then. The energy in the capacitor at that time is

$$U(t) = \frac{q_{max\,1}^2}{2C} = \frac{Q^2}{2C}e^{-Rt/L},$$

where

$$q_{max\,1} = Q\,e^{-Rt/2L}$$

was used. Here Q is the charge at $t = 0$. One cycle later the maximum charge is

$$q_{max\,2} = Q\,e^{-R(t+T)/2L}$$

and the energy is

$$U(t+T) = \frac{q_{max\,2}^2}{2C} = \frac{Q^2}{2C}e^{-R(t+T)/L},$$

where T is the period of oscillation. The fractional loss in energy is

$$\frac{\Delta U}{U} = \frac{U(t) - U(t+T)}{U(t)} = \frac{e^{-Rt/L} - e^{-R(t+T)/L}}{e^{-Rt/L}} = 1 - e^{-RT/L}.$$

Assume that RT/L is small compared to 1 (the resistance is small) and use the binomial theorem to expand the exponential. The first two terms are:

$$e^{-RT/L} \approx 1 - \frac{RT}{L}.$$

Replace T with $2\pi/\omega$, where ω is the angular frequency of oscillation. Thus

$$\frac{\Delta U}{U} \approx 1 - \left(1 - \frac{RT}{L}\right) = \frac{RT}{L} = \frac{2\pi R}{\omega L}.$$

Chapter 36

3E

(a) The current amplitude I is given by $I = V_L/X_L$, where V_L is the voltage amplitude across the inductor and X_L is the inductive reactance. The reactance is given by $X_L = \omega L = 2\pi f L$, where ω is the angular frequency, f is the frequency, and L is the inductance. Since the circuit contains only the inductor and a sinusoidal generator, $V_L = \mathcal{E}_m$, where $\mathcal{E}_m$ is the generator emf amplitude. Thus

$$I = \frac{V_L}{X_L} = \frac{\mathcal{E}_m}{2\pi f L} = \frac{30.0\,\text{V}}{2\pi(1.00 \times 10^3\,\text{Hz})(50.0 \times 10^{-3}\,\text{H})} = 0.0955\,\text{A}.$$

(b) The frequency is now eight times larger than in part (a), so the inductive reactance is eight times larger and the current is one-eighth as much, or $= (0.0955\,\text{A})/8 = 0.0119\,\text{A}$.

7E

(a) The inductive reactance for angular frequency ω is given by $X_L = \omega L$, where L is the inductance, and the capacitive reactance is given by $X_C = 1/\omega C$, where C is the capacitance. The two reactances are equal if $\omega L = 1/\omega C$, or $\omega = 1/\sqrt{LC}$. The frequency is

$$f = \frac{\omega}{2\pi} = \frac{1}{2\pi\sqrt{LC}} = \frac{1}{2\pi\sqrt{(6.0 \times 10^{-3}\,\text{H})(10 \times 10^{-6}\,\text{F})}} = 650\,\text{Hz}.$$

(b) The inductive reactance is $X_L = \omega L = 2\pi f L = 2\pi(650\,\text{Hz})(6.0 \times 10^{-3}\,\text{H}) = 24\,\Omega$. The capacitive reactance has the same value for this frequency.

(c) The natural frequency for free LC oscillations is $f = \omega/2\pi = 1/2\pi\sqrt{LC}$, the same as that for which the reactances are equal.

10P

(a) The generator emf is a maximum when $\sin(\omega t - \pi/4) = 1$ or $\omega t - \pi/4 = \pi/2 \pm 2n\pi$, where n is an integer, including zero. The first time this occurs after $t = 0$ is when $\omega t - \pi/4 = \pi/2$ or

$$t = \frac{3\pi}{4\omega} = \frac{3\pi}{4(350\,\text{s}^{-1})} = 6.73 \times 10^{-3}\,\text{s}.$$

(b) The current is a maximum when $\sin(\omega t - 3\pi/4) = 1$, or $\omega t - 3\pi/4 = \pi/2 \pm 2n\pi$. The first time this occurs after $t = 0$ is when

$$t = \frac{5\pi}{4\omega} = \frac{5\pi}{4(350\,\text{s}^{-1})} = 1.12 \times 10^{-2}\,\text{s}.$$

(c) The current lags the inductor by $\pi/2\,\text{rad}$, so the circuit element must be an inductor.

(d) The current amplitude I is related to the voltage amplitude V_L by $V_L = IX_L$, where X_L is the inductive reactance, given by $X_L = \omega L$. Furthermore, since there is only one element in the circuit the amplitude of the potential difference across the element must be the same as the amplitude of the generator emf: $V_L = \mathcal{E}_m$. Thus $\mathcal{E}_m = I\omega L$ and

$$L = \frac{\mathcal{E}_m}{I\omega} = \frac{30.0\,\text{V}}{(620 \times 10^{-3}\,\text{A})(350\,\text{rad/s})} = 0.138\,\text{H}.$$

15E

(a) The capacitive reactance is

$$X_C = \frac{1}{\omega C} = \frac{1}{2\pi f C} = \frac{1}{2\pi(60.0\,\text{Hz})(70.0 \times 10^{-6}\,\text{F})} = 37.9\,\Omega.$$

The inductive reactance is unchanged, $86.7\,\Omega$. The new impedance is

$$Z = \sqrt{R^2 + (X_L - X_C)^2} = \sqrt{(160\,\Omega)^2 + (37.9\,\Omega - 86.7\,\Omega)^2} = 167\,\Omega.$$

The current amplitude is

$$I = \frac{\mathcal{E}_m}{Z} = \frac{36.0\,\text{V}}{167\,\Omega} = 0.216\,\text{A}.$$

The phase angle is

$$\phi = \tan^{-1}\left(\frac{X_L - X_C}{R}\right) = \tan^{-1}\left(\frac{86.7\,\Omega - 37.9\,\Omega}{160\,\Omega}\right) = 17.0°.$$

(b) The voltage amplitudes are

$$V_R = IR = (0.216\,\text{A})(160\,\Omega) = 34.6\,\text{V},$$

$$V_L = IX_L = (0.216\,\text{A})(86.7\,\Omega) = 18.7\,\text{V},$$

and

$$V_C = IX_C = (0.216\,\text{A})(37.9\,\Omega) = 8.19\,\text{V}.$$

Note that $X_L > X_C$, so that $\mathcal{E}_m$ leads I. The phasor diagram is drawn to scale on the right.

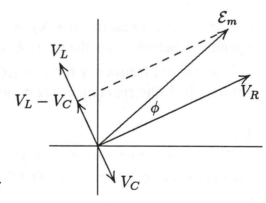

18P

The amplitude of the voltage across the inductor in an RLC series circuit is given by $V_L = IX_L$, where $X_L \, (= \omega L)$ is the inductive reactance. At resonance $\omega = 1/\sqrt{LC}$, where L is the inductance and C is the capacitance. For the given circuit

$$X_L = \frac{L}{\sqrt{LC}} = \frac{1.0\,\text{H}}{\sqrt{(1.0\,\text{H})(1.0 \times 10^{-6}\,\text{F})}} = 1000\,\Omega.$$

At resonance the capacitive reactance has the same value as the inductive reactance, so $X_C = 1000\,\Omega$. For $X_L = X_C$, Eq. 36–23 gives $Z = R$. Hence

$$I = \frac{\mathcal{E}_m}{R} = \frac{10\,\text{V}}{10\,\Omega} = 1.0\,\text{A}\,.$$

Thus

$$V_L = IX_L = (1.0\,\text{A})(1000\,\Omega) = 1000\,\text{V}\,.$$

This is much larger than the amplitude of the generator emf (10 V).

21P

(a) For a given amplitude $\mathcal{E}_m$ of the generator emf, the current amplitude is given by

$$I = \frac{\mathcal{E}_m}{Z} = \frac{\mathcal{E}_m}{\sqrt{R^2 + (\omega L - 1/\omega C)^2}}\,,$$

where R is the resistance, L is the inductance, C is the capacitance, and ω is the angular frequency. To find the maximum set the derivative with respect to ω equal to zero and solve for ω. The derivative is

$$\frac{dI}{d\omega} = -\mathcal{E}_m \left[R^2 + (\omega L - 1/\omega C)^2 \right]^{-3/2} \left[\omega L - \frac{1}{\omega C} \right] \left[L + \frac{1}{\omega^2 C} \right]\,.$$

The only factor that can equal zero is $\omega L - (1/\omega C)$ and it does for $\omega = 1/\sqrt{LC}$. For the given circuit

$$\omega = \frac{1}{\sqrt{LC}} = \frac{1}{\sqrt{(1.00\,\text{H})(20.0 \times 10^{-6}\,\text{F})}} = 224\,\text{rad/s}\,.$$

(b) For this value of the angular frequency the impedance is $Z = R$ and the current amplitude is

$$I = \frac{\mathcal{E}_m}{R} = \frac{30.0\,\text{V}}{5.00\,\Omega} = 6.00\,\text{A}\,.$$

(c) You want to find the values of ω for which $I = \mathcal{E}_m/2R$. This means

$$\frac{\mathcal{E}_m}{\sqrt{R^2 + (\omega L - 1/\omega C)^2}} = \frac{\mathcal{E}_m}{2R}\,.$$

Cancel the factors $\mathcal{E}_m$ that appear on both sides, square both sides, and set the reciprocals of the two sides equal to each other to obtain

$$R^2 + \left(\omega L - \frac{1}{\omega C} \right)^2 = 4R^2\,.$$

Thus

$$\left(\omega L - \frac{1}{\omega C} \right)^2 = 3R^2\,.$$

Now take the square root of both sides and multiply by ωC to obtain

$$\omega^2 (LC) \pm \omega \left(\sqrt{3} CR \right) - 1 = 0,$$

where the symbol $\pm$ indicates the two possible signs for the square root. The last equation is a quadratic equation for ω. Its solutions are

$$\omega = \frac{\pm\sqrt{3}CR \pm \sqrt{3C^2 R^2 + 4LC}}{2LC}.$$

You want the two positive solutions. The smaller of these is

$$\omega_2 = \frac{-\sqrt{3}CR + \sqrt{3C^2R^2 + 4LC}}{2LC}$$

$$= \frac{-\sqrt{3}(20.0 \times 10^{-6}\,\text{F})(5.00\,\Omega)}{2(1.00\,\text{H})(20.0 \times 10^{-6}\,\text{F})}$$

$$+ \frac{\sqrt{3(20.0 \times 10^{-6}\,\text{F})^2(5.00\,\Omega)^2 + 4(1.00\,\text{H})(20.0 \times 10^{-6}\,\text{F})}}{2(1.00\,\text{H})(20.0 \times 10^{-6}\,\text{F})}$$

$$= 219\,\text{rad/s}$$

and the larger is

$$\omega_1 = \frac{+\sqrt{3}CR + \sqrt{3C^2R^2 + 4LC}}{2LC}$$

$$= \frac{+\sqrt{3}(20.0 \times 10^{-6}\,\text{F})(5.00\,\Omega)}{2(1.00\,\text{H})(20.0 \times 10^{-6}\,\text{F})}$$

$$+ \frac{\sqrt{3(20.0 \times 10^{-6}\,\text{F})^2(5.00\,\Omega)^2 + 4(1.00\,\text{H})(20.0 \times 10^{-6}\,\text{F})}}{2(1.00\,\text{H})(20.0 \times 10^{-6}\,\text{F})}$$

$$= 228\,\text{rad/s}.$$

(*d*) The fractional width is

$$\frac{\omega_1 - \omega_2}{\omega_0} = \frac{228\,\text{rad/s} - 219\,\text{rad/s}}{224\,\text{rad/s}} = 0.04.$$

27P

Use the results of Problem 21:

$$\omega_1 = \frac{+\sqrt{3}CR + \sqrt{3C^2R^2 + 4LC}}{2LC}$$

and

$$\omega_2 = \frac{-\sqrt{3}CR + \sqrt{3C^2R^2 + 4LC}}{2LC}.$$

Also use

$$\omega_0 = \frac{1}{\sqrt{LC}}.$$

Thus

$$\frac{\Delta\omega}{\omega_0} = \frac{\omega_1 - \omega_2}{\omega_0} = \frac{2\sqrt{3}CR\sqrt{LC}}{2LC} = R\sqrt{\frac{3C}{L}}.$$

For the data of Problem 21

$$\frac{\Delta\omega}{\omega_0} = (5.00\,\Omega)\sqrt{\frac{3(20.0 \times 10^{-6}\,\text{F})}{1.00\,\text{H}}} = 3.87 \times 10^{-2}.$$

This is in agreement with the result of Problem 21. The method of Problem 21, however, gives only one significant figure since two numbers that are close in value (ω_1 and ω_2) are subtracted. Here the subtraction is done algebraically and 3 significant figures are obtained.

30E

The average power dissipated in resistance R when the current is alternating is given by $P_{av} = I_{rms}^2 R$, where I_{rms} is the root-mean-square current. Since $I_{rms} = I/\sqrt{2}$, where I is the current amplitude, this can be written $P_{av} = I^2 R/2$. The power dissipated in the same resistor when the current is direct is given by $P = i^2 R$, where i is the current. Set the two powers equal to each other and solve for i. You should get

$$i = \frac{I}{\sqrt{2}} = \frac{2.60\,\text{A}}{\sqrt{2}} = 1.84\,\text{A}.$$

34E

(a) The impedance is given by

$$Z = \sqrt{R^2 + (X_L - X_C)^2},$$

where R is the resistance, X_L is the inductive reactance, and X_C is the capacitive reactance. Thus

$$Z = \sqrt{(12.0\,\Omega)^2 + (1.30\,\Omega - 0)^2} = 12.1\,\Omega.$$

(b) The average rate at which energy is supplied to the air conditioner is given by

$$P_{av} = \frac{\mathcal{E}_{rms}^2}{Z} \cos\phi,$$

where $\cos\phi$ is the power factor. Now

$$\cos\phi = \frac{R}{Z} = \frac{12\,\Omega}{12.1\,\Omega} = 0.992,$$

so

$$P_{av} = \left[\frac{(120\,\text{V})^2}{12.1\,\Omega}\right](0.992) = 1.18 \times 10^3\,\text{W}.$$

40P

(a) The power factor is $\cos\phi$, where ϕ is the phase angle when the current is written $i = I\sin(\omega t - \phi)$. Thus $\phi = -42.0°$ and $\cos\phi = \cos(-42.0°) = 0.743$.

(b) Since $\phi < 0$, $\omega t - \phi > \omega t$ and the current leads the emf.

(c) The phase angle is given by $\tan\phi = (X_L - X_C)/R$, where X_L is the inductive reactance, X_C is the capacitive reactance, and R is the resistance. Now $\tan\phi = \tan(-42.0°) = -0.900$, a negative number. This means $X_L - X_C$ is negative, or $X_C > X_L$. The circuit in the box is predominantly capacitive.

(d) If the circuit were in resonance X_L would be the same as X_C, $\tan\phi$ would be zero, and ϕ would be zero. Since ϕ is not zero, we conclude the circuit is not in resonance.

(e) Since $\tan\phi$ is negative and finite, neither the capacitive reactance nor the resistance are zero. This means the box must contain a capacitor and a resistor. The inductive reactance may be zero, so there need not be an inductor. If there is an inductor its reactance must be less than that of the capacitor at the operating frequency.

(f) The average power is

$$P_{av} = \frac{1}{2}\mathcal{E}_m I \cos\phi = \frac{1}{2}(75.0\,\text{V})(1.20\,\text{A})(0.743) = 33.4\,\text{W}.$$

(g) The answers above depend on the frequency only through the phase angle ϕ, which is given. If values were given for R, L and C then the value of the frequency would also be needed to compute the power factor.

41P

(a) The average power is given by

$$P_{av} = \mathcal{E}_{rms} I_{rms} \cos\phi,$$

where $\mathcal{E}_{rms}$ is the root-mean-square emf of the generator, I_{rms} is the root-mean-square current, and $\cos\phi$ is the power factor. Now

$$I_{rms} = \frac{I}{\sqrt{2}} = \frac{\mathcal{E}_m}{\sqrt{2}Z},$$

where I is the current amplitude, $\mathcal{E}_m$ is the maximum emf of the generator, and Z is the impedance of the circuit. $I = \mathcal{E}_m/Z$ was used. In addition, $\mathcal{E}_{rms} = \mathcal{E}_m/\sqrt{2}$ and $\cos\phi = R/Z$, where R is the resistance. Thus

$$P_{av} = \frac{\mathcal{E}_m^2 R}{2Z^2} = \frac{\mathcal{E}_m^2 R}{2\left[R^2 + (\omega L - 1/\omega C)^2\right]}.$$

Here the expression $Z = \sqrt{R^2 + (\omega L - 1/\omega C)^2}$ for the impedance in terms of the angular frequency was substituted.

Considered as a function of C, P_{av} has its largest value when the factor $R^2 + (\omega L - 1/\omega C)^2$ has the smallest possible value. This occurs for $\omega L = 1/\omega C$, or

$$C = \frac{1}{\omega^2 L} = \frac{1}{(2\pi)^2(60.0\,\text{Hz})^2(60.0 \times 10^{-3}\,\text{H})} = 1.17 \times 10^{-4}\,\text{F}.$$

The circuit is then at resonance.

(b) Now you want Z^2 to be as large as possible. Notice that it becomes large without bound as C becomes small. Thus the smallest average power occurs for $C = 0$.

(c) When $\omega L = 1/\omega C$ the expression for the average power becomes

$$P_{av} = \frac{\mathcal{E}_m^2}{2R},$$

so the maximum average power is

$$P_{av} = \frac{(30.0\,\text{V})^2}{2(5.00\,\Omega)} = 90.0\,\text{W}.$$

The minimum average power is $P_{av} = 0$.

(d) At maximum power $X_L = X_C$, where X_L is the inductive reactance and X_C is the capacitive reactance. The phase angle ϕ is

$$\tan\phi = \frac{X_L - X_C}{R} = 0,$$

so $\phi = 0$. At minimum power X_C is infinite, so $\tan\phi = -\infty$ and $\phi = -90°$.

(e) At maximum power the power factor is $\cos\phi = \cos 0° = 1$ and at minimum power it is $\cos\phi = \cos(-90°) = 0$.

45E

(a) If N_p is the number of primary turns and N_s is the number of secondary turns, then

$$V_s = \frac{N_s}{N_p} V_p = \left(\frac{10}{500}\right)(120\,\text{V}) = 2.4\,\text{V}.$$

(b) The current in the secondary is given by Ohm's law:

$$I_s = \frac{V_s}{R_s} = \frac{2.4\,\text{V}}{15\,\Omega} = 0.16\,\text{A}.$$

The current in the primary is

$$I_p = \frac{N_s}{N_p} I_s = \left(\frac{10}{500}\right)(0.16\,\text{A}) = 3.2 \times 10^{-3}\,\text{A}.$$

48P

The amplifier is connected across the primary windings of a transformer and the resistor R is connected across the secondary windings. If I_s is the rms current in the secondary coil then the average power delivered to R is $P_{av} = I_s^2 R$. Now $I_s = (N_p/N_s)I_p$, where N_p is the number of turns in the primary coil, N_s is the number of turns in the secondary coil, and I_p is the rms current in the primary coil. Thus

$$P_{av} = \left(\frac{I_p N_p}{N_s}\right)^2 R.$$

Now find the current in the primary circuit. It acts like a circuit consisting of a generator and two resistors in series. One resistance is the resistance r of the amplifier and the other is the equivalent resistance R_{eq} of the secondary circuit. Thus $I_p = \mathcal{E}/(r + R_{eq})$, where $\mathcal{E}$ is the rms emf of the amplifier. According to Eq. 36–38 $R_{eq} = (N_p/N_s)^2 R$, so

$$I_p = \frac{\mathcal{E}}{r + (N_p/N_s)^2 R}$$

and

$$P_{av} = \frac{\mathcal{E}^2 (N_p/N_s)^2 R}{\left[r + (N_p/N_s)^2 R\right]^2} .$$

You wish to find the value of N_p/N_s so that P_{av} is a maximum.
Let $x = (N_p/N_s)^2$. Then

$$P_{av} = \frac{\mathcal{E}^2 R x}{(r + xR)^2}$$

and the derivative with respect to x is

$$\frac{dP_{av}}{dx} = \frac{\mathcal{E}^2 R(r - xR)}{(r + xR)^3} .$$

This is zero for $x = r/R = (1000\,\Omega)/(10\,\Omega) = 100$. Notice that for small x, P_{av} increases linearly with x and for large x it decreases in proportion to $1/x$. Thus $x = r/R$ is indeed a maximum, not a minimum.

Since $x = (N_p/N_s)^2$, maximum power is achieved for $(N_p/N_s)^2 = 100$, or $N_p/N_s = 10$.

Chapter 37

3E

Let R be the radius of a capacitor plate and r be the distance from axis of the capacitor. For points with $r \leq R$ the magnitude of the magnetic field is given by

$$B = \frac{\mu_0 \epsilon_0 r}{2} \frac{dE}{dt}$$

and for $r \geq R$ it is

$$B = \frac{\mu_0 \epsilon_0 R^2}{2r} \frac{dE}{dt}.$$

The maximum magnetic field occurs at points for which $r = R$ and its value is given by either of the formulas above:

$$B_{max} = \frac{\mu_0 \epsilon_0 R}{2} \frac{dE}{dt}.$$

There are two values of r for which $B = B_{max}/2$: one less than R and one greater. To find the one that is less than R, solve

$$\frac{\mu_0 \epsilon_0 r}{2} \frac{dE}{dt} = \frac{\mu_0 \epsilon_0 R}{4} \frac{dE}{dt}$$

for r. The result is $r = R/2 = (55.0\,\text{mm})/2 = 27.5\,\text{mm}$.

To find the one that is greater than R, solve

$$\frac{\mu_0 \epsilon_0 R^2}{2r} \frac{dE}{dt} = \frac{\mu_0 \epsilon_0 R}{4} \frac{dE}{dt}$$

for r. The result is $r = 2R = 2(55.0\,\text{mm}) = 110\,\text{mm}$.

6E

The displacement current is given by

$$i_d = \epsilon_0 A \frac{dE}{dt},$$

where A is the area of a plate and E is the magnitude of the electric field between the plates. The field between the plates is uniform, so $E = V/d$, where V is the potential difference across the plates and d is the plate separation. Thus

$$i_d = \frac{\epsilon_0 A}{d} \frac{dV}{dt}.$$

Now $\epsilon_0 A/d$ is the capacitance C of a parallel-plate capacitor without a dielectric, so

$$i_d = C \frac{dV}{dt}.$$

8E

Consider an area A, normal to a uniform electric field **E**. The displacement current density is uniform and normal to the area. Its magnitude is given by $J_d = i_d/A$. For this situation

$$i_d = \epsilon_0 A \frac{dE}{dt},$$

so

$$J_d = \frac{1}{A} \epsilon_0 A \frac{dE}{dt} = \epsilon_0 \frac{dE}{dt}.$$

14P

(a) Use the results of Exercise 6, with $V = V_m \sin(2\pi ft)$. The derivative with respect to time is $dV/dt = 2\pi f V_m \cos(2\pi ft)$, so $i_d = 2\pi f C V_m \cos(2\pi ft)$ and the maximum displacement current is

$$i_{d\,max} = 2\pi f C V_m = 2\pi (50.0\,\text{Hz})(100 \times 10^{-12}\,\text{F})(174 \times 10^3\,\text{V}) = 5.47 \times 10^{-3}\,\text{A}.$$

(b) The maximum displacement current is directly proportional to the maximum potential difference applied. A large value of V_m produces a more easily measurable value of $i_{d\,max}$ than a smaller value of V_m.

15P

(a) At any instant the displacement current i_d in the gap between the plates equals the conduction current i in the wires. Thus $i_{max} = i_{d\,max} = 7.60\,\mu\text{A}$.

(b) Since $i_d = \epsilon_0\,(d\Phi_E/dt)$,

$$\left(\frac{d\Phi_E}{dt}\right)_{max} = \frac{i_{d\,max}}{\epsilon_0} = \frac{7.60 \times 10^{-6}\,\text{A}}{8.85 \times 10^{-12}\,\text{F/m}} = 8.59 \times 10^5\,\text{V}\cdot\text{m/s}.$$

(c) According to Exercise 6

$$i_d = \frac{\epsilon_0 A}{d} \frac{dV}{dt}.$$

Now the potential difference across the capacitor is the same in magnitude as the emf of the generator, so $V = \mathcal{E}_m \sin \omega t$ and $dV/dt = \omega \mathcal{E}_m \cos \omega t$. Thus

$$i_d = \frac{\epsilon_0 A \omega \mathcal{E}_m}{d} \cos \omega t$$

and

$$i_{d\,max} = \frac{\epsilon_0 A \omega \mathcal{E}_m}{d}.$$

This means

$$d = \frac{\epsilon_0 A \omega \mathcal{E}_m}{i_{d\,max}} = \frac{(8.85 \times 10^{-12}\,\text{F/m})\pi(0.180\,\text{m})^2(130\,\text{rad/s})(220\,\text{V})}{7.60 \times 10^{-6}\,\text{A}}$$
$$= 3.39 \times 10^{-3}\,\text{m},$$

where $A = \pi R^2$ was used.

(*d*) Use the Ampere-Maxwell law in the form $\oint \mathbf{B} \cdot d\mathbf{s} = \mu_0 I_d$, where the path of integration is a circle of radius r between the plates and parallel to them. I_d is the displacement current through the area bounded by the path of integration. Since the displacement current density is uniform between the plates $I_d = (r^2/R^2)i_d$, where i_d is the total displacement current between the plates and R is the plate radius. The field lines are circles centered on the axis of the plates, so $\mathbf{B}$ is parallel to d**s**. The field has constant magnitude around the circular path, so $\oint \mathbf{B} \cdot d\mathbf{s} = 2\pi r B$. Thus

$$2\pi r B = \mu_0 \left(\frac{r^2}{R^2} \right) i_d$$

and

$$B = \frac{\mu_0 i_d r}{2\pi R^2}.$$

The maximum magnetic field is given by

$$B_{\text{max}} = \frac{\mu_0 i_{d\ \text{max}} r}{2\pi R^2} = \frac{(4\pi \times 10^{-7}\,\text{T} \cdot \text{m/A})(7.6 \times 10^{-6}\,\text{A})(0.110\,\text{m})}{2\pi(0.180\,\text{m})^2} = 5.16 \times 10^{-12}\,\text{T}.$$

20P*

(*a*) At time t the charge on the right face is given by

$$q = \int_0^t i\,dt = \int_0^t \alpha t\,dt = \frac{1}{2}\alpha t^2.$$

At that time the charge on the left face is $-\frac{1}{2}\alpha t^2$.

(*b*) Use a Gaussian surface in the shape of a cylinder, concentric with the rod, with one end in the gap and the other in the rod to the left of the gap, as shown in the diagram to the right. The electric field is in the positive x direction so we need to consider only the faces of the cylinder. The magnitude of the electric field at the left face is given by ρJ, where ρ is the resistivity of the rod and J is the current density.

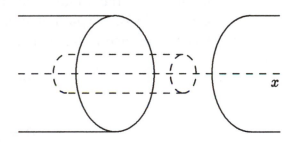

Let E be the magnitude of the field at the right face. Assume the current density is uniform on the left face and the electric field is uniform on the right face. Then

$$\oint \mathbf{E} \cdot d\mathbf{A} = -\rho J A + E A,$$

where A is the area of a face. We suppose the resistivity is so small that the first term can be neglected. Gauss' law becomes $EA = Q/\epsilon_0$, where Q is the charge on the rod end, within the Gaussian surface. The area of the face of the Gaussian cylinder is $A = \pi r^2$, where r it is radius, and the charge enclosed is $Q = (r^2/R^2)q$, where q is the charge on the face of the rod. Thus

$$E = \frac{q}{\pi \epsilon_0 R^2} = \frac{\alpha t^2}{2\pi \epsilon_0 R^2},$$

where the result of part (*a*), $q = \frac{1}{2}\alpha t^2$, was used.

(c) The magnetic field lines form circles that are concentric with the rod axis (the x axis) and lie in planes that are parallel to the rod faces.

(d) Use the Ampere-Maxwell law:

$$\oint \mathbf{B} \cdot \mathbf{ds} = \mu_0 i + \mu_0 \epsilon_0 \frac{d\Phi_E}{dt}.$$

For the path of integration choose a circle that coincides with a magnetic field line. Suppose its radius is r (with $r < R$) and that B is the magnitude of the field at points on it. Then $\oint \mathbf{B} \cdot \mathbf{ds} = 2\pi r B$. In the gap the current is zero and only the displacement current contributes to the right side of the Ampere-Maxwell equation. Since

$$\frac{d\Phi_E}{dt} = A \frac{dE}{dt} = \pi r^2 \frac{\alpha t}{\pi \epsilon_0 R^2} = \frac{\alpha t r^2}{\epsilon_0 R^2},$$

the Ampere-Maxwell law yields

$$2\pi r B = \mu_0 \epsilon_0 \frac{\alpha t r^2}{\epsilon_0 R^2}.$$

Thus

$$B = \frac{\mu_0 \alpha t r}{2\pi R^2}.$$

The magnetic field inside the rod, a distance r from its axis, is given by exactly the same expression. Now only the conduction current contributes to the right side of the Ampere-Maxwell equation. Take the path of integration to be a circle centered at the axis and parallel to the rod faces. The current through the circle is $(r^2/R^2)i$ and the Ampere-Maxwell equation yields

$$2\pi r B = \mu_0 (r^2/R^2)i,$$

so

$$B = \frac{\mu_0 i r}{2\pi R^2} = \frac{\mu_0 \alpha t r}{2\pi R^2},$$

where αt was substituted for i.

Chapter 38

2E

The time for light to travel a distance d in free space is $t = d/c$, where c is the speed of light $(3.00 \times 10^8\,\text{m/s})$.

(*a*) Take d to be $150\,\text{km} = 150 \times 10^3\,\text{m}$. Then

$$t = \frac{d}{c} = \frac{150 \times 10^3\,\text{m}}{3.00 \times 10^8\,\text{m/s}} = 5.00 \times 10^{-4}\,\text{s}.$$

(*b*) At full moon the moon and the sun are on opposite sides of the Earth, so the distance traveled by the light is $d = (1.50 \times 10^8\,\text{km}) + 2(3.82 \times 10^5\,\text{km}) = 1.508 \times 10^8\,\text{km} = 1.508 \times 10^{11}\,\text{m}$. The time taken by light to travel this distance is

$$t = \frac{d}{c} = \frac{1.508 \times 10^{11}\,\text{m}}{3.00 \times 10^8\,\text{m/s}} = 503\,\text{s} = 8.38\,\text{min}.$$

The distances are given in Appendix C.

(*c*) Take d to be $2(1.3 \times 10^9\,\text{km}) = 2.6 \times 10^{12}\,\text{m}$. Then

$$t = \frac{d}{c} = \frac{2.6 \times 10^{12}\,\text{m}}{3.00 \times 10^8\,\text{m/s}} = 8.7 \times 10^3\,\text{s} = 2.4\,\text{h}.$$

(*d*) Take d to be $6500\,\text{ly}$ and the speed of light to be $1.00\,\text{ly/y}$. Then

$$t = \frac{d}{c} = \frac{6500\,\text{ly}}{1.00\,\text{ly/y}} = 6500\,\text{y}.$$

The explosion took place in the year $1054 - 6500 = -5446$ or B.C. 5446.

7P

(*a*) Suppose that at time t_1 the moon is starting a revolution (on the verge of going behind Jupiter, say) and that at this instant the distance between Jupiter and the Earth is ℓ_1. The time of the start of the revolution as seen on Earth is $t_1^* = t_1 + \ell_1/c$. Suppose the moon starts the next revolution at time t_2 and at that instant the Earth-Jupiter distance is ℓ_2. The start of the revolution as seen on Earth is $t_2^* = t_2 + \ell_2/c$. Now the actual period of the moon is given by $T = t_2 - t_1$ and the period as measured on Earth is

$$T^* = t_2^* - t_1^* = t_2 - t_1 + \frac{\ell_2}{c} - \frac{\ell_2}{c} = T + \frac{\ell_2 - \ell_1}{c}.$$

The period as measured on Earth is longer than the actual period because the Earth moves during a revolution and light takes a finite time to travel from Jupiter to the Earth. For the situation depicted in the diagram, light emitted at the end of a revolution travels a longer distance to get to the Earth than light emitted at the beginning.

Suppose the position of the Earth is given by the angle θ, measured from x. Let R be the radius of the Earth's orbit and d be the distance from the sun to Jupiter. Then the law of cosines, applied to the triangle with the sun, the Earth, and Jupiter at the vertices, yields $\ell^2 = d^2 + R^2 - 2dR \cos \theta$. This expression can be used to calculate ℓ_1 and ℓ_2. Since the Earth does not move very far during one revolution of the moon, we may approximate $\ell_2 - \ell_1$ by $(d\ell/dt)T$ and T^* by $T + (d\ell/dt)(T/c)$. Now

$$\frac{d\ell}{dt} = \frac{2Rd \sin \theta}{\sqrt{d^2 + R^2 - 2dR \cos \theta}} \frac{d\theta}{dt} = \frac{2vd \sin \theta}{\sqrt{d^2 + R^2 - 2dR \cos \theta}},$$

where $v = R(d\theta/dt)$ is the speed of the Earth in its orbit. For $\theta = 0$, $(d\ell/dt) = 0$ and $T^* = T$. Since the Earth is then moving perpendicularly to the line from the sun to Jupiter its distance from the planet does not change much during a revolution of the moon. On the other hand, when $\theta = 90°$, $d\ell/dt = vd/\sqrt{d^2 + R^2}$ and

$$T^* = T \left(1 + \frac{vd}{c\sqrt{d^2 + R^2}} \right).$$

The Earth is now moving parallel to the line from the sun to Jupiter and its distance from the planet changes during a revolution of the moon.

(b) Let t be the actual time for the moon to make N revolutions and t^* the time for N revolutions to be observed on Earth. Then

$$t^* = t + \frac{\ell_2 - \ell_2}{c},$$

where ℓ_1 is the Earth-Jupiter distance at the beginning of the interval and ℓ_2 is the Earth-Jupiter distance at the end. Suppose the Earth is at x at the beginning of the interval and at y at the end. Then $\ell_1 = d - R$ and $\ell_2 = \sqrt{d^2 + R^2}$. Thus

$$t^* = t + \frac{\sqrt{d^2 + R^2} - (d - R)}{c}.$$

A value can be found for t by measuring the observed period of revolution when the Earth is at x and multiplying by N. Notice that the observed period is the true period when the Earth is at x. Now measure the time interval as the Earth moves from x to y. This is t^*. The difference is

$$t^* - t = \frac{\sqrt{d^2 + R^2} - (d - R)}{c}.$$

If the radii of the orbits of Jupiter and the Earth are known, the value for $t^* - t$ can be used to compute c.

Since Jupiter is much further from the sun than the Earth, $\sqrt{d^2 + R^2}$ may be approximated by d and $t^* - t$ may be approximated by R/c. In this approximation only the radius of the Earth's orbit need be known.

9E

If f is the frequency and λ is the wavelength of an electromagnetic wave then $f\lambda = c$. The frequency is the same as the frequency of oscillation of the current in the LC circuit of the generator. That is, $f = 1/2\pi\sqrt{LC}$, where C is the capacitance and L is the inductance. Thus

$$\frac{\lambda}{2\pi\sqrt{LC}} = c.$$

The solution for L is

$$L = \frac{\lambda^2}{4\pi^2 C c^2} = \frac{(550 \times 10^{-9}\,\mathrm{m})^2}{4\pi^2(17 \times 10^{-12}\,\mathrm{F})(3.00 \times 10^8\,\mathrm{m/s})^2} = 5.00 \times 10^{-21}\,\mathrm{H}.$$

This is exceedingly small.

16E

If P is the power and Δt is the time interval of one pulse, then the energy in a pulse is

$$E = P\,\Delta t = (100 \times 10^{12}\,\mathrm{W})(1.0 \times 10^{-9}\,\mathrm{s}) = 1.0 \times 10^5\,\mathrm{J}.$$

20E

The region illuminated on the moon is a circle with radius $R = r\theta/2$, where r is the Earth-moon distance (3.82×10^8 m) and θ is the full-angle beam divergence in radians. The area A illuminated is

$$A = \pi R^2 = \frac{\pi r^2 \theta^2}{4} = \frac{\pi(3.82 \times 10^8\,\mathrm{m})^2(0.880 \times 10^{-6}\,\mathrm{rad})^2}{4} = 8.88 \times 10^4\,\mathrm{m}^2.$$

26P

At any point and time the energy density associated with the electric field is given by $u_E = \epsilon_0 E^2/2$, where E is the magnitude of the electric field at that point and that time. The energy density associated with the magnetic field is given by $u_B = B^2/2\mu_0$, where B is the magnitude of the magnetic field. The electric and magnetic fields in an electromagnetic wave are related by $E = cB$, where c is the speed of light. Substitute cB for E and $1/\sqrt{\mu_0\epsilon_0}$ for c in the expression for the electric-field energy density. You should obtain

$$u_E = \frac{1}{2}\epsilon_0 c^2 B^2 = \frac{\epsilon_0 B^2}{2\mu_0\epsilon_0} = \frac{B^2}{2\mu_0}.$$

Since this is the energy density associated with the magnetic field, $u_E = u_B$.

29P

(a) Take the electric field to be $\mathbf{E} = E\,\mathbf{j}$ and the magnetic field to be $\mathbf{B} = B\,\mathbf{i}$, where E and B are the magnitudes of the fields. The Poynting vector is

$$\mathbf{S} = \frac{1}{\mu_0}\mathbf{E} \times \mathbf{B} = \frac{1}{\mu_0}(E\,\mathbf{j}) \times (B\,\mathbf{i}) = -\frac{EB}{\mu_0}\,\mathbf{k},$$

where $\mathbf{j} \times \mathbf{i} = -\mathbf{k}$ was used. The Poynting vector is in the negative z direction. If the z axis is upward, the Poynting vector view predicts that energy flows into the top face of the cube at the rate

$$P = SA = \frac{EBa^2}{\mu_0},$$

where A $(= a^2)$ is the area of a face. It also predicts that energy flows out of the bottom face at the same rate. The Poynting vector is parallel to all other faces of the cube, so the view predicts that no energy flows through these faces.

(b) The net rate with which energy in the cube is changing is the difference between the rate with which it is flowing into the top face and the rate with which it is flowing out of the bottom face. Since these two rates are the same, the energy in the cube is not changing.

The Poynting vector may not give the actual rate of energy flow through any given portion of a surface. Note that the fields given above may be static, in which case no energy actually flows through any face of the cube. Nevertheless, $\oint \mathbf{S} \cdot d\mathbf{A}$, around a *closed* surface, correctly gives the rate of change of energy within the surface, whether the fields are static or not.

32P

(a) The average rate of energy flow per unit area, or intensity, is related to the electric field amplitude E_m by $I = E_m^2/2\mu_0 c$, so

$$E_m = \sqrt{2\mu_0 c I} = \sqrt{2(4\pi \times 10^{-7}\,\text{H/m})(3.00 \times 10^8\,\text{m/s})(10 \times 10^{-6}\,\text{W/m}^2)}$$
$$= 8.7 \times 10^{-2}\,\text{V/m}.$$

(b) The amplitude of the magnetic field is given by

$$B_m = \frac{E_m}{c} = \frac{8.7 \times 10^{-2}\,\text{V/m}}{3.00 \times 10^8\,\text{m/s}} = 2.9 \times 10^{-10}\,\text{T}.$$

(c) At a distance r from the transmitter the intensity is $I = P/4\pi r^2$, where P is the power of the transmitter. Thus

$$P = 4\pi r^2 I = 4\pi (10 \times 10^3\,\text{m})^2 (10 \times 10^{-6}\,\text{W/m}^2) = 1.3 \times 10^4\,\text{W}.$$

35E

The plasma completely reflects all the energy incident on it, so the radiation pressure is given by $p_r = 2I/c$, where I is the intensity. The intensity is $I = P/A$, where P is the power and A is the area intercepted by the radiation. Thus

$$p_r = \frac{2P}{Ac} = \frac{2(1.5 \times 10^9\,\text{W})}{(1.00 \times 10^{-6}\,\text{m}^2)(3.00 \times 10^8\,\text{m/s})} = 1.0 \times 10^7\,\text{Pa} = 10\,\text{MPa}.$$

38E

Since the surface is perfectly absorbing the radiation pressure is given by $p_r = I/c$, where I is the intensity. Since the bulb radiates uniformly in all directions the intensity a distance r from it is given by $I = P/4\pi r^2$, where P is the power of the bulb. Thus

$$p_r = \frac{P}{4\pi r^2 c} = \frac{500\,\text{W}}{4\pi(1.5\,\text{m})^2(3.00 \times 10^8\,\text{m/s})} = 5.9 \times 10^{-8}\,\text{Pa}\,.$$

39P

(a) Since $c = \lambda f$, where λ is the wavelength and f is the frequency of the wave,

$$f = \frac{c}{\lambda} = \frac{3.00 \times 10^8\,\text{m/s}}{3.0\,\text{m}} = 1.0 \times 10^8\,\text{Hz}\,.$$

(b) The magnetic field amplitude is

$$B_m = \frac{E_m}{c} = \frac{300\,\text{V/m}}{3.00 \times 10^8\,\text{m/s}} = 1.00 \times 10^{-6}\,\text{T}\,.$$

$\mathbf{B}$ must be in the positive z direction when $\mathbf{E}$ is in the positive y direction in order for $\mathbf{E} \times \mathbf{B}$ to be in the positive x direction (the direction of propagation).

(c) The angular wave number is

$$k = \frac{2\pi}{\lambda} = \frac{2\pi}{3.0\,\text{m}} = 2.1\,\text{rad/m}\,.$$

The angular frequency is

$$\omega = 2\pi f = 2\pi(1.0 \times 10^8\,\text{Hz}) = 6.3 \times 10^8\,\text{rad/s}\,.$$

(d) The intensity of the wave is

$$I = \frac{E_m^2}{2\mu_0 c} = \frac{(300\,\text{V/m})^2}{2(4\pi \times 10^{-7}\,\text{H/m})(3.00 \times 10^8\,\text{m/s})} = 119\,\text{W/m}^2\,.$$

(e) Since the sheet is perfectly absorbing, the rate per unit area with which momentum is delivered to it is I/c, so

$$\frac{dp}{dt} = \frac{IA}{c} = \frac{(119\,\text{W/m}^2)(2.0\,\text{m}^2)}{3.00 \times 10^8\,\text{m/s}} = 8.0 \times 10^{-7}\,\text{N}\,.$$

The radiation pressure is

$$p_r = \frac{dp/dt}{A} = \frac{8.0 \times 10^{-7}\,\text{N}}{2.0\,\text{m}^2} = 4.0 \times 10^{-7}\,\text{Pa}\,.$$

43P

Let f be the fraction of the incident beam intensity that is reflected. The fraction absorbed is $1 - f$. The reflected portion exerts a radiation pressure of

$$p_r = \frac{2fI_0}{c}$$

and the absorbed portion exerts a radiation pressure of

$$p_a = \frac{(1-f)I_0}{c},$$

where I_0 is the incident intensity. The factor 2 enters the first expression because the momentum of the reflected portion is reversed. The total radiation pressure is the sum of the two contributions:

$$p_{total} = p_r + p_a = \frac{2fI_0 + (1-f)I_0}{c} = \frac{(1+f)I_0}{c}.$$

To relate the intensity and energy density, consider a tube with length ℓ and cross-sectional area A, lying with its axis along the propagation direction of an electromagnetic wave. The electromagnetic energy inside is $U = uA\ell$, where u is the energy density. All this energy will pass through the end in time $t = \ell/c$ so the intensity is

$$I = \frac{U}{At} = \frac{uA\ell c}{\ell} = uc.$$

Thus $u = I/c$. The intensity and energy density are inherently positive, regardless of the propagation direction.

For the partially reflected and partially absorbed wave the intensity just outside the surface is $I = I_0 + fI_0 = (1+f)I_0$, where the first term is associated with the incident beam and the second is associated with the reflected beam. The energy density is therefore

$$u = \frac{I}{c} = \frac{(1+f)I_0}{c},$$

the same as radiation pressure.

45P

If the beam carries energy U away from the spaceship then it also carries momentum $p = U/c$ away. Since the total momentum of the spaceship and light is conserved, this is the magnitude of the momentum acquired by the spaceship. If P is the power of the laser, then the energy carried away in time t is $U = Pt$. Thus $p = Pt/c$ and, if m is mass of the spaceship, its speed is

$$v = \frac{p}{m} = \frac{Pt}{mc} = \frac{(10 \times 10^3 \,\text{W})(1\,\text{d})(8.64 \times 10^4 \,\text{s/d})}{(1.5 \times 10^3 \,\text{kg})(3.00 \times 10^8 \,\text{m/s})} = 1.9 \times 10^{-3} \,\text{m/s} = 1.9\,\text{mm/s}.$$

47P

(a) Let r be the radius and ρ be the density of the particle. Since its volume is $(4\pi/3)r^3$, its mass is $m = (4\pi/3)\rho r^3$. Let R be the distance from the sun to the particle and let M be the mass of the sun. Then the gravitational force of attraction of the sun on the particle has magnitude

$$F_g = \frac{GMm}{R^2} = \frac{4\pi GM\rho r^3}{3R^2}.$$

If P is the power output of the sun, then at the position of the particle the radiation intensity is $I = P/4\pi R^2$ and since the particle is perfectly absorbing the radiation pressure on it is

$$p_r = \frac{I}{c} = \frac{P}{4\pi R^2 c}.$$

All of the radiation that passes through a circle of radius r and area $A = \pi r^2$, perpendicular to the direction of propagation, is absorbed by the particle, so the force of the radiation on the particle has magnitude

$$F_r = p_r A = \frac{\pi P r^2}{4\pi R^2 c} = \frac{P r^2}{4R^2 c}.$$

The force is radially outward from the sun. Notice that both the force of gravity and the force of the radiation are inversely proportional to R^2. If one of these forces is larger than the other at some distance from the sun, then that force is larger at all distances.

The two forces depend on the particle radius r differently: F_g is proportional to r^3 and F_r is proportional to r^2. We expect a small radius particle to be blown away by the radiation pressure and a large radius particle with the same density to be pulled inward toward the sun. The critical value for the radius is the value for which the two forces are equal. Equate the expressions for F_g and F_r, then solve for r. You should obtain

$$r = \frac{3P}{16\pi GM\rho c}.$$

(b) According to Appendix C, $M = 1.99 \times 10^{30}$ kg and $P = 3.90 \times 10^{26}$ W. Thus,

$$r = \frac{3(3.90 \times 10^{26}\text{ W})}{16\pi(6.67 \times 10^{-11}\text{ N}\cdot\text{m}^2/\text{kg}^2)(1.99 \times 10^{30}\text{ kg})(1.0 \times 10^3\text{ kg/m}^3)(3.00 \times 10^8\text{ m/s})}$$
$$= 5.8 \times 10^{-7}\text{ m}.$$

49E

(a) Since the incident light is unpolarized half the intensity is transmitted and half is absorbed. Thus the transmitted intensity is $I = 5.0\text{ mW/m}^2$. The intensity and the electric field amplitude are related by $I = E_m^2/2\mu_0 c$, so

$$E_m = \sqrt{2\mu_0 cI} = \sqrt{2(4\pi \times 10^{-7}\text{ H/m})(3.00 \times 10^8\text{ m/s})(5.0 \times 10^{-3}\text{ W/m}^2)}$$
$$= 1.9\text{ V/m}.$$

(b) The radiation pressure is $p_r = I_a/c$, where I_a is the absorbed intensity. Thus

$$p_r = \frac{5.0 \times 10^{-3}\text{ W/m}^2}{3.00 \times 10^8\text{ m/s}} = 1.7 \times 10^{-11}\text{ Pa}.$$

58P

Let I_0 be the intensity of the incident beam and f be the fraction that is polarized. Thus the intensity of the polarized portion is $f I_0$. After transmission this portion contributes $f I_0 \cos^2 \theta$ to the intensity of the transmitted beam. Here θ is the angle between the direction of polarization of the radiation and the polarizing direction of the filter. The intensity of the unpolarized portion of the incident beam is $(1 - f) I_0$ and after transmission this portion contributes $(1 - f) I_0 / 2$ to the transmitted intensity. Thus the transmitted intensity is

$$I = f I_0 \cos^2 \theta + \frac{1}{2}(1 - f) I_0.$$

As the filter is rotated $\cos^2 \theta$ varies from a minimum of 0 to a maximum of 1, so the transmitted intensity varies from a minimum of

$$I_{min} = \frac{1}{2}(1 - f) I_0$$

to a maximum of

$$I_{max} = f I_0 + \frac{1}{2}(1 - f) I_0 = \frac{1}{2}(1 + f) I_0.$$

The ratio of I_{max} to I_{min} is

$$\frac{I_{max}}{I_{min}} = \frac{1 + f}{1 - f}.$$

Set the ratio equal to 5.0 and solve for f. You should get $f = 0.67$.

59P

(a) The rotation cannot be done with a single sheet. If a sheet is placed with its polarizing direction at an angle of $90°$ to the direction of polarization of the incident radiation, no radiation is transmitted.

It can be done with two sheets. Place the first sheet with its polarizing direction at some angle θ, between 0 and $90°$, to the direction of polarization of the incident radiation. Place the second sheet with its polarizing direction at $90°$ to the polarization direction of the incident radiation. The transmitted radiation is then polarized at $90°$ to the incident polarization direction. The intensity is $I_0 \cos^2 \theta \cos^2(90° - \theta) = I_0 \cos^2 \theta \sin^2 \theta$, where I_0 is the incident radiation. If θ is not 0 or $90°$ the transmitted intensity is not zero.

(b) Consider n sheets, with the polarizing direction of the first sheet making an angle of $\theta = 90°/n$ with the direction of polarization of the incident radiation and with the polarizing direction of each successive sheet rotated $90°/n$ in the same direction from the polarizing direction of the previous sheet. The transmitted radiation is polarized with its direction of polarization making an angle of $90°$ with the direction of polarization of the incident radiation. The intensity is $I = I_0 \cos^{2n}(90°/n)$. You want the smallest integer value of n for which this is greater than $0.60 I_0$.

Start with $n = 2$ and calculate $\cos^{2n}(90°/n)$. If the result is greater than 0.60 you have obtained the solution. If it is less, increase n by 1 and try again. Repeat this process, increasing n by 1 each time, until you have a value for which $\cos^{2n}(90°/n)$ is greater than 0.60. The first one will be $n = 5$.

Chapter 39

2E

Use the law of refraction:

$$n_1 \sin \theta_1 = n_2 \sin \theta_2 \,.$$

Take medium 1 to be the vacuum, with $n_1 = 1$ and $\theta_1 = 32.0°$. Medium 2 is the glass, with $\theta_2 = 21.0°$. Solve for n_2:

$$n_2 = n_1 \frac{\sin \theta_1}{\sin \theta_2} = (1.00) \left(\frac{\sin 32.0°}{\sin 21.0°} \right) = 1.48 \,.$$

3E

Note that the normal to the refracting surface is vertical in the diagram. The angle of refraction is $\theta_2 = 90°$ and the angle of incidence is given by $\tan \theta_1 = w/h$, where h is the height of the tank and w is its width. Thus

$$\theta_1 = \tan^{-1} \left(\frac{w}{h} \right) = \tan^{-1} \left(\frac{1.10\,\text{m}}{0.850\,\text{m}} \right) = 52.31° \,.$$

The law of refraction yields

$$n_1 = n_2 \frac{\sin \theta_2}{\sin \theta_1} = (1.00) \left(\frac{\sin 90°}{\sin 52.31°} \right) = 1.26 \,,$$

where the index of refraction of air was taken to be unity.

5P

Consider a ray that grazes the top of the pole, as shown in the diagram to the right. Here $\theta_1 = 35°$, $\ell_1 = 0.50\,\text{m}$, and $\ell_2 = 1.50\,\text{m}$. The length of the shadow is $x + L$. x is given by $x = \ell_1 \tan \theta_1 = (0.50\,\text{m}) \tan 35° = 0.35\,\text{m}$. According to the law of refraction $n_2 \sin \theta_2 = n_1 \sin \theta_1$. Take $n_1 = 1$ and $n_2 = 1.33$ (from Table 39–1). Then

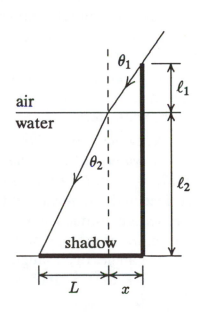

$$\theta_2 = \sin^{-1} \left(\frac{\sin \theta_1}{n_2} \right) = \sin^{-1} \left(\frac{\sin 35.0°}{1.33} \right) = 25.55° \,.$$

L is given by

$$L = \ell_2 \tan \theta_2 = (1.50\,\text{m}) \tan 25.55° = 0.72\,\text{m} \,.$$

The length of the shadow is $0.35\,\text{m} + 0.72\,\text{m} = 1.07\,\text{m}$.

Let θ be the angle of incidence and θ_2 be the angle of refraction at the left face of the plate. Let n be the index of refraction of the glass. Then the law of refraction yields $\sin\theta = n\sin\theta_2$. The angle of incidence at the right face is also θ_2. If θ_3 is the angle of emergence there, then $n\sin\theta_2 = \sin\theta_3$. Thus $\sin\theta_3 = \sin\theta$ and $\theta_3 = \theta$. The emerging ray is parallel to the incident ray.

You wish to derive an expression for x in terms of θ. If D is the length of the ray in the glass, then $D\cos\theta_2 = t$ and $D = t/\cos\theta_2$. The angle α in the diagram equals $\theta - \theta_2$ and $x = D\sin\alpha = D\sin(\theta-\theta_2)$. Thus

$$x = \frac{t\sin(\theta - \theta_2)}{\cos\theta_2}.$$

If all the angles θ, θ_2, θ_3, and $\theta - \theta_2$ are small and measured in radians then $\sin\theta \approx \theta$, $\sin\theta_2 \approx \theta_2$, $\sin(\theta - \theta_2) \approx \theta - \theta_2$, and $\cos\theta_2 \approx 1$. Thus $x \approx t(\theta - \theta_2)$. The law of refraction applied to the point of incidence at the left face of the plate is now $\theta \approx n\theta_2$, so $\theta_2 \approx \theta/n$ and

$$x \approx t\left(\theta - \frac{\theta}{n}\right) = \frac{(n-1)t\theta}{n}.$$

12P

(a) A ray diagram is shown to the right. The incident ray is normal to the water surface and so is not refracted. The angle of incidence at the first mirror is $\theta_1 = 45°$. According to the law of reflection the angle of reflection is also $45°$. This means the ray is horizontal after reflection and the angle of incidence at the second mirror is $\theta_2 = 45°$. Since the angle of reflection at the second mirror

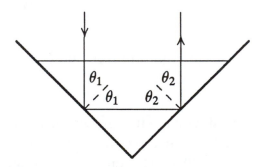

is also $45°$ the ray leaves that mirror normal to the water surface. Again there is no refraction at the water surface and the emerging ray is parallel to the incident ray.

(b) A ray diagram is shown below. On incidence the ray makes the angle θ_1 with the normal to the water surface. The angle of refraction θ_2 can be found using the law of refraction: $\sin\theta_1 = n\sin\theta_2$, where n is the index of refraction of the water. The normal to the water surface and the normal to the first mirror make an angle of $45°$. If the normal to the water surface is continued downward until it meets the normal to the first mirror the triangle formed has an interior angle of $180° - 45° = 135°$ at the vertex formed by the normal. Since the interior angles of a triangle must sum to $180°$, the angle of incidence at the first mirror satisfies $\theta_3 + \theta_2 + 135° = 180°$, so $\theta_3 = 45° - \theta_2$. The angle of reflection at the first mirror is also $45° - \theta_2$.

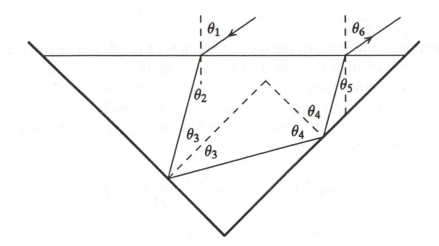

Now look at the triangle formed by the ray and the normals to the two mirrors. It is a right triangle, so $\theta_3 + \theta_4 + 90° = 180°$ and $\theta_4 = 90° - \theta_3 = 90° - 45° + \theta_2 = 45° + \theta_2$. The angle of reflection at the second mirror is also $45° + \theta_2$.

Now continue the normal to the water surface downward from the exit point of the ray to the second mirror. It makes an angle of $45°$ with the mirror. Consider the triangle formed by the second mirror, the ray, and the normal to the water surface. The angle at the intersection of the normal and the mirror is $180° - 45° = 135°$. The angle at the intersection of the ray and the mirror is $90° - \theta_4 = 90° - (45° + \theta_2) = 45° - \theta_2$. The angle at the intersection of the ray and the water surface is θ_5. These three angles must sum to $180°$, so $135° + 45° - \theta_2 + \theta_5 = 180°$. This means $\theta_5 = \theta_2$.

Now use the law of refraction to find θ_6: $\sin \theta_6 = n \sin \theta_5$. But since $\theta_5 = \theta_2$, this means $\sin \theta_6 = n \sin \theta_2$. Finally, since $\sin \theta_1 = n \sin \theta_2$, we conclude that $\sin \theta_6 = \sin \theta_1$ and $\theta_6 = \theta_1$. The exiting ray is parallel to the incident ray.

17E

Let θ_1 $(= 45°)$ be the angle of incidence at the first surface and θ_2 be the angle of refraction there. Let θ_3 be the angle of incidence at the second surface. The condition for total internal reflection at the second surface is $n \sin \theta_3 \geq 1$. You want to find the smallest value of the index of refraction n for which this inequality holds.

The law of refraction, applied to the first surface, yields $n \sin \theta_2 = \sin \theta_1$. Consideration of the triangle formed by the surface of the slab and the ray in the slab tells us that $\theta_3 = 90° - \theta_2$. Thus the condition for total internal reflection becomes $1 \leq n \sin(90° - \theta_2) = n \cos \theta_2$. Square this equation and use $\sin^2 \theta_2 + \cos^2 \theta_2 = 1$ to obtain $1 \leq n^2(1 - \sin^2 \theta_2)$. Now substitute $\sin \theta_2 = (1/n) \sin \theta_1$ to obtain

$$1 \leq n^2 \left(1 - \frac{\sin^2 \theta_1}{n^2} \right) = n^2 - \sin^2 \theta_1 .$$

The largest value of n for which this equation is true is the value for which $1 = n^2 - \sin^2 \theta_1$. Solve for n:

$$n = \sqrt{1 + \sin^2 \theta_1} = \sqrt{1 + \sin^2 45°} = 1.22 .$$

18E

(a) No refraction occurs at the surface ab, so the angle of incidence at surface ac is $90° - \phi$. For total internal reflection at the second surface, $n_g \sin(90° - \phi)$ must be greater than n_a. Here n_g is the index of refraction for the glass and n_a is the index of refraction for air. Since $\sin(90° - \phi) = \cos\phi$, you want the largest value of ϕ for which $n_g \cos\phi \geq n_a$. Recall that $\cos\phi$ decreases as ϕ increases from zero. When ϕ has the largest value for which total internal reflection occurs, then $n_g \cos\phi = n_a$, or

$$\phi = \cos^{-1}\left(\frac{n_a}{n_g}\right) = \cos^{-1}\left(\frac{1}{1.52}\right) = 48.9° .$$

The index of refraction for air was taken to be unity.

(b) Replace the air with water. If n_w ($= 1.33$) is the index of refraction for water, then the largest value of ϕ for which total internal reflection occurs is

$$\phi = \cos^{-1}\left(\frac{n_w}{n_g}\right) = \cos^{-1}\left(\frac{1.33}{1.52}\right) = 29.0° .$$

21P

(a) The diagram on the right shows a cross section, through the center of the cube and parallel to a face. L is the length of a cube edge and S labels the spot. A portion of a ray from the source to a cube face is also shown. Light leaving the source at a small angle θ is refracted at the face and leaves the cube; light leaving at a sufficiently large angle is totally reflected. The light that passes through the cube face forms a circle, the radius r being associated with the critical angle for total internal reflection. If θ_c is that angle then

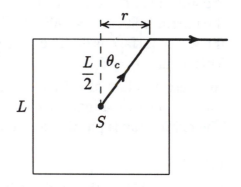

$$\sin\theta_c = \frac{1}{n} ,$$

where n is the index of refraction for the glass. As the diagram shows, the radius of the circle is given by $r = (L/2)\tan\theta_c$. Now

$$\tan\theta_c = \frac{\sin\theta_c}{\cos\theta_c} = \frac{\sin\theta_c}{\sqrt{1 - \sin^2\theta_c}} = \frac{1/n}{\sqrt{1 - (1/n)^2}} = \frac{1}{\sqrt{n^2 - 1}}$$

and the radius of the circle is

$$r = \frac{L}{2\sqrt{n^2 - 1}} = \frac{10\,\text{mm}}{2\sqrt{(1.5)^2 - 1}} = 4.47\,\text{mm} .$$

If an opaque circular disk with this radius is pasted at the center of each cube face the spot will not be seen (provided internally reflected light can be ignored).

(b) There must be six opaque disks, one for each face. The total area covered by disks is $6\pi r^2$ and the total surface area of the cube is $6L^2$. The fraction of the surface area that must covered by disks is

$$f = \frac{6\pi r^2}{6L^2} = \frac{\pi r^2}{L^2} = \frac{\pi(4.47\,\text{mm})^2}{(10\,\text{mm})^2} = 0.63\,.$$

23P

(a) A ray diagram is shown to the right. Let θ_1 be the angle of incidence and θ_2 be the angle of refraction at the first surface. Let θ_3 be the angle of incidence at the second surface. The angle of refraction there is $\theta_4 = 90°$. The law of refraction, applied to the second surface, yields $n \sin\theta_3 = \sin\theta_4 = 1$. As shown in the diagram, the normals to the surfaces at P and Q make an angle of 90° with each other. The interior angles of the triangle formed by the ray and the two normals

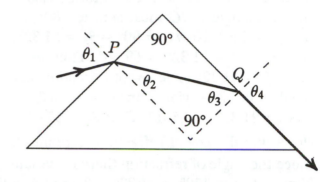

must sum to 180°, so $\theta_3 = 90° - \theta_2$ and $\sin\theta_3 = \sin(90° - \theta_2) = \cos\theta_2 = \sqrt{1 - \sin^2\theta_2}$.

According to the law of refraction, applied at Q, $n\sqrt{1 - \sin^2\theta_2} = 1$.

The law of refraction, applied to point P, yields $\sin\theta_1 = n\sin\theta_2$, so $\sin\theta_2 = (\sin\theta_1)/n$ and

$$n\sqrt{1 - \frac{\sin^2\theta_1}{n^2}} = 1\,.$$

Square both sides and solve for n. You should get

$$n = \sqrt{1 + \sin^2\theta_1}\,.$$

(b) The greatest possible value of $\sin^2\theta_1$ is 1, so the greatest possible value of n is $n_\text{max} = \sqrt{2} = 1.41$.

(c) For a given value of n, if the angle of incidence at the first surface is greater than θ_1, the angle of refraction there is greater than θ_2 and the angle of incidence at the second face is less than θ_3 ($= 90° - \theta_2$). That is, it is less than the critical angle for total internal reflection, so light leaves the second surface and emerges into the air.

(d) If the angle of incidence at the first surface is less than θ_1, the angle of refraction there is less than θ_2 and the angle of incidence at the second surface is greater than θ_3. This is greater than the critical angle for total internal reflection, so all the light is reflected at Q.

24P

(a) When the incident ray is at the minimum angle for which light exits the prism, the light exits along the second face. That is, the angle of refraction at the second face is 90° and the angle of incidence there is the critical angle for total internal reflection. Let θ_1 be the angle of incidence and θ_2 be the angle of refraction at the first face and let θ_3 be the angle of incidence at the second face. You want to solve for θ_1.

A ray diagram is shown to the right. The law of refraction, applied to point C, yields $n \sin \theta_3 = 1$, so $\sin \theta_3 = 1/n = 1/1.60 = 0.625$ and $\theta_3 = 38.68°$. The interior angles of the triangle ABC must sum to 180°, so $\alpha + \beta = 120°$. Now $\alpha = 90° - \theta_3 = 51.32°$, so $\beta = 120° - 51.32° = 69.68°$. Thus, $\theta_2 = 90° - \beta = 21.32°$. The law of refraction, applied to point A, yields $\sin \theta_1 = n \sin \theta_2 = 1.60 \sin 21.32° = 0.5817$. Thus $\theta_1 = 35.6°$.

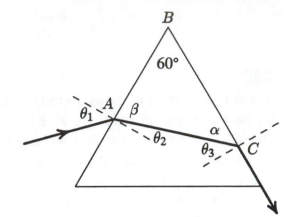

(b) Apply the law of refraction to point C.

Since the angle of refraction there is the same as the angle of incidence at A, $n \sin \theta_3 = \sin \theta_1$. Now $\alpha + \beta = 120°$, $\alpha = 90° - \theta_3$, and $\beta = 90° - \theta_2$, as before. This means $\theta_2 + \theta_3 = 60°$. Thus the law of refraction becomes $\sin \theta_1 = n \sin(60° - \theta_2)$, or $\sin \theta_1 = n \sin 60° \cos \theta_2 - n \cos 60° \sin \theta_2$, where the trigonometric identity $\sin(A - B) = \sin A \cos B - \cos A \sin B$ was used.

Now apply the law of refraction to point A: $\sin \theta_1 = n \sin \theta_2$. This means $\sin \theta_2 = (1/n) \sin \theta_1$ and $\cos \theta_2 = \sqrt{1 - \sin^2 \theta_2} = \sqrt{1 - (1/n^2) \sin^2 \theta_1}$. Thus

$$\sin \theta_1 = n \sin 60° \sqrt{1 - (1/n)^2 \sin^2 \theta_1} - \cos 60° \sin \theta_1$$

or

$$(1 + \cos 60°) \sin \theta_1 = \sin 60° \sqrt{n^2 - \sin^2 \theta_1} \, .$$

Square both sides and solve for $\sin \theta_1$. You should get

$$\sin \theta_1 = \frac{n \sin 60°}{\sqrt{(1 + \cos 60°)^2 + \sin^2 60°}} = \frac{1.60 \sin 60°}{\sqrt{(1 + \cos 60°)^2 + \sin^2 60°}} = 0.80$$

and $\theta_1 = 53.1°$.

25P

(a) The light that passes through the surface of the lake is within a cone of apex angle $2\theta_c$, where θ_c is the critical angle for total internal reflection. Imagine a sphere with the light source at its center and suppose that surface area A of the sphere is inside the cone. Since the intensity is the same at every point on the sphere, the fraction of the total energy emitted that passes through the surface is $frac = A/4\pi R^2$, where R is the radius of the sphere.

You now need to develop an expression for A in terms of the critical angle θ_c. Consider a ring of angular thickness $d\theta$ on the sphere surface. As you can see on the diagram, its radius is $r = R \sin\theta$, so its circumference is $2\pi R \sin\theta$. Its thickness is $R\,d\theta$, so its area is $dA = 2\pi R^2 \sin\theta\,d\theta$. Integrate from zero to θ_c to obtain

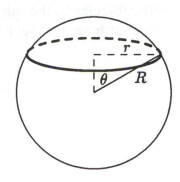

$$A = 2\pi R^2 \int_0^{\theta_c} \sin\theta\,d\theta = -2\pi R^2 \cos\theta \Big|_0^{\theta_c}$$
$$= 2\pi R^2 \left(1 - \cos\theta_c\right).$$

The critical angle is given by $\sin\theta_c = 1/n$, so $\cos\theta_c = \sqrt{1 - \sin^2\theta_c} = \sqrt{1 - 1/n^2}$ and

$$frac = \frac{2\pi R^2 \left(1 - \cos\theta_c\right)}{4\pi R^2} = \frac{1}{2}\left[1 - \sqrt{1 - \frac{1}{n^2}}\right].$$

(b) For $n = 1.33$,

$$frac = \frac{1}{2}\left[1 - \sqrt{1 - \frac{1}{(1.33)^2}}\right] = 0.170.$$

29E

The angle of incidence θ_B for which reflected light is fully polarized is given by Brewster's law, Eq. 39–4 of the text. If n_1 is the index of refraction for the medium of incidence and n_2 is the index of refraction for the second medium, then $\theta_B = \tan^{-1}(n_2/n_1) = \tan^{-1}(1.53/1.33) = 63.8°$.

34E

The image is 10 cm behind the mirror and you are 30 cm in front of the mirror. You must focus your eyes for a distance of 10 cm + 30 cm = 40 cm.

38E

(a) There are three images. Two are formed by single reflections from each of the mirrors and the third is formed by successive reflections from both mirrors.

(b) The positions of the images are shown on the two diagrams below. The diagram on the left shows the image I_1, formed by reflections from the left-hand mirror. It is the same distance behind the mirror as the object O is in front and is on the line that is perpendicular to the mirror and through the object. Image I_2 is formed by light that is reflected from both mirrors. You may consider I_2 to be the image of I_1 formed by the right-hand mirror, extended. I_2 is the same distance behind the line of the right-hand mirror as I_1 is in front and it is on the line that is perpendicular to the line of the mirror.

The diagram on the right shows image I_3, formed by reflections from the right-hand mirror. It is the same distance behind the mirror as the object is in front and is on the line that is perpendicular to the mirror and through the object. As the diagram shows, light that is first reflected from the right-hand mirror and then from the left-hand mirror forms an image at I_2.

41P

Consider a single ray from the source to the mirror and let θ be the angle of incidence. The angle of reflection is also θ and the reflected ray makes an angle of 2θ with the incident ray. Now rotate the mirror through the angle α so the angle of incidence increases to $\theta + \alpha$. The reflected ray now makes an angle of $2(\theta + \alpha)$ with the incident ray. The reflected ray has been rotated through an angle of 2α. If the mirror is rotated so the angle of incidence is decreased, then the reflected ray makes an angle of $2(\theta - \alpha)$ with the incident ray. Again it has been rotated through 2α. The diagrams show the situation for $\alpha = 45°$. The ray from the object to the mirror is the same in both cases and the reflected rays are 90° apart.

43P

The intensity of light from a point source varies as the inverse of the square of the distance from the source. Before the mirror is in place, the intensity at the center of the screen is given by $I_0 = A/d^2$, where A is a constant of proportionality. After the mirror is in place, the light that goes directly to the screen contributes intensity I_0, as before. Reflected light also reaches the screen. This light appears to come from the image of the source, a distance d behind the mirror and a distance $3d$ from the screen. Its contribution to the intensity at the center of the screen is

$$I_r = \frac{A}{(3d)^2} = \frac{A}{9d^2} = \frac{I_0}{9}.$$

The total intensity at the center of the screen is

$$I = I_0 + I_r = I_0 + \frac{I_0}{9} = \frac{10}{9}I_0.$$

The ratio of the new intensity to the original intensity is $I/I_0 = 10/9$.

44P

You may think of the final image as the image formed by the bottom mirror of the image in the top mirror. Consider an object a distance ℓ in front of a plane mirror that is tilted at 45° to the horizontal, as shown. The image is on the normal to the mirror, a distance ℓ behind it. It is on the vertical line shown. If L is the horizontal distance from the object to the mirror then the vertical distance from the mirror to the image is also L. If the object is extended its image has the same orientation and size.

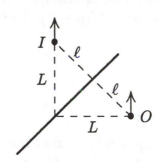

(a) The final image is behind the bottom mirror. No light passes through it, so it is virtual.

(b) The image in a plane mirror of an upright object is itself upright. The image in the top mirror is upright and this becomes an upright object for the bottom mirror. The final image is therefore upright.

(c) The image in a plane mirror is the same size as the object. The image in the top mirror is the same size as the penguin and the image in the bottom mirror is the same size as the first image and therefore the same size as the penguin. The magnification is $+1$.

(d) The image formed by the top mirror is a distance D directly above it. This image, which is the object for the bottom mirror, is therefore a distance $L + D$ from the bottom mirror. The final image is directly in front of the bottom mirror, a distance $L + D$ from it.

49P

(a) Suppose one end of the object is a distance p from the mirror and the other end is a distance $p + L$. The position i_1 of the image of the first end is given by

$$\frac{1}{p} + \frac{1}{i_1} = \frac{1}{f},$$

where f is the focal length of the mirror. Thus

$$i_1 = \frac{fp}{p - f}.$$

The image of the other end is at

$$i_2 = \frac{f(p + L)}{p + L - f},$$

so the length of the image is

$$L' = i_1 - i_2 = \frac{fp}{p - f} - \frac{f(p + L)}{p + L - f} = \frac{f^2 L}{(p - f)(p + L - f)}.$$

Since the object is short compared to $p - f$ we may neglect the L in the denominator and write

$$L' = L \left(\frac{f}{p - f} \right)^2.$$

(b) The lateral magnification is $m = -i/p$ and since $i = fp/(p - f)$, this can be written $m = -f/(p - f)$. The longitudinal magnification is

$$m' = \frac{L'}{L} = \left(\frac{f}{p - f}\right)^2 = m^2.$$

53P

Light refracted at the left surface of the sphere passes through and is refracted again at the right surface. Apply Eq. 39–14 twice, using the image formed by the first surface as the object for the second surface. For refraction at the first surface the object distance is $p_1 = \infty$ and the radius of curvature is $r = R$, where R is the radius of the sphere. The surface is convex to the incoming light so the radius of curvature is positive. The medium of incidence is air (or vacuum) and the medium of refraction is glass, so $n_1 = 1$ and $n_2 = n$, where n is the index of refraction of the glass. If i_1 is the image distance for the image formed by the left surface, then

$$\frac{n}{i_1} = \frac{n - 1}{R},$$

so

$$i_1 = \frac{nR}{n - 1}.$$

The image distance is measured from the left edge of the sphere. Since n is less than 2, i_1 is greater than $2R$ and the image is formed to the right of the sphere, a distance $i_1 - 2R$ from its right edge.

The object distance for refraction at the right surface is

$$p_2 = 2R - i_1 = 2R - \frac{nR}{n - 1} = \frac{n - 2}{n - 1}R.$$

This is a negative number, indicating that the light is converging as it strikes the right surface of the sphere. Now the medium of incidence is glass and the medium of the refracted light is air, so $n_1 = n$ and $n_2 = 1$. The surface is concave to the incoming light so the radius of curvature is $r = -R$. Eq. 39–14 yields

$$\frac{n(n - 1)}{(n - 2)R} + \frac{1}{i_2} = \frac{n - 1}{R},$$

where i_2 is the image distance for the second refraction. Thus

$$\frac{1}{i_2} = \frac{n - 1}{R} - \frac{n(n - 1)}{(n - 2)R} = \frac{2(n - 1)}{(2 - n)R}.$$

The image distance is

$$i_2 = \frac{2 - n}{2(n - 1)}R.$$

This is positive, indicating that the final image is to the right of the sphere.

Solve Eq. 39–15 for the image distance i: $i = pf/(p - f)$. The lens is diverging, so its focal length is $f = -30$ cm. The object distance is $p = 20$ cm. Thus

$$i = \frac{(20\,\text{cm})(-30\,\text{cm})}{(20\,\text{cm}) - (-30\,\text{cm})} = -12\,\text{cm}.$$

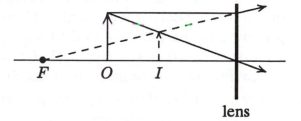

The negative sign indicates that the image is virtual and is on the same side of the lens as the object. The ray diagram, drawn to scale, is shown on the right.

lens

57E

Use the lens maker's equation, Eq. 39–16:

$$\frac{1}{f} = (n - 1)\left(\frac{1}{r_1} - \frac{1}{r_2}\right),$$

where f is the focal length, n is the index of refraction, r_1 is the radius of curvature of the first surface encountered by the light and r_2 is the radius of curvature of the second surface. Since one surface has twice the radius of the other and since one surface is convex to the incoming light while the other is concave, set $r_2 = -2r_1$ to obtain

$$\frac{1}{f} = (n - 1)\left(\frac{1}{r_1} + \frac{1}{2r_1}\right) = \frac{3(n - 1)}{2r_1}.$$

Solve for r_1:

$$r_1 = \frac{3(n - 1)f}{2} = \frac{3(1.5 - 1)(60\,\text{mm})}{2} = 45\,\text{mm}.$$

The radii are 45 mm and 90 mm.

64P

For a thin lens $(1/p) + (1/i) = (1/f)$, where p is the object distance, i is the image distance, and f is the focal length. Solve for i:

$$i = \frac{fp}{p - f}.$$

Let $p = f + x$, where x is positive if the object is outside the focal point and negative if it is inside. Then

$$i = \frac{f(f + x)}{x}.$$

Now let $i = f + x'$, where x' is positive if the image is outside the focal point and negative if it is inside. Then

$$x' = i - f = \frac{f(f + x)}{x} - f = \frac{f^2}{x}$$

and $xx' = f^2$.

71P

Place an object far away from the composite lens and find the image distance i. Since the image is at a focal point $i = f$, the effective focal length of the composite. The final image is produced by two lenses, with the image of the first lens being the object for the second. For the first lens $(1/p_1) + (1/i_1) = (1/f_1)$, where f_1 is the focal length of this lens and i_1 is the image distance for the image it forms. Since $p_1 = \infty$, $i_1 = f_1$.

The thin lens equation, applied to the second lens, is $(1/p_2) + (1/i_2) = (1/f_2)$, where p_2 is the object distance, i_2 is the image distance, and f_2 is the focal length. If the thicknesses of the lenses can be ignored the object distance for the second lens is $p_2 = -i_1$. The negative sign must be used since the image formed by the first lens is beyond the second lens if i_1 is positive. This means the object for the second lens is virtual and the object distance is negative. If i_1 is negative the image formed by the first lens is in front of the second lens and p_2 is positive. In the thin lens equation, replace p_2 with $-f_1$ and i_2 with f to obtain

$$-\frac{1}{f_1} + \frac{1}{f} = \frac{1}{f_2}$$

or

$$\frac{1}{f} = \frac{1}{f_1} + \frac{1}{f_2} = \frac{f_1 + f_2}{f_1 f_2}.$$

Thus

$$f = \frac{f_1 f_2}{f_1 + f_2}.$$

74P

For an object in front of a thin lens, the object distance p and the image distance i are related by $(1/p) + (1/i) = (1/f)$, where f is the focal length of the lens. For the situation described by the problem all quantities are positive, so the distance x between the object and image is $x = p + i$. Substitute $i = x - p$ into the thin lens equation and solve for x. You should get

$$x = \frac{p^2}{p - f}.$$

To find the minimum value of x, set $dx/dp = 0$ and solve for p. Since

$$\frac{dx}{dp} = \frac{p(p - 2f)}{(p - f)^2},$$

the result is $p = 2f$. The minimum distance is

$$x_{min} = \frac{p^2}{p - f} = \frac{(2f)^2}{2f - f} = 4f.$$

This is a minimum, rather than a maximum, since the image distance i becomes large without bound as the object approaches the focal point.

75P

(a) If the object distance is x, then the image distance is $D - x$ and the thin lens equation becomes

$$\frac{1}{x} + \frac{1}{D-x} = \frac{1}{f}.$$

Multiply each term in the equation by $fx(D-x)$ to obtain $x^2 - Dx + Df = 0$. Solve for x. The two object distances for which images are formed on the screen are

$$x_1 = \frac{D - \sqrt{D(D-4f)}}{2}$$

and

$$x_2 = \frac{D + \sqrt{D(D-4f)}}{2}.$$

The distance between the two object positions is

$$d = x_2 - x_1 = \sqrt{D(D-4f)}.$$

(b) The ratio of the image sizes is the same as the ratio of the lateral magnifications. If the object is at $p = x_1$ the lateral magnification is

$$m_1 = \frac{i_1}{p_1} = \frac{D - x_1}{x_1}.$$

Now $x_1 = \frac{1}{2}(D - d)$, where $d = \sqrt{D(D-f)}$, so

$$m_1 = \frac{D - (D-d)/2}{(D-d)/2} = \frac{D+d}{D-d}.$$

Similarly, when the object is at x_2 the magnification is

$$m_2 = \frac{I_2}{p_2} = \frac{D - x_2}{x_2} = \frac{D - (D+d)/2}{(D+d)/2} = \frac{D-d}{D+d}.$$

The ratio of the magnifications is

$$\frac{m_2}{m_1} = \frac{(D-d)/(D+d)}{(D+d)/(D-d)} = \left(\frac{D-d}{D+d}\right)^2.$$

76E

(a) If L is the distance between the lenses, then according to Fig. 39–26, the tube length is $s = L - f_{ob} - f_{ey} = 25.0\,\text{cm} - 4.00\,\text{cm} - 8.00\,\text{cm} = 13.0\,\text{cm}$.

(b) Solve $(1/p) + (1/i) = (1/f_{ob})$ for p. The image distance is $i = f_{ob} + s = 4.00\,\text{cm} + 13.0\,\text{cm} = 17.0\,\text{cm}$, so

$$p = \frac{if}{i-f} = \frac{(17.0\,\text{cm})(4.00\,\text{cm})}{17.0\,\text{cm} - 4.00\,\text{cm}} = 5.23\,\text{cm}.$$

(c) The magnification of the objective is

$$m = -\frac{i}{p} = -\frac{17.0\,\text{cm}}{5.23\,\text{cm}} = -3.25\,.$$

(d) The angular magnification of the eyepiece is

$$m_\theta = \frac{15\,\text{cm}}{f_{ey}} = \frac{15\,\text{cm}}{8.00\,\text{cm}} = 1.88\,.$$

(e) The overall magnification of the microscope is

$$M = mm_\theta = (-3.25)(1.88) = -6.09\,.$$

82P

(a) As far as magnification is concerned the mirror M has exactly the same effect as a lens with the same focal length, at the same position. The only difference is that the image produced by the mirror is in front while the image produced by the lens is behind. Thus Eq. 39–21 is directly applicable and $m_\theta = -f_{ob}/f_{ey}$.

(b) Solve the mirror equation $(1/p) + (1/i) = (1/f)$ for i. Here the object distance is $p = 2.0 \times 10^3\,\text{m}$ and the focal length is $f = 16.8\,\text{m}$. The result is

$$i = \frac{pf}{p-f} = \frac{(2.0 \times 10^3\,\text{m})(16.8\,\text{m})}{2.0 \times 10^3\,\text{m} - 16.8\,\text{m}} = 16.94\,\text{m}\,.$$

The magnification is

$$m = -\frac{i}{p} = -\frac{16.94\,\text{m}}{2.0 \times 10^3\,\text{m}} = -8.47 \times 10^{-3}\,.$$

Since the length of the object is $\ell = 1.0\,\text{m}$, the length of its image is

$$\ell' = |m|\ell = (8.47 \times 10^{-3})(1.0\,\text{m}) = 8.47 \times 10^{-3}\,\text{m}\,.$$

(c) Use $m_\theta = -f_{ob}/f_{ey}$. If R is the radius of curvature of the objective mirror, its focal length is $f_{ob} = R/2 = 5.0\,\text{m}$. Take the angular magnification to be -200. Then

$$f_{ey} = -\frac{f_{ob}}{m_\theta} = \frac{5.0\,\text{m}}{200} = 0.025\,\text{m} = 2.5\,\text{cm}\,.$$

Chapter 40

11P

(*a*) Take the phases of both waves to be zero at the front surfaces of the layers. The phase of the first wave at the back surface of the glass is given by $\phi_1 = k_1 L - \omega t$, where $k_1 (= 2\pi/\lambda_1)$ is the angular wave number and λ_1 is the wavelength in glass. Similarly, the phase of the second wave at the back surface of the plastic is given by $\phi_2 = k_2 L - \omega t$, where $k_2 (= 2\pi/\lambda_2)$ is the angular wave number and λ_2 is the wavelength in plastic. The angular frequencies are the same since the waves have the same wavelength in air and the frequency of a wave does not change when the wave enters another medium. The phase difference is

$$\phi_1 - \phi_2 = (k_1 - k_2)L = 2\pi \left(\frac{1}{\lambda_1} - \frac{1}{\lambda_2}\right) L.$$

Now $\lambda_1 = \lambda_{air}/n_1$, where λ_{air} is the wavelength in air and n_1 is the index of refraction of the glass. Similarly $\lambda_2 = \lambda_{air}/n_2$, where n_2 is the index of refraction of the plastic. This means that the phase difference is $\phi_1 - \phi_2 = (2\pi/\lambda_{air})(n_1 - n_2)L$. The value of L that makes this 5.65 rad is

$$L = \frac{(\phi_1 - \phi_2)\lambda_{air}}{2\pi(n_1 - n_2)} = \frac{5.65(400 \times 10^{-9}\,\text{m})}{2\pi(1.60 - 1.50)} = 3.60 \times 10^{-6}\,\text{m}.$$

(*b*) 5.65 rad is less than 2π rad ($= 6.28$ rad), the phase difference for completely constructive interference, and greater than π rad ($= 3.14$ rad), the phase difference for completely destructive interference. The interference is therefore intermediate, neither completely constructive nor completely destructive. It is, however, closer to completely constructive than to completely destructive.

18E

The condition for a maximum in the two-slit interference pattern is $d\sin\theta = m\lambda$, where d is the slit separation, λ is the wavelength, m is an integer, and θ is the angle made by the interfering rays with the forward direction. If θ is small, $\sin\theta$ may be approximated by θ in radians. Then $d\theta = m\lambda$ and the angular separation of adjacent maxima, one associated with the integer m and the other associated with the integer $m + 1$, is given by $\Delta\theta = \lambda/d$. The separation on a screen a distance D away is given by $\Delta y = D\,\Delta\theta = \lambda D/d$. Thus

$$\Delta y = \frac{(500 \times 10^{-9}\,\text{m})(5.40\,\text{m})}{1.20 \times 10^{-3}\,\text{m}} = 2.25 \times 10^{-3}\,\text{m} = 2.25\,\text{mm}.$$

20E

The angular positions of the maxima of a two-slit interference pattern are given by $d\sin\theta = m\lambda$, where d is the slit separation, λ is the wavelength, and m is an integer. If θ is small,

$\sin\theta$ may be approximated by θ in radians. Then $d\theta = m\lambda$. The angular separation of two adjacent maxima is $\Delta\theta = \lambda/d$. Let λ' be the wavelength for which the angular separation is 10.0% greater. Then $1.10\lambda/d = \lambda'/d$ or $\lambda' = 1.10\lambda = 1.10(589\,\text{nm}) = 648\,\text{nm}$.

24P

The maxima of a two-slit interference pattern are at angles θ given by $d\sin\theta = m\lambda$, where d is the slit separation, λ is the wavelength, and m is an integer. If θ is small, $\sin\theta$ may be replaced by θ in radians. Then $d\theta = m\lambda$. The angular separation of two maxima associated with different wavelengths but the same value of m is $\Delta\theta = (m/d)(\lambda_2 - \lambda_1)$ and the separation on a screen a distance D away is

$$\Delta y = D\tan\Delta\theta \approx D\,\Delta\theta = \left[\frac{mD}{d}\right](\lambda_2 - \lambda_1)$$

$$= \left[\frac{3(1.0\,\text{m})}{5.0 \times 10^{-3}\,\text{m}}\right](600 \times 10^{-9}\,\text{m} - 480 \times 10^{-9}\,\text{m}) = 7.2 \times 10^{-5}\,\text{m}.$$

The small angle approximation $\tan\Delta\theta \approx \Delta\theta$ was made. $\Delta\theta$ must be in radians.

28P

Consider the two waves, one from each slit, that produce the seventh bright fringe in the absence of the mica. They are in phase at the slits and travel different distances to the seventh bright fringe, where they are out of phase by $2\pi m = 14\pi$. Now a piece of mica with thickness x is placed in front of one of the slits and the waves are no longer in phase at the slits. In fact, their phases at the slits differ by

$$\frac{2\pi x}{\lambda_m} - \frac{2\pi x}{\lambda} = \frac{2\pi x}{\lambda}(n-1),$$

where λ_m is the wavelength in the mica and n is the index of refraction of the mica. The relationship $\lambda_m = \lambda/n$ was used to substitute for λ_m. Since the waves are now in phase at the screen

$$\frac{2\pi x}{\lambda}(n-1) = 14\pi$$

or

$$x = \frac{7\lambda}{n-1} = \frac{7(550 \times 10^{-9}\,\text{m})}{1.58 - 1} = 6.64 \times 10^{-6}\,\text{m}.$$

30P

Interference maxima occur at angles θ such that $d\sin\theta = m\lambda$, where d is the separation of the sources, λ is the wavelength, and m is an integer. Since $d = 2.0\,\text{m}$ and $\lambda = 0.50\,\text{m}$, this means that $\sin\theta = 0.25m$. You want all values of m (positive and negative) for which $|0.25m| \leq 1$. These are $-4, -3, -2, -1, 0, +1, +2, +3$, and $+4$. For each of these except -4 and $+4$ there are 2 different values for θ. A single value of θ ($-90°$) is associated with $m = -4$ and a single value ($-90°$) is associated with $m = +4$. There are 16 different angles in all and therefore 16 maxima.

39P

The phasor diagram is shown to the right. Here $E_1 = 1.00$, $E_2 = 2.00$, and $\phi = 60°$. The resultant amplitude E_m is given by the trigonometric law of cosines:

$$E_m^2 = E_1^2 + E_2^2 - 2E_1E_2 \cos(180° - \phi),$$

so

$$E_m = \sqrt{(1.00)^2 + (2.00)^2 - 2(1.00)(2.00)\cos 120°} = 2.65.$$

40P

(a) To get to the detector the wave from A travels a distance x and the wave from B travels a distance $\sqrt{d^2 + x^2}$. The difference in phase of the two waves is

$$\Delta\phi = \frac{2\pi}{\lambda}\left[\sqrt{d^2 + x^2} - x\right],$$

where λ is the wavelength. For a maximum in intensity this must be a multiple of 2π. Solve

$$\sqrt{d^2 + x^2} - x = m\lambda$$

for x. Here m is an integer. Write the equation as $\sqrt{d^2 + x^2} = x + m\lambda$, then square both sides to obtain $d^2 + x^2 = x^2 + m^2\lambda^2 + 2m\lambda x$. The solution is

$$x = \frac{d^2 - m^2\lambda^2}{2m\lambda}.$$

The largest value of m that produces a positive value for x is $m = 3$. This corresponds to the maximum that is nearest A, at

$$x = \frac{(4.00\,\text{m})^2 - 9(1.00\,\text{m})^2}{(2)(3)(1.00\,\text{m})} = 1.17\,\text{m}.$$

For the next maximum $m = 2$ and $x = 3.00\,\text{m}$. For the third maximum $m = 1$ and $x = 7.50\,\text{m}$.

(b) Minima in intensity occur where the phase difference is π rad; the intensity at a minimum, however, is not zero because the amplitudes of the waves are different. Although the amplitudes are the same at the sources, the waves travel different distances to get to the points of minimum intensity and each amplitude decreases in inverse proportion to the distance traveled.

42P

According to Eqs. 40–18 and 40–19, the intensity is given by

$$I = 4I_0 \cos^2(\phi/2),$$

where

$$\phi = \left(\frac{2\pi d}{\lambda}\right)\sin\theta.$$

Here d is the slit separation and λ is the wavelength. The intensity at the center of the interference pattern is $4I_0$, so you want the value of θ for which $I = 2I_0$. First solve $2I_0 = 4I_0\cos^2(\frac{1}{2}\phi)$ for ϕ:

$$\phi = 2\cos^{-1}\left(\frac{1}{\sqrt{2}}\right) = \frac{\pi}{2}\,\text{rad}.$$

Now solve

$$\frac{\pi}{2} = \left(\frac{2\pi d}{\lambda}\right)\sin\theta$$

for θ. Since θ is small, $\sin\theta \approx \theta$, provided θ is measured in radians. Then

$$\frac{\pi}{2} = \frac{2\pi d\theta}{\lambda}$$

and

$$\theta = \frac{\lambda}{4d}.$$

Another point of half intensity, at $\theta = -\lambda/4d$, is symmetrically placed relative to the central point of the pattern, so the width of the pattern at half intensity is $2(\lambda/4d) = \lambda/2d$.

43P*

Take the electric field of one wave, at the screen, to be

$$E_1 = E_0\sin(\omega t)$$

and the electric field of the other to be

$$E_2 = 2E_0\sin(\omega t + \phi),$$

where the phase difference is given by

$$\phi = \left(\frac{2\pi d}{\lambda}\right)\sin\theta.$$

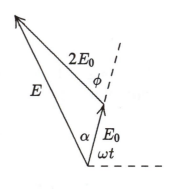

Here d is the center-to-center slit separation and λ is the wavelength. The resultant wave can be written $E = E_1 + E_2 = E\sin(\omega t + \alpha)$, where α is a phase constant. The phasor diagram is shown above.

The resultant amplitude E is given by the trigonometric law of cosines:

$$E^2 = E_0^2 + (2E_0)^2 - 4E_0^2\cos(180° - \phi) = E_0^2(5 + 4\cos\phi).$$

The intensity is given by $I = I_0(5 + 4\cos\phi)$, where I_0 is the intensity that would be produced by the first wave if the second were not present. Since $\cos\phi = 2\cos^2(\phi/2) - 1$, this may also be written $I = I_0\left[1 + 8\cos^2(\phi/2)\right]$.

47E

The wave reflected from the front surface suffers a phase change of π rad since it is incident in air on a medium of higher index of refraction. The phase of the wave reflected from the back surface does not change on reflection since the medium beyond the soap film is air and has a lower index of refraction than the film. If L is the thickness of the film this wave travels a distance $2L$ further than the wave reflected from the front surface. The phase difference of the two waves is $2L(2\pi/\lambda_f) + \pi$, where λ_f is the wavelength in the film. If λ is the wavelength in vacuum and n is the index of refraction of the soap film, then $\lambda_f = \lambda/n$ and the phase difference is

$$2nL\left(\frac{2\pi}{\lambda}\right) + \pi = 2(1.33)(1.21 \times 10^{-6}\,\text{m})\left(\frac{2\pi}{585 \times 10^{-9}\,\text{m}}\right) + \pi = 12\pi\,\text{rad}.$$

Since the phase difference is an even multiple of π the interference is completely constructive.

52E

Light reflected from the front surface of the coating suffers a phase change of π rad while light reflected from the back surface does not change phase. If L is the thickness of the coating, light reflected from the back surface travels a distance $2L$ further than light reflected from the front surface. The difference in phase of the two waves is $2L(2\pi/\lambda_c) - \pi$, where λ_c is the wavelength in the coating. If λ is the wavelength in vacuum, then $\lambda_c = \lambda/n$, where n is the index of refraction of the coating. Thus the phase difference is $2nL(2\pi/\lambda) - \pi$. For fully constructive interference this should be a multiple of 2π. Solve

$$2nL\left(\frac{2\pi}{\lambda}\right) - \pi = 2m\pi$$

for L. Here m is an integer. This equation is a slightly rearranged version of Eq. 40–27. The solution is

$$L = \frac{(2m + 1)\lambda}{4n}.$$

To find the smallest coating thickness, take $m = 0$. Then

$$L = \frac{\lambda}{4n} = \frac{560 \times 10^{-9}\,\text{m}}{4(2.00)} = 7.00 \times 10^{-8}\,\text{m}.$$

53E

For complete destructive interference, you want the waves reflected from the front and back of the coating to differ in phase by an odd multiple of π rad. Each wave is incident on a medium of higher index of refraction from a medium of lower index, so both suffer phase changes of π rad on reflection. If L is the thickness of the coating, the wave reflected from the back surface travels a distance $2L$ further than the wave reflected from the front. The phase difference is $2L(2\pi/\lambda_c)$, where λ_c is the wavelength in the coating. If n is the index of refraction of the coating, $\lambda_c = \lambda/n$, where λ is the wavelength in vacuum, and the phase difference is $2nL(2\pi/\lambda)$. Solve

$$2nL\left(\frac{2\pi}{\lambda}\right) = (2m+1)\pi$$

for L. Here m is an integer. The result is

$$L = \frac{(2m+1)\lambda}{4n}.$$

To find the least thickness for which destructive interference occurs, take $m = 0$. Then

$$L = \frac{\lambda}{4n} = \frac{600 \times 10^{-9}\,\text{m}}{4(1.25)} = 1.2 \times 10^{-7}\,\text{m}.$$

60P

Light reflected from the upper oil surface (in contact with air) changes phase by π rad. Light reflected from the lower surface (in contact with glass) changes phase by π rad if the index of refraction of the oil is less than that of the glass and does not change phase if the index of refraction of the oil is greater than that of the glass.

First suppose the index of refraction of the oil is greater than the index of refraction of the glass. The condition for fully destructive interference is $2n_o d = m\lambda$, where d is the thickness of the oil film, n_o is the index of refraction of the oil, λ is the wavelength in vacuum, and m is an integer. For the shorter wavelength $2n_o d = m_1 \lambda_1$ and for the longer $2n_o d = m_2 \lambda_2$. Since λ_1 is less than λ_2, m_1 is greater than m_2 and since fully destructive interference does not occur for any wavelengths between, $m_1 = m_2 + 1$. Solve $(m_2 + 1)\lambda_1 = m_2 \lambda_2$ for m_2. The result is

$$m_2 = \frac{\lambda_1}{\lambda_2 - \lambda_1} = \frac{500\,\text{nm}}{700\,\text{nm} - 500\,\text{nm}} = 2.50.$$

Since m_2 must be an integer the oil cannot have an index of refraction that is greater than that of the glass.

Now suppose the index of refraction of the oil is less than that of the glass. The condition for fully destructive interference is then $2n_o d = (2m+1)\lambda$. For the shorter wavelength $2m_o d = (2m_1 + 1)\lambda_1$ and for the longer $2n_o d = (2m_2 + 1)\lambda_2$. Again $m_1 = m_2 + 1$, so $(2m_2 + 3)\lambda_1 = (2m_2 + 1)\lambda_2$. This means the value of m_2 is

$$m_2 = \frac{3\lambda_1 - \lambda_2}{2(\lambda_2 - \lambda_1)} = \frac{3(500\,\text{nm}) - 700\,\text{nm}}{2(700\,\text{nm} - 500\,\text{nm})} = 2.00.$$

This is an integer. Thus the index of refraction of the oil is less than that of the glass.

62P

Since the index of refraction of water is greater than that of the oil, which in turn is greater than that of air, both the wave reflected from the upper surface of the oil drop and the wave reflected from the lower surface undergo phase changes of π rad on reflection. At a place where the thickness of the drop is L the condition for a bright fringe is $2nL = m\lambda$ and the condition for a dark fringe is $2nL = (m + \frac{1}{2})\lambda$, where λ is the wavelength in vacuum, n is the index of refraction of the oil, and m is an integer. Near the rim of the drop, where $L < \lambda/4$, only the condition for a bright fringe (with $m = 0$) can be met, so the rim is bright.

(b) Take λ to be 475 nm and m to be 3. Then

$$L = \frac{3\lambda}{2n} = \frac{3(475\,\text{nm})}{2(1.20)} = 594\,\text{nm}.$$

(c) At places where the drop is thick, fractionally small variations in thickness from place to place may amount to variations of many wavelengths. The fringes are quite close together and cannot be distinguished by an unaided eye.

65P

Consider the interference of waves reflected from the top and bottom surfaces of the air film. The wave reflected from the upper surface does not change phase on reflection but the wave reflected from the bottom surface changes phase by π rad. At a place where the thickness of the air film is L the condition for fully constructive interference is $2L = (m + \frac{1}{2})\lambda$, where λ ($= 680$ nm) is the wavelength and m is an integer. The largest value of m for which L is less than $48.0\,\mu$m is 140. Note that for $m = 140$,

$$L = \frac{(m + \frac{1}{2})\lambda}{2} = \frac{(140.5)(680 \times 10^{-9}\,\text{m})}{2} = 4.78 \times 10^{-5}\,\text{m} = 47.8\,\mu\text{m}$$

and for $m = 141$,

$$L = \frac{(141.5)(680 \times 10^{-9}\,\text{m})}{2} = 4.81 \times 10^{-5}\,\text{m} = 48.1\,\mu\text{m}.$$

At the thin end of the air film there is a bright fringe associated with $m = 0$. There are therefore 141 bright fringes in all.

69P

Assume the wedge-shaped film is in air, so the wave reflected from one surface undergoes a phase change of π rad while the wave reflected from the other surface does not. At a place where the film thickness is L the condition for fully constructive interference is $2nL = (m + \frac{1}{2})\lambda$, where n is the index of refraction of the film, λ is the wavelength in vacuum, and m is an integer. The ends of the film are bright. Suppose the end where the film is narrow has thickness L_1 and the bright fringe there corresponds to $m = m_1$. Suppose the end where the film is thick has thickness L_2 and the bright fringe there corresponds to $m = m_2$. Since there are 10 bright fringes $m_2 = m_1 + 9$. Subtract $2nL_1 = (m_1 + \frac{1}{2})\lambda$ from $2nL_2 = (m_1 + 9 + \frac{1}{2})\lambda$ to obtain $2n\,\Delta L = 9\lambda$, where $\Delta L = L_2 - L_1$ is the change in the film thickness over its length. Thus

$$\Delta L = \frac{9\lambda}{2n} = \frac{9(630 \times 10^{-9}\,\text{m})}{2(1.50)} = 1.89 \times 10^{-6}\,\text{m}.$$

72P

Consider the interference pattern formed by waves reflected from the upper and lower surfaces of the air wedge. The wave reflected from the lower surface undergoes a π rad phase change while the wave reflected from the upper surface does not. At a place where the thickness of the wedge is d the condition for a maximum in intensity is $2d = (m + \frac{1}{2})\lambda$, where λ is the wavelength in air and m is an integer. Thus $d = (2m + 1)\lambda/4$. As the geometry of Fig. 40–31 shows, $d = R - \sqrt{R^2 - r^2}$, where R is the radius of curvature of the lens and r is the radius of a Newton's ring. Thus $(2m + 1)\lambda/4 = R - \sqrt{R^2 - r^2}$. Solve for r. First rearrange the terms so the equation becomes

$$\sqrt{R^2 - r^2} = R - \frac{(2m + 1)\lambda}{4}.$$

Now square both sides and solve for r^2. When you take the square root you should get

$$r = \sqrt{\frac{(2m + 1)R\lambda}{2} - \frac{(2m + 1)^2\lambda^2}{16}}.$$

If R is much larger than a wavelength the first term dominates the second and

$$r = \sqrt{\frac{(2m + 1)R\lambda}{2}}.$$

78E

A shift of one fringe corresponds to a change in the optical path length of one wavelength. When the mirror moves a distance d the path length changes by $2d$ since the light traverses the mirror arm twice. Let N be the number of fringes shifted. Then $2d = N\lambda$ and

$$\lambda = \frac{2d}{N} = \frac{2(0.233 \times 10^{-3}\,\text{m})}{792} = 5.88 \times 10^{-7}\,\text{m} = 588\,\text{nm}.$$

80P

Let ϕ_1 be the phase difference of the waves in the two arms when the tube has air in it and let ϕ_2 be the phase difference when the tube is evacuated. These are different because the wavelength in air is different from the wavelength in vacuum. If λ is the wavelength in vacuum then the wavelength in air is λ/n, where n is the index of refraction of air. This means

$$\phi_1 - \phi_2 = 2L\left[\frac{2\pi n}{\lambda} - \frac{2\pi}{\lambda}\right] = \frac{4\pi(n - 1)L}{\lambda},$$

where L is the length of the tube. The factor 2 arises because the light traverses the tube twice, once on the way to a mirror and once after reflection from the mirror.

Each shift by 1 fringe corresponds to a change in phase of 2π rad so if the interference pattern shifts by N fringes as the tube is evacuated,

$$\frac{4\pi(n - 1)L}{\lambda} = 2N\pi$$

and

$$n - 1 = \frac{N\lambda}{2L} = \frac{60(500 \times 10^{-9}\,\text{m})}{2(5.0 \times 10^{-2}\,\text{m})} = 3.0 \times 10^{-4}.$$

Thus $n = 1.00030$.

Chapter 41

3E

The condition for a minimum of a single-slit diffraction pattern is

$$a \sin \theta = m\lambda,$$

where a is the slit width, λ is the wavelength, and m is an integer. The angle θ is measured from the forward direction, so for the situation described in the problem it is $0.60°$, for $m = 1$. Thus

$$a = \frac{m\lambda}{\sin \theta} = \frac{633 \times 10^{-9}\,\text{m}}{\sin 0.60°} = 6.04 \times 10^{-5}\,\text{m}.$$

4E

(a) The condition for a minimum in a single-slit diffraction pattern is given by $a \sin \theta = m\lambda$, where a is the slit width, λ is the wavelength, and m is an integer. For $\lambda = \lambda_a$ and $m = 1$ the angle θ is the same as for $\lambda = \lambda_b$ and $m = 2$. Thus $\lambda_a = 2\lambda_b$.

(b) Let m_a be the integer associated with a minimum in the pattern produced by light with wavelength λ_a and let m_b be the integer associated with a minimum in the pattern produced by light with wavelength λ_b. A minimum in one pattern coincides with a minimum in the other if they occur at the same angle. This means $m_a \lambda_a = m_b \lambda_b$. Since $\lambda_a = 2\lambda_b$, the minima coincide if $2m_a = m_b$. Thus every other minimum of the λ_b pattern coincides with a minimum of the λ_a pattern.

7E

(a) A plane wave is incident on the lens so it is brought to focus in the focal plane of the lens, a distance of 70 cm from the lens.

(b) Waves leaving the lens at an angle θ to the forward direction interfere to produce an intensity minimum if $a \sin \theta = m\lambda$, where a is the slit width, λ is the wavelength, and m is an integer. The distance on the screen from the center of the pattern to the minimum is given by $y = D \tan \theta$, where D is the distance from the lens to the screen. For the conditions of this problem

$$\sin \theta = \frac{m\lambda}{a} = \frac{(1)(590 \times 10^{-9}\,\text{m})}{0.40 \times 10^{-3}\,\text{m}} = 1.475 \times 10^{-3}.$$

This means $\theta = 1.475 \times 10^{-3}\,\text{rad}$ and $y = (70 \times 10^{-2}\,\text{m}) \tan(1.475 \times 10^{-3}\,\text{rad}) = 1.03 \times 10^{-3}\,\text{m}$.

8P

The condition for a minimum of intensity in a single-slit diffraction pattern is $a \sin \theta = m\lambda$, where a is the slit width, λ is the wavelength, and m is an integer. To find the angular position of the first minimum to one side of the central maximum set $m = 1$:

$$\theta_1 = \sin^{-1}\left(\frac{\lambda}{a}\right) = \sin^{-1}\left(\frac{589 \times 10^{-9}\,\text{m}}{1.00 \times 10^{-3}\,\text{m}}\right) = 5.89 \times 10^{-4}\,\text{rad}.$$

If D is the distance from the slit to the screen, the distance on the screen from the center of the pattern to the minimum is $y_1 = D\tan\theta_1 = (3.00\,\text{m})\tan(5.89 \times 10^{-4}\,\text{rad}) = 1.767 \times 10^{-3}\,\text{m}$. To find the second minimum set $m = 2$:

$$\theta_2 = \sin^{-1}\left(\frac{2(589 \times 10^{-9}\,\text{m})}{1.00 \times 10^{-3}\,\text{m}}\right) = 1.178 \times 10^{-3}\,\text{rad}.$$

The distance from the pattern center to the minimum is $y_2 = D\tan\theta_2 = (3.00\,\text{m})\tan(1.178\times 10^{-3}\,\text{rad}) = 3.534 \times 10^{-3}\,\text{m}$. The separation of the two minima is $\Delta y = y_2 - y_1 = 3.534\,\text{mm} - 1.767\,\text{mm} = 1.77\,\text{mm}$.

13P

If you divide the original slit into N strips and represent the light from each strip, when it reaches the screen, by a phasor, then at the central maximum in the diffraction pattern you add N phasors, all in the same direction and each with the same amplitude. The intensity there is proportional to N^2. If you double the slit width you need $2N$ phasors if they are each to have the amplitude of the phasors you used for the narrow slit. The intensity at the central maximum is proportional to $(2N)^2$ and is therefore 4 times the intensity for the narrow slit. The energy reaching the screen per unit time, however, is only twice the energy reaching it per unit time when the narrow slit is in place. The energy is simply redistributed. For example, the central peak is now half as wide and the integral of the intensity over the peak is only twice the analogous integral for the narrow slit.

14P

Think of the Huygens' explanation of diffraction phenomenon. When A is in place only the Huygens' wavelets that pass through the hole get to point P. Suppose they produce a resultant electric field E_A. When B is in place the light that was blocked by A gets to P and the light that passed through the hole in A is blocked. Suppose the electric field at P is now E_B. The sum $E_A + E_B$ is the resultant of all waves that get to P when neither A nor B are present. Since P is in the geometric shadow this is zero. Thus $E_A = -E_B$ and since the intensity is proportional to the square of the electric field, the intensity at P is the same when A is present as when B is present.

15P

(a) The intensity for a single-slit diffraction pattern is given by

$$I = I_m \frac{\sin^2 \alpha}{\alpha^2},$$

where $\alpha = (\pi a/\lambda)\sin\theta$, a is the slit width and λ is the wavelength. The angle θ is measured from the forward direction. You want $I = I_m/2$, so

$$\sin^2\alpha = \frac{1}{2}\alpha^2.$$

(b) Evaluate $\sin^2\alpha$ and $\alpha^2/2$ for $\alpha = 1.39\,\text{rad}$ and compare the results. To be sure that $1.39\,\text{rad}$ is closer to the correct value for α than any other value with 3 significant digits, you should also try $1.385\,\text{rad}$ and $1.395\,\text{rad}$.

(c) Since $\alpha = (\pi a/\lambda)\sin\theta$,

$$\theta = \sin^{-1}\left(\frac{\alpha\lambda}{\pi a}\right).$$

Now $\alpha/\pi = 1.39/\pi = 0.442$, so

$$\theta = \sin^{-1}\left(\frac{0.442\lambda}{a}\right).$$

The angular separation of the two points of half intensity, one on either side of the center of the diffraction pattern, is

$$\Delta\theta = 2\theta = 2\sin^{-1}\left(\frac{0.442\lambda}{a}\right).$$

(d) For $a/\lambda = 1.0$,

$$\Delta\theta = 2\sin^{-1}(0.442/1.0) = 0.916\,\text{rad},$$

for $a/\lambda = 5.0$,

$$\Delta\theta = 2\sin^{-1}(0.442/5.0) = 0.177\,\text{rad},$$

and for $a/\lambda = 10$,

$$\Delta\theta = 2\sin^{-1}(0.442/10) = 0.0884\,\text{rad}.$$

16P

(a) The intensity for a single-slit diffraction pattern is given by

$$I = I_m\frac{\sin^2\alpha}{\alpha^2},$$

where $\alpha = (\pi a/\lambda)\sin\theta$. Here a is the slit width and λ is the wavelength. To find the maxima and minima, set the derivative of I with respect to α equal to zero and solve for α. The derivative is

$$\frac{dI}{d\alpha} = 2I_m\frac{\sin\alpha}{\alpha^3}(\alpha\cos\alpha - \sin\alpha).$$

The derivative vanishes if $\alpha \neq 0$ but $\sin\alpha = 0$. This yields $\alpha = m\pi$, where m is an integer. Except for $m = 0$ these are the intensity minima: $I = 0$ for $\alpha = m\pi$.

The derivative also vanishes for $\alpha\cos\alpha - \sin\alpha = 0$. This condition can be written $\tan\alpha = \alpha$. These are the maxima.

(b) The values of α that satisfy $\tan\alpha = \alpha$ can be found by trial and error on a pocket calculator or computer. Each of them is slightly less than one of the values $(m + \frac{1}{2})\pi$ rad, so start with these values. The first few are 0, 4.4934, 7.7252, 10.9041, 14.0662, and 17.2207. They can also be found graphically. As in the diagram to the right, plot $y = \tan\alpha$ and $y = \alpha$ on the same graph. The intersections of the line with the $\tan\alpha$ curves are the solutions. The first two solutions listed above are shown on the diagram.

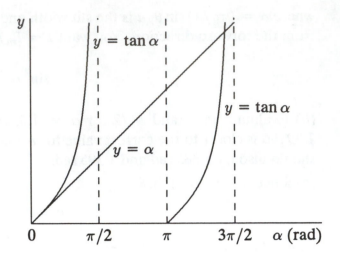

(c) Write $\alpha = (m + \frac{1}{2})\pi$ for the maxima. For the central maximum, $\alpha = 0$ and $m = -\frac{1}{2}$. For the next, $\alpha = 4.4934$ and $m = 0.930$. For the next $\alpha = 7.7252$ and $m = 1.959$.

17P*

Since the slit width is much less than the wavelength of the light, the central peak of the single-slit diffraction pattern is spread across the screen and the diffraction envelope can be ignored. Consider 3 waves, one from each slit. Since the slits are evenly spaced the phase difference for waves from the first and second slits is the same as the phase difference for waves from the second and third slits. The electric fields of the waves at the screen can be written $E_1 = E_0\sin(\omega t)$, $E_2 = E_0\sin(\omega t + \phi)$, and $E_3 = E_0\sin(\omega t + 2\phi)$, where $\phi = (2\pi d/\lambda)\sin\theta$. Here d is the separation of adjacent slits and λ is the wavelength.

The phasor diagram is shown to the right. It yields

$$E = E_0\cos\phi + E_0 + E_0\cos\phi = E_0(1 + 2\cos\phi)$$

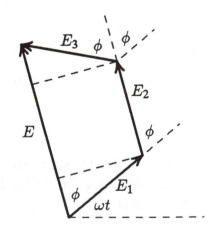

for the amplitude of the resultant wave. Since the intensity of a wave is proportional to the square of the electric field, we may write $I = AE_0^2(1 + 2\cos\phi)^2$, where A is a constant of proportionality. If I_m is the intensity at the center of the pattern, for which $\phi = 0$, then $I_m = 9AE_0^2$. Take A to be $I_m/9E_0^2$ and obtain

$$I = \frac{I_m}{9}(1 + 2\cos\phi)^2 = \frac{I_m}{9}(1 + 4\cos\phi + 4\cos^2\phi).$$

18E

(a) Use the Rayleigh criteria. To resolve two point sources the central maximum of the diffraction pattern of one must lie at or beyond the first minimum of the diffraction pattern of the other. This means the angular separation of the sources must be at least $\theta_R = 1.22\lambda/d$, where λ is the wavelength and d is the diameter of the aperture. For the headlights of this problem

$$\theta_R = \frac{1.22(550 \times 10^{-9}\,\text{m})}{5.0 \times 10^{-3}\,\text{m}} = 1.34 \times 10^{-4}\,\text{rad}\,.$$

(b) If D is the distance from the headlights to the eye when the headlights are just resolvable and ℓ is the separation of the headlights, then $\ell = D \tan \theta_R \approx D\theta_R$, where the small angle approximation $\tan \theta_R \approx \theta_R$ was made. This is valid if θ_R is measured in radians. Thus $D = \ell/\theta_R = (1.4\,\text{m})/(1.34 \times 10^{-4}\,\text{rad}) = 1.0 \times 10^4\,\text{m} = 10\,\text{km}$.

23E

(a) Use the Rayleigh criteria: two objects can be resolved if their angular separation at the observer is greater than $\theta_R = 1.22\lambda/d$, where λ is the wavelength of the light and d is the diameter of the aperture (the eye or mirror). If D is the distance from the observer to the objects then the smallest separation ℓ they can have and still be resolvable is $\ell = D \tan \theta_R \approx D\theta_R$, where θ_R is measured in radians. The small angle approximation $\tan \theta_R \approx \theta_R$ was made. Thus

$$\ell = \frac{1.22D\lambda}{d} = \frac{1.22(8.0 \times 10^{10}\,\text{m})(550 \times 10^{-9}\,\text{m})}{5.0 \times 10^{-3}\,\text{m}} = 1.1 \times 10^7\,\text{m} = 1.1 \times 10^4\,\text{km}\,.$$

This distance is greater than the diameter of Mars. One part of the planet's surface cannot be resolved from another part.

(b) Now $d = 5.1\,\text{m}$ and

$$\ell = \frac{1.22(8.0 \times 10^{10}\,\text{m})(550 \times 10^{-9}\,\text{m})}{5.1\,\text{m}} = 1.1 \times 10^4\,\text{m} = 11\,\text{km}\,.$$

31P

(a) The first minimum in the diffraction pattern is at an angular position θ, measured from the center of the pattern, such that $\sin \theta = 1.22\lambda/d$, where λ is the wavelength and d is the diameter of the antenna. If f is the frequency then the wavelength is $\lambda = c/f = (3.00 \times 10^8\,\text{m/s})/(220 \times 10^9\,\text{Hz}) = 1.36 \times 10^{-3}\,\text{m}$. Thus

$$\theta = \sin^{-1}\left(\frac{1.22\lambda}{d}\right) = \sin^{-1}\left(\frac{1.22(1.36 \times 10^{-3}\,\text{m})}{55.0 \times 10^{-2}\,\text{m}}\right) = 3.02 \times 10^{-3}\,\text{rad}\,.$$

The angular width of the central maximum is twice this, or $6.04 \times 10^{-3}\,\text{rad}$ ($0.346°$).

(b) Now $\lambda = 1.6\,\text{cm}$ and $d = 2.3\,\text{m}$, so

$$\theta = \sin^{-1}\left(\frac{1.22(1.6 \times 10^{-2}\,\text{m})}{2.3\,\text{m}}\right) = 8.5 \times 10^{-3}\,\text{rad}\,.$$

The angular width of the central maximum is $1.7 \times 10^{-2}\,\text{rad}$ ($0.97°$).

35E

Bright interference fringes occur at angles θ given by $d \sin\theta = m\lambda$, where d is the slit separation, λ is the wavelength, and m is an integer. For the slits of this problem $d = 11a/2$, so $a \sin\theta = 2m\lambda/11$ (see Sample Problem 41–7). The first minimum of the diffraction pattern occurs at the angle θ_1 given by $a \sin\theta_1 = \lambda$ and the second occurs at the angle θ_2 given by $a \sin\theta_2 = 2\lambda$, where a is the slit width. You want to count the values of m for which $\theta_1 < \theta < \theta_2$, or what is the same, the values of m for which $\sin\theta_1 < \sin\theta < \sin\theta_2$. This means $1 < (2m/11) < 2$. The values are $m = 6, 7, 8, 9$, and 10. There are five bright fringes in all.

40P

(a) The angular positions θ of the bright interference fringes are given by $d \sin\theta = m\lambda$, where d is the slit separation, λ is the wavelength, and m is an integer. The first diffraction minimum occurs at the angle θ_1 given by $a \sin\theta_1 = \lambda$, where a is the slit width. The diffraction peak extends from $-\theta_1$ to $+\theta_1$, so you want to count the number of values of m for which $-\theta_1 < \theta < +\theta_1$, or what is the same, the number of values of m for which $-\sin\theta_1 < \sin\theta < +\sin\theta_1$. This means $-1/a < m/d < 1/a$ or $-d/a < m < +d/a$. Now $d/a = (0.150 \times 10^{-3}\,\text{m})/(30.0 \times 10^{-6}\,\text{m}) = 5.00$, so the values of m are $m = -4, -3, -2, -1, 0, +1, +2, +3$, and $+4$. There are 9 fringes.

(b) The intensity at the screen is given by

$$I = I_m \left(\cos^2\beta\right) \left(\frac{\sin\alpha}{\alpha}\right)^2,$$

where $\alpha = (\pi a/\lambda)\sin\theta$, $\beta = (\pi d/\lambda)\sin\theta$, and I_m is the intensity at the center of the pattern. For the third bright interference fringe $d \sin\theta = 3\lambda$, so $\beta = 3\pi$ rad and $\cos^2\beta = 1$. Similarly, $\alpha = 3\pi a/d = 3\pi/5.00 = 0.600\pi$ rad and $(\sin\alpha)^2/\alpha^2 = (\sin 0.600\pi)^2/(0.600\pi)^2 = 0.255$. The intensity ratio is $I/I_m = 0.255$.

46E

The ruling separation is $d = 1/(400\,\text{mm}^{-1}) = 2.5 \times 10^{-3}\,\text{mm}$. Diffraction lines occur at angles θ such that $d \sin\theta = m\lambda$, where λ is the wavelength and m is an integer. Notice that for a given order the line associated with a long wavelength is produced at a greater angle than the line associated with a shorter wavelength. Take λ to be the longest wavelength in the visible spectrum (700 nm) and find the greatest integer value of m such that θ is less than 90°. That is, find the greatest integer value of m for which $m\lambda < d$. Since $d/\lambda = (2.5 \times 10^{-6}\,\text{m})/(700 \times 10^{-9}\,\text{m}) = 3.57$ that value is $m = 3$. There are 3 complete orders on each side of the $m = 0$ order. As stated in Problem 56, the second and third orders overlap.

49P

(a) Maxima of a two-slit interference pattern occur at angles θ given by $d \sin\theta = m\lambda$, where d is the slit separation, λ is the wavelength, and m is an integer. The two lines are adjacent so their order numbers differ by unity. Let m be the order number for the line with $\sin\theta = 0.2$ and $m + 1$ be the order number for the line with $\sin\theta = 0.3$. Then $0.2d = m\lambda$ and $0.3d =$

$(m + 1)\lambda$. Subtract the first equation from the second to obtain $0.1d = \lambda$, or $d = \lambda/0.1 = (600 \times 10^{-9}\,\text{m})/0.1 = 6.0 \times 10^{-6}\,\text{m}$.

(b) Minima of the single-slit diffraction pattern occur at angles θ given by $a \sin\theta = m\lambda$, where a is the slit width. Since the fourth order interference maximum is missing it must fall at one of these angles. If a is the smallest slit width for which this order is missing the angle must be given by $a \sin\theta = \lambda$. It is also given by $d \sin\theta = 4\lambda$, so $a = d/4 = (6.0 \times 10^{-6}\,\text{m})/4 = 1.5 \times 10^{-6}\,\text{m}$.

(c) First set $\theta = 90°$ and find the largest value of m for which $m\lambda < d \sin\theta$. This is the highest order that is diffracted toward the screen. The condition is the same as $m < d/\lambda$ and since $d/\lambda = (6.0 \times 10^{-6}\,\text{m})/(600 \times 10^{-9}\,\text{m}) = 10.0$, the highest order seen is the $m = 9$ order. The fourth and eighth orders are missing so the observable orders are $m = 0, 1, 2, 3, 5, 6, 7$, and 9.

51P

The angular positions of the first-order diffraction lines are given by $d \sin\theta = \lambda$, where d is the slit separation and λ is the wavelength. Let λ_1 be the shorter wavelength (430 nm) and θ be the angular position of the line associated with it. Let λ_2 be the longer wavelength (680 nm) and let $\theta + \Delta\theta$ be the angular position of the line associated with it. Here $\Delta\theta = 20°$. Then $d \sin\theta = \lambda_1$ and $d \sin(\theta + \Delta\theta) = \lambda_2$. Use a trigonometric identity to replace $\sin(\theta + \Delta\theta)$ with $\sin\theta \cos\Delta\theta + \cos\theta \sin\Delta\theta$, then use the equation for the first line to replace $\sin\theta$ with λ_1/d and $\cos\theta$ with $\sqrt{1 - \lambda_1^2/d^2}$. After multiplying by d you should obtain $\lambda_1 \cos\Delta\theta + \sqrt{d^2 - \lambda_1^2}\sin\Delta\theta = \lambda_2$. Rearrange to get $\sqrt{d^2 - \lambda_1^2}\sin\Delta\theta = \lambda_2 - \lambda_1 \cos\Delta\theta$. Square both sides and solve for d. You should get

$$d = \sqrt{\frac{(\lambda_2 - \lambda_1 \cos\Delta\theta)^2 + (\lambda_1 \sin\Delta\theta)^2}{\sin^2 \Delta\theta}}$$

$$= \sqrt{\frac{[(680\,\text{nm}) - (430\,\text{nm})\cos 20°]^2 + [(430\,\text{nm})\sin 20°]^2}{\sin^2 20°}}$$

$$= 914\,\text{nm} = 9.14 \times 10^{-4}\,\text{mm}.$$

There are $1/d = 1/(9.14 \times 10^{-4}\,\text{mm}) = 1090$ rulings per mm.

59P

The derivation is similar to that used to obtain Eq. 41–21. At the first minimum beyond the mth principal maximum two waves from adjacent slits have a phase difference of $\Delta\phi = 2\pi m + (2\pi/N)$, where N is the number of slits. This implies a difference in path length of $\Delta L = (\Delta\phi/2\pi)\lambda = m\lambda + (\lambda/N)$. If θ_m is the angular position of the mth maximum then the difference in path length is also given by $\Delta L = d \sin(\theta_m + \Delta\theta)$. Thus $d \sin(\theta_m + \Delta\theta) = m\lambda + (\lambda/N)$. Use the trigonometric identity $\sin(\theta_m + \Delta\theta) = \sin\theta_m \cos\Delta\theta + \cos\theta_m \sin\Delta\theta$. Since $\Delta\theta$ is small, we may approximate $\sin\Delta\theta$ by $\Delta\theta$ in radians and $\cos\Delta\theta$ by unity. Thus $d \sin\theta_m + $

$d \, \Delta\theta \cos\theta_m = m\lambda + (\lambda/N)$. Use the condition $d \sin\theta_m = m\lambda$ to obtain $d \, \Delta\theta \cos\theta_m = \lambda/N$ and

$$\Delta\theta = \frac{\lambda}{N d \cos\theta_m}.$$

62E

If a grating just resolves two wavelengths whose mean is λ and whose separation is $\Delta\lambda$ then its resolving power is defined by $R = \lambda/\Delta\lambda$. The text shows this is Nm, where N is the number of rulings in the grating and m is the order of the lines. Thus $\lambda/\Delta\lambda = Nm$ and

$$N = \frac{\lambda}{m \, \Delta\lambda} = \frac{656.3 \, \text{nm}}{(1)(0.18 \, \text{nm})} = 3650 \, \text{rulings}.$$

65E

The dispersion of a grating is given by $D = d\theta/d\lambda$, where θ is the angular position of a line associated with wavelength λ. The angular position and wavelength are related by $d \sin\theta = m\lambda$, where d is the slit separation and m is an integer. Differentiate this with respect to θ to obtain $(d\theta/d\lambda) \, d\cos\theta = m$ or

$$D = \frac{d\theta}{d\lambda} = \frac{m}{d \cos\theta}.$$

Now $m = (d/\lambda)\sin\theta$, so

$$D = \frac{d \sin\theta}{d\lambda \cos\theta} = \frac{\tan\theta}{\lambda}.$$

The trigonometric identity $\tan\theta = \sin\theta/\cos\theta$ was used.

69P

(a) Since the resolving power of a grating is given by $R = \lambda/\Delta\lambda$ and by Nm, the range of wavelengths that can just be resolved in order m is $\Delta\lambda = \lambda/Nm$. Here N is the number of rulings in the grating. The frequency f is related to the wavelength by $f\lambda = c$, where c is the speed of light. This means $f \, \Delta\lambda + \lambda \, \Delta f = 0$, so $\Delta\lambda = -(\lambda/f)\Delta f = -(\lambda^2/c)\Delta f$, where $f = c/\lambda$ was used. The negative sign means simply that an increase in frequency corresponds to a decrease in wavelength. If we interpret Δf as the range of frequencies that can be resolved we may take it to be positive. Then

$$\frac{\lambda^2}{c}\Delta f = \frac{\lambda}{Nm}$$

and

$$\Delta f = \frac{c}{Nm\lambda}.$$

(b) The difference in travel time for waves traveling along the two extreme rays is $\Delta t = \Delta L/c$, where ΔL is the difference in path length. The waves originate at slits that are separated by $(N-1)d$, where d is the slit separation and N is the number of slits, so the path difference is $\Delta L = (N-1)d \sin\theta$ and the time difference is

$$\Delta t = \frac{(N-1)d \sin\theta}{c}.$$

If N is large this may be approximated by $\Delta t = (Nd/c)\sin\theta$. The lens does not affect the travel time.

(c) Substitute the expressions you derived for Δt and Δf to obtain

$$\Delta f \, \Delta t = \left(\frac{c}{Nm\lambda}\right)\left(\frac{Nd\sin\theta}{c}\right) = \frac{d\sin\theta}{m\lambda} = 1.$$

The condition $d\sin\theta = m\lambda$ for a diffraction line was used to obtain the last result.

70E

Bragg's law gives the condition for a diffraction maximum:

$$2d\sin\theta = m\lambda,$$

where d is the spacing of the crystal planes and λ is the wavelength. The angle θ is measured from the normal to the planes. For a second order reflection $m = 2$, so

$$d = \frac{m\lambda}{2\sin\theta} = \frac{2(0.12\times 10^{-9}\,\text{m})}{2\sin 28^\circ} = 2.56\times 10^{-10}\,\text{m} = 256\,\text{pm}.$$

76P

There are two unknowns, the x-ray wavelength λ and the plane separation d, so data for scattering at two angles from the same planes should suffice. The observations obey Bragg's law, so

$$2d\sin\theta_1 = m_1\lambda$$

and

$$2d\sin\theta_2 = m_2\lambda.$$

However, these cannot be solved for the unknowns. For example, use first equation to eliminate λ from the second. You obtain $m_2\sin\theta_1 = m_1\sin\theta_2$, an equation that does not contain either of the unknowns.

79P

(a) The sets of planes with the next five smaller interplanar spacings (after a_0) are shown in the diagram to the right. In terms of a_0 the spacings are:

(i): $a_0/\sqrt{2} = 0.7071a_0$

(ii): $a_0/\sqrt{5} = 0.4472a_0$

(iii): $a_0/\sqrt{10} = 0.3162a_0$

(iv): $a_0/\sqrt{13} = 0.2774a_0$

(v): $a_0/\sqrt{17} = 0.2425a_0$

(*b*) Since any crystal plane passes through lattice points its slope can be written as the ratio of two integers. Consider a set planes with slope m/n, as shown in the diagram to the right. The first and last planes shown pass through adjacent lattice points along a horizontal line and there are $m-1$ planes between. If h is the separation of the first and last planes, then the interplanar spacing is $d = h/m$. If the planes make the angle θ with the horizontal, then the normal to planes (shown dotted) makes the angle $\phi = 90° - \theta$. The distance h is given by $h = a_0 \cos \phi$ and the interplanar spacing is

$d = h/m = (a_0/m) \cos \phi$. Since $\tan \theta = m/n$, $\tan \phi = n/m$ and $\cos \phi = 1/\sqrt{1 + \tan^2 \phi} = m/\sqrt{n^2 + m^2}$. Thus

$$d = \frac{h}{m} = \frac{a_0 \cos \phi}{m} = \frac{a_0}{\sqrt{n^2 + m^2}}.$$

Chapter 42

2E

(a) The time an electron with a horizontal component of velocity v takes to travel a horizontal distance L is

$$t = \frac{L}{v} = \frac{20 \times 10^{-2}\,\text{m}}{(0.992)(3.00 \times 10^8\,\text{m/s})} = 6.72 \times 10^{-10}\,\text{s}\,.$$

(b) During this time it falls a vertical distance $y = \frac{1}{2}gt^2 = \frac{1}{2}(9.8\,\text{m/s}^2)(6.72 \times 10^{-10}\,\text{s})^2 = 2.21 \times 10^{-18}\,\text{m}$.

5E

Use the time dilation equation $\Delta t = \gamma \Delta t_0$, where Δt_0 is the proper time interval, $\gamma = 1/\sqrt{1 - \beta^2}$, and $\beta = v/c$. Thus $\Delta t = \Delta t_0/\sqrt{1 - \beta^2}$ and the solution for β is

$$\beta = \sqrt{1 - \left(\frac{\Delta t_0}{\Delta t}\right)^2}\,.$$

The proper time interval is measured by a clock at rest relative to the muon. That is, $\Delta t_0 = 2.2\,\mu\text{s}$ and $\Delta t = 16\,\mu\text{s}$. This means

$$\beta = \sqrt{1 - \left(\frac{2.2\,\mu\text{s}}{16\,\mu\text{s}}\right)^2} = 0.9905\,.$$

The muon speed is $v = \beta c = 0.9905(3.00 \times 10^8\,\text{m/s}) = 2.97 \times 10^8\,\text{m/s}$.

9E

The length L of the rod, as measured in a frame in which it is moving with speed v parallel to its length, is related to its rest length L_0 by $L = L_0/\gamma$, where $\gamma = 1/\sqrt{1 - \beta^2}$ and $\beta = v/c$. Since γ must be greater than 1, L is less than L_0. For this problem $L_0 = 1.70\,\text{m}$ and $\beta = 0.630$, so $L = (1.70\,\text{m})\sqrt{1 - (0.630)^2} = 1.32\,\text{m}$.

13E

(a) The rest length L_0 (= 130 m) of the spaceship and its length L as measured by the timing station are related by $L = L_0/\gamma = L_0\sqrt{1 - \beta^2}$, where $\gamma = 1/\sqrt{1 - \beta^2}$ and $\beta = v/c$. Thus $L = (130\,\text{m})\sqrt{1 - (0.740)^2} = 87.4\,\text{m}$.

(b) The time interval for the passage of the spaceship is

$$\Delta t = \frac{L}{v} = \frac{87.4\,\text{m}}{(0.740)(3.00 \times 10^8\,\text{m/s})} = 3.94 \times 10^{-7}\,\text{s}\,.$$

14P

(a) The speed of the traveler is $v = 0.99c$ which is the same as $0.99\,\mathrm{ly/y}$. Let d be the distance traveled. Then the time for the trip, as measured in the frame of the Earth, is $\Delta t = d/v = (26\,\mathrm{ly})/(0.99\,\mathrm{ly/y}) = 26.3\,\mathrm{y}$.

(b) The signal, presumed to be a radio wave, travels with speed c and so takes $26.0\,\mathrm{y}$ to reach Earth. The total time elapsed, in the frame of the Earth, is $26.3\,\mathrm{y} + 26.0\,\mathrm{y} = 52.3\,\mathrm{y}$.

(c) The proper time interval is measured by a clock in the spaceship, so $\Delta t_0 = \Delta t/\gamma$. Now $\gamma = 1/\sqrt{1-\beta^2} = 1/\sqrt{1-(0.99)^2} = 7.09$. Thus $\Delta t_0 = (26.3\,\mathrm{y})/(7.09) = 3.7\,\mathrm{y}$.

17E

The proper time is not measured by clocks in either frame S or frame S' since a single clock at rest in either frame cannot be present at the origin and at the event. The full Lorentz transformation must be used:

$$x' = \gamma[x - vt]$$
$$t' = \gamma[t - \beta x/c],$$

where $\beta = v/c = 0.950$ and $\gamma = 1/\sqrt{1-\beta^2} = 1/\sqrt{1-(0.950)^2} = 3.2026$. Thus

$$x' = (3.2026)\left[100 \times 10^3\,\mathrm{m} - (0.950)(3.00 \times 10^8\,\mathrm{m/s})(200 \times 10^{-6}\,\mathrm{s}\right]$$
$$= 1.38 \times 10^5\,\mathrm{m} = 138\,\mathrm{km}$$

and

$$t' = (3.2026)\left[200 \times 10^{-6}\,\mathrm{s} - \frac{(0.950)(100 \times 10^3\,\mathrm{m})}{3.00 \times 10^8\,\mathrm{m/s}}\right] = -3.74 \times 10^{-4}\,\mathrm{s} = -374\,\mu\mathrm{s}\,.$$

20E

(a) Take the flashbulbs to be at rest in frame S and let frame S' be the rest frame of the second observer. Clocks in neither frame measure the proper time interval between the flashes, so the full Lorentz transformation must be used. Let t_b be the time and x_b be the coordinate of the blue flash, as measured in frame S. Then the time of the blue flash, as measured in frame S', is

$$t'_b = \gamma\left[t_b - \frac{\beta x_b}{c}\right],$$

where $\beta = v/c = 0.250$ and $\gamma = 1/\sqrt{1-\beta^2} = 1/\sqrt{1-(0.250)^2} = 1.0328$. Similarly, let t_r be the time and x_r be the coordinate of the red flash, as measured in frame S. Then the time of the red flash, as measured in frame S', is

$$t'_r = \gamma\left[t_r - \frac{\beta x_r}{c}\right].$$

Now subtract the first Lorentz transformation equation from the second. Recognize that $t_b = t_r$ since the flashes are simultaneous in S. Let $\Delta x = x_r - x_b = 30.0\,\mathrm{km}$ and let $\Delta t' = t'_r - t'_b$. Then

$$\Delta t' = -\frac{\gamma \beta \Delta x}{c} = -\frac{(1.0328)(0.250)(30 \times 10^3\,\mathrm{m})}{3.00 \times 10^8\,\mathrm{m/s}} = -2.58 \times 10^{-5}\,\mathrm{s}\,.$$

(b) Since $\Delta t'$ is negative, t'_b is greater than t'_r. The red bulb flashes first in S'.

22P

(a) The Lorentz factor is

$$\gamma = \frac{1}{\sqrt{1 - \beta^2}} = \frac{1}{\sqrt{1 - (0.600)^2}} = 1.25 \, .$$

(b) In the unprimed frame the time for the clock to travel from the origin to $x = 180 \, \text{m}$ is

$$t = \frac{x}{v} = \frac{180 \, \text{m}}{(0.600)(3.00 \times 10^8 \, \text{m/s})} = 1.00 \times 10^{-6} \, \text{s} \, .$$

The proper time interval between the two events (clock at origin and clock at $x = 180 \, \text{m}$) is measured by the clock itself. The reading on the clock at the beginning of the interval is zero, so the reading at the end is

$$t' = \frac{t}{\gamma} = \frac{1.00 \times 10^{-6} \, \text{s}}{1.25} = 8.00 \times 10^{-7} \, \text{s} \, .$$

25E

There are two possible solutions, depending on the relative directions of the velocity of the particle and the velocity of frame S, both as measured in S'. First suppose both are in the positive x direction. Then $v = 0.40c$ and $u = -0.60c$. According to the velocity transformation equation (Eq. 42–23)

$$v = \frac{v' + u}{1 + uv'/c^2} = \frac{0.40c - 0.60c}{1 + (0.40c)(-0.60c)/c^2} = -0.263c \, .$$

Notice that in the equation u is the velocity of S' relative to S. Since S is moving with speed $0.60c$ in the positive x direction according to S', S' is moving with the same speed but in the negative x direction according to S.

Now suppose the velocity of frame S is in the negative x direction when viewed from S'. Then $u = +0.60c$ and

$$v = \frac{0.40c + 0.60c}{1 + (0.40c)(0.60c)/c^2} = 0.81c \, .$$

28E

(a) Let frame S' be attached to us and frame S be attached to Galaxy A. Take the positive axis to be in the direction of motion of Galaxy A, as seen by us. In S' our velocity is $v' = 0$ and the velocity of Galaxy A is $0.35c$ in the positive x direction. This means $u = -0.35c$. Our velocity, as observed from Galaxy A, is

$$v = \frac{v' + u}{1 + uv'/c^2} = u = -0.35c \, .$$

The negative sign indicates motion in the negative x direction.

(b) In frame S' the velocity of Galaxy B is $v' = -0.35c$, so in S it is

$$v = \frac{v' + u}{1 + uv'} = \frac{-0.35c - 0.35c}{1 + (-0.35c)(-0.35c)/c^2} = -0.62c.$$

The negative sign again indicates motion in the negative x direction.

30P

Calculate the speed of the micrometeorite relative to the spaceship. Let S' be the reference frame for which the data is given and attach frame S to the spaceship. Suppose the micrometeorite is going in the positive x direction and the spaceship is going in the negative x direction, both as viewed from S'. Then in Eq. 42–23, $v' = 0.82c$ and $u = 0.82c$. Notice that u in the equation is the velocity of S' relative to S. Thus the velocity of the micrometeorite in the frame of the spaceship is

$$v = \frac{v' + u}{1 + uv'/c^2} = \frac{0.82c + 0.82c}{1 + (0.82c)(0.82c)/c^2} = 0.9806c.$$

The time for the micrometeorite to pass the spaceship is

$$\Delta t = \frac{L}{v} = \frac{350\,\text{m}}{(0.9806)(3.00 \times 10^8\,\text{m/s})} = 1.19 \times 10^{-6}\,\text{s}.$$

33E

The spaceship is moving away from the Earth, so the frequency received is given by Eq. 42–26:

$$f = f_0 \sqrt{\frac{1 - \beta}{1 + \beta}},$$

where f_0 is the frequency in the frame of the spaceship, $\beta = v/c$, and v is the speed of the spaceship relative to the Earth. Thus

$$f = (100\,\text{MHz})\sqrt{\frac{1 - 0.9000}{1 + 0.9000}} = 22.9\,\text{MHz}.$$

36P

The spaceship is moving away from the Earth, so the frequency received is given by Eq. 42–26:

$$f = f_0 \sqrt{\frac{1 - \beta}{1 + \beta}},$$

where f_0 is the frequency in the frame of the spaceship, $\beta = v/c$, and v is the speed of the spaceship relative to the Earth. The frequency f and wavelength λ are related by $f\lambda = c$, so if λ_0 is the wavelength of the light as seen on the spaceship and λ is the wavelength detected on Earth, then

$$\lambda = \lambda_0 \sqrt{\frac{1+\beta}{1-\beta}} = (450\,\text{nm})\sqrt{\frac{1+0.20}{1-0.20}} = 550\,\text{nm}.$$

This is in the yellow-green portion of the visible spectrum.

42E

Use the two expressions for the total energy: $E = mc^2 + K$ and $E = \gamma mc^2$, where m is the mass of an electron, K is the kinetic energy, and $\gamma = 1/\sqrt{1-\beta^2}$. Thus $mc^2 + K = \gamma mc^2$ and $\gamma = (mc^2 + K)/mc^2$. This means $\sqrt{1-\beta^2} = (mc^2)/(mc^2 + K)$ and

$$\beta = \sqrt{1 - \left(\frac{mc^2}{mc^2 + K}\right)^2}.$$

Now $mc^2 = 0.511\,\text{MeV}$ so

$$\beta = \sqrt{1 - \left(\frac{0.511\,\text{MeV}}{0.511\,\text{MeV} + 100\,\text{MeV}}\right)^2} = 0.999987.$$

The speed of the electron is $0.999987c$ or 99.9987% the speed of light.

45E

Since the rest energy E_0 and the mass m of the quasar are related by $E_0 = mc^2$, the rate P of energy radiation and the rate of mass loss are related by $P = dE_0/dt = (dm/dt)c^2$. Thus

$$\frac{dm}{dt} = \frac{P}{c^2} = \frac{10^{41}\,\text{W}}{(3.00 \times 10^8\,\text{m/s})^2} = 1.11 \times 10^{24}\,\text{kg/s}.$$

Since a solar mass is $2.0 \times 10^{30}\,\text{kg}$ and a year is $3.156 \times 10^7\,\text{s}$,

$$\frac{dm}{dt} = (1.11 \times 10^{24}\,\text{kg/s})\left(\frac{3.156 \times 10^7\,\text{s/y}}{2.0 \times 10^{30}\,\text{kg/smu}}\right) = 17.5\,\text{smu/y}.$$

49P

(a) In general, the momentum of a particle with mass m and speed v is given by $p = \gamma mv$, where γ is the Lorentz factor. Since the particle of the problem has momentum mc, $mc = \gamma mv$ or $1 = \gamma\beta$, where $\beta = v/c$. Now find an expression for β in terms of γ. Since $\gamma = 1/\sqrt{1-\beta^2}$, $\beta^2 = 1 - (1/\gamma^2) = (\gamma^2 - 1)/\gamma^2$. Thus $\beta = \sqrt{\gamma^2 - 1}/\gamma$ and $1 = \sqrt{\gamma^2 - 1}$. The solution is $\gamma = \sqrt{2} = 1.41$.

(b) The speed parameter is

$$\beta = \frac{\sqrt{\gamma^2 - 1}}{\gamma} = \frac{\sqrt{2-1}}{\sqrt{2}} = 0.707.$$

The speed of the particle is $v = \beta c = 0.707c$.

(c) The kinetic energy is $K = (\gamma - 1)mc^2 = (\sqrt{2} - 1)mc^2 = 0.414mc^2$.

52P

The energy equivalent of one tablet is $mc^2 = (320 \times 10^{-6}\,\text{kg})(3.00 \times 10^8\,\text{m/s})^2 = 2.88 \times 10^{13}\,\text{J}$. This provides the same energy as $(2.88 \times 10^{13}\,\text{J})/(1.30 \times 10^8\,\text{J/gal}) = 2.21 \times 10^5$ gal of gasoline. The distance the car can go is $d = (2.21 \times 10^5\,\text{gal})(30.0\,\text{mi/gal}) = 6.65 \times 10^6$ mi.

54P

The distance traveled by the pion in the frame of the Earth is $d = v\,\Delta t$, where v is the speed of the pion and Δt is the pion lifetime, both as measured in that frame. The proper time interval Δt_0 is measured in the rest frame of the pion, so $\Delta t = \gamma \Delta t_0$. We must calculate the speed of the pion and the Lorentz factor γ.

Since the total energy of the pion is given by $E = \gamma mc^2$,

$$\gamma = \frac{E}{mc^2} = \frac{1.35 \times 10^5\,\text{MeV}}{139.6\,\text{MeV}} = 967.1\,.$$

Since $\gamma = 1/\sqrt{1 - \beta^2}$,

$$\beta = \frac{\sqrt{\gamma^2 - 1}}{\gamma} = \frac{\sqrt{(967.1)^2 - 1}}{967.1} = 0.9999995\,.$$

The speed of the pion is extremely close to the speed of light and we may approximate β as 1. Thus $v = 3.00 \times 10^8$ m/s.

The pion lifetime as measured in the frame of the Earth is $\Delta t = (967.1)(35.0\,\text{ns}) = 3.385 \times 10^4\,\text{ns} = 3.385 \times 10^{-5}$ s. The distance traveled is $d = (3.00 \times 10^8\,\text{m/s})(3.385 \times 10^{-5}\,\text{s}) = 1.02 \times 10^4\,\text{m} = 10.2\,\text{km}$. The altitude at which the pion decays is $120\,\text{km} - 10.2\,\text{km} = 110\,\text{km}$.

58P

The radius r of the path is given in Problem 57 as

$$r = \frac{mv}{qB\sqrt{1 - \beta^2}}\,,$$

so

$$m = \frac{qBr\sqrt{1 - \beta^2}}{v} = \frac{2(1.60 \times 10^{-19}\,\text{C})(1.00\,\text{T})(6.28\,\text{m})\sqrt{1 - (0.710)^2}}{(0.710)(3.00 \times 10^8\,\text{m/s})}$$
$$= 6.64 \times 10^{-27}\,\text{kg}\,.$$

Since $1.00\,\text{u} = 1.66 \times 10^{-27}\,\text{u}$, $m = 4.00\,\text{u}$. The nuclear particle contains 4 nucleons. Since there must be 2 protons to provide the charge $2e$, the nuclear particle is a helium nucleus with 2 protons and 2 neutrons.

Chapter 43

1E

The energy of a photon is given by $E = hf$, where h is the Planck constant and f is the frequency. The wavelength λ is related to the frequency by $\lambda f = c$, so $E = hc/\lambda$. Since $h = 6.63 \times 10^{-34}\,\text{J} \cdot \text{s}$ and $c = 3.00 \times 10^{8}\,\text{m/s}$,

$$hc = 1.99 \times 10^{-25}\,\text{J} \cdot \text{m} = \frac{1.99 \times 10^{-25}\,\text{J} \cdot \text{m}}{(1.602 \times 10^{-19}\,\text{J/eV}) \times (10^{-9}\,\text{m/nm})} = 1240\,\text{eV} \cdot \text{nm}.$$

Thus

$$E = \frac{1240\,\text{eV} \cdot \text{nm}}{\lambda}.$$

9P

(a) Let R be the rate of photon emission (number of photons emitted per unit time) and let E be the energy of a single photon. Then the power output of a bulb is given by $P = RE$ if all the power goes into photon production. Now $E = hf = hc/\lambda$, where h is the Planck constant, f is the frequency of the light emitted, and λ is the wavelength. Thus

$$P = \frac{Rhc}{\lambda}$$

and

$$R = \frac{\lambda P}{hc}.$$

The bulb emitting light with the longer wavelength (the 700 nm bulb) emits more photons per unit time. The energy of each photon is less so it must emit photons at a greater rate.

(b) Let R_ℓ be the rate of photon production at the longer wavelength (λ_ℓ) and let R_s be the rate of production at the shorter wavelength (λ_s). Then

$$R_\ell - R_s = \frac{(\lambda_\ell - \lambda_s)P}{hc} = \frac{(700\,\text{nm} - 400\,\text{nm})(400\,\text{J/s})}{(1.60 \times 10^{-19}\,\text{J/eV})(1240\,\text{eV} \cdot \text{nm})} = 6.0 \times 10^{20}\,\text{photon/s}.$$

The result $hc = 1240\,\text{eV} \cdot \text{nm}$, developed in Exercise 1, was used.

13P

(a) Assume all the power results in photon production at the wavelength $\lambda = 589\,\text{nm}$. Let R be the rate of photon production and E be the energy of a single photon. Then $P = RE = Rhc/\lambda$, where $E = hf$ and $f = c/\lambda$ were used. Here h is the Planck constant, f is the frequency of the emitted light, and λ is its wavelength. Thus

$$R = \frac{\lambda P}{hc} = \frac{(589 \times 10^{-9}\,\text{m})(100\,\text{W})}{(6.63 \times 10^{-34}\,\text{J} \cdot \text{s})(3.00 \times 10^{8}\,\text{m/s})} = 2.96 \times 10^{20}\,\text{photon/s}.$$

(b) Let I be the photon flux a distance r from the source. Since photons are emitted uniformly in all directions $R = 4\pi r^2 I$ and

$$r = \sqrt{\frac{R}{4\pi I}} = \sqrt{\frac{2.96 \times 10^{20}\,\text{photon/s}}{4\pi(1.00 \times 10^4\,\text{photon/m}^2 \cdot \text{s})}} = 4.85 \times 10^7\,\text{m}.$$

(c) Let n be the photon density and consider a narrow column with cross-sectional area A and length ℓ, with its axis along the direction of photon motion. At any instant the number of photons in the column is $N = nA\ell$. All of them will move through one end in time $t = \ell/c$, so the photon flux through the end is $I = N/At = nA\ell c/A\ell = nc$. Now $I = R/4\pi r^2$, so $R/4\pi r^2 = nc$ and

$$r = \sqrt{\frac{R}{4\pi nc}} = \sqrt{\frac{2.96 \times 10^{20}\,\text{photon/s}}{4\pi(1.00 \times 10^6\,\text{photon/m}^3)(3.00 \times 10^8\,\text{m/s})}} = 280\,\text{m}.$$

(d) The photon flux is

$$I = \frac{R}{4\pi r^2} = \frac{2.96 \times 10^{20}\,\text{photon/s}}{4\pi(2.00\,\text{m})^2} = 5.89 \times 10^{18}\,\text{photon/m}^2 \cdot \text{s}$$

and the photon density is

$$n = \frac{I}{c} = \frac{5.89 \times 10^{18}\,\text{photon/m}^2 \cdot \text{s}}{3.00 \times 10^8\,\text{m/s}} = 1.96 \times 10^{10}\,\text{photon/m}^3.$$

18E

The energy of an incident photon is $E = hf = hc/\lambda$, where h is the Planck constant, f is the frequency of the electromagnetic radiation, and λ is its wavelength. The kinetic energy of the most energetic electron emitted is $K_m = E - \phi = (hc/\lambda) - \phi$, where ϕ is the work function for sodium. The stopping potential V_0 is related to the maximum kinetic energy by $eV_0 = K_m$, so $eV_0 = (hc/\lambda) - \phi$ and

$$\lambda = \frac{hc}{eV_0 + \phi} = \frac{1240\,\text{eV} \cdot \text{nm}}{5.0\,\text{eV} + 2.2\,\text{eV}} = 170\,\text{nm}.$$

Here $eV_0 = 5.0\,\text{eV}$ and $hc = 1240\,\text{ev} \cdot \text{nm}$ were used. See Exercise 1.

20E

(a) The kinetic energy K_m of the fastest electron emitted is given by $K_m = hf - \phi = (hc/\lambda) - \phi$, where ϕ is the work function of aluminum, f is the frequency of the incident radiation, and λ is its wavelength. The relationship $f = c/\lambda$ was used to obtain the second form. Thus

$$K_m = \frac{1240\,\text{eV} \cdot \text{nm}}{200\,\text{nm}} - 4.20\,\text{eV} = 2.00\,\text{eV},$$

where the result of Exercise 1 was used.

(b) The slowest electron just breaks free of the surface and so has zero kinetic energy.

(c) The stopping potential V_0 is given by $K_m = eV_0$, so $V_0 = K_m/e = (2.00\,\text{eV})/e = 2.00\,\text{V}$.

(d) The value of the cutoff wavelength is such that $K_m = 0$. Thus $hc/\lambda = \phi$ or

$$\lambda = \frac{hc}{\phi} = \frac{1240\,\text{eV}\cdot\text{nm}}{4.2\,\text{eV}} = 295\,\text{nm}\,.$$

If the wavelength is longer the photon energy is less and a photon does not have sufficient energy to knock even the most energetic electron out of the aluminum sample.

23P

(a) Use the photoelectric effect equation (Eq. 43–8) in the form $hc/\lambda = \phi + K_m$. The work function depends only on the material and the condition of the surface and not on the wavelength of the incident light. Let λ_1 be first wavelength described and λ_2 be the second. Let K_{m1} (= 0.710 eV) be the maximum kinetic energy of electrons ejected by light with the first wavelength and K_{m2} (= 1.43 eV) be the maximum kinetic energy of electron ejected by light with the second wavelength. Then

$$\frac{hc}{\lambda_1} = \phi + K_{m1}$$

and

$$\frac{hc}{\lambda_2} = \phi + K_{m2}\,.$$

Solve these equations simultaneously for λ_2.

The first equation yields $\phi = (hc/\lambda_1) - K_{m1}$. When this is used to substitute for ϕ in the second equation the result is $(hc/\lambda_2) = (hc/\lambda_1) - K_{m1} + K_{m2}$. The solution for λ_2 is

$$\lambda_2 = \frac{hc\lambda_1}{hc + \lambda_1(K_{m2} - K_{m1})} = \frac{(1240\,\text{eV}\cdot\text{nm})(491\,\text{nm})}{1240\,\text{eV}\cdot\text{nm} + (491\,\text{nm})(1.43\,\text{eV} - 0.710\,\text{eV})}$$

$$= 382\,\text{nm}\,.$$

Here $hc = 1240\,\text{eV}\cdot\text{nm}$, calculated in Exercise 1, was used.

(b) The first equation displayed above yields

$$\phi = \frac{hc}{\lambda_1} - K_{m1} = \frac{1240\,\text{eV}\cdot\text{nm}}{491\,\text{nm}} - 0.710\,\text{eV} = 1.82\,\text{eV}\,.$$

28P*

The momentum of the photon is given by hf/c, where f is the frequency of the radiation. The momentum of the electron is initially zero. After the collision the momentum of the electron is $mv/\sqrt{1-v^2/c^2}$, where v is its velocity. Assume the photon is absorbed. Then conservation of momentum yields

$$\frac{hf}{c} = \frac{mv}{\sqrt{1-v^2/c^2}}\,.$$

Initially the photon has energy hf and the electron has energy mc^2, its rest energy. After the collision the electron energy is $mc^2/\sqrt{1 - v^2/c^2}$. Conservation of energy yields

$$hf + mc^2 = \frac{mc^2}{\sqrt{1 - v^2/c^2}}.$$

Solve the momentum conservation equation for hf and substitute the result into the energy conservation equation. You should get

$$v = c\left[1 - \sqrt{1 - v^2/c^2}\right].$$

There are two solutions for v: either $v = 0$ or $v = c$. If $v = 0$ then the momentum conservation equation tells us that $f = 0$. There really is no interaction since no radiation energy is incident on the electron. The solution $v = c$ is prohibited since this means the electron has infinite momentum and energy after the collision and the photon has infinite momentum and energy before.

29E

(a) When a photon scatters from an electron initially at rest the change in wavelength is given by

$$\Delta\lambda = \frac{h}{mc}(1 - \cos\phi),$$

where m is the mass of an electron and ϕ is the scattering angle. Now $h/mc = 2.43 \times 10^{-12}$ m $= 2.43$ pm, so $\Delta\lambda = (2.43\,\text{pm})(1 - \cos 30°) = 0.326$ pm. The final wavelength is $\lambda' = \lambda + \Delta\lambda = 2.4\,\text{pm} + 0.326\,\text{pm} = 2.73\,\text{pm}$.

(b) Now $\Delta\lambda = (2.43\,\text{pm})(1 - \cos 120°) = 3.645$ pm and $\lambda' = 2.4\,\text{pm} + 3.645\,\text{pm} = 6.05\,\text{pm}$.

32E

(a) Since the mass of an electron is $m = 9.109 \times 10^{-31}$ kg, its Compton wavelength is

$$\lambda_C = \frac{h}{mc} = \frac{6.626 \times 10^{-34}\,\text{J}\cdot\text{s}}{(9.109 \times 10^{-31}\,\text{kg})(2.998 \times 10^8\,\text{m/s})} = 2.426 \times 10^{-12}\,\text{m} = 2.43\,\text{pm}.$$

(b) Since the mass of a proton is $m = 1.673 \times 10^{-27}$ kg, its Compton wavelength is

$$\lambda_C = \frac{6.626 \times 10^{-34}\,\text{J}\cdot\text{s}}{(1.673 \times 10^{-27}\,\text{kg})(2.998 \times 10^8\,\text{m/s})} = 1.321 \times 10^{-15}\,\text{m} = 1.32\,\text{fm}.$$

(c) Use the formula developed in Exercise 1: $E = (1240\,\text{eV}\cdot\text{nm})/\lambda$, where E is the energy and λ is the wavelength. Thus for the electron

$$E = \frac{1240\,\text{eV}\cdot\text{nm}}{2.426 \times 10^{-3}\,\text{nm}} = 5.11 \times 10^5\,\text{eV} = 0.511\,\text{MeV}.$$

(d) For the proton

$$E = \frac{1240\,\text{eV}\cdot\text{nm}}{1.321 \times 10^{-6}\,\text{nm}} = 9.39 \times 10^8\,\text{eV} = 939\,\text{MeV}\,.$$

34E

If E is the original energy of the photon and E' is the energy after scattering, then the fractional energy loss is

$$frac = \frac{E - E'}{E}\,.$$

Sample Problem 43–6 shows that this is

$$frac = \frac{\Delta\lambda}{\lambda + \Delta\lambda}\,.$$

Thus

$$\frac{\Delta\lambda}{\lambda} = \frac{frac}{1 - frac} = \frac{0.75}{1 - 0.75} = 3\,.$$

A 300% increase in the wavelength leads to a 75% decrease in the energy of the photon.

36P

(a) Use the result developed in Sample Problem 43–6:

$$\frac{E - E'}{E} = \frac{\Delta\lambda}{\lambda + \Delta\lambda}\,,$$

where E is the initial energy of the photon, E' is its final energy, λ is the initial wavelength of the wave associated with the photon, and $\Delta\lambda$ is the change in wavelength. Substitute $E = hf$ and $E' = hf'$ to obtain

$$\frac{f - f'}{f} = \frac{\Delta\lambda}{\lambda + \Delta\lambda}\,.$$

If $(f - f')/f = frac$, the solution for $\Delta\lambda$ is

$$\Delta\lambda = \frac{\lambda frac}{1 - frac}\,.$$

The wavelength of a 6.2-keV photon is $\lambda = (hc)/E = (1240\,\text{eV}\cdot\text{nm})/(6.2 \times 10^3\,\text{eV}) = 0.200\,\text{nm}$, where the result of Exercise 1 was used. Thus

$$\Delta\lambda = \frac{(0.200\,\text{nm})(1.0 \times 10^{-4})}{1 - 1.0 \times 10^{-4}} = 2.0 \times 10^{-5}\,\text{nm} = 2.0 \times 10^{-2}\,\text{pm}\,.$$

Now use the Compton scattering equation $\Delta\lambda = \lambda_C(1 - \cos\phi)$, where λ_C is the Compton wavelength for an electron (2.43 pm) and ϕ is the scattering angle. This means

$$\phi = \cos^{-1}\left(1 - \frac{\Delta\lambda}{\lambda_C}\right) = \cos^{-1}\left(1 - \frac{2.0 \times 10^{-2}\,\text{pm}}{2.43\,\text{pm}}\right) = 7.4°\,.$$

(b) The kinetic energy imparted to the electron equals the energy lost by the photon. That is,

$$K = \frac{hc}{\lambda} - \frac{hc}{\lambda'} = \frac{hc(\lambda' - \lambda)}{\lambda\lambda'} = \frac{hc\,\Delta\lambda}{\lambda(\lambda + \Delta\lambda)}$$

$$= \frac{(1240\,\text{eV} \cdot \text{nm})(2.00 \times 10^{-5}\,\text{nm})}{(0.200\,\text{nm})(0.200\,\text{nm} + 2.00 \times 10^{-5}\,\text{nm})} = 0.62\,\text{eV}.$$

42E

Since $\lambda_{\max} T = 2898\,\mu\text{m} \cdot \text{K}$,

$$\lambda_{\max} = \frac{2898\,\mu\text{m} \cdot \text{K}}{5800\,\text{K}} = 0.500\,\mu\text{m} = 500\,\text{nm}.$$

This is in the blue-green portion of the visible spectrum. The sun appears yellow because the human eye is more sensitive to light at about 550 nm than to light at 500 nm and the sun has a high intensity at the somewhat redder wavelengths. See Fig. 38–2.

44E

The Wien displacement law (see Exercise 43–42) gives

$$T = \frac{2898\,\mu\text{m} \cdot \text{K}}{\lambda_{\max}} = \frac{2.898 \times 10^6\,\text{nm} \cdot \text{K}}{550\,\text{nm}} = 5270\,\text{K}.$$

49P

(a) If $S(\lambda)$ is the spectral radiancy then the power emitted per unit area is

$$P = \int_0^\infty S(\lambda)\,\mathrm{d}\lambda = (2\pi c^2 h) \int_0^\infty \frac{\lambda^{-5}\,\mathrm{d}\lambda}{e^{hc/\lambda kT} - 1}.$$

Let $x = hc/\lambda kT$; then $\lambda = hc/kTx$ and $\mathrm{d}\lambda = -(hc/kTx^2)\,\mathrm{d}x$. When $\lambda = \infty$ then $x = 0$ and when $\lambda = \infty$ then $x = 0$. The negative sign in $\mathrm{d}\lambda$ can be used to reverse the limits on the integral. Hence

$$P = (2\pi c^2 h)\left(\frac{kT}{hc}\right)^4 \int_0^\infty \frac{x^3\,\mathrm{d}x}{e^x - 1} = (2\pi c^2 h)\left(\frac{k}{hc}\right)^4 \left(\frac{\pi^4}{15}\right) T^4 = \left(\frac{2\pi^5 k^4}{15h^3 c^2}\right) T^4.$$

This can be written $P = \sigma T^4$, where $\sigma = (2\pi^5 k^4)/(15h^3 c^2)$.

(b) Use values of the fundamental constants given in Appendix B to obtain

$$\sigma = \frac{2\pi^5 (1.3807 \times 10^{-23}\,\text{J/K})^4}{15(6.6261 \times 10^{-34}\,\text{J} \cdot \text{s})^3 (2.9979 \times 10^8\,\text{m/s})^2} = 5.67 \times 10^{-8}\,\text{W/m}^2 \cdot \text{K}^4.$$

54P

Let T_o be the temperature of the oven in kelvins ($227 + 273 = 500\,\text{K}$) and T_r be the temperature of the room in kelvins ($27 + 273 = 300\,\text{K}$). The oven radiates energy into the room at the rate $\sigma A T_o^4$ and the room radiates energy into the oven at the rate $\sigma A T_r^4$, where A is the area of the opening and σ is the Stefan-Boltzmann constant, developed in Problem 49. The net rate at which energy is transferred from the oven to the room is

$$P = \sigma A \left(T_o^4 - T_r^4\right) = (5.67 \times 10^{-8}\,\mathrm{W/m^2 \cdot K^4})(5.0 \times 10^{-4}\,\mathrm{m^2})\left[(500\,\mathrm{K})^4 - (300\,\mathrm{K})^4\right]$$
$$= 1.5\,\mathrm{W}\,.$$

57E

Since energy is conserved the energy E of the emitted photon equals $E_i - E_f$, where E_i is the energy of the initial state of the gold atom and E_f is the energy of the final state. Now $E = hf = hc/\lambda = (1240\,\mathrm{eV \cdot nm})/(18.5 \times 10^{-3}\,\mathrm{nm}) = 6.70 \times 10^4\,\mathrm{eV} = 67.0\,\mathrm{keV}$. Hence

$$E_f = E_i - hf = -13.7\,\mathrm{keV} - 67.0\,\mathrm{keV} = -80.7\,\mathrm{keV}\,.$$

60E

(a) Since energy is conserved the energy E of the photon is given by $E = E_i - E_f$, where E_i is the initial energy of the electron and E_f is the final energy. The electron energy is given by $(-13.6\,\mathrm{eV})/n^2$, where n is the principal quantum number. Thus

$$E = E_i - E_f = \frac{-13.6\,\mathrm{eV}}{(3)^2} - \frac{-13.6\,\mathrm{eV}}{(1)^2} = 12.1\,\mathrm{eV}\,.$$

(b) The photon momentum is given by

$$p = \frac{E}{c} = \frac{(12.1\,\mathrm{eV})(1.60 \times 10^{-19}\,\mathrm{J/eV})}{3.00 \times 10^8\,\mathrm{m/s}} = 6.45 \times 10^{-27}\,\mathrm{kg \cdot m/s}\,.$$

(c) The wavelength is given by

$$\lambda = \frac{1240\,\mathrm{eV \cdot nm}}{12.1\,\mathrm{eV}} = 102\,\mathrm{nm}\,,$$

where the result of Exercise 1 was used.

62E

Use Eq. 43–24 for the wavelengths of the electromagnetic radiation emitted by hydrogen:

$$\frac{1}{\lambda} = R\left(\frac{1}{\ell^2} - \frac{1}{u^2}\right)\,,$$

where R is the Rydberg constant, ℓ is the principal quantum number associated with the lower electron state, and u is the principal quantum number associated with the upper electron state. A series limit is obtained if $u = \infty$. For the Balmer series $\ell = 2$ and $\lambda_B = 4/R$. For the Lyman series $\ell = 1$ and $\lambda_L = 1/R$. The ratio is

$$\frac{\lambda_B}{\lambda_L} = \frac{4/R}{1/R} = 4\,.$$

64P

(a) Take the electrostatic potential energy to be zero when the electron and proton are far removed from each other. Then the final energy of the atom is zero and the work done in pulling it apart is $W = -E_i$, where E_i is the energy of the initial state. The energy of the initial state is given by $E_i = (-13.6\,\text{eV})/n^2$, where n is the principal quantum number of the state. For the ground state $n = 1$ and $W = 13.6\,\text{eV}$.

(b) For the state with $n = 2$, $W = (13.6\,\text{eV})/(2)^2 = 3.40\,\text{eV}$.

70P

If kinetic energy is not conserved some of the neutron's initial kinetic energy is used to excite the hydrogen atom. The least energy that the hydrogen atom can accept is the difference between the first excited state ($n = 2$) and the ground state ($n = 1$). Since the energy of a state with principal quantum number n is $-(13.6\,\text{eV})/n^2$, the smallest excitation energy is $13.6\,\text{eV} - (13.6\,\text{eV})/(2)^2 = 10.2\,\text{eV}$. The neutron does not have sufficient kinetic energy to excite the hydrogen atom, so the hydrogen atom is left in its ground state and all the initial kinetic energy of the neutron ends up as the final kinetic energies of the neutron and atom. The collision must be elastic.

78P

(a) If angular momentum is quantized as in the Bohr model of the hydrogen atom, then $L = nh/2\pi$, where n is the principal quantum number and h is the Planck constant. Since $L = I\omega$, where I is the rotational inertia and ω is the angular velocity, $I\omega = nh/2\pi$ and $\omega = nh/2\pi I$. The rotational inertia of the two-particle system shown is $I = 2m(d/2)^2 = md^2/2$, so

$$\omega = \frac{nh}{2\pi}\frac{2}{md^2} = \frac{nh}{\pi md^2}.$$

(b) The kinetic energy of rotation is

$$E = \frac{1}{2}I\omega^2 = \frac{1}{2}\left(\frac{md^2}{2}\right)\left(\frac{nh}{\pi md^2}\right)^2 = \frac{n^2h^2}{4\pi^2md^2}.$$

Chapter 44

2E

(a) Substitute the classical relationship between momentum p and velocity v, $v = p/m$ into the classical definition of kinetic energy, $K = \frac{1}{2}mv^2$, to obtain $K = p^2/2m$. Here m is the mass of an electron. Solve for p: $p = \sqrt{2mK}$. The relationship between the momentum and the de Broglie wavelength λ is $\lambda = h/p$, where h is the Planck constant. Thus

$$\lambda = \frac{h}{\sqrt{2mK}}.$$

If K is given in electron volts then

$$\lambda = \frac{6.626 \times 10^{-34}\,\text{J}\cdot\text{s}}{\sqrt{2(9.109 \times 10^{-31}\,\text{kg})}\sqrt{(1.602 \times 10^{-19}\,\text{J/eV})K}} = \frac{1.226 \times 10^{-9}\,\text{m}\cdot\text{eV}^{1/2}}{\sqrt{K}}$$

$$= \frac{1.226\,\text{nm}\cdot\text{eV}^{1/2}}{\sqrt{K}}.$$

(b) Replace K with eV, where V is the accelerating potential and e is the charge on the electron, to obtain

$$\lambda = \frac{h}{\sqrt{2meV}} = \frac{6.626 \times 10^{-34}\,\text{J}\cdot\text{s}}{\sqrt{2(9.109 \times 10^{-31}\,\text{kg})(1.602 \times 10^{-19}\,\text{C})V}}$$

$$= \sqrt{\frac{1.50 \times 10^{-18}\,\text{nm}^2\cdot\text{V}}{V}} = \sqrt{\frac{1.50\,\text{nm}^2\cdot\text{V}}{V}}.$$

3E

Use the result of Exercise 2:

$$\lambda = \sqrt{\frac{1.50\,\text{nm}^2\cdot\text{V}}{V}} = \sqrt{\frac{1.50\,\text{nm}^2\cdot\text{V}}{25.0 \times 10^3\,\text{V}}} = 7.75 \times 10^{-3}\,\text{nm} = 7.75\,\text{pm}.$$

6E

Use the result of Exercise 2: $\lambda = (1.226\,\text{nm}\cdot\text{eV}^{1/2})/\sqrt{K}$, where K is the kinetic energy. Thus

$$K = \left(\frac{1.226\,\text{nm}\cdot\text{eV}^{1/2}}{\lambda}\right)^2 = \left(\frac{1.226\,\text{nm}\cdot\text{eV}^{1/2}}{590\,\text{nm}}\right)^2 = 4.32 \times 10^{-6}\,\text{eV}.$$

10P

(a) The momentum of the photon is given by $p = E/c$, where E is its energy. Its wavelength is

$$\lambda = \frac{h}{p} = \frac{hc}{E} = \frac{1240\,\text{eV}\cdot\text{nm}}{1.00\,\text{eV}} = 1240\,\text{nm}.$$

See Exercise 1 of Chapter 43. The momentum of the electron is given by $p = \sqrt{2mK}$, where K is its kinetic energy and m is its mass. Its wavelength is

$$\lambda = \frac{h}{p} = \frac{h}{\sqrt{2mK}}.$$

According to Exercise 2 if K is in electron volts this is

$$\lambda = \frac{1.226\,\text{nm}}{\sqrt{K}} = \frac{1.226\,\text{nm}}{\sqrt{1.00}} = 1.23\,\text{nm}.$$

(b) For the photon

$$\lambda = \frac{hc}{E} = \frac{1240\,\text{eV} \cdot \text{nm}}{1.00 \times 10^9\,\text{eV}} = 1.24 \times 10^{-6}\,\text{nm}.$$

Relativity theory must be used to calculate the wavelength for the electron. According to Eq. 42–40 the momentum p and kinetic energy K are related by $(pc)^2 = K^2 + 2Kmc^2$. Thus

$$pc = \sqrt{K^2 + 2Kmc^2} = \sqrt{(1.00 \times 10^9\,\text{eV})^2 + 2(1.00 \times 10^9\,\text{eV})(0.511 \times 10^6\,\text{eV})}$$
$$= 1.00 \times 10^9\,\text{eV}.$$

The wavelength is

$$\lambda = \frac{h}{p} = \frac{hc}{pc} = \frac{1240\,\text{eV} \cdot \text{nm}}{1.00 \times 10^9\,\text{eV}} = 1.24 \times 10^{-12}\,\text{nm}.$$

12P

(a) The kinetic energy acquired is $K = qV$, where q is the charge on an ion and V is the accelerating potential. Thus $K = (1.60 \times 10^{-19}\,\text{C})(300\,\text{V}) = 4.80 \times 10^{-17}\,\text{J}$. The mass of a single sodium atom is, from Appendix D, $m = (22.9898\,\text{g/mol})/(6.02 \times 10^{23}\,\text{atom/mol}) = 3.819 \times 10^{-23}\,\text{g} = 3.819 \times 10^{-26}\,\text{kg}$. Thus the momentum of an ion is

$$p = \sqrt{2mK} = \sqrt{2(3.819 \times 10^{-26}\,\text{kg})(4.80 \times 10^{-17}\,\text{J})} = 1.91 \times 10^{-21}\,\text{kg} \cdot \text{m/s}.$$

(b) The de Broglie wavelength is

$$\lambda = \frac{h}{p} = \frac{6.63 \times 10^{-34}\,\text{J} \cdot \text{s}}{1.91 \times 10^{-21}\,\text{kg} \cdot \text{m/s}} = 3.47 \times 10^{-13}\,\text{m}.$$

15P

The wavelength associated with the unknown particle is $\lambda_p = h/p_p = h/(m_p v_p)$, where p_p is its momentum, m_p is its mass, and v_p is its speed. The classical relationship $p_p = m_p v_p$ was used. Similarly, the wavelength associated with the electron is $\lambda_e = h/(m_e v_e)$, where m_e is its mass and v_e is its speed. The ratio of the wavelengths is $\lambda_p/\lambda_e = (m_e v_e)/(m_p v_p)$, so

$$m_p = \frac{v_e \lambda_e}{v_p \lambda_p} m_e = \frac{9.109 \times 10^{-31}\,\text{kg}}{3(1.813 \times 10^{-4})} = 1.675 \times 10^{-27}\,\text{kg}.$$

According to Appendix B this is the mass of a neutron.

17P

The same resolution requires the same wavelength and since the wavelength and particle momentum are related by $p = h/\lambda$, this means the same particle momentum. The momentum of a 100-keV photon is $p = E/c = (100 \times 10^3 \, \text{eV})(1.60 \times 10^{-19} \, \text{J/eV})/(3.00 \times 10^8 \, \text{m/s}) = 5.33 \times 10^{-23} \, \text{kg·m/s}$. This is also the magnitude of the momentum of the electron. The kinetic energy of the electron is

$$K = \frac{p^2}{2m} = \frac{(5.33 \times 10^{-23} \, \text{kg} \cdot \text{m/s})^2}{2(9.11 \times 10^{-31} \, \text{kg})} = 1.56 \times 10^{-15} \, \text{J}.$$

The accelerating potential is

$$V = \frac{K}{e} = \frac{1.56 \times 10^{-15} \, \text{J}}{1.60 \times 10^{-19} \, \text{C}} = 9.76 \times 10^3 \, \text{V}.$$

23E

(a) Use Eq. 44–10, with $n = 1$ and L equal to the atomic diameter, to estimate the energy:

$$E = n^2 \frac{h^2}{8mL^2} = \frac{(6.63 \times 10^{-34} \, \text{J} \cdot \text{s})^2}{8(9.11 \times 10^{-31} \, \text{kg})(1.4 \times 10^{-14} \, \text{m})^2} = 3.07 \times 10^{-10} \, \text{J} = 1920 \, \text{MeV}.$$

(b) The available binding energy is only a few MeV, so we cannot expect to find an electron bound inside a nucleus.

26P

(a) The allowed values of the energy for a particle of mass m trapped in a box of length L are given by

$$E = n^2 \frac{h^2}{8mL^2},$$

where n is an integer. Since $(2)^2 - (1)^2 = 3$, the difference in energy of the two lowest energy states is

$$\Delta E = \frac{3h^2}{8mL^2}.$$

The mass of an argon atom is $m = (39.9 \, \text{g/mol})/(6.02 \times 10^{23} \, \text{atoms/mol}) = 6.63 \times 10^{-23} \, \text{g} = 6.63 \times 10^{-26} \, \text{kg}$. Thus

$$\Delta E = \frac{3(6.63 \times 10^{-34} \, \text{J} \cdot \text{s})^2}{8(6.63 \times 10^{-26} \, \text{kg})(20 \times 10^{-2} \, \text{m})^2} = 6.2 \times 10^{-41} \, \text{J} = 3.9 \times 10^{-22} \, \text{eV}.$$

(b) The thermal energy (average kinetic energy) at absolute temperature T is given by $\overline{K} = \frac{3}{2}kT$, where k is the Boltzmann constant. Thus

$$\overline{K} = \frac{3}{2}(1.38 \times 10^{-23} \, \text{J/K})(300 \, \text{K}) = 6.21 \times 10^{-21} \, \text{J} = 3.88 \times 10^{-2} \, \text{eV},$$

greater than the separation of the two lowest levels by a factor of about 10^{20}.

(c) Set $\frac{3}{2}kT$ equal to ΔE and solve for T:

$$T = \frac{2\Delta E}{3k} = \frac{2(3.88 \times 10^{-22}\,\text{eV})(1.60 \times 10^{-19}\,\text{J/eV})}{3(1.38 \times 10^{-23}\,\text{J/K})} = 3.0 \times 10^{-18}\,\text{K}.$$

28P

First evaluate the integral

$$I = \int_0^L |\psi(x)|^2\,dx = A^2 \int_0^L \sin^2(\pi x/L)\,dx.$$

Change the variable of integration to $u = \pi x/L$. Then $dx = (L/\pi)\,du$, the lower limit is $u = 0$, and the upper limit is $u = \pi L/L = \pi$. Since $\cos(2u) = \cos^2 u - \sin^2 u = 1 - 2\sin^2 u$, $\sin^2 u = \frac{1}{2}[1 - \cos(2u)]$, and the integral becomes

$$I = \frac{LA^2}{2\pi} \int_0^\pi [1 - \cos(2u)]\,du = \frac{LA^2}{2\pi} \left[u - \frac{1}{2}\sin(2u)\right]_0^\pi = \frac{LA^2}{2\pi}\,\pi = \frac{LA^2}{2}.$$

To normalize the wave function, set $I = 1$ and solve for A. The result is $A = \sqrt{2/L}$.

31E

According to Sample Problem 44–6 the probability the electron in the ground state of a hydrogen atom will be found inside a sphere of radius r is given by

$$p(r) = 1 - e^{-2x}\left(1 + 2x + 2x^2\right),$$

where $x = r/r_B$ and r_B is the Bohr radius. You want $r = r_B$, so $x = 1$ and

$$p(r_B) = 1 - e^{-2}\left(1 + 2 + 2\right) = 1 - 5e^{-2} = 0.323.$$

36P

Use the equation given in Sample Problem 44–6: the probability an electron in the ground state of a hydrogen atom will be found inside a sphere of radius r is given by

$$p(r) = 1 - e^{-2x}\left(1 + 2x + 2x^2\right),$$

where $x = r/r_B$ and r_B is the Bohr radius. You want $r = 1.1 \times 10^{-15}\,\text{m}$. Since $r_B = 5.292 \times 10^{-11}\,\text{m}$, $x = (1.1 \times 10^{-15}\,\text{m})/(5.292 \times 10^{-11}\,\text{m}) = 2.08 \times 10^{-5}$. Since x is small and $p(r)$ is the difference between 1 and a number very close to 1, your calculator probably does not carry enough significant digits for you to obtain an accurate answer. It is better to use a power series expansion. The expansion of e^{-2x} for small x is $1 - 2x + 2x^2 - (4/3)x^3 + \ldots$, so

$$p(r) \approx 1 - \left(1 - 2x + 2x^2 - \frac{4}{3}x^3\right)\left(1 + 2x + 2x^3\right)$$

$$= 1 - \left(1 - \frac{4}{3}x^3\right) = \frac{4}{3}x^3 = \frac{4}{3}\left(2.08 \times 10^{-5}\right)^3 = 1.2 \times 10^{-14}.$$

When multiplying the quantities in parentheses, terms of the order x^3 and lower were retained; higher order terms were discarded.

37E

The probability T that a particle of mass m and energy E tunnels through a barrier of height U and width L is given by

$$T = e^{-2kL},$$

where

$$k = \sqrt{\frac{8\pi^2 m(U - E)}{h^2}}.$$

For the proton

$$k = \sqrt{\frac{8\pi^2(1.6726 \times 10^{-27}\,\text{kg})(10\,\text{MeV} - 3.0\,\text{MeV})(1.6022 \times 10^{-13}\,\text{J/MeV})}{(6.6261 \times 10^{-34}\,\text{J} \cdot \text{s})^2}}$$

$$= 5.8082 \times 10^{14}\,\text{m}^{-1},$$

$kL = (5.8082 \times 10^{14}\,\text{m}^{-1})(10 \times 10^{-15}\,\text{m}) = 5.8082$, and

$$T = e^{-2 \times 5.8082} = 9.02 \times 10^{-6}.$$

The value of k was computed to a greater number of significant digits than usual because an exponential is quite sensitive to the value of the exponent.

The mass of a deuteron is $2.0141\,\text{u} = 3.3454 \times 10^{-27}\,\text{kg}$, so

$$k = \sqrt{\frac{8\pi^2(3.3454 \times 10^{-27}\,\text{kg})(10\,\text{MeV} - 3.0\,\text{MeV})(1.6022 \times 10^{-13}\,\text{J/MeV})}{(6.6261 \times 10^{-34}\,\text{J} \cdot \text{s})^2}}$$

$$= 8.2143 \times 10^{14}\,\text{m}^{-1},$$

$kL = (8.2143 \times 10^{14}\,\text{m}^{-1})(10 \times 10^{-15}\,\text{m}) = 8.2143$, and

$$T = e^{-2 \times 8.2143} = 7.33 \times 10^{-8}.$$

40P

(a) If m is the mass of the particle and E is its energy then the probability it will tunnel through a barrier of height U and width L is given by

$$T = e^{-2kL},$$

where

$$k = \sqrt{\frac{8\pi^2 m(U - E)}{h^2}}.$$

If the change ΔU in U is small (as it is) the change in the tunneling probability is given by

$$\Delta T = \frac{\mathrm{d}T}{\mathrm{d}U}\,\Delta U = -2LT\frac{\mathrm{d}k}{\mathrm{d}U}\,\Delta U\,.$$

Now

$$\frac{\mathrm{d}k}{\mathrm{d}U} = \frac{1}{2\sqrt{U-E}}\sqrt{\frac{8\pi^2 m}{h^2}} = \frac{1}{2(U-E)}\sqrt{\frac{8\pi^2 m(U-E)}{h^2}} = \frac{k}{2(U-E)}\,.$$

Thus

$$\Delta T = -LTk\frac{\Delta U}{U-E}\,.$$

For the data of Sample Problem 44–7, $2kL = 10.0$, so $kL = 5.0$ and

$$\frac{\Delta T}{T} = -kL\frac{\Delta U}{U-E} = -(5.0)\frac{(0.0100)(6.8\,\mathrm{eV})}{6.8\,\mathrm{eV} - 5.1\,\mathrm{eV}} = -0.20\,.$$

There is a 20% decrease in the tunneling probability.

(b) The change in the tunneling probability is given by

$$\Delta T = \frac{\mathrm{d}T}{\mathrm{d}L}\,\Delta L = -2ke^{-2kL}\,\Delta L = -2kT\,\Delta L$$

and

$$\frac{\Delta T}{T} = -2k\,\Delta L = -2(6.67\times 10^9\,\mathrm{m}^{-1})(0.0100)(750\times 10^{-12}\,\mathrm{m}) = -0.10\,.$$

There is a 10% decrease in the tunneling probability.

(c) The change in the tunneling probability is given by

$$\Delta T = \frac{\mathrm{d}T}{\mathrm{d}E}\,\Delta E = -2Le^{-2kL}\frac{\mathrm{d}k}{\mathrm{d}E}\,\Delta E = -2LT\frac{\mathrm{d}k}{\mathrm{d}E}\,\Delta E\,.$$

Now $\mathrm{d}k/\mathrm{d}E = -\mathrm{d}k/\mathrm{d}U = -k/2(U-E)$, so

$$\frac{\Delta T}{T} = kL\frac{\Delta E}{U-E} = (5.0)\frac{(0.01000)(5.1\,\mathrm{eV})}{6.8\,\mathrm{eV} - 5.1\,\mathrm{eV}} = 0.15\,.$$

There is a 15% increase in the tunneling probability.

46P

If Δx is the uncertainty in the position and Δp is the uncertainty in the momentum of a particle, then the uncertainty principle tells us that $\Delta x\,\Delta p \geq h$, where h is the Planck constant. Now $p = mv$, where v is the velocity and m is the mass, so $\Delta p = m\,\Delta v$ and the uncertainty principle becomes $m\,\Delta x\,\Delta v \geq h$. Put $\Delta x = \lambda$, where λ is the de Broglie wavelength, and use $\lambda = h/p = h/mv$ to obtain $\Delta x = h/mv$. The uncertainty principle now becomes $m(h/mv)(\Delta v) \geq h$ or $\Delta v \geq v$. The minimum uncertainty in the velocity is the velocity itself.

Chapter 45

4E

(a) For a given value of the principal quantum number n, the orbital quantum number ℓ ranges from 0 to $n - 1$. For $n = 3$ there are 3 possible values: 0, 1, and 2.

(b) For a given value of ℓ, the magnetic quantum number m_ℓ ranges from $-\ell$ to $+\ell$. For $\ell = 1$, there are 3 possible values: -1, 0, and $+1$.

7E

The principal quantum number n must be greater than 3. The magnetic quantum number m_ℓ can have any of the values -3, -2, -1, 0, $+1$, $+2$, or $+3$. The spin quantum number can have either of the values $-\frac{1}{2}$ or $+\frac{1}{2}$.

10P

(a) In the Bohr model of the hydrogen atom the electron travels in a circular orbit. Let r be the radius of the orbit and v be the speed of the electron. The magnitude of the magnetic dipole moment is given by $\mu = iA$, where i is the current and A ($= \pi r^2$) is the area of the orbit. The current is $i = e/T$, where T is the period of the motion. Since $T = 2\pi r/v$, $i = ev/2\pi r$ and

$$\mu = \left(\frac{ev}{2\pi r}\right)\left(\pi r^2\right) = \frac{erv}{2}.$$

The orbital angular momentum of the electron is given by $L = mrv$, so $rv = L/m$ and

$$\mu = \frac{eL}{2m}.$$

Bohr assumed the angular momentum could have only the values $n\hbar$, where n is a positive integer and $\hbar$ is the Planck constant divided by 2π (see Chapter 43 of the text). This means

$$\mu = \frac{en\hbar}{2m} = n\mu_B,$$

where μ_B ($= e\hbar/2m$) is the Bohr magneton.

(b) The Bohr model predicts that the orbital angular momentum of the electron is a multiple of the Bohr magneton, which is correct, but it gives an incorrect relationship between the orbital angular momentum and the principal quantum number n. For most values of n the magnetic moment can have any of a number of values, not just one.

11P

Use $L_z = m_\ell\hbar$ to calculate the z component of the angular momentum, $\mu_z = -m_\ell\mu_B$ to calculate the z component of the magnetic dipole moment, and $\cos\theta = m_\ell/\sqrt{\ell(\ell+1)}$ to calculate the angle between the angular momentum vector and the z axis. Here $\hbar$ is the Planck constant divided by 2π ($\hbar = 1.055 \times 10^{-34}$ J·s) and μ_B is the Bohr magneton ($\mu_B = 9.274 \times 10^{-24}$ J/T). Spin was ignored in computing the dipole moment. For $\ell = 3$ the magnetic quantum number m_ℓ can take on the values $-3, -2, -1, 0, +1, +2, +3$. Results are tabulated below.

m_ℓ	L_z (kg·m^2/s)	μ_z (J/T)	θ (degrees)
-3	-3.16×10^{-34}	2.78×10^{-23}	$150.0°$
-2	-2.10×10^{-34}	1.85×10^{-23}	$125.3°$
-1	-1.05×10^{-34}	9.27×10^{-24}	$106.8°$
0	0	0	$90.0°$
1	1.05×10^{-34}	-9.27×10^{-24}	$73.2°$
2	2.10×10^{-34}	-1.85×10^{-23}	$54.7°$
3	3.16×10^{-34}	-2.78×10^{-23}	$30.0°$

The magnitude of $\mathbf{L}$ is $\sqrt{\ell(\ell+1)}\,\hbar = \sqrt{3 \times 4}\,(1.055 \times 10^{-34} \text{ J·s}) = 3.655 \times 10^{-34}$ kg·m^2/s. The magnitude of the dipole moment is $\mu_B\sqrt{\ell(\ell+1)} = (9.274 \times 10^{-24} \text{ J/T})\sqrt{3 \times 4} = 3.21 \times 10^{-23}$ J/T.

15P

Since $L^2 = L_x^2 + L_y^2 + L_z^2$, $\sqrt{L_x^2 + L_y^2} = \sqrt{L^2 - L_z^2}$. Replace L^2 with $\ell(\ell+1)\hbar^2$ and L_z with $m_\ell \hbar$ to obtain

$$\sqrt{L_x^2 + L_y^2} = \hbar\sqrt{\ell(\ell+1) - m_\ell^2}.$$

For a given value of ℓ, the greatest that m_ℓ can be is ℓ, so the smallest that $\sqrt{L_x^2 + L_y^2}$ can be is $\hbar\sqrt{\ell(\ell+1) - \ell^2} = \hbar\sqrt{\ell}$. The smallest possible magnitude of m_ℓ is zero, so the largest $\sqrt{L_x^2 + L_y^2}$ can be is $\hbar\sqrt{\ell(\ell+1)}$. Thus

$$\hbar\sqrt{\ell} \le \sqrt{L_x^2 + L_y^2} \le \hbar\sqrt{\ell(\ell+1)}.$$

19E

The radial probability density for the ground state of hydrogen is given by Eq. 45–13:

$$P(r) = \left(\frac{4r^2}{r_B^3}\right) e^{-2r/r_B},$$

where r_B is the Bohr radius. The probability that the electron is in a spherical shell with inner radius r_1 and outer radius r_2 is

$$p = \int_{r_1}^{r_2} P(r)\,dr = \frac{4}{r_B^3} \int_{r_1}^{r_2} r^2\, e^{-2r/r_B}\,dr\,.$$

The indefinite integral $\int x^2\, e^{-ar}\,dx$ is given in Appendix G as $-(1/a^3)(a^2 x^2 + 2ax + 2)e^{-ax}$. Let $x = r$ and $a = 2/r_B$ to obtain

$$p = -\frac{4}{r_B^3}\,\frac{r_B^3}{8}\left[\frac{4r^2}{r_B^2} + \frac{4r}{r_B} + 2\right]e^{-2r/r_B}\bigg|_{r_1}^{r_2}$$

$$= \frac{1}{2}\left[\frac{4r_1^2}{r_B^2} + \frac{4r_1}{r_B} + 2\right]e^{-2r_1/r_B} - \frac{1}{2}\left[\frac{4r_2^2}{r_B^2} + \frac{4r_2}{r_B} + 2\right]e^{-2r_2/r_B}\,.$$

Let $r_1 = 1.00r_B$ and $r_2 = 1.01r_B$. Then

$$p = \frac{1}{2}\left[4(1.00)^2 + 4(1.00) + 2\right]e^{-2\times1.00} - \frac{1}{2}\left[4(1.01)^2 + 4(1.00) + 2\right]e^{-2\times1.01}$$

$$= 0.00541\,.$$

24P

The radial probability density for the $n = 2$, $\ell = 0$, $m_\ell = 0$ state of hydrogen is given by Eq. 45–14:

$$P(r) = \left(\frac{r^2}{8r_B^3}\right)\left(2 - \frac{r}{r_B}\right)^2 e^{-r/r_B}\,,$$

where r_B is the Bohr radius. Replace r/r_B with x to obtain

$$8r_B P = x^2\,(2-x)^2\,e^{-x}\,.$$

Differentiate with respect to x:

$$8r_B\,\frac{dP}{dx} = 2x(2-x)^2\,e^{-x} - 2x^2(2-x)\,e^{-x} - x^2(2-x)^2\,e^{-x} = x(2-x)(4 - 6x + x^2)e^{-x}\,.$$

There are 5 zeros. Three occur at $x = 0$, $x = 2$, and $x = \infty$. For each of these $P = 0$. They are minima. For the maxima, $4 - 6x + x^2 = 0$. The solutions to this quadratic equation are $x = 3 - \sqrt{5} = 0.7639$ and $x = 3 + \sqrt{5} = 5.236$. Thus the maxima of the radial distribution function occur at $r = 0.7639r_B$ and $r = 5.236r_B$.

29E

The magnitude of the spin angular momentum is $S = \sqrt{s(s+1)}\,\hbar = (\sqrt{3}/2)\hbar$, where $s = \frac{1}{2}$ was used. The z component is either $S_z = \hbar/2$ or $-\hbar/2$. If $S_z = +\hbar/2$ the angle θ between the spin angular momentum vector and the positive z axis is

$$\theta = \cos^{-1}\left(\frac{S_z}{S}\right) = \cos^{-1}\left(\frac{1}{\sqrt{3}}\right) = 54.7°\,.$$

If $S_z = -\hbar/2$ the angle is $\theta = 180° - 54.7° = 125.3°$.

30E

The acceleration is

$$a = \frac{F}{M} = \frac{(\mu \cos \theta)(dB/dz)}{M},$$

where M is the mass of a silver atom, μ is its magnetic dipole moment, B is the magnetic field, and θ is the angle between the dipole moment and the magnetic field. Take the moment and the field to be parallel ($\cos \theta = 1$) and use the data given in Sample Problem 45–3 to obtain

$$a = \frac{(9.27 \times 10^{-24}\,\text{J/T})(1.4 \times 10^3\,\text{T/m})}{1.8 \times 10^{-25}\,\text{kg}} = 7.21 \times 10^4\,\text{m/s}^2.$$

33E

The energy of a magnetic dipole in an external magnetic field $\mathbf{B}$ is $U = -\boldsymbol{\mu} \cdot \mathbf{B} = -\mu_z B$, where $\boldsymbol{\mu}$ is the dipole moment and μ_z is its component along the field. The energy required to change the moment direction from parallel to antiparallel is $\Delta E = 2\mu_z B$. Since the z component of the spin magnetic moment of an electron is the Bohr magneton μ_B, $\Delta E = 2\mu_B B = 2(9.274 \times 10^{-24}\,\text{J/T})(0.200\,\text{T}) = 3.71 \times 10^{-24}\,\text{J}$. The photon wavelength is

$$\lambda = \frac{c}{f} = \frac{hc}{\Delta E} = \frac{(6.626 \times 10^{-34}\,\text{J} \cdot \text{s})(3.00 \times 10^8\,\text{m/s})}{3.71 \times 10^{-24}\,\text{J}} = 5.36 \times 10^{-2}\,\text{m},$$

where $f = \Delta E/h$ was used.

39E

(a) All states with principal quantum number $n = 1$ are filled. The next lowest states have $n = 2$. The orbital quantum number can have the values $\ell = 0$ or 1 and of these the $\ell = 0$ states have the lowest energy. The magnetic quantum number must be $m_\ell = 0$ since this is the only possibility if $\ell = 0$. The spin quantum number can have either of the values $m_s = -\frac{1}{2}$ or $+\frac{1}{2}$. Since there is no external magnetic field the energies of these two states are the same. Thus in the ground state the quantum numbers of the third electron are either $n = 2$, $\ell = 0$, $m_\ell = 0$, $m_s = -\frac{1}{2}$ or $n = 2$, $\ell = 0$, $m_\ell = 0$, $m_s = +\frac{1}{2}$.

(b) The next lowest state in energy is an $n = 2$, $\ell = 1$ state. All $n = 3$ states are higher in energy. The magnetic quantum number can be $m_\ell = -1, 0$, or $+1$; the spin quantum number can be $m_s = -\frac{1}{2}$ or $+\frac{1}{2}$. If both external and internal magnetic fields can be neglected, all these states have the same energy.

41P

For a given value of the principal quantum number n there are n possible values of the orbital quantum number ℓ, ranging from 0 to $n - 1$. For any value of ℓ there are $2\ell + 1$ possible values of the magnetic quantum number m_ℓ, ranging from $-\ell$ to $+\ell$. Finally, for each set of values of ℓ and m_ℓ there are two states, one corresponding to the spin quantum number $m_s = -\frac{1}{2}$

and the other corresponding to $m_s = +\frac{1}{2}$. Hence the total number of states with principal quantum number n is

$$N = 2 \sum_0^{n-1} (2\ell + 1).$$

Now

$$\sum_0^{n-1} 2\ell = 2 \sum_0^{n-1} \ell = 2\frac{n}{2}(n-1) = n(n-1),$$

since there are n terms in the sum and the average term is $(n-1)/2$. Furthermore,

$$\sum_0^{n-1} 1 = n.$$

Thus

$$N = 2\left[n(n-1) + n\right] = 2n^2.$$

45E

The kinetic energy of the electron is eV, where V is the accelerating potential. A photon with the minimum wavelength is produced when all of the electron's kinetic energy goes to a single photon in a bremsstrahlung event. Thus

$$eV = \frac{hc}{\lambda_{\min}} = \frac{1240\,\mathrm{eV}\cdot\mathrm{nm}}{0.10\,\mathrm{nm}} = 1.24 \times 10^4\,\mathrm{eV},$$

where the result of Exercise 1 of Chapter 43 was used. The accelerating potential is $V = 1.24 \times 10^4\,\mathrm{V} = 12.4\,\mathrm{kV}$.

47P

The initial kinetic energy of the electron is $50.0\,\mathrm{keV}$. After the first collision the kinetic energy is $25\,\mathrm{keV}$, after the second it is $12.5\,\mathrm{keV}$, and after the third it is zero. The energy of the photon produced in the first collision is $50.0\,\mathrm{keV} - 25.0\,\mathrm{keV} = 25.0\,\mathrm{keV}$. The wavelength associated with the photon is

$$\lambda = \frac{1240\,\mathrm{eV}\cdot\mathrm{nm}}{25.0 \times 10^3\,\mathrm{eV}} = 4.96 \times 10^{-2}\,\mathrm{nm} = 49.6\,\mathrm{pm},$$

where the result of Exercise 1 of Chapter 43 was used. The energies of the photons produced in the second and third collisions are each $12.5\,\mathrm{keV}$ and their wavelengths are

$$\lambda = \frac{1240\,\mathrm{eV}\cdot\mathrm{nm}}{12.5 \times 10^3\,\mathrm{eV}} = 9.92 \times 10^{-2}\,\mathrm{nm} = 99.2\,\mathrm{pm}.$$

48P

Suppose an electron with total energy E and momentum $\mathbf{p}$ spontaneously changes into a photon. If energy is conserved, the energy of the photon is E and its momentum has magnitude

E/c. Now the energy and momentum of the electron are related by $E^2 = (pc)^2 + (mc^2)^2$, so $pc = \sqrt{E^2 - (mc^2)^2}$. Since the electron has non-zero mass E/c and p cannot have the same value. Hence momentum cannot be conserved. A third particle must participate in the interaction, primarily to conserve momentum. It does, however, carry off some energy.

50E

(a) The cut-off wavelength λ_{min} is characteristic of the incident electrons, not of the target material. This wavelength is the wavelength of a photon with energy equal to the kinetic energy of an incident electron. According to the result of Exercise 1 of Chapter 43,

$$\lambda_{min} = \frac{1240\,\text{eV} \cdot \text{nm}}{35 \times 10^3\,\text{eV}} = 3.54 \times 10^{-2}\,\text{nm} = 35.4\,\text{pm}.$$

(b) A K_α photon results when an electron in a target atom jumps from the L-shell to the K-shell. The energy of this photon is $25.51\,\text{keV} - 3.56\,\text{keV} = 21.95\,\text{keV}$ and its wavelength is $\lambda_{K\alpha} = (1240\,\text{eV} \cdot \text{nm})/(21.95 \times 10^3\,\text{eV}) = 5.65 \times 10^{-2}\,\text{nm} = 56.5\,\text{pm}$.

(c) A K_β photon results when an electron in a target atom jumps from the M-shell to the K-shell. The energy of this photon is $25.51\,\text{keV} - 0.53\,\text{keV} = 24.98\,\text{keV}$ and its wavelength is $\lambda_{K\beta} = (1240\,\text{eV} \cdot \text{nm})/(24.98 \times 10^3\,\text{eV}) = 4.96 \times 10^{-2}\,\text{nm} = 49.6\,\text{pm}$.

53E

As shown in Sample Problem 45–7, the ratio of the wavelength λ_{Nb} for the K_α line of niobium to the wavelength λ_{Ga} for the K_α line of gallium is given by

$$\frac{\lambda_{Nb}}{\lambda_{Ga}} = \left[\frac{Z_{Ga} - 1}{Z_{Nb} - 1} \right]^2,$$

where Z_{Nb} is the atomic number of niobium (41) and the Z_{Ga} is the atomic number of gallium (31). Thus

$$\frac{\lambda_{Nb}}{\lambda_{Ga}} = \left[\frac{30}{40} \right]^2 = \frac{9}{16}.$$

55P

(a) An electron must be removed from the K-shell, so that an electron from a higher energy shell can drop. This requires an energy of 69.5 keV. The accelerating potential must be at least 69.5 kV.

(b) After it is accelerated the kinetic energy of the bombarding electron is 69.5 keV. The energy of a photon associated with the minimum wavelength is 69.5 keV, so its wavelength is

$$\lambda_{min} = \frac{1240\,\text{eV} \cdot \text{nm}}{69.5 \times 10^3\,\text{eV}} = 1.78 \times 10^{-2}\,\text{nm} = 17.8\,\text{pm}.$$

(c) The energy of a photon associated with the K_α line is $69.5\,\text{keV} - 11.3\,\text{keV} = 58.2\,\text{keV}$ and its wavelength is $\lambda_{K\alpha} = (1240\,\text{eV} \cdot \text{nm})/(58.2 \times 10^3\,\text{eV}) = 2.13 \times 10^{-2}\,\text{nm} = 21.3\,\text{pm}$.

The energy of a photon associated with the K_β line is $69.5 \text{ keV} - 2.30 \text{ keV} = 67.2 \text{ keV}$ and its wavelength is $\lambda_{K_\beta} = (1240 \text{ eV} \cdot \text{nm})/(67.2 \times 10^3 \text{ eV}) = 1.85 \times 10^{-2} \text{ nm} = 18.5 \text{ pm}$.

60E

The number of atoms in a state with energy E is proportional to $e^{-E/kT}$, where T is the temperature on the Kelvin scale and k is the Boltzmann constant. Thus the ratio of the number of atoms in the thirteenth excited state to the number in the eleventh excited state is

$$\frac{n_{13}}{n_{11}} = e^{-\Delta E/kT},$$

where ΔE is the difference in the energies: $\Delta E = E_{13} - E_{11} = 2(1.2 \text{ eV}) = 2.4 \text{ eV}$. For the given temperature $kT = (8.62 \times 10^{-2} \text{ eV/K})(2000 \text{ K}) = 0.1724 \text{ eV}$. Hence

$$\frac{n_{13}}{n_{11}} = e^{-2.4/0.1724} = 9.0 \times 10^{-7}.$$

64E

(a) If t is the time interval over which the pulse is emitted, the length of the pulse is $L = ct = (3.00 \times 10^8 \text{ m/s})(1.20 \times 10^{-11} \text{ s}) = 3.60 \times 10^{-3} \text{ m}$.

(b) If E_p is the energy of the pulse, E is the energy of a single photon in the pulse, and N is the number of photons in the pulse, then $E_p = NE$. The energy of the pulse is $E_p = (0.150 \text{ J})/(1.602 \times 10^{-19} \text{ J/eV}) = 9.36 \times 10^{17} \text{ eV}$ and the energy of a single photon is $E = (1240 \text{ eV} \cdot \text{nm})/694.4 \text{ nm}) = 1.786 \text{ eV}$. Hence

$$N = \frac{E_p}{E} = \frac{9.36 \times 10^{17} \text{ eV}}{1.786 \text{ eV}} = 5.24 \times 10^{17} \text{ photons}.$$

68P

(a) If both mirrors are perfectly reflecting there is a node at each end of the crystal. With one end partially silvered there is a node very close to that end. Assume nodes at both ends, so there are an integer number of half-wavelengths in the length of the crystal. The wavelength in the crystal is $\lambda_c = \lambda/n$, where λ is the wavelength in vacuum and n is the index of refraction of ruby. Thus $N(\lambda/2n) = L$, where N is the number of standing wave nodes, so

$$N = \frac{2nL}{\lambda} = \frac{2(1.75)(0.0600 \text{ m})}{694 \times 10^{-9} \text{ m}} = 3.03 \times 10^5.$$

(b) Since $\lambda = c/f$, where f is the frequency, $N = 2nLf/c$ and $\Delta N = (2nL/c)\Delta f$. Hence

$$\Delta f = \frac{c \Delta N}{2nL} = \frac{(3.00 \times 10^8 \text{ m/s})(1)}{2(1.75)(0.0600 \text{ m})} = 1.43 \times 10^9 \text{ Hz}.$$

The speed of light in the crystal is c/n and the round-trip distance is $2L$, so the round-trip travel time is $2nL/c$. This is the same as the reciprocal of the change in frequency.

(c) The frequency is $f = c/\lambda = (3.00 \times 10^8 \text{ m/s})/(694 \times 10^{-9} \text{ m}) = 4.32 \times 10^{14} \text{ Hz}$ and the fractional change in the frequency is $\Delta f/f = (1.43 \times 10^9 \text{ Hz})/(4.32 \times 10^{14} \text{ Hz}) = 3.31 \times 10^{-6}$.

72P

(a) The intensity at the target is given by $I = P/A$, where P is the power output of the source and A is the area of the beam at the target. You want to compute I and compare the result with 10^8 W/m^2.

The beam spreads because diffraction occurs at the aperture of the laser. Consider the part of the beam that is within the central diffraction maximum. The angular position of the edge is given by $\sin\theta = 1.22\lambda/d$, where λ is the wavelength and d is the diameter of the aperture (see Exercise 67). At the target, a distance D away, the radius of the beam is $r = D\tan\theta$. Since θ is small we may approximate both $\sin\theta$ and $\tan\theta$ by θ, in radians. Then $r = D\theta = 1.22D\lambda/d$ and

$$I = \frac{P}{\pi r^2} = \frac{Pd^2}{\pi(1.22D\lambda)^2} = \frac{(5.0 \times 10^6 \text{ W})(4.0 \text{ m})^2}{\pi\left[(1.22(3000 \times 10^3 \text{ m})(3.0 \times 10^{-6} \text{ m})\right]^2}$$

$$= 2.1 \times 10^5 \text{ W/m}^2,$$

not great enough to destroy the missile.

(b) Solve for the wavelength in terms of the intensity and substitute $I = 1.0 \times 10^8 \text{ W/m}^2$:

$$\lambda = \frac{d}{1.22D}\sqrt{\frac{P}{\pi I}} = \frac{4.0 \text{ m}}{1.22(3000 \times 10^3 \text{ m})}\sqrt{\frac{5.0 \times 10^6 \text{ W}}{\pi(1.0 \times 10^8 \text{ W/m}^2)}}$$

$$= 1.4 \times 10^{-7} \text{ m} = 140 \text{ nm}.$$

Chapter 46

2E

The number of atoms per unit volume is given by $n = d/M$, where d is the mass density of gold and M is the mass of a single gold atom. Since each atom contributes one conduction electron, n is also the number of conduction electrons per unit volume. Since the molar mass of gold is $A = 197\,\text{g/mol}$, $M = A/N_A = (197\,\text{g/mol})/(6.022 \times 10^{23}\,\text{mol}^{-1}) = 3.271 \times 10^{-22}\,\text{g}$. Thus

$$n = \frac{19.3\,\text{g/cm}^3}{3.271 \times 10^{-22}\,\text{g}} = 5.90 \times 10^{22}\,\text{cm}^{-3}.$$

6E

(a) Eq. 46–3 gives

$$n(E) = \frac{8\sqrt{2}\pi m^{3/2}}{h^3} E^{1/2}$$

for the density of states associated with the conduction electrons of a metal. This can be written

$$n(E) = CE^{1/2},$$

where

$$C = \frac{8\sqrt{2}\pi m^{3/2}}{h^3} = \frac{8\sqrt{2}\pi(9.109 \times 10^{-31}\,\text{kg})^{3/2}}{(6.626 \times 10^{-34}\,\text{J} \cdot \text{s})^3} = 1.062 \times 10^{56}\,\text{kg}^{3/2}/\text{J}^3 \cdot \text{s}^3.$$

Now $1\,\text{J} = 1\,\text{kg} \cdot \text{m}^2/\text{s}^2$ (think of the equation for kinetic energy $K = \frac{1}{2}mv^2$), so $1\,\text{kg} = 1\,\text{J} \cdot \text{s}^2 \cdot \text{m}^{-2}$. Thus the units of C can be written $(\text{J} \cdot \text{s}^2)^{3/2} \cdot (\text{m}^{-2})^{3/2} \cdot \text{J}^{-3} \cdot \text{s}^{-3} = \text{J}^{-3/2} \cdot \text{m}^{-3}$. This means

$$C = (1.062 \times 10^{56}\,\text{J}^{-3/2} \cdot \text{m}^{-3})(1.602 \times 10^{-19}\,\text{J/eV})^{3/2} = 6.81 \times 10^{27}\,\text{m}^{-3} \cdot \text{eV}^{-3/2}.$$

(b) If $E = 5.00\,\text{eV}$ then

$$n(E) = (6.81 \times 10^{27}\,\text{m}^{-3} \cdot \text{eV}^{-3/2})(5.00\,\text{eV})^{1/2} = 1.52 \times 10^{28}\,\text{eV}^{-1} \cdot \text{m}^{-3}.$$

8E

(a) At absolute temperature $T = 0$, the probability is zero that any state with energy above the Fermi energy is occupied.

(b) The probability that a state with energy E is occupied at temperature T is given by

$$p(E) = \frac{1}{e^{(E-E_F)/kT} + 1},$$

where k is the Boltzmann constant and E_F is the Fermi energy. Now $E - E_F = 0.062\,eV$ and $(E - E_F)/kT = (0.062\,eV)/(8.62 \times 10^{-5}\,eV/K)(320\,K) = 2.248$, so

$$p(E) = \frac{1}{e^{2.248} + 1} = 0.0956.$$

See Sample Problem 46–4 for the value of k.

10E

According to Eq. 46–6 the Fermi energy is given by

$$E_F = \left(\frac{3}{16\sqrt{2\pi}}\right)^{2/3} \frac{h^2}{m} n^{2/3},$$

where n is the number of conduction electrons per unit volume, m is the mass of an electron, and h is the Planck constant. This can be written $E_F = An^{2/3}$, where

$$A = \left(\frac{3}{16\sqrt{2\pi}}\right)^{2/3} \frac{h^2}{m} = \left(\frac{3}{16\sqrt{2\pi}}\right)^{2/3} \frac{(6.626 \times 10^{-34}\,J \cdot s)^2}{9.109 \times 10^{-31}\,kg} = 5.842 \times 10^{-38}\,J^2 \cdot s^2/kg.$$

Since $1\,J = 1\,kg \cdot m^2/s^2$, the units of A can be taken to be $m^2 \cdot J$. Divide by $1.602 \times 10^{-19}\,J/eV$ to obtain $A = 3.65 \times 10^{-19}\,m^2 \cdot eV$.

15E

Let N be the number of atoms per unit volume and n be the number of free electrons per unit volume. Then the number of free electrons per atom is n/N. Use the result of Exercise 10 to find n: $E_F = An^{2/3}$, where $A = 3.65 \times 10^{-19}\,m^2 \cdot eV$. Thus

$$n = \left(\frac{E_F}{A}\right)^{3/2} = \left(\frac{11.6\,eV}{3.65 \times 10^{-19}\,m^2 \cdot eV}\right)^{3/2} = 1.79 \times 10^{29}\,m^{-3}.$$

If M is the mass of a single aluminum atom and d is the mass density of aluminum then $N = d/M$. Now $M = (27.0\,g/mol)/(6.022 \times 10^{23}\,mol^{-1}) = 4.48 \times 10^{-23}\,g$, so $N = (2.70\,g/cm^3)/(4.48 \times 10^{-23}\,g) = 6.03 \times 10^{22}\,cm^{-3} = 6.03 \times 10^{28}\,m^{-3}$. Thus the number of free electrons per atom is

$$\frac{n}{N} = \frac{1.79 \times 10^{29}\,m^{-3}}{6.03 \times 10^{28}\,m^{-3}} = 2.97.$$

20P

Use the result of Exercise 10: at temperature $T = 0$ the Fermi energy is given by $E_F = An^{2/3}$, where n is the number of electrons per unit volume and $A = 3.65 \times 10^{-19}\,m^2 \cdot eV$. If N is the number of atoms per unit volume and each atom contributes Z electrons, then $n = NZ$. If M_S is the mass of the star, R is its radius, and M is the mass of a single iron atom, then $N = (3M_S)/(4\pi R^3 M)$. Now $M_S = 1.99 \times 10^{30}\,kg$, $R = 6.37 \times 10^6\,m$, and

$M = (55.847 \, \text{g/mol})/(6.022 \times 10^{23} \, \text{mol}^{-1}) = 9.27 \times 10^{-23} \, \text{g} = 9.27 \times 10^{-26} \, \text{kg}$. The data used was found in appendices of the text. This means

$$N = \frac{3(1.99 \times 10^{30} \, \text{kg})}{4\pi(6.37 \times 10^6 \, \text{m})^3(9.27 \times 10^{-26} \, \text{kg})} = 1.98 \times 10^{34} \, \text{m}^{-3} \,.$$

The atomic number of iron is 26, so $n = 26(1.98 \times 10^{34} \, \text{m}^{-3}) = 5.15 \times 10^{35} \, \text{m}^{-3}$. The Fermi energy is

$$E_F = (3.65 \times 10^{-19} \, \text{m}^2 \cdot \text{eV}) \left(5.15 \times 10^{35} \, \text{m}^{-3} \right)^{2/3} = 2.35 \times 10^5 \, \text{eV} \,.$$

25P

The fraction f of electrons with energies greater than the Fermi energy is given in Problem 23:

$$f = \frac{3kT/2}{E_F} \,,$$

where T is the temperature on the Kelvin scale, k is the Boltzmann constant, and E_F is the Fermi energy. Solve for T:

$$T = \frac{2fE_F}{3k} = \frac{2(0.013)(4.7 \, \text{eV}))}{3(8.62 \times 10^{-5} \, \text{eV/K})} = 473 \, \text{K} \,.$$

27P

The average energy of the conduction electrons is given by

$$\overline{E} = \frac{1}{n} \int_0^\infty E n(E) p(E) \, \mathrm{d}E \,,$$

where n is the number of free electrons per unit volume, $n(E)$ is the density of states, and $p(E)$ is the probability function. The density of states is proportional to $E^{1/2}$, so we may write $n(E) = CE^{1/2}$, where C is a constant of proportionality. The probability function is unity for energies below the Fermi energy and 0 for energies above. Thus

$$\overline{E} = \frac{C}{n} \int_0^{E_F} E^{3/2} \, \mathrm{d}E = \frac{2C}{5n} E_F^{5/2} \,.$$

Now

$$n = \int_0^\infty n(E) p(E) \, \mathrm{d}E = C \int_0^{E_F} E^{1/2} \, \mathrm{d}E = \frac{2C}{3} E_F^{3/2} \,.$$

Substitute this expression into the formula for $\overline{E}$ to obtain

$$\overline{E} = \left(\frac{2C}{5} \right) E_F^{5/2} \left(\frac{3}{2C E_F^{3/2}} \right) = \frac{3}{5} E_F \,.$$

33P

(a) The probability that a state with energy E is occupied is given by

$$p(E) = \frac{1}{e^{(E-E_F)/kT} + 1},$$

where E_F is the Fermi energy, T is the temperature on the Kelvin scale, and k is the Boltzmann constant. If energies are measured from the top of the valence band, then the energy associated with a state at the bottom of the conduction band is $E = 1.11\,\text{eV}$. Furthermore, $kT = (8.62 \times 10^{-5}\,\text{eV/K})(300\,\text{K}) = 0.02586\,\text{eV}$. For pure silicon $E_F = 0.555\,\text{eV}$ and $(E - E_F)/kT = (0.555\,\text{eV})/(0.02586\,\text{eV}) = 21.46$. Thus

$$p(E) = \frac{1}{e^{21.46} + 1} = 4.79 \times 10^{-10}.$$

For the doped semiconductor $(E - E_F)/kT = (0.11\,\text{eV})/(0.02586\,\text{eV}) = 4.254$ and

$$p(E) = \frac{1}{e^{4.254} + 1} = 1.40 \times 10^{-2}.$$

(b) The energy of the donor state, relative to the top of the valence band, is $1.11\,\text{eV} - 0.15\,\text{eV} = 0.96\,\text{eV}$. The Fermi energy is $1.11\,\text{eV} - 0.11\,\text{eV} = 1.00\,\text{eV}$. Hence $(E - E_F)/kT = (0.96\,\text{eV} - 1.00\,\text{eV})/(0.2586\,\text{eV}) = -1.547$ and

$$p(E) = \frac{1}{e^{-1.547} + 1} = 0.824.$$

36E

To find the maximum number of electrons that can be excited across the gap, assume that the energy received by each electron is exactly the difference in energy between the top of the conduction band and the bottom of the valence band (1.1 eV). Then the number that can be excited is

$$N = \frac{662 \times 10^3\,\text{eV}}{1.1\,\text{eV}} = 6.0 \times 10^5.$$

Since each electron that jumps the gap leaves a hole behind, this is also the maximum number of electron-hole pairs that can be created.

39E

The valence band is essentially filled and the conduction band is essentially empty. If an electron in the valence band is to absorb a photon, the energy it receives must be sufficient to excite it across the band gap. Photons with energies less than the gap width are not absorbed and the semiconductor is transparent to this radiation. Photons with energies greater than the gap width are absorbed and the semiconductor is opaque to this radiation. Thus the width of the band gap is the same as the energy of a photon associated with a wavelength of 295 nm. Use the result of Exercise 1 of Chapter 43 to obtain

$$E_{\text{gap}} = \frac{1240\,\text{eV} \cdot \text{nm}}{\lambda} = \frac{1240\,\text{eV} \cdot \text{nm}}{295\,\text{nm}} = 4.20\,\text{eV}.$$

Chapter 47

13E

The binding energy is given by $E = [Zm_H + (A - Z)m_n - M_{Pu}]c^2$, where Z is the atomic number (number of protons), A is the mass number (number of nucleons), m_H is the mass of a hydrogen atom, m_n is the mass of a neutron, and M_{Pu} is the mass of a $^{239}_{94}$Pu atom. In principal, nuclear masses should have been used, but the mass of the Z electrons included in ZM_H is canceled by the mass of the Z electrons included in M_{Pu}, so the result is the same. First calculate the mass difference in atomic mass units: $\Delta m = (94)(1.00783\,u) + (239 - 94)(1.00867\,u) - (239.05216\,u) = 1.94101\,u$. Since $1\,u$ is equivalent to $931.5\,MeV$, $E = (1.94101\,u)(931.5\,MeV/u) = 1808\,MeV$. Since there are 239 nucleons, the binding energy per nucleon is $E_n = (1808\,MeV)/(239) = 7.56\,MeV$.

16E

(a) The de Broglie wavelength is given by $\lambda = h/p$, where p is the magnitude of the momentum. The kinetic energy K and momentum are related by Eq. 42–40, which yields

$$pc = \sqrt{K^2 + 2Kmc^2} = \sqrt{(200\,MeV)^2 + 2(200\,MeV)(0.511\,MeV)} = 200.5\,MeV.$$

Thus

$$\lambda = \frac{hc}{pc} = \frac{1240\,eV \cdot nm}{200.5 \times 10^6\,eV} = 6.18 \times 10^{-6}\,nm = 6.18\,fm.$$

(b) The diameter of a copper nucleus, for example, is about $8.6\,fm$, just a little larger than the de Broglie wavelength of a 200-MeV electron. To resolve detail the wavelength should be smaller than the target, ideally a tenth of the diameter or less. 200-MeV electrons are perhaps at the lower limit in energy for useful probes.

19P

(a) Since the nuclear force has a short range, any nucleon interacts only with its nearest neighbors, not with more distant nucleons in the nucleus. Let N be the number of neighbors that interact with any nucleon. It is independent of the number A of nucleons in the nucleus. The number of interactions in a nucleus is approximately NA, so the energy associated with the strong nuclear force is proportional to NA and therefore proportional to A itself.

(b) Each proton in a nucleus interacts electrically with every other proton. The number of pairs of protons is $Z(Z - 1)/2$, where Z is the number of protons. The Coulomb energy is therefore proportional to $Z(Z - 1)$.

(c) As A increases, Z increases at a slightly slower rate but Z^2 increases at a faster rate than A and the energy associated with Coulomb interactions increases faster than the energy associated with strong nuclear interactions.

20P

Let f_{24} be the abundance of ^{24}Mg, let f_{25} be the abundance of ^{25}Mg, and let f_{26} be the abundance of ^{26}Mg. Then the entry in the periodic table for Mg is $24.312 = 23.98504 f_{24} + 24.98584 f_{25} + 25.98259 f_{26}$. Since there are only three isotopes, $f_{24} + f_{25} + f_{26} = 1$. Solve for f_{25} and f_{26}. The second equation gives $f_{26} = 1 - f_{24} - f_{25}$. Substitute this expression and $f_{24} = 0.7899$ into the first equation to obtain $24.312 = (23.98504)(0.7899) + 24.98584 f_{25} + 25.98259 - (25.98259)(0.7899) - 25.98259 f_{25}$. The solution is $f_{25} = 0.09303$. Then $f_{26} = 1 - 0.7899 - 0.09303 = 0.1171$. 78.99% of naturally occurring magnesium is ^{24}Mg, 9.30% is ^{25}Mg, and 11.71% is ^{26}Mg.

25P

If a nucleus contains Z protons and N neutrons, its binding energy is $E = (Zm_H + Nm_n - m)c^2$, where m_H is the mass of a hydrogen atom, m_n is the mass of a neutron, and m is the mass of the atom containing the nucleus of interest. If the masses are given in atomic mass units then mass excesses are defined by $\Delta_H = (m_H - 1)c^2$, $\Delta_n = (m_n - 1)c^2$, and $\Delta = (m - A)c^2$. This means $m_H c^2 = \Delta_H + c^2$, $m_n c^2 = \Delta_n + c^2$, and $mc^2 = \Delta + Ac^2$. Thus $E = (Z\Delta_H + N\Delta_n - \Delta) + (Z + N - A)c^2 = Z\Delta_H + N\Delta_n - \Delta$, where $A = Z + N$ was used.

For $^{197}_{79}$Au, $Z = 79$ and $N = 197 - 79 = 118$. Hence

$$E = (79)(7.29 \, \text{MeV}) + (118)(8.07 \, \text{MeV}) - (-31.2 \, \text{MeV}) = 1560 \, \text{MeV}.$$

This means the binding energy per nucleon is $E_n = (1560 \, \text{MeV})/(197) = 7.92 \, \text{MeV}$.

29E

(a) The decay rate is given by $R = \lambda N$, where λ is the disintegration constant and N is the number of undecayed nuclei. Initially $R = R_0 = \lambda N_0$, where N_0 is the number of undecayed nuclei at that time. You must find values for both N_0 and λ. The disintegration constant is related to the half-life τ by $\lambda = (\ln 2)/\tau = (\ln 2)/(78 \, \text{h}) = 8.89 \times 10^{-3} \, \text{h}^{-1}$. If M is the mass of the sample and m is the mass of a single atom of gallium, then $N_0 = M/m$. Now $m = (67 \, \text{u})(1.661 \times 10^{-24} \, \text{g/u}) = 1.113 \times 10^{-22} \, \text{g}$ and $N_0 = (3.4 \, \text{g})/(1.113 \times 10^{-22} \, \text{g}) = 3.05 \times 10^{22}$. Thus $R_0 = (8.89 \times 10^{-3} \, \text{h}^{-1})(3.05 \times 10^{22}) = 2.71 \times 10^{20} \, \text{h}^{-1} = 7.53 \times 10^{16} \, \text{s}^{-1}$.

(b) The decay rate at any time t is given by

$$R = R_0 \, e^{-\lambda t}.$$

where R_0 is the decay rate at $t = 0$. At $t = 48 \, \text{h}$, $\lambda t = (8.89 \times 10^{-3} \, \text{h}^{-1})(48 \, \text{h}) = 0.427$ and

$$R = (7.53 \times 10^{16} \, \text{s}^{-1}) \, e^{-0.427} = 4.91 \times 10^{16} \, \text{s}^{-1}.$$

30E

(a) The half-life τ and the disintegration constant are related by $\tau = (\ln 2)/\lambda$, so $\tau = (\ln 2)/(0.0108 \, \text{h}^{-1}) = 64.2 \, \text{h}$.

(b) At time t the number of undecayed nuclei remaining is given by

$$N = N_0 e^{-\lambda t} = N_0 e^{-(\ln 2)t/\tau}.$$

Substitute $t = 3\tau$ to obtain

$$\frac{N}{N_0} = e^{-3\ln 2} = 0.125.$$

In each half-life the number of undecayed nuclei is reduced by half. At the end of one half-life $N = N_0/2$, at the end of two half-lives $N = N_0/4$, and at the end of three half-lives $N = N_0/8 = 0.125N_0$.

(c) Use

$$N = N_0 e^{-\lambda t}.$$

10.0 d is 240 h, so $\lambda t = (0.0108\,\text{h}^{-1})(240\,\text{h}) = 2.592$ and

$$\frac{N}{N_0} = e^{-2.592} = 0.0749.$$

33E

The rate of decay is given by $R = \lambda N$, where λ is the disintegration constant and N is the number of undecayed nuclei. In terms of the half-life τ the disintegration constant is $\lambda = (\ln 2)/\tau$, so

$$N = \frac{R}{\lambda} = \frac{R\tau}{\ln 2} = \frac{(6000\,\text{Ci})(3.7 \times 10^{10}\,\text{s}^{-1}/\text{Ci})(5.27\,\text{y})(3.16 \times 10^7\,\text{s/y})}{\ln 2}$$
$$= 5.33 \times 10^{22}\,\text{nuclei}.$$

34P

(a) Assume that the chlorine in the sample had the naturally occurring isotopic mixture, so the average mass number was 35.453, as given in Appendix D. Then the mass of ^{226}Ra was

$$m = \frac{226}{226 + 2(35.453)}(0.10\,\text{g}) = 76.1 \times 10^{-3}\,\text{g}.$$

The mass of a ^{226}Ra nucleus is $(226\,\text{u})(1.661 \times 10^{-24}\,\text{g/u}) = 3.75 \times 10^{-22}\,\text{g}$, so the number of ^{226}Ra nuclei present was $N = (76.1 \times 10^{-3}\,\text{g})/(3.75 \times 10^{-22}\,\text{g}) = 2.03 \times 10^{20}$.

(b) The decay rate is given by $R = N\lambda = (N \ln 2)/\tau$, where λ is the disintegration constant, τ is the half-life, and N is the number of nuclei. The relationship $\lambda = (\ln 2)/\tau$ was used. Thus

$$R = \frac{(2.03 \times 10^{20})\ln 2}{(1600\,\text{y})(3.156 \times 10^7\,\text{s/y})} = 2.79 \times 10^9\,\text{s}^{-1}.$$

38P

The number N of undecayed nuclei present at any time and the rate of decay R at that time are related by $R = \lambda N$, where λ is the disintegration constant. The disintegration constant is related to the half-life τ by $\lambda = (\ln 2)/\tau$, so $R = (N \ln 2)/\tau$ and $\tau = (N \ln 2)/R$. Since 15.0% by mass of the sample is ^{147}Sm, the number of ^{147}Sm nuclei present in the sample is

$$N = \frac{(0.150)(1.00\,\mathrm{g})}{(147\,\mathrm{u})(1.661 \times 10^{-24}\,\mathrm{g/u})} = 6.143 \times 10^{20}\,.$$

Thus

$$\tau = \frac{(6.143 \times 10^{20})\ln 2}{120\,\mathrm{s}^{-1}} = 3.55 \times 10^{18}\,\mathrm{s} = 1.12 \times 10^{11}\,\mathrm{y}\,.$$

43P

If N is the number of undecayed nuclei present at time t, then

$$\frac{dN}{dt} = R - \lambda N\,,$$

where R is the rate of production by the cyclotron and λ is the disintegration constant. The second term gives the rate of decay. Rearrange the equation slightly and integrate:

$$\int_{N_0}^{N} \frac{dN}{R - \lambda N} = \int_0^t dt\,,$$

where N_0 is the number of undecayed nuclei present at time $t = 0$. This yields

$$-\frac{1}{\lambda} \ln \frac{R - \lambda N}{R - \lambda N_0} = t\,.$$

Solve for N:

$$N = \frac{R}{\lambda} + \left(N_0 - \frac{R}{\lambda}\right) e^{-\lambda t}\,.$$

After many half-lives the exponential is small and the second term can be neglected. Then $N = R/\lambda$, regardless of the initial value N_0. At times that are long compared to the half-life the rate of production equals the rate of decay and N is a constant.

44P

(a) The sample is in secular equilibrium with the source and the decay rate equals the production rate. Let R be the rate of production of ^{56}Mn and let λ be the disintegration constant. According the result of Problem 43, $R = \lambda N$ after a long time has passed. Now $\lambda N = 8.88 \times 10^{10}\,\mathrm{s}^{-1}$, so $R = 8.88 \times 10^{10}\,\mathrm{s}^{-1}$.

(b) They decay at the same rate as they are produced, $8.88 \times 10^{10}\,\mathrm{s}^{-1}$.

(c) Use $N = R/\lambda$. If τ is the half-life, then the disintegration constant is $\lambda = (\ln 2)/\tau = (\ln 2)/(2.58\,\mathrm{h}) = 0.269\,\mathrm{h}^{-1} = 7.46 \times 10^{-5}\,\mathrm{s}^{-1}$, so $N = (8.88 \times 10^{10}\,\mathrm{s}^{-1})/(7.46 \times 10^{-5}\,\mathrm{s}^{-1}) = 1.19 \times 10^{15}$.

(d) The mass of a ^{56}Mn nucleus is $(56\,u)(1.661 \times 10^{-24}\,g/u) = 9.30 \times 10^{-23}\,g$ and the total mass of ^{56}Mn in the sample at the end of the bombardment is $Nm = (1.19 \times 10^{15})(9.30 \times 10^{-23}\,g) = 1.11 \times 10^{-7}\,g$.

47E

The fraction of undecayed nuclei remaining after time t is given by

$$\frac{N}{N_0} = e^{-\lambda t} = e^{-(\ln 2)t/\tau},$$

where λ is the disintegration constant and $\tau\,(= (\ln 2)/\lambda)$ is the half-life. The time for half the original ^{238}U nuclei to decay is $4.5 \times 10^9\,y$. For ^{244}Pu at that time

$$\frac{(\ln 2)t}{\tau} = \frac{(\ln 2)(4.5 \times 10^9\,y)}{8.2 \times 10^7\,y} = 38.0$$

and

$$\frac{N}{N_0} = e^{-38.0} = 3.1 \times 10^{-17}.$$

For ^{248}Cm at that time

$$\frac{(\ln 2)t}{\tau} = \frac{(\ln 2)(4.5 \times 10^9\,y)}{3.4 \times 10^5\,y} = 9170$$

and

$$\frac{N}{N_0} = e^{-9170} = 3.31 \times 10^{-3983}.$$

For any reasonably sized sample this is less than 1 nucleus and may be taken to be zero. Your calculator probably cannot evaluate e^{-9170} directly. Treat it as $(e^{-91.70})^{100}$.

48P

Energy and momentum are conserved. Assume the residual thorium nucleus is in its ground state. Let K_α be the kinetic energy of the alpha particle and K_{Th} be the kinetic energy of the thorium nucleus. Then $Q = K_\alpha + K_{Th}$. Assume the uranium nucleus is initially at rest. Then conservation of momentum yields $0 = p_\alpha + p_{Th}$, where p_α is the momentum of the alpha particle and p_{Th} is the momentum of the thorium nucleus.

Both particles travel slowly enough that the classical relationship between momentum and energy can be used. Thus $K_{Th} = p_{Th}^2/2m_{Th}$, where m_{Th} is the mass of the thorium nucleus. Substitute $p_{Th} = -p_\alpha$ and use $K_\alpha = p_\alpha^2/2m_\alpha$ to obtain $K_{Th} = (m_\alpha/m_{Th})K_\alpha$. Thus

$$Q = K_\alpha + \frac{m_\alpha}{m_{Th}}K_\alpha = \left(1 + \frac{m_\alpha}{m_{Th}}\right)K_\alpha = \left(1 + \frac{4.00\,u}{234\,u}\right)(4.196\,MeV) = 4.27\,MeV.$$

52E

Let $^A_Z X$ represent the unknown nuclide. The reaction equation is

$$^A_Z X + ^1_0 n \rightarrow ^0_{-1}e + 2\,^4_2\text{He}.$$

Conservation of charge yields $Z + 0 = -1 + 4$ or $Z = 3$. Conservation of mass number yields $A + 1 = 0 + 8$ or $A = 7$. According to the periodic table in Appendix E, lithium has atomic number 3, so the nuclide must be ^7_3Li.

53E

Let M_{Cs} be the mass of one atom of $^{137}_{55}\text{Cs}$ and M_{Ba} be the mass of one atom of $^{137}_{56}\text{Ba}$. To obtain the nuclear masses we must subtract the mass of 55 electrons from M_{Cs} and the mass of 56 electrons from M_{Ba}. The energy released is $Q = [(M_{\text{Cs}} - 55m) - (M_{\text{Ba}} - 56m) - m] c^2$, where m is the mass of an electron. Once cancellations have been made, $Q = (M_{\text{Cs}} - M_{\text{Ba}})c^2$ is obtained. Thus

$$Q = [136.9073\,\text{u} - 136.9058\,\text{u}] c^2 = (0.0015\,\text{u})c^2 = (0.0015\,\text{u})(932\,\text{MeV/u}) = 1.40\,\text{MeV}\,.$$

55E

The decay scheme is $n \rightarrow p + e^- + \nu$. The electron kinetic energy is a maximum if no neutrino is emitted. Then $K_{\text{max}} = (m_n - m_p - m_e)c^2$, where m_n is the mass of a neutron, m_p is the mass of a proton, and m_e is the mass of an electron. Since $m_p + m_e = m_H$, where m_H is the mass of a hydrogen atom, this can be written $K_{\text{max}} = (m_n - m_H)c^2$. Hence $K_{\text{max}} = (840 \times 10^{-6}\,\text{u})c^2 = (840 \times 10^{-6}\,\text{u})(932\,\text{MeV/u}) = 0.783\,\text{MeV}$.

61P*

Since the electron has the maximum possible kinetic energy, no neutrino is emitted. Since momentum is conserved, the momentum of the electron and the momentum of the residual sulfur nucleus are equal in magnitude and opposite in direction. If p_e is the momentum of the electron and p_S is the momentum of the sulfur nucleus, then $p_S = -p_e$. The kinetic energy K_S of the sulfur nucleus is $K_S = p_S^2/2M_S = p_e^2/2M_S$, where M_S is the mass of the sulfur nucleus. Now the electron's kinetic energy K_e is related to its momentum by the relativistic equation $(p_e c)^2 = K_e^2 + 2K_e mc^2$, where m is the mass of an electron. See Eq. 42–40. Thus

$$K_S = \frac{(p_e c)^2}{2M_S c^2} = \frac{K_e^2 + 2K_e mc^2}{2M_S c^2} = \frac{(1.71\,\text{MeV})^2 + 2(1.71\,\text{MeV})(0.511\,\text{MeV})}{2(32\,\text{u})(931.5\,\text{MeV/u})}$$

$$= 7.83 \times 10^{-5}\,\text{MeV} = 78.3\,\text{eV}\,,$$

where $mc^2 = 0.511\,\text{MeV}$ was used.

62E

(a) The mass of a ^{238}U atom is $(238\,\text{u})(1.661 \times 10^{-24}\,\text{g/u}) = 3.95 \times 10^{-22}\,\text{g}$, so the number of uranium atoms in the rock is $N_U = (4.20 \times 10^{-3}\,\text{g})/(3.95 \times 10^{-22}\,\text{g}) = 1.06 \times 10^{19}$. The mass of a ^{206}Pb atom is $(206\,\text{u})(1.661 \times 10^{-24}\,\text{g}) = 3.42 \times 10^{-22}\,\text{g}$, so the number of lead atoms in the rock is $N_{\text{Pb}} = (2.135 \times 10^{-3}\,\text{g})/(3.42 \times 10^{-22}\,\text{g}) = 6.24 \times 10^{18}$.

(b) If no lead was lost there was originally one uranium atom for each lead atom formed by decay, in addition to the uranium atoms that did not yet decay. Thus the original number of uranium atoms was $N_{U0} = N_U + N_{\text{Pb}} = 1.06 \times 10^{19} + 6.24 \times 10^{18} = 1.68 \times 10^{19}$.

(c) Use

$$N_U = N_{U0}\, e^{-\lambda t},$$

where λ is the disintegration constant for the decay. It is related to the half-life τ by $\lambda = (\ln 2)/\tau$. Thus

$$t = -\frac{1}{\lambda} \ln\left(\frac{N_U}{N_{U0}}\right) = -\frac{\tau}{\ln 2} \ln\left(\frac{N_U}{N_{U0}}\right) = -\frac{4.47 \times 10^9\,\mathrm{y}}{\ln 2} \ln\left(\frac{1.06 \times 10^{19}}{1.68 \times 10^{19}}\right) = 2.97 \times 10^9\,\mathrm{y}.$$

67E

The decay rate R is related to the number of nuclei N by $R = \lambda N$, where λ is the disintegration constant. The disintegration constant is related to the half-life τ by $\lambda = (\ln 2)/\tau$, so $N = R/\lambda = R\tau/\ln 2$. Since $1\,\mathrm{Ci} = 3.7 \times 10^{10}$ disintegrations/s,

$$N = \frac{(250\,\mathrm{Ci})(3.7 \times 10^{10}\,\mathrm{s}^{-1}/\mathrm{Ci})(2.7\,\mathrm{d})(8.64 \times 10^4\,\mathrm{s/d})}{\ln 2} = 3.11 \times 10^{18}.$$

The mass of a ^{198}Au atom is $M = (198\,\mathrm{u})(1.661 \times 10^{-24}\,\mathrm{g/u}) = 3.29 \times 10^{-22}\,\mathrm{g}$ so the mass required is $NM = (3.11 \times 10^{18})(3.29 \times 10^{-22}\,\mathrm{g}) = 1.02 \times 10^{-3}\,\mathrm{g} = 1.02\,\mathrm{mg}$.

70P

The dose equivalent is the product of the absorbed dose and the RBE factor, so the absorbed dose is (dose equivalent)/(RBE) $= (25 \times 10^{-3}\,\mathrm{rem})/(0.85) = 2.94 \times 10^{-2}\,\mathrm{rad}$. But $1\,\mathrm{rad} = 10\,\mathrm{mJ/kg}$, so the absorbed dose is $(2.94 \times 10^{-2})(10 \times 10^{-3}\,\mathrm{J/kg}) = 2.94 \times 10^{-4}\,\mathrm{J/kg}$. To obtain the total energy received, multiply this by the mass receiving the energy: $E = (2.94 \times 10^{-4}\,\mathrm{J/kg})(44\,\mathrm{kg}) = 1.29 \times 10^{-2}\,\mathrm{J}$.

75P

A generalized formation reaction can be written $\mathrm{X} + \mathrm{x} \to \mathrm{Y}$, where X is the target nucleus, x is the incident light particle, and Y is the excited compound nucleus (^{20}Ne). Assume X is initially at rest. Then conservation of energy yields

$$m_X c^2 + m_x c^2 + K_x = m_Y c^2 + K_Y + E_Y,$$

where m_X, m_x, and M_Y are masses, K_x and K_Y are kinetic energies, and E_Y is the excitation energy of Y. Conservation of momentum yields

$$p_x = p_Y.$$

Now $K_Y = p_Y^2/2m_Y = p_x^2/2m_Y = (m_x/m_Y)K_x$, so

$$m_X c^2 + m_x c^2 + K_x = m_Y c^2 + (m_x/m_Y)K_x + E_Y$$

and

$$K_x = \frac{m_Y}{m_Y - m_x}\left[(m_Y - m_X - m_x)c^2 + E_Y\right].$$

(a) Let x represent the alpha particle and X represent the ^{16}O nucleus. Then $(m_Y - m_X - m_x)c^2 = (19.99244\,u - 15.99491\,u - 4.00260\,u)(931.5\,MeV/u) = -4.722\,MeV$ and

$$K_\alpha = \frac{19.99244\,u}{19.99244\,u - 4.00260\,u}(-4.722\,MeV + 25.0\,MeV) = 25.35\,MeV\,.$$

(b) Let x represent the proton and X represent the ^{19}F nucleus. Then $(m_Y - m_X - m_x)c^2 = (19.99244\,u - 18.99841\,u - 1.00783\,u)(931.5\,MeV/u) = -12.85\,MeV$ and

$$K_\alpha = \frac{19.99244\,u}{19.99244\,u - 1.00783\,u}(-12.85\,MeV + 25.0\,MeV) = 12.80\,MeV\,.$$

(c) Let x represent the photon and X represent the ^{20}Ne nucleus. Since the mass of the photon is zero we must rewrite the conservation of energy equation: if E_γ is the energy of the photon, then $E_\gamma + m_X c^2 = m_Y c^2 + K_Y + E_Y$. Since $m_X = m_Y$, this equation becomes $E_\gamma = K_Y + E_Y$. Since the momentum and energy of a photon are related by $p_\gamma = E_\gamma/c$, the conservation of momentum equation becomes $E_\gamma/c = p_Y$. The kinetic energy of the compound nucleus is $K_Y = p_Y^2/2m_Y = E_\gamma^2/2m_Y c^2$. Substitute this result into the conservation of energy equation to obtain

$$E_\gamma = \frac{E_\gamma^2}{2m_Y c^2} + E_Y\,.$$

This quadratic equation has the solutions

$$E_\gamma = m_Y c^2 \pm \sqrt{(m_Y c^2)^2 - 2m_Y c^2 E_Y}\,.$$

If the problem is solved using the relativistic relationship between the energy and momentum of the compound nucleus only one solution would be obtained, the one corresponding to the negative sign above. Since $M_Y c^2 = (19.99244\,u)(931.5\,MeV/u) = 1.862 \times 10^4\,MeV$,

$$E_\gamma = (1.862 \times 10^4\,MeV) - \sqrt{(1.862 \times 10^4\,MeV)^2 - 2(1.862 \times 10^4\,MeV)(25.0\,MeV)}$$
$$= 25.0\,MeV\,.$$

The kinetic energy of the compound nucleus is very small; essentially all of the photon energy goes to excite the nucleus.

Chapter 48

1E

(a) The mass of a single atom of ^{235}U is $(235\,\mathrm{u})(1.661 \times 10^{-27}\,\mathrm{kg/u}) = 3.90 \times 10^{-25}\,\mathrm{kg}$, so the number of atoms in $1.0\,\mathrm{kg}$ is $(1.0\,\mathrm{kg})/(3.90 \times 10^{-25}\,\mathrm{kg}) = 2.56 \times 10^{24}$.

(b) The energy released by N fission events is given by $E = NQ$, where Q is the energy released in each event. For $1.0\,\mathrm{kg}$ of ^{235}U, $E = (2.56 \times 10^{24})(200 \times 10^{6}\,\mathrm{eV})(1.60 \times 10^{-19}\,\mathrm{J/eV}) = 8.19 \times 10^{13}\,\mathrm{J}$.

(c) If P is the power requirement of the lamp, then $t = E/P = (8.19 \times 10^{13}\,\mathrm{J})/(100\,\mathrm{W}) = 8.19 \times 10^{11}\,\mathrm{s} = 26{,}000\,\mathrm{y}$. The conversion factor $3.156 \times 10^{7}\,\mathrm{s/y}$ was used to obtain the last result.

3E

If R is the fission rate, then the power output is $P = RQ$, where Q is the energy released in each fission event. Hence $R = P/Q = (1.0\,\mathrm{W})/(200 \times 10^{6}\,\mathrm{eV})(1.60 \times 10^{-19}\,\mathrm{J/eV}) = 3.12 \times 10^{10}\,\mathrm{fissions/s}$.

6E

If M_{Cr} is the mass of a ^{52}Cr nucleus and M_{Mg} is the mass of a ^{26}Mg nucleus, then the disintegration energy is $Q = \left(M_{\mathrm{Cr}} - 2M_{\mathrm{Mg}}\right)c^2 = [51.94051\,\mathrm{u} - 2(25.98259\,\mathrm{u})](931.5\,\mathrm{MeV/u}) = -23.0\,\mathrm{MeV}$.

12P

(a) If X represents the unknown fragment, then the reaction can be written

$$^{235}_{92}\mathrm{U} + {}^{1}_{0}\mathrm{n} \rightarrow {}^{83}_{32}\mathrm{Ge} + {}^{A}_{Z}X ,$$

where A is the mass number and Z is the atomic number of the fragment. Conservation of charge yields $92 + 0 = 32 + Z$, so $Z = 60$. Conservation of mass number yields $235 + 1 = 83 + A$, so $A = 153$. Look in Appendix D or E for nuclides with $Z = 60$. You should find that the unknown fragment is $^{153}_{60}$Nd.

(b) Ignore the small kinetic energy and momentum carried by the neutron that triggers the fission event. Then $Q = K_{\mathrm{Ge}} + K_{\mathrm{Nd}}$, where K_{Ge} is the kinetic energy of the germanium nucleus and K_{Nd} is the kinetic energy of the neodymium nucleus. Conservation of momentum yields $p_{\mathrm{Ge}} + p_{\mathrm{Nd}} = 0$, where p_{Ge} is the momentum of the germanium nucleus and p_{Nd} is the momentum of the neodymium nucleus. Since $p_{\mathrm{Nd}} = -p_{\mathrm{Ge}}$, the kinetic energy of the neodymium nucleus is

$$K_{\mathrm{Nd}} = \frac{p_{\mathrm{Nd}}^2}{2M_{\mathrm{Nd}}} = \frac{p_{\mathrm{Ge}}^2}{2M_{\mathrm{Nd}}} = \frac{M_{\mathrm{Ge}}}{M_{\mathrm{Nd}}} K_{\mathrm{Ge}} .$$

Thus the energy equation becomes

$$Q = K_{Ge} + \frac{M_{Ge}}{M_{Nd}}K_{Ge} = \frac{M_{Nd} + M_{Ge}}{M_{Nd}}K_{Ge}$$

and

$$K_{Ge} = \frac{M_{Nd}}{M_{Nd} + M_{Ge}}Q = \frac{153\,u}{153\,u + 83\,u}(170\,\text{MeV}) = 110\,\text{MeV}.$$

Similarly,

$$K_{Nd} = \frac{M_{Ge}}{M_{Nd} + M_{Ge}}Q = \frac{83\,u}{153\,u + 83\,u}(170\,\text{MeV}) = 60\,\text{MeV}.$$

(c) The initial speed of the germanium nucleus is

$$v_{Ge} = \sqrt{\frac{2K_{Ge}}{M_{Ge}}} = \sqrt{\frac{2(110 \times 10^6\,\text{eV})(1.60 \times 10^{-19}\,\text{J/eV})}{(83\,u)(1.661 \times 10^{-27}\,\text{kg/u})}} = 1.60 \times 10^7\,\text{m/s}.$$

The initial speed of the neodymium nucleus is

$$v_{Nd} = \sqrt{\frac{2K_{Nd}}{M_{Nd}}} = \sqrt{\frac{2(60 \times 10^6\,\text{eV})(1.60 \times 10^{-19}\,\text{J/eV})}{(153\,u)(1.661 \times 10^{-27}\,\text{kg/u})}} = 8.69 \times 10^6\,\text{m/s}.$$

13P

(a) The electrostatic potential energy is given by

$$U = \frac{1}{4\pi\epsilon_0}\frac{Z_{Xe}Z_{Sr}e^2}{R_{Xe} + R_{Sr}},$$

where Z_{Xe} is the atomic number of xenon, Z_{Sr} is the atomic number of strontium, R_{Xe} is the radius of a xenon nucleus, and R_{Sr} is the radius of a strontium nucleus. Atomic numbers can be found in Appendix D. The radii are given by $R = (1.2\,\text{fm})A^{1/3}$, where A is the mass number, also found in Appendix D. Thus $R_{Xe} = (1.2\,\text{fm})(140)^{1/3} = 6.23\,\text{fm} = 6.23 \times 10^{-15}\,\text{m}$ and $R_{Sr} = (1.2\,\text{fm})(96)^{1/3} = 5.49\,\text{fm} = 5.46 \times 10^{-15}\,\text{m}$. Hence the potential energy is

$$U = (8.99 \times 10^9\,\text{m/F})\frac{(54)(38)(1.60 \times 10^{-19}\,\text{C})^2}{6.23 \times 10^{-15}\,\text{m} + 5.49 \times 10^{-15}\,\text{m}} = 4.08 \times 10^{-11}\,\text{J}.$$

This is 251 MeV.

(b) The energy released in a typical fission event is about 200 MeV, roughly the same as the electrostatic potential energy when the fragments are touching. The energy appears as kinetic energy of the fragments and neutrons produced by fission.

15E

If P is the power output, then the energy E produced in the time interval $\Delta t\ (=\ 3\,\text{y})$ is $E = P\,\Delta t = (200 \times 10^6\,\text{W})(3\,\text{y})(3.156 \times 10^7\,\text{s/y}) = 1.89 \times 10^{16}\,\text{J}$, or $(1.89 \times 10^{16}\,\text{J})/(1.60 \times 10^{-19}\,\text{J/eV}) = 1.18 \times 10^{35}\,\text{eV} = 1.18 \times 10^{29}\,\text{MeV}$. At 200 MeV per event, this means $(1.18 \times$

$10^{29})/(200\,\mathrm{MeV}) = 5.92 \times 10^{26}$ fission events occurred. This must be half the number of fissionable nuclei originally available. Thus there were $2(5.92 \times 10^{26}) = 1.18 \times 10^{27}$ nuclei. The mass of a ^{235}U nucleus is $(235\,\mathrm{u})(1.661 \times 10^{-27}\,\mathrm{kg/u}) = 3.90 \times 10^{-25}\,\mathrm{kg}$, so the total mass of ^{235}U originally present was $(1.18 \times 10^{27})(3.90 \times 10^{-25}\,\mathrm{kg}) = 462\,\mathrm{kg}$.

18P

If R is the decay rate then the power output is $P = RQ$, where Q is the energy produced by each alpha decay. Now $R = \lambda N = N\ln 2/\tau$, where λ is the disintegration constant and τ is the half-life. The relationship $\lambda = (\ln 2)/\tau$ was used. If M is the total mass of material and m is the mass of a single ^{238}Pu nucleus, then

$$ N = \frac{M}{m} = \frac{1.00\,\mathrm{kg}}{(238\,\mathrm{u})(1.661 \times 10^{-27}\,\mathrm{kg/u})} = 2.53 \times 10^{24}. $$

Thus

$$ P = \frac{NQ\ln 2}{\tau} = \frac{(2.53 \times 10^{24})(5.50 \times 10^{6}\,\mathrm{eV})(1.60 \times 10^{-19}\,\mathrm{J/eV})(\ln 2)}{(87.7\,\mathrm{y})(3.156 \times 10^{7}\,\mathrm{s/y})} = 558\,\mathrm{W}. $$

22P

(a) The energy yield of the bomb is $E = (66 \times 10^{-3}\,\mathrm{Mton})(2.6 \times 10^{28}\,\mathrm{MeV/Mton}) = 1.72 \times 10^{27}\,\mathrm{MeV}$. At $200\,\mathrm{MeV}$ per fission event, $(1.72 \times 10^{27}\,\mathrm{MeV})/(200\,\mathrm{MeV}) = 8.58 \times 10^{24}$ fission events take place. Since only 4.0% of the ^{235}U nuclei originally present undergo fission, there must have been $(8.58 \times 10^{24})/(0.040) = 2.14 \times 10^{26}$ nuclei originally present. The mass of ^{235}U originally present was $(2.14 \times 10^{26})(235\,\mathrm{u})(1.661 \times 10^{-27}\,\mathrm{kg/u}) = 83.7\,\mathrm{kg}$.

(b) Two fragments are produced in each fission event so the total number of fragments is $2(8.58 \times 10^{24}) = 1.72 \times 10^{25}$.

(c) One neutron produced in a fission event is used to trigger the next fission event, so the average number of neutrons released to the environment in each event is 1.5. The total number released is $(8.58 \times 10^{24})(1.5) = 1.29 \times 10^{25}$.

24P

Let P_0 be the initial power output, P be the final power output, k be the multiplication factor, t be the time for the power reduction, and t_{gen} be the neutron generation time. Then, according to the result of Problem 23,

$$ P = P_0\, k^{t/t_{\mathrm{gen}}}. $$

Divide by P_0, then take the natural logarithm of both sides of the equation and solve for $\ln k$. You should obtain

$$ \ln k = \frac{t_{\mathrm{gen}}}{t}\ln\frac{P}{P_0}. $$

Hence

$$ k = e^{\alpha}, $$

where
$$\alpha = \frac{t_{gen}}{t} \ln \frac{P}{P_0} = \frac{1.3 \times 10^{-3} \, s}{2.6 \, s} \ln \frac{350 \, MW}{1200 \, MW} = -6.161 \times 10^{-4}.$$

This yields $k = .99938$.

30P

Let t be the present time and $t = 0$ be the time when the ratio of ^{235}U to ^{238}U was 3.0%. Let N_{235} be the number of ^{235}U nuclei present in a sample now and $N_{235, \, 0}$ be the number present at $t = 0$. Let N_{238} be the number of ^{238}U nuclei present in the sample now and $N_{238, \, 0}$ be the number present at $t = 0$. The law of radioactive decay holds for each specie, so

$$N_{235} = N_{235, \, 0} \, e^{-\lambda_{235} t}$$

and

$$N_{238} = N_{238, \, 0} \, e^{-\lambda_{238} t}.$$

Divide the first equation by the second to obtain

$$r = r_0 \, e^{-(\lambda_{235} - \lambda_{238})t},$$

where $r = N_{235}/N_{238} \, (= 0.0072)$ and $r_0 = N_{235, \, 0}/N_{238, \, 0} \, (= 0.030)$. Solve for t:

$$t = -\frac{1}{\lambda_{235} - \lambda_{238}} \ln \frac{r}{r_0}.$$

Now use $\lambda_{235} = (\ln 2)/\tau_{235}$ and $\lambda_{238} = (\ln 2)/\tau_{238}$, where τ_{235} and τ_{238} are the half-lives, to obtain

$$t = -\frac{\tau_{235}\tau_{238}}{(\tau_{238} - \tau_{235}) \ln 2} \ln \frac{r}{r_0} = -\frac{(7.0 \times 10^8 \, y)(4.5 \times 10^9 \, y)}{(4.5 \times 10^9 \, y - 7.0 \times 10^8 \, y) \ln 2} \ln \frac{0.0072}{0.030} = 1.71 \times 10^9 \, y.$$

32E

The height of the Coulomb barrier is taken to be the value of the kinetic energy K each nucleus must initially have if they are to come to rest when their surfaces touch (see Sample Problem 48–4). If R is the radius of a nucleus, conservation of energy yields

$$2K = \frac{1}{4\pi\epsilon_0} \frac{e^2}{2R},$$

so

$$K = \frac{1}{4\pi\epsilon_0} \frac{e^2}{4R} = (8.99 \times 10^9 \, m/F) \frac{(1.60 \times 10^{-19} \, C)^2}{4(0.80 \times 10^{-15} \, m)} = 7.19 \times 10^{-14} \, J.$$

This is $0.450 \, MeV$.

40E

If M_{He} is the mass of an atom of helium and M_C is the mass of an atom of carbon, then the energy released in a single fusion event is

$$Q = [3M_{He} - M_C] c^2 = [3(4.0026 \, u) - (12.0000 \, u)] (931.5 \, MeV/u) = 7.27 \, MeV.$$

Note that $3M_{He}$ contains the mass of 6 electrons and so does M_C. The electron masses cancel and the mass difference calculated is the same as the mass difference of the nuclei.

44P

(a) Let M be the mass of the sun at time t and E be the energy radiated to that time. Then the power output is $P = dE/dt = (dM/dt)c^2$, where $E = Mc^2$ was used. At the present time

$$\frac{dM}{dt} = \frac{P}{c^2} = \frac{3.9 \times 10^{26}\,\text{W}}{(3.00 \times 10^8\,\text{m/s})^2} = 4.33 \times 10^9\,\text{kg/s}\,.$$

(b) Assume the rate of mass loss remained constant. Then the total mass loss is $\Delta M = (dM/dt)\,\Delta t = (4.33 \times 10^9\,\text{kg/s})(4.5 \times 10^9\,\text{y})(3.156 \times 10^7\,\text{s/y}) = 6.15 \times 10^{26}\,\text{kg}$. The fraction lost is

$$\frac{\Delta M}{M + \Delta M} = \frac{6.15 \times 10^{26}\,\text{kg}}{2.0 \times 10^{30}\,\text{kg} + 6.15 \times 10^{26}\,\text{kg}} = 3.07 \times 10^{-4}\,.$$

46P

(a) The mass of a carbon atom is $(12.0\,\text{u})(1.661 \times 10^{-27}\,\text{kg/u}) = 1.99 \times 10^{-26}\,\text{kg}$, so the number of carbon atoms in $1.00\,\text{kg}$ of carbon is $(1.00\,\text{kg})/(1.99 \times 10^{-26}\,\text{kg}) = 5.02 \times 10^{25}$. The heat of combustion per atom is $(3.3 \times 10^7\,\text{J/kg})/(5.02 \times 10^{25}\,\text{atom/kg}) = 6.58 \times 10^{-19}\,\text{J/atom}$. This is $4.11\,\text{eV/atom}$.

(b) In each combustion event two oxygen atoms combine with one carbon atom, so the total mass involved is $2(16.0\,\text{u}) + (12.0\,\text{u}) = 44\,\text{u}$. This is $(44\,\text{u})(1.661 \times 10^{-27}\,\text{kg/u}) = 7.31 \times 10^{-26}\,\text{kg}$. Each combustion event produces $6.58 \times 10^{-19}\,\text{J}$ so the energy produced per unit mass of reactants is $(6.58 \times 10^{-19}\,\text{J})/(7.31 \times 10^{-26}\,\text{kg}) = 9.00 \times 10^6\,\text{J/kg}$.

(c) If the sun were composed of the appropriate mixture of carbon and oxygen, the number of combustion events that could occur before the sun burns out would be $(2.0 \times 10^{30}\,\text{kg})/(7.31 \times 10^{-26}\,\text{kg}) = 2.74 \times 10^{55}$. The total energy released would be $E = (2.74 \times 10^{55})(6.58 \times 10^{-19}\,\text{J}) = 1.80 \times 10^{37}\,\text{J}$. If P is the power output of the sun, the burn time would be $t = E/P = (1.80 \times 10^{37}\,\text{J})/(3.9 \times 10^{26}\,\text{W}) = 4.62 \times 10^{10}\,\text{s}$. This is $1460\,\text{y}$.

50P

Since the mass of a helium atom is $(4.00\,\text{u})(1.661 \times 10^{-27}\,\text{kg/u}) = 6.64 \times 10^{-27}\,\text{kg}$, the number of helium nuclei originally in the star is $(4.6 \times 10^{32}\,\text{kg})/(6.64 \times 10^{-27}\,\text{kg}) = 6.92 \times 10^{58}$. Since each fusion event requires three helium nuclei, the number of fusion events that can take place is $N = 6.92 \times 10^{58}/3 = 2.31 \times 10^{58}$. If Q is the energy released in each event and t is the conversion time, then the power output is $P = NQ/t$ and

$$t = \frac{NQ}{P} = \frac{(2.31 \times 10^{58})(7.27 \times 10^6\,\text{eV})(1.60 \times 10^{-19}\,\text{J/eV})}{5.3 \times 10^{30}\,\text{W}} = 5.07 \times 10^{15}\,\text{s}\,.$$

This is $1.6 \times 10^8\,\text{y}$.

54P

Since 1.00 L of water has a mass of 1.00 kg, the mass of the heavy water in 1.00 L is 0.0150×10^{-2} kg $= 1.50 \times 10^{-4}$ kg. Since a heavy water molecule contains one oxygen atom, one hydrogen atom and one deuterium atom, its mass is $(16.0 \, u + 1.00 \, u + 2.00 \, u) = 19.0 \, u$ or $(19.0 \, u)(1.661 \times 10^{-27} \, kg/u) = 3.16 \times 10^{-26}$ kg. The number of heavy water molecules in a liter of water is $(1.50 \times 10^{-4} \, kg)/(3.16 \times 10^{-26} \, kg) = 4.75 \times 10^{21}$. Since each fusion event requires two deuterium nuclei, the number of fusion events that can occur is $N = 4.75 \times 10^{21}/2 = 2.38 \times 10^{21}$. Each event releases energy $Q = (3.27 \times 10^6 \, eV)(1.60 \times 10^{-19} \, J/eV) = 5.23 \times 10^{-13}$ J. Since all events take place in a day, the power output is

$$P = \frac{NQ}{t} = \frac{(2.38 \times 10^{21})(5.23 \times 10^{-13} \, J)}{8.64 \times 10^4 \, s} = 1.44 \times 10^4 \, W = 14.4 \, kW \, .$$

56E

(a) The most probable speed is given by Eq. 21–25:

$$v_P = \sqrt{\frac{2RT}{M}} \, ,$$

where R is the universal gas constant (8.31 J/mol · K), T is the temperature on the Kelvin scale, and M is the molar mass (2.0×10^{-3} kg/mol). Hence

$$v_P = \sqrt{\frac{2(8.31 \, J/mol \cdot K)(1.0 \times 10^8 \, K)}{2.0 \times 10^{-3} \, kg}} = 9.12 \times 10^5 \, m/s \, .$$

(b) The confinement time for the conditions of Sample Problem 48–7 is $\tau = 1.0 \times 10^{-12}$ s. In this time a deuteron with speed v_P moves a distance $d = v_P \tau = (9.12 \times 10^5 \, m/s)(1.0 \times 10^{-12} \, s) = 9.12 \times 10^{-7} \, m = 912 \, nm$.

Chapter 49

2E

Conservation of momentum requires that the gamma-ray particles move in opposite directions with momenta of the same magnitude. Since the magnitude p of the momentum of a gamma-ray particle is related to its energy by $p = E/c$, the particles have the same energy E. Conservation of energy yields $m_\pi c^2 = 2E$, where m_π is the mass of a neutral pion. According to Table 49–5 the rest energy of a neutral pion is $m_\pi c^2 = 135.0\,\text{MeV}$. Hence $E = (135.0\,\text{MeV})/2 = 67.5\,\text{MeV}$. Use the result of Exercise 1 of Chapter 43 to obtain the wavelength of the gamma-ray waves:

$$\lambda = \frac{1240\,\text{eV} \cdot \text{nm}}{67.5 \times 10^6\,\text{eV}} = 1.84 \times 10^{-5}\,\text{nm} = 18.4\,\text{fm}\,.$$

5E

The energy released would be twice the rest energy of the Earth, or $E = 2mc^2 = 2(5.98 \times 10^{24}\,\text{kg})(3.00 \times 10^8\,\text{m/s})^2 = 1.08 \times 10^{42}\,\text{J}$.

10P

Table 49–5 gives the rest energy of each pion as $139.6\,\text{MeV}$. The magnitude of the momentum of each pion is $p_\pi = (358.3\,\text{MeV})/c$. Use the relativistic relationship between energy and momentum (Eq. 42–41) to find the total energy of each pion:

$$E_\pi = \sqrt{(p_\pi c)^2 + (m_\pi c^2)^2} = \sqrt{(358.3\,\text{MeV})^2 + (139.6\,\text{MeV})^2} = 384.5\,\text{MeV}\,.$$

Conservation of energy yields $m_\rho c^2 = 2E_\pi = 2(384.5\,\text{MeV}) = 769\,\text{MeV}$.

13E

(a) The conservation laws considered so far are associated with energy, momentum, angular momentum, charge, and baryon number. The rest energy of the muon is $105.7\,\text{MeV}$, the rest energy of the electron is $0.511\,\text{MeV}$, and rest energy of the neutrino is zero. Thus the total rest energy before the decay is greater than the total rest energy after. The excess energy can be carried away as the kinetic energies of the decay products and energy can be conserved. Momentum is conserved if the electron and neutrino move away from the decay in opposite directions with equal magnitudes of momenta. Since the orbital angular momentum is zero, we consider only spin angular momentum. All the particles have spin $\hbar/2$. The total angular momentum after the decay must be either $\hbar$ (if the spins are aligned) or zero (if the spins are antialigned). Since the spin before the decay is $\hbar/2$, angular momentum cannot be conserved. The muon has charge $-e$, the electron has charge $-e$, and the neutrino has charge zero, so the total charge before the decay is $-e$ and the total charge after is $-e$. Charge is conserved. All particles have baryon number zero, so baryon number is conserved. The only conservation law not obeyed is the conservation of angular momentum.

(b) Analyze the decay in the same way. You should find that only charge is not conserved.

(c) Here you should find that energy and angular momentum cannot be conserved.

15E

For purposes of deducing the properties of the antineutron, cancel a proton from each side of the reaction and write the equivalent reaction as

$$\pi^+ \to p + \bar{n}.$$

Particle properties can be found in Tables 49–4 and 5. The pion and proton each have charge $+e$, so the antineutron must be neutral. The pion has baryon number zero (it is a meson) and the proton has baryon number $+1$, so the baryon number of the antineutron must be -1. The pion and the proton each have strangeness zero, so the strangeness of the antineutron must also be zero. In summary, $Q = 0$, $B = -1$, and $S = 0$ for the antineutron.

17E

(a) See the solution to Exercise 13 for the quantities to be considered, but add strangeness to the list. The lambda has a rest energy of 1115.6 MeV, the proton has a rest energy of 938.3 MeV, and the kaon has a rest energy of 493.7 MeV. The rest energy before the decay is less than the total rest energy after, so energy cannot be conserved. Momentum can be conserved. The lambda and proton each have spin $\hbar/2$ and the kaon has spin zero, so angular momentum can be conserved. The lambda has charge zero, the proton has charge $+e$, and the kaon has charge $-e$, so charge is conserved. The lambda and proton each have baryon number $+1$ and the kaon has baryon number zero, so baryon number is conserved. The lambda and kaon each have strangeness -1 and the proton has strangeness zero, so strangeness is conserved. Only energy cannot be conserved.

(b) The omega has a rest energy of 1680 MeV, the sigma has a rest energy of 1197.3 MeV, and the pion has a rest energy of 135 MeV. The rest energy before the decay is greater than the total rest energy after, so energy can be conserved. Momentum can be conserved. The omega and sigma each have spin $\hbar/2$ and the pion has spin zero, so angular momentum can be conserved. The omega has charge $-e$, the sigma has charge $-e$, and the pion has charge zero, so charge is conserved. The omega and sigma have baryon number $+1$ and the pion has baryon number , so baryon number is conserved. The omega has strangeness -3, the sigma has strangeness -1, and the pion has strangeness zero, so strangeness is not conserved.

(c) The kaon and proton can bring kinetic energy to the reaction, so energy can be conserved even though the total rest energy after the collision is greater than the total rest energy before. Momentum can be conserved. The proton and lambda each have spin $\hbar/2$ and the kaon and pion each have spin zero, so angular momentum can be conserved. The kaon has charge $-e$, the proton has charge $+e$, the lambda has charge zero, and the pion has charge $+e$, so charge is not conserved. The proton and lambda each have baryon number $+1$ and the kaon and pion each have baryon number zero, so baryon number is conserved. The kaon has strangeness -1, the proton and pion each have strangeness zero, and the lambda has strangeness -1, so strangeness is conserved. Only charge is not conserved.

20P

(a) As far as the conservation laws are concerned, we may cancel a proton from each side of the reaction equation and write the reaction as p $\rightarrow \Lambda^0 +$ x. Since the proton and the lambda each have a spin angular momentum of $\hbar/2$, the spin angular momentum of x must be either zero or $\hbar$. Since the proton has charge $+e$ and the lambda is neutral, x must have charge $+e$. Since the proton and the lambda each have a baryon number of $+1$, the baryon number of x is zero. Since the strangeness of the proton is zero and the strangeness of the lambda is -1, the strangeness of x is $+1$. Take the unknown particle to be a spin zero meson with a charge of $+e$ and a strangeness of $+1$. Look at Table 49–5 to identify it as a K^+ particle.

(b) Similar analysis tells us that x is a spin-$\frac{1}{2}$ antibaryon ($B = -1$) with charge and strangeness both zero. Inspection of Table 49–4 reveals it is an antineutron.

(c) Here x is a spin-0 (or spin-1) meson with charge zero and strangeness -1. According to Table 49–5 it could be a $\overline{K}^0$ particle.

26E

(a) Look at the first three lines of Table 49–6. Since the particle is a baryon it must consist of 3 quarks. To obtain a strangeness of -2, two of them must be s quarks. Each of these has a charge of $-e/3$, so the sum of their charges is $-2e/3$. To obtain a total charge of e, the charge on the third quark must be $5e/3$. There is no quark with this charge, so the particle cannot be constructed. In fact, such a particle has never been observed.

(b) Again the particle consists of three quarks (and no antiquarks). To obtain a strangeness of zero, none of them may be s quarks. We must find a combination of three u and d quarks with a total charge of $2e$. The only such combination consists of three u quarks.

31E

Apply Eq. 18–55 for the Doppler shift in wavelength:

$$\frac{\Delta\lambda}{\lambda} = \frac{v}{c},$$

where v is the recessional speed of the galaxy. (The speed is denoted by u in Eq. 18-55.) Use Hubble's law to find the recessional speed: $v = Hr$, where H is the Hubble constant and r is the distance to the galaxy. Thus $v = [17 \times 10^{-3}\,\mathrm{m/(s \cdot ly)}](2.40 \times 10^8\,\mathrm{ly}) = 4.08 \times 10^6\,\mathrm{m/s}$ and

$$\Delta\lambda = \frac{v}{c}\lambda = \left(\frac{4.08 \times 10^6\,\mathrm{m/s}}{3.00 \times 10^8\,\mathrm{m/s}}\right)(656.3\,\mathrm{nm}) = 8.9\,\mathrm{nm}.$$

Since the galaxy is receding the observed wavelength is longer than the wavelength in the rest frame of the galaxy. Its value is $656.3\,\mathrm{nm} + 8.9\,\mathrm{nm} = 665.2\,\mathrm{nm}$.

37P

(a) The mass M within the Earth's orbit is used to calculate the gravitational force on the Earth. If r is the radius of the orbit, R is the radius of the new sun, and M_S is the mass of the sun, then

$$M = \left(\frac{r}{R}\right)^3 M_S = \left(\frac{1.50 \times 10^{11}\,\text{m}}{5.90 \times 10^{12}\,\text{m}}\right)^3 (1.99 \times 10^{30}\,\text{kg}) = 3.27 \times 10^{25}\,\text{kg}.$$

The gravitational force on the Earth is given by GMm/r^2, where m is the mass of the Earth and G is the universal gravitational constant. Since the centripetal acceleration is given by v^2/r, where v is the speed of the Earth, $GMm/r^2 = mv^2/r$ and

$$v = \sqrt{\frac{GM}{r}} = \sqrt{\frac{(6.67 \times 10^{-11}\,\text{m}^3/\text{s}^2 \cdot \text{kg})(3.27 \times 10^{25}\,\text{kg})}{1.50 \times 10^{11}\,\text{m}}} = 1.21 \times 10^2\,\text{m/s}.$$

(*b*) The period of revolution is

$$T = \frac{2\pi r}{v} = \frac{2\pi(1.50 \times 10^{11}\,\text{m})}{1.21 \times 10^2\,\text{m/s}} = 7.82 \times 10^9\,\text{s}.$$

This is 248 y.

40E

(*a*) Substitute $\lambda = (2898\,\mu\text{m}\cdot\text{K})/T$ into the result of Exercise 1 of Chapter 43: $E = (1240\,\text{nm}\cdot\text{eV})/\lambda$. First convert units: $2898\,\mu\text{m}\cdot\text{K} = 2.898 \times 10^6\,\text{nm}\cdot\text{K}$ and $1240\,\text{nm}\cdot\text{eV} = 1.240 \times 10^{-3}\,\text{nm}\cdot\text{MeV}$. Hence

$$E = \frac{(1.240 \times 10^{-3}\,\text{nm}\cdot\text{MeV})T}{2.898 \times 10^6\,\text{nm}\cdot\text{K}} = (4.28 \times 10^{-10}\,\text{MeV/K})T.$$

(*b*) The minimum energy required to create an electron-positron pair is twice the rest energy of an electron, or $2(0.511\,\text{MeV}) = 1.022\,\text{MeV}$. Hence

$$T = \frac{E}{4.28 \times 10^{-10}\,\text{MeV/K}} = \frac{1.022\,\text{MeV}}{4.28 \times 10^{-10}\,\text{MeV/K}} = 2.39 \times 10^9\,\text{K}.$$